# Comprehensive Organometallic Chemistry II

A Review of the Literature 1982–1994

# Comprehensive Organometallic Chemistry II

## A Review of the Literature 1982–1994

*Editors-in-Chief*

Edward W. Abel
*University of Exeter, UK*

F. Gordon A. Stone
*Baylor University, Waco, TX, USA*

Geoffrey Wilkinson
*Imperial College of Science, Technology and Medicine, London, UK*

**Volume 6**
MANGANESE GROUP

Volume Editor

Charles P. Casey
*University of Wisconsin–Madison, WI, USA*

PERGAMON

UK         Elsevier Science Ltd., The Boulevard, Langford Lane, Kidlington, Oxford OX5 1GB, UK

USA      Elsevier Science Inc., 660 White Plains Road, Tarrytown, New York 10591-5153, USA

JAPAN   Elsevier Science Japan, Tsunashima Building Annex, 3-20-12 Yushima, Bunkyo-ku, Tokyo 113, Japan

---

First edition 1995

**Library of Congress Cataloging in Publication Data**
Comprehensive organometallic chemistry II : a review of the literature 1982–1994 / editors-in-chief, Edward W. Abel, F. Gordon A. Stone, Geoffrey Wilkinson
    p.  cm.
   Includes indexes.
   1. Organometallic chemistry.   I. Abel, Edward W.   II. Stone, F. Gordon A.   III.  Wilkinson, Geoffrey.
QD411.C652   1995
547'.05—dc20                             95–7030

**British Library Cataloguing in Publication Data**
A catalogue record for this book is available from the British Library.

ISBN 0–08–040608–4 (set : alk. paper)
ISBN 0–08–042313–2 (Volume 6)

**Important note**
For safety reasons, readers should always consult the list of abbreviations on p. xi before making use of the experimental details provided.

∞™   The paper used in this publication meets the minimum requirements of the American National Standard for Information Sciences—Permanence of Paper for Printed Library Materials, ANSI Z39.48–1984.

Chemical structures drawn by Synopsys Scientific Systems Ltd., Leeds, UK.

Printed and bound in Great Britain by BPC Wheatons Ltd., Exeter, UK.

# Contents

# Preface

'Comprehensive Organometallic Chemistry', published in 1982, was well received and remains very highly cited in the primary journal literature. Since its publication, studies on the chemistry of molecules with carbon–metal bonds have continued to expand rapidly. This is due to many factors, ranging from the sheer intellectual challenge and excitement provided by the continuing production of novel results, which demand new ideas, through to the successful application of organometallic species in organic syntheses, the generation of living catalysts for polymerization, and the synthesis of precursors for materials employed in the electronic and ceramic industries. For many reasons, therefore, we judged it timely to update 'Comprehensive Organometallic Chemistry' with a new work.

Due to the scope and depth of this area of chemistry, to have merely revised each of the original nine volumes did not seem the most user-friendly or cost-effective procedure to follow. As a consequence of the sheer bulk of the literature of the subject, a revised edition would necessarily require either the elimination of much chemistry of archival value but which is still important, or the production of a set of volumes significantly larger in number than the original nine. Accordingly, we decided it would be best to use the original work as a basis for new volumes focusing on organometallic chemistry reported since 1982, with reference back to the original work when necessary. For ease of use the new volumes maintain the same general structure as employed previously but reflect the changes in substance and direction the field has undergone in the last ten years. Thus it is not surprising that the largest volume in the new work concerns the role of the transition elements in metal-mediated organic syntheses.

The expansion of organometallic chemistry since the early 1980s also led us to decide that an updating of 'Comprehensive Organometallic Chemistry' would be more effectively accomplished if each volume had one or two editors who would be responsible both for recruiting experts for the Herculean task of writing the many chapters of each volume and for overseeing the content. We are deeply indebted to the volume editors and their authors for the time and effort they have given to the project.

As with the original 'Comprehensive Organometallic Chemistry', published some thirteen years ago, we hope this new version will serve as a pivotal reference point for new work and will function to generate new ideas and perceptions for the continued advance of what will surely continue as a vibrant area of chemistry.

Edward W. Abel
*Exeter, UK*

F. Gordon A. Stone
*Waco, Texas, USA*

Geoffrey Wilkinson
*London, UK*

# Preface to 'Comprehensive Organometallic Chemistry'

Although the discovery of the platinum complex that we now know to be the first π-alkene complex, $K[PtCl_3(C_2H_4)]$, by Zeise in 1827 preceded Frankland's discovery (1849) of diethylzinc, it was the latter that initiated the rapidly developing interest during the latter half of the nineteenth century in compounds with organic groups bound to the elements. This era may be considered to have reached its apex in the discovery by Grignard of the magnesium reagents which occupy a special place because of their ease of synthesis and reactivity. With the exception of trimethylplatinum chloride discovered by Pope, Peachy and Gibson in 1907 by use of the Grignard reagent, attempts to make stable transition metal alkyls and aryls corresponding to those of main group elements met with little success, although it is worth recalling that even in 1919 Hein and his co-workers were describing the 'polyphenyl-chromium' compounds now known to be arene complexes.

The other major area of organometallic compounds, namely metal compounds of carbon monoxide, originated in the work starting in 1868 of Schützenberger and later of Mond and his co-workers and was subsequently developed especially by Hieber and his students. During the first half of this century, aided by the use of magnesium and, later, lithium reagents the development of main group organo chemistry was quite rapid, while from about 1920 metal carbonyl chemistry and catalytic reactions of carbon monoxide began to assume importance.

In 1937 Krause and von Grosse published their classic book 'Die Chemie der Metallorganischen Verbindungen'. Almost 1000 pages in length, it listed scores of compounds, mostly involving metals of the main groups of the periodic table. Compounds of the transition elements could be dismissed in 40 pages. Indeed, even in 1956 the stimulating 197-page monograph 'Organometallic Compounds' by Coates adequately reviewed organo transition metal complexes within 27 pages.

Although exceedingly important industrial processes in which transition metals were used for catalysis of organic reactions were developed in the 1930s, mainly in Germany by Reppe, Koch, Roelen, Fischer and Tropsch and others, the most dramatic growth in our knowledge of organometallic chemistry, particularly of transition metals, has stemmed from discoveries made in the middle years of this century. The introduction in the same period of physical methods of structure determination (infrared, nuclear magnetic resonance, and especially single-crystal X-ray diffraction) as routine techniques to be used by preparative chemists allowed increasingly sophisticated exploitation of discoveries. Following the recognition of the structure of ferrocene, other major advances quickly followed, including the isolation of a host of related π-complexes, the synthesis of a plethora of organometallic compounds containing metal–metal bonds, the characterization of low-valent metal species in which hydrocarbons are the only ligands, and the recognition from dynamic NMR spectra that ligand site exchange and tautomerism were common features in organometallic and metal carbonyl chemistry. The discovery of alkene polymerization using aluminium alkyl–titanium chloride systems by Ziegler and Natta and of the Wacker palladium–copper catalysed ethylene oxidation led to enormous developments in these areas.

In the last two decades, organometallic chemistry has grown more rapidly in scope than have the classical divisions of chemistry, leading to publications in journals of all national chemical societies, the appearance of primary journals specifically concerned with the topic, and the growth of annual review volumes designed to assist researchers to keep abreast of accelerating developments.

Organometallic chemistry has become a mature area of science which will obviously continue to grow. We believe that this is an appropriate time to produce a comprehensive review of the subject, treating organo derivatives in the widest sense of both main group and transition elements. Although advances in transition metal chemistry have appeared to dominate progress in recent years, spectacular progress has, nevertheless, also been made in our knowledge of organo compounds of main group elements such as aluminium, boron, lithium and silicon.

In these Volumes we have assembled a compendium of knowledge covering contemporary organometallic and carbon monoxide chemistry. In addition to reviewing the chemistry of the elements individually, two Volumes survey the use of organometallic species in organic synthesis and in catalysis, especially of industrial utility. Within the other Volumes are sections devoted to such diverse topics as the nature of carbon–metal bonds, the dynamic behaviour of organometallic compounds in solution, heteronuclear metal–metal bonded compounds, and the impact of organometallic compounds on the environment. The Volumes provide a unique record, especially of the intensive studies conducted during the past 25 years. The last Volume of indexes of various kinds will assist readers seeking information on the properties and synthesis of compounds and on earlier reviews.

As Editors, we are deeply indebted to all those who have given their time and effort to this project. Our Contributors are among the most active research workers in those areas of the subject that they have reviewed and they have well justified international reputations for their scholarship. We thank them sincerely for their cooperation.

Finally, we believe that 'Comprehensive Organometallic Chemistry', as well as providing a lasting source of information, will provide the stimulus for many new discoveries since we do not believe it possible to read any of the articles without generating ideas for further research.

E. W. ABEL
*Exeter*

F. G. A. STONE
*Bristol*

G. WILKINSON
*London*

# Contributors to Volume 6

Dr. J. C. Bryan
Oak Ridge National Laboratory, PO Box 2008, Oak Ridge, TN 37831, USA

Professor T. C. Flood
Department of Chemistry, University of Southern California, Los Angeles, CA 90089, USA

Professor G. S. Girolami
Department of Chemistry, University of Illinois, 505 South Matthews Avenue, Urbana, IL 61801, USA

Professor D. M. Hoffman
Department of Chemistry, University of Houston, Houston, TX 77204, USA

Professor K. F. McDaniel
Department of Chemistry, Ohio University, Athens, OH 45701, USA

Professor R. J. Morris
Ball State University, 2000 University Avenue, Muncie, IN 47306, USA

Professor J. M. O'Connor
Department of Chemistry, University of California, San Diego, 9500 Gilman Drive, La Jolla, CA 92093-0358, USA

Dr. A. P. Sattelberger
Isotope and Nuclear Chemistry Division, Los Alamos National Laboratory, Mail Stop J515, Los Alamos, NM 87545, USA

Professor P. M. Treichel
Department of Chemistry, University of Wisconsin–Madison, 1101 University Avenue, Madison, WI 53706, USA

# Abbreviations

The abbreviations used throughout 'Comprehensive Organometallic Chemistry II' are consistent with those used in 'Comprehensive Organometallic Chemistry' and with other standard texts in this area. The abbreviations in some instances may differ from those commonly used in other branches of chemistry.

| | |
|---|---|
| Ac | acetyl |
| acac | acetylacetonate |
| AIBN | 2,2'-azobisisobutyronitrile |
| Ar | aryl |
| arphos | 1-(diphenylphosphino)-2-(diphenylarsino)ethane |
| Azb | azobenzene |
| | |
| 9-BBN | 9-borabicyclo[3.3.1]nonyl |
| 9-BBN-H | 9-borabicyclo[3.3.1]nonane |
| BHT | 2,6-di-*t*-butyl-4-methylphenol (butylated hydroxytoluene) |
| bipy | 2,2'-bipyridyl |
| t-BOC | *t*-butoxycarbonyl |
| bsa | *N,O*-bis(trimethylsilyl)acetamide |
| bstfa | *N,O*-bis(trimethylsilyl)trifluoroacetamide |
| btaf | benzyltrimethylammonium fluoride |
| Bz | benzyl |
| | |
| can | ceric ammonium nitrate |
| cbd | cyclobutadiene |
| 1,5,9-cdt | cyclododeca-1,5,9-triene |
| chd | cyclohexadiene |
| chpt | cycloheptatriene |
| [Co] | cobalamin |
| (Co) | cobaloxime [Co(DMG)$_2$] derivative |
| cod | 1,5-cyclooctadiene |
| cot | cyclooctatetraene |
| Cp | $\eta^5$-cyclopentadienyl |
| Cp* | pentamethylcyclopentadienyl |
| 18-crown-6 | 1,4,7,10,13,16-hexaoxacyclooctadecane |
| CSA | camphorsulfonic acid |
| csi | chlorosulfonyl isocyanate |
| Cy | cyclohexyl |
| | |
| dabco | 1,4-diazabicyclo[2.2.2]octane |
| dba | dibenzylideneacetone |
| dbn | 1,5-diazabicyclo[4.3.0]non-5-ene |
| dbu | 1,8-diazabicyclo[5.4.0]undec-7-ene |
| dcc | dicyclohexylcarbodiimide |
| dcpe | 1,2-bis(dicyclohexylphosphino)ethane |
| ddq | 2,3-dichloro-5,6-dicyano-1,4-benzoquinone |
| deac | diethylaluminum chloride |
| dead | diethyl azodicarboxylate |
| depe | 1,2-bis(diethylphosphino)ethane |
| depm | 1,2-bis(diethylphosphino)methane |
| det | diethyl tartrate (+ or −) |

| DHP | dihydropyran |
|-----|--------------|
| diars | 1,2-bis(dimethylarsino)benzene |
| dibal-H | diisobutylaluminum hydride |
| dien | diethylenetriamine |
| DIGLYME | bis(2-methoxyethyl)ether |
| diop | 2,3-$O$-isopropylidene-2,3-dihydroxy-1,4-bis(diphenylphosphino)butane |
| dipt | diisopropyl tartrate (+ or −) |
| dma | dimethylacetamide |
| dmac | dimethylaluminum chloride |
| DMAD | dimethyl acetylenedicarboxylate |
| dmap | 4-dimethylaminopyridine |
| DME | dimethoxyethane |
| DMF | $N,N'$-dimethylformamide |
| DMG | dimethylglyoximate |
| DMI | $N,N'$-dimethylimidazalone |
| dmpe | 1,2-bis(dimethylphosphino)ethane |
| dmpm | bis(dimethylphosphino)methane |
| DMSO | dimethyl sulfoxide |
| dmtsf | dimethyl(methylthio)sulfonium fluoroborate |
| dpam | bis(diphenylarsino)methane |
| dppb | 1,4-bis(diphenylphosphino)butane |
| dppe | 1,2-bis(diphenylphosphino)ethane |
| dppf | 1,1'-bis(diphenylphosphino)ferrocene |
| dpph | 1,6-bis(diphenylphosphino)hexane |
| dppm | bis(diphenylphosphino)methane |
| dppp | 1,3-bis(diphenylphosphino)propane |
| | |
| eadc | ethylaluminum dichloride |
| edta | ethylenediaminetetraacetate |
| eedq | $N$-ethoxycarbonyl-2-ethoxy-1,2-dihydroquinoline |
| en | ethylene-1,2-diamine |
| Et$_2$O | diethyl ether |
| | |
| F$_6$acac | hexafluoroacetylacetonate |
| Fc | ferrocenyl |
| Fp | Fe(CO)$_2$Cp |
| | |
| HFA | hexafluoroacetone |
| hfacac | hexafluoroacetylacetonate |
| hfb | hexafluorobut-2-yne |
| HMPA | hexamethylphosphoramide |
| hobt | hydroxybenzotriazole |
| | |
| IpcBH$_2$ | isopinocampheylborane |
| Ipc$_2$BH | diisopinocampheylborane |
| | |
| kapa | potassium 3-aminopropylamide |
| K-selectride | potassium tri-$s$-butylborohydride |
| | |
| LAH | lithium aluminum hydride |
| LDA | lithium diisopropylamide |
| LICA | lithium isopropylcyclohexylamide |
| LITMP | lithium tetramethylpiperidide |
| L-selectride | lithium tri-$s$-butylborohydride |
| LTA | lead tetraacetate |
| | |
| mcpba | $m$-chloroperbenzoic acid |
| MeCN | acetonitrile |
| MEM | methoxyethoxymethyl |
| MEM-Cl | β-methoxyethoxymethyl chloride |

| | |
|---|---|
| Mes | mesityl |
| mma | methyl methacrylate |
| mmc | methylmagnesium carbonate |
| MOM | methoxymethyl |
| Ms | methanesulfonyl |
| MSA | methanesulfonic acid |
| MsCl | methanesulfonyl chloride |
| | |
| nap | 1-naphthyl |
| nbd | norbornadiene |
| NBS | *N*-bromosuccinimide |
| NCS | *N*-chlorosuccinimide |
| nmo | *N*-methylmorpholine *N*-oxide |
| NMP | *N*-methyl-2-pyrrolidone |
| Nu⁻ | nucleophile |
| | |
| ox | oxalate |
| | |
| pcc | pyridinium chlorochromate |
| pdc | pyridinium dichromate |
| phen | 1,10-phenanthroline |
| phth | phthaloyl |
| ppa | polyphosphoric acid |
| ppe | polyphosphate ester |
| [PPN]⁺ | [(Ph$_3$P)$_2$N]⁺ |
| ppts | pyridinium *p*-toluenesulfonate |
| py | pyridine |
| pz | pyrazolyl |
| | |
| Red-Al | sodium bis(2-methoxyethoxy)aluminum dihydride |
| | |
| sal | salicylaldehyde |
| salen | *N,N'*-bis(salicylaldehydo)ethylenediamine |
| SEM | β-trimethylsilylethoxymethyl |
| | |
| tas | tris(diethylamino)sulfonium |
| tasf | tris(diethylamino)sulfonium difluorotrimethylsilicate |
| tbaf | tetra-*n*-butylammonium fluoride |
| TBDMS | *t*-butyldimethylsilyl |
| TBDMS-Cl | *t*-butyldimethylsilyl chloride |
| TBDPS | *t*-butyldiphenylsilyl |
| tbhp | *t*-butyl hydroperoxide |
| TCE | 2,2,2-trichloroethanol |
| TCNE | tetracyanoethene |
| TCNQ | 7,7,8,8-tetracyanoquinodimethane |
| terpy | 2,2':6',2"-terpyridyl |
| tes | triethylsilyl |
| Tf | triflyl (trifluoromethanesulfonyl) |
| TFA | trifluoracetic acid |
| TFAA | trifluoroacetic anhydride |
| tfacac | trifluoroacetylacetonate |
| THF | tetrahydrofuran |
| THP | tetrahydropyranyl |
| tipbs-Cl | 2,4,6-triisopropylbenzenesulfonyl chloride |
| tips-Cl | 1,3-dichloro-1,1,3,3-tetraisopropyldisiloxane |
| TMEDA | tetramethylethylenediamine [1,2-bis(dimethylamino)ethane] |
| TMS | trimethylsilyl |
| TMS-Cl | trimethylsilyl chloride |
| TMS-CN | trimethylsilyl cyanide |
| Tol | tolyl |
| tpp | *meso*-tetraphenylporphyrin |

| | |
|---|---|
| Tr | trityl (triphenylmethyl) |
| tren | 2,2',2"-triaminotriethylamine |
| trien | triethylenetetraamine |
| triphos | 1,1,1-tris(diphenylphosphinomethyl)ethane |
| Ts | tosyl |
| TsMIC | tosylmethyl isocyanide |
| ttfa | thallium trifluoroacetate |

# Contents of All Volumes

# 1

# Manganese Carbonyls and Manganese Carbonyl Halides

## PAUL M. TREICHEL
### University of Wisconsin–Madison, WI, USA

## 1.1 INTRODUCTION

### 1.1.1 Review of Earlier Work

A brief abstract of the chemistry known prior to 1982 is given below, to provide a logical introduction and to place recent studies in a clearer context. Further details are available in *COMC-I*.

Dimanganese decacarbonyl, [Mn$_2$(CO)$_{10}$], was a latecomer in the metal carbonyl field, details of its synthesis being reported in 1954. Sodium pentacarbonylmanganate, Na[Mn(CO)$_5$], and manganese pentacarbonyl iodide, [MnI(CO)$_5$], were reported in the same year, while the cationic species hexacarbonylmanganese(I), $[Mn(CO)_6]^+$, was described in 1961. Extensive synthetic efforts have produced many related complexes with carbonyl groups replaced by other ligands. In addition, synthetic methodology to convert between these groups of compounds was developed and refined. Known complexes, virtually all 18-electron species, have been characterized by conventional techniques.

Interest in mechanisms of carbonyl substitution of $[Mn_2(CO)_{10}]$ is evident in research prior to 1982. Two mechanistic pathways exist, with the first step being either carbonyl loss or metal–metal bond scission. There has been considerable attention paid to highly reactive radical species such as $[Mn(CO)_5]$, partly because 17-electron organometallic complexes are rare. Several $[Mn(CO)_3(L)_2]$ species (L = phosphine) are stable in solution relative to dimerization, but in 1982 none had been isolated. These species are reactive toward halogen and hydrogen donors. Mechanisms for reactions of radical species have remained an active area of study and, as described in this chapter, the use of sophisticated techniques to study the process of metal–metal bond cleavage vs. carbonyl loss is an important part of recent research.

The commercial availability of $[Mn_2(CO)_{10}]$ was a factor in its use in the synthesis of many other organometallic complexes, including those with nitrosyl groups and with hydrocarbon ligands. In addition, $Na[Mn(CO)_5]$ was found to be a useful reactant in the synthesis of hydrido–manganese carbonyls as well as many metal alkyl, aryl, and acyl species and derivatives of metalloidal elements. A more highly reduced species, $Na_3[Mn(CO)_4]$, was reported, and a number of polynuclear complexes and mixed-metal species were known. Some data on redox chemistry was in place by the early 1980s.

## 1.1.2 Other Literature

Review articles provide detailed coverage of specific topics included in this chapter and are referred to in the text as appropriate. These include: electron-transfer processes;[1] photochemistry of compounds with metal–metal bonds;[2] nonhomolytic cleavage of metal–metal bonds on photolysis;[3] light-induced disproportionation of metal carbonyls;[4] 19-electron species in organometallic chemistry;[5] electron transfer and radicals in organometallic chemistry;[6] organometallic ions and ion pairs;[7] ion pairing in transition metal anions;[8] highly reduced metal carbonyl anions;[9] synthetic routes to metal carbonyl cations;[10] photochemical behavior of phosphine substituted dimanganese carbonyls;[11] catalyst- and reagent-assisted carbonyl substitution;[12] dppm compounds;[13-14] photochemistry of diimine complexes;[15-16] redox chemistry applied to the synthesis of transition metal complexes;[17] organometallic radical processes (symposium papers);[18] and manganese carbonyl–diene photochemistry.[19]

Tested synthetic procedures for several manganese carbonyl and manganese carbonyl halides can be found in recent issues of *Inorganic Syntheses* and *Organometallic Syntheses*. Procedures are given for the synthesis of the following compounds: $[MnX(CO)_5]$ (X = Cl, Br, I;[20] $O_3SCF_3$);[21] $[Mn_2(\mu\text{-X})_2(CO)_8]$ (X = Cl, Br, I);[22] and $Na_3[Mn(CO)_4]$.[23]

## 1.2 DIMANGANESE DECACARBONYL AND DERIVATIVES

### 1.2.1 Dimanganese Decacarbonyl $[Mn_2(CO)_{10}]$

#### 1.2.1.1 Physical properties

Attention in this section is narrowly directed to $[Mn_2(CO)_{10}]$. Separate sections are set aside to cover details arising from studies involving measurement of metal–metal bond energies and the chemistry of the mixed-metal species $[MnRe(CO)_{10}]$. A survey on $[Mn_2(CO)_{10-n}(L)_n]$ chemistry is found in Section 1.2.2 and radical species arising from homolytic metal–metal bond cleavage, $[Mn(CO)_{5-n}(L)_n]$, are discussed more fully in Section 1.2.3.

The physical properties of $[Mn_2(CO)_{10}]$ were mostly determined in the years following its discovery. A recent redetermination of the solid-state structure of $[Mn_2(CO)_{10}]$ at 74 K has resulted in a more accurate metal–metal length (0.2895(1) nm). Experimental accuracy was sufficient to distinguish between axial and equatorial Mn–C and C–O bond lengths, with the data suggesting a higher degree of backbonding to axial carbonyl groups.[24] Microcalorimetric techniques (on bromination of $[Mn_2(CO)_{10}]$ and sublimation of this compound) produced values for the enthalpy of formation of this compound ($\Delta H_f^\circ = -1675 \pm 8$ kJ mol$^{-1}$) and the heat of sublimation ($\Delta H^\circ = +92.3 \pm 2.1$ kJ mol$^{-1}$). Both results compare favorably with results from previous studies.[25]

Reactivity studies on $[Mn_2(CO)_{10}]$ have provided both new information and a more detailed perspective of known processes. Particularly important studies are described below.

Photolysis of $[Mn_2(CO)_{10}]$ results in two reactions, loss of a carbonyl group and Mn–Mn bond scission. Gas-phase and condensed-phase (matrix) studies are discussed separately.

A simple bonding model describes a Mn–Mn single bond in $[Mn_2(CO)_{10}]$. Photolysis into the lowest excited singlet state ($\sigma \to \sigma^*$ transition), decreases the Mn–Mn bond order to zero and results in dissociation of $[Mn_2(CO)_{10}]$ into $[Mn(CO)_5]$ radicals. Most early studies focused on this process, but as techniques were refined it became apparent that carbonyl loss is a concurrent process. Weitz and co-workers[26] detected both $[Mn(CO)_5]$ and $[Mn_2(CO)_9]$ as products of photolysis using time-resolved spectroscopic methods; infrared spectra of these species were recorded and recombination kinetics were measured. The observation that the $[Mn(CO)_5]/[Mn_2(CO)_9]$ ratio decreased at lower wavelength was studied further by Prinslow and Vaida[27] who subjected the complex, in the gas phase, to eximer laser photolysis at 350 nm, 248 nm, and 193 nm, identifying the photoproducts using a quadrupole mass spectrometer with electron-impact ionization. With 350 nm radiation (the wavelength required for the $\sigma \to \sigma^*$ transition), products of both photoprocesses were detected, while at the two shorter wavelengths only the product of ligand loss was observed.

Condensed-phase photolysis has been studied more intensively. As the early work in this area is described in a review,[2] only a summary of pertinent information is given here. Early studies identified radical species as intermediates in kinetic schemes.[28] More direct evidence for radical intermediates was provided by conventional flash photolysis experiments, from which the spectra of transient species and kinetic data for their very fast recombination were obtained.[29] The presence of a second long-lived intermediate was confirmed by flash photolysis experiments in ethanol[30] or in cyclohexane;[31] $[Mn_2(CO)_9(EtOH)]$ was named as the intermediate in the former instance, while in the latter the intermediate was assumed to be $[Mn_2(CO)_9]$ because hydrocarbons rarely serve as ligands. In the latter study, rate data were consistent with recombination of the ligand. The structure of this species was assigned in a separate matrix-isolation experiment at 77 K, a temperature at which it does not degrade.[32] Based on a low-frequency carbonyl stretch at 1760 cm$^{-1}$, a structure with a semibridging carbonyl group (1) was proposed.

(1)

In 1984, Brown and co-workers described flash-photolysis experiments using $[Mn_2(CO)_8(L)_2]$ (L = CO, phosphines, phosphites).[33-4] For $[Mn_2(CO)_{10}]$ in hexane, the rates of decay of the two photoproducts were determined. The rate for $[Mn(CO)_5]$, a very fast process, was monitored by the decrease in intensity of an absorption at 850 nm. The second species, $[Mn_2(CO)_9]$, has an absorption at 510 nm; this decreases at a considerably slower rate which is dependent on [CO].

More recently, Sullivan and Brown described reactions of $[Mn_2(CO)_{10-n}(L)_n]$ (L = CO, phosphines) with Bu$_3$SnH and Et$_3$SiH, using both continuous photolysis[35] and flash photolysis (in hexane, 316 nm or 366 nm radiation under either CO or argon).[36] With $[Mn_2(CO)_{10}]$, $[MnH(CO)_5]$, and $[Mn(AR_3)(CO)_5]$ (AR$_3$ = SnBu$_3$, SiEt$_3$) are formed. These reactions are believed to involve the loss of a carbonyl group followed by oxidative addition of the main group complex to the metal; reductive elimination of $[MnH(CO)_5]$ from this site and readdition of a carbonyl group gives the second product. In the [Bu$_3$SnH] reaction under low CO pressure, a second product, $[MnH(SnBu_3)_2(CO)_4]$, is observed which is formed by oxidative addition of a second mole of the tin reagent to a coordinatively unsaturated intermediate $[Mn(SnBu_3)(CO)_4]$.

Recent work adds another dimension to this story. Using ultrafast techniques to study the photodissociation of CO from $[Mn_2(CO)_{10}]$ in hydrocarbons, Zhang and Harris[37] detected a very short-lived intermediate, $[Mn_2(CO)_9]$, during the formation. They speculated that this might be a nonbridged species. However, Brown and co-workers[38] subsequently pointed out that the rate of isomerization appeared to be too fast for this simple rearrangement, considering that the activation energy would probably be of the order of 12–21 kJ. Instead, they suggested that this intermediate is a solvated species, $[Mn_2(CO)_9(S)]$. When photolysis at 93 K in glassy 3-methylpentane was carried out, infrared data could be obtained; $\nu(CO)$ values are indicative of the solvated species. These new absorptions disappear over about 10 min at this temperature, with bands corresponding to the semibridged complex growing over this period.

Three mass spectrometry studies on $[Mn_2(CO)_{10}]$ merit mention. Ion cluster fragments, $[Mn_x(CO)_y]^+$ ($x = 1, 2; y = 0–10$), are formed from $[Mn_2(CO)_{10}]$ using laser–ion beam photodissociation. Upper limits for $D(Mn–Mn)$ and $D(Mn–CO)$ in these ionic species were established to be 184 kJ mol$^{-1}$ and 105 kJ mol$^{-1}$, respectively.[39] In another study, metal cluster ions with up to eight metal atoms were observed to be formed by the combination of electron-impact fragments with $[Mn_2(CO)_{10}]$ (and also with $[MnRe(CO)_{10}]$) and rate constants for these aggregation reactions were measured.[40] Gas-phase reactions of $[Mn_2(CO)_{10}]$, $[MnRe(CO)_{10}]$, and $[Re_2(CO)_{10}]$ with thermal electrons (negative-ion mass spectrometry) gave $[M_2(CO)_{10}]^-$ which decomposes by splitting off either CO or $[M(CO)_5]$ fragments. From these experiments, values of $2.43 \pm 0.21$ eV for the electron affinity of $[Mn(CO)_5]$ and $774 \pm 13$ kJ mol$^{-1}$ for the proton affinity of $[Mn_2(CO)_{10}]$ were determined.[41]

### 1.2.1.2 Studies defining bond energies

The metal–metal bond energy in $[Mn_2(CO)_{10}]$ ($142 \pm 54$ kJ mol$^{-1}$) was estimated from calorimetric data in 1960.[42] Studies over the next 20 years involved the use of a variety of techniques in an attempt to refine this value. One value that gained early acceptance ($153.8 \pm 1.6$ kJ mol$^{-1}$) was obtained from kinetic data, where it was assumed that metal–metal bond breaking was the primary factor defining activation energy for reactions with alkyl halides, oxygen, and other species.

During the 1980s, further studies have been reported. Rate-constant data from flash-photolysis experiments were used along with rate parameters obtained for the thermal decomposition of $[Mn_2(CO)_{10}]$ in decalin, to give the values $\Delta H = 151$ kJ mol$^{-1}$ and $\Delta S = 134$ J mol$^{-1}$ deg$^{-1}$.[43]

In 1984, Beauchamp and co-workers[44] used PE spectroscopy and ICR data on $[Mn(Bz)(CO)_5]$ to calculate a value for $\Delta H_f^\circ$ for $[Mn(CO)_5]$ (g.) ($-707$ kJ mol$^{-1}$). This, in conjunction with the heat of formation of $[Mn_2(CO)_{10}]$ (g.), yields a value for $\Delta H$ of $172 \pm 37$ kJ mol$^{-1}$ for metal–metal scission. The following year, Vaida and co-workers[45] carried out photoacoustic time-resolved calorimetry studies on $[Mn_2(CO)_{10}]$ in several solvents to obtain a value of $159 \pm 20$ kJ mol$^{-1}$.

The most recent effort to obtain a value for the metal–metal bond energy was an electrochemical study using $[Mn_2(CO)_{10}]$ (in MeCN, with $Bu_4NPF_6$ as supporting electrolyte).[46] From electrochemical potentials for $[Mn_2(CO)_{10}] + 2e \rightleftharpoons 2[Mn(CO)_5]^-$ ($E = -0.69$ V vs. SCCE) and $[Mn(CO)_5] + e \rightleftharpoons [Mn(CO)_5^-]$ ($E = -0.15$ V vs. SCCE), a value for $\Delta G^\circ$ of $117 \pm 20$ kJ mol$^{-1}$ was determined for the process $[Mn_2(CO)_{10}] \rightarrow 2[Mn(CO)_5]$ in solution. Combining this value with a reasonable estimate for the entropy of this process (+134 J mol$^{-1}$ deg$^{-1}$, the value obtained from flash photolysis experiments)[43] gave 159 kJ mol$^{-1}$ for the Mn–Mn bond energy.

For the energy required for ligand dissociation, the value of 155 kJ mol$^{-1}$ for $D(Mn–CO)$, obtained from a 1969 study on the kinetics of the $Mn_2(CO)_{10} + PPh_3$ reaction,[47] is generally quoted.[27,48]

### 1.2.1.3 The mixed-metal species [MnRe(CO)₁₀]

Considerable attention has been directed to the mixed-metal species $[MnRe(CO)_{10}]$. Its properties are expected to be intermediate between the two homonuclear species; in addition, the rhenium atom can act as a label in some mechanistic studies.

It is possible to prepare $[MnRe(CO)_{10}]$ by thermal or photolytic disproportionation of $[Mn_2(CO)_{10}]$ and $[Re_2(CO)_{10}]$. The need for quite high temperatures for the thermal process was first noted in crossover experiments.[49] The compound $[MnRe(CO)_9(PPh_3)]$ forms in a reaction of $[MnRe(CO)_{10}]$ and $PPh_3$ at 130 °C, whereas this species is not detected when mixtures of $[Mn_2(CO)_{10}]$ and $[Re_2(CO)_{10}]$ are heated with the phosphine under similar conditions. In addition,[50-1] it was shown that the disproportionation reaction forming $[MnRe(CO)_{10}]$ occurs fairly rapidly at 170–190 °C. A value for the equilibrium constant of $2.13 \pm 0.13$ (confirmed in a separate study)[52] indicates a near statistical distribution of products. Two bimolecular pathways are postulated, neither of which involves metal–metal bond cleavage.[50] Both processes are thought to involve formation of tetrametallic clusters as intermediates. In the slower process, a rate-determining dissociation of a carbonyl group is proposed.

A second synthetic route to $[MnRe(CO)_{10}]$ involves displacement of halide ion in $[Re(X)(CO)_5]$ by $[Mn(CO)_5]^-$. Using $[Re(O_3SCF_3)(CO)_5]$ enriched with $^{13}CO$ in this reaction provides a route to $[(OC)_5MnRe(^{13}CO)_5]$ (Equation (1)).[53] Beck et al.[54] extended this route to prepare several phosphine substituted complexes, $[(OC)_5MnRe(CO)_4(L)]$.

$$K[Mn(CO)_5] + Re(O_3SCF_3)(^{13}CO)_5 \longrightarrow (OC)_5MnRe(^{13}CO)_5 + K[O_3SCF_3] \qquad (1)$$

The $^{13}$CO labeled species made other experiments possible. The label does not scramble readily under ambient conditions; however, carbonyl exchange between metals was observed at 65–85 °C.[54] Using $^{13}$C NMR, the activation energy parameters for this process were determined: $\Delta H^{\ddagger} = 53 \pm 6$ kJ mol$^{-1}$ and $\Delta S^{\ddagger} = -170$ J mol$^{-1}$ deg$^{-1}$. The large negative entropy is consistent with a mechanism involving pairwise migration of carbonyl groups.

By mass spectrometry, it was found that a carbonyl group attached to manganese in [MnRe(CO)$_{10}$] is preferentially lost.[55] Near-UV irradiation of this species at 77 K in a glassy matrix also causes loss of a carbonyl group from manganese.[56] Spectroscopic data suggest that the intermediate, [MnRe(CO)$_9$], has Structure (2). At low temperature, scrambling of the $^{13}$CO label between rhenium and manganese in (2) does not occur. If photolysis is carried out with PPh$_3$ present, and the matrix is then warmed to 298 K, [(PPh$_3$)(CO)$_4$MnRe($^{13}$CO)$_5$] forms.[57]

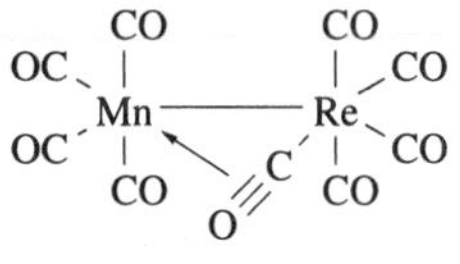

(2)

Formation of *eq*-[(OC)$_5$MnRe(CO)$_4$(CNR)] (R = Pr$^i$, Bu$^t$, Cy, 2,6-dimethylphenyl (*eq* = equatorial)) occurs in reactions between [MnRe(CO)$_{10}$] and these isocyanides using a Pd$^0$ catalyst.[58] However, an unexpected result was obtained in this reaction using [(OC)$_5$MnRe($^{13}$CO)$_5$]. The product, *eq*-[(OC)$_5$MnRe(CO)$_4$(CNBu$^t$)], retains the label on rhenium, with little or no transfer to manganese.[59] This is evidence that the loss of CO and addition of CNBu$^t$ occurs on the same metal in the presence of Pd$^0$, which is striking in view of evidence showing preferential CO loss from manganese.[55-7]

A single-crystal x-ray diffraction study on [MnRe(CO)$_{10}$] reveals a Mn–Re distance of 0.2909(1) nm, which is 0.005 nm shorter than the length predicted based on metal–metal distances in the two parent species.[60] Subsequently, molecular structures were determined for [(OC)$_5$MnRe(CO)$_4$(L)] (L = *eq*-CNBu$^t$, *ax*-PBz$_3$) and [(OC)$_5$MnRe(CO)$_3$(*eq*-CNBu$^t$) (*ax*-PPh$_3$)] (*ax* = axial). In these examples, metal–metal bond distances did not vary greatly, the average of the three values (0.2963 ± 0.0007 nm) being very close to the sum of the manganese and rhenium covalent radii.[61]

### 1.2.1.4 Chemical properties of [Mn$_2$(CO)$_{10}$]

The mechanism of ligand substitution in [Mn$_2$(CO)$_{10}$] under thermal conditions was shown to be fully consistent with a scheme not involving metal–metal bond scission (Scheme 1).[48] With L = $^{13}$CO at 550–650 torr and at 120 °C in octane, a half-life of 45 ± 10 min was established for this reaction (very similar to the rate of the reaction with PPh$_3$). Lack of metal–metal bond scission in this reaction was shown when the reaction with CO was run using a mixture of [Mn$_2$(CO)$_{10}$] and Mn$_2$($^{13}$CO)$_{10}$ (94% enriched). Mass-spectrometric analysis of the products did not show the buildup of [Mn$_2$(CO)$_{4-6}$($^{13}$CO)$_{6-4}$] expected for metal–metal bond cleavage. A similar result was obtained when this mixture of [Mn$_2$(CO)$_{10}$] and [Mn$_2$($^{13}$CO)$_{10}$] was reacted with PPh$_3$.

$$[Mn_2(CO)_{10}] \xrightleftharpoons{\pm CO} [Mn_2(CO)_9] \xrightarrow{+ PPh_3} [Mn_2(CO)_9(PPh_3)]$$

**Scheme 1**

A value for the equilibrium constant for the reaction between [Mn$_2$(CO)$_{10}$] and dihydrogen[26] (Equation (2)) has been obtained.[62] This study used supercritical CO$_2$ as a solvent and [Co$_2$(CO)$_8$] as a catalyst, at 165–220 °C and under CO and H$_2$ pressures of 142–300 atm. NMR ($^1$H and $^{55}$Mn) was used to monitor equilibration. These experiments yielded the thermodynamic parameters $\Delta H^{\circ} = +36.4 \pm 1.2$ kJ mol$^{-1}$ and $\Delta S^{\circ} = 35.6 \pm 4$ J mol$^{-1}$ deg$^{-1}$.

$$Mn_2(CO)_{10} + H_2 \rightleftharpoons 2 MnH(CO)_5 \qquad (2)$$

Some reactions between [Mn$_2$(CO)$_{10}$] and various ligands result in ionic products [Mn(CO)$_{6-n}$(L)$_n$]$^+$ [Mn(CO)$_5$]$^-$ (*n* is usually 3 or 6). Tyler and co-workers[3,63-4] studied photochemical reactions with amines (py, NEt$_3$) and other ligands. These reactions are said to occur by a radical chain mechanism. Metal–metal bond cleavage is the first step in the process, followed by rapid substitution of two carbonyl ligands. At this point, the addition of another ligand produces a 19-electron species, [Mn(CO)$_3$(L)$_3$], which undergoes electron transfer with [Mn$_2$(CO)$_{10}$], giving the products.

Further support for a 19-electron intermediate comes from the reaction of the monomer–dimer mixture, [Mn(CO)$_3$(depe)] ⇌ [Mn$_2$(CO)$_6$(depe)$_2$] with [Mn$_2$(CO)$_{10}$] and PMe$_3$ in the dark, producing [Mn(CO)$_3$(depe)(PMe$_3$)]$^+$ and [Mn(CO)$_5$]$^-$.[64] Related chemistry is encountered with [(OC)$_5$MnMn(CO)$_3$(diimine)][65–6] and [(OC)$_4$CoMn(CO)$_3$(diimine)][67] complexes (diimine = bipy, *o*-phen, R-dab; where R-dab is RN=CHCH=NR, R = Pr$^i$, Bu$^t$, *o*-anisyl). Photolysis of these complexes in the presence of PBu$_3$ causes disproportionation, giving [Mn(CO)$_3$(diimine)(PBu$_3$)]$^+$ and [Mn(CO)$_5$]$^-$ or [Co(CO)$_4$]$^-$. Two review articles summarize additional work in this area.[15–16]

Several similar reactions with mixed metal complexes have been reported.[68] Photolysis of [Mn$_2$(CO)$_{10}$], dppe, and [Fe(CO)$_2$(η-Cp)]$_2$ gives [Fe(CO)(dppe)(η-Cp)]$^+$ and [Mn(CO)$_5$]$^-$, while the reaction of [Mn$_2$(CO)$_{10}$], dien, and [Mo(CO)$_3$(η-Cp)]$_2$ under similar conditions yields [Mn(CO)$_3$(dien)]$^+$ and [Mo(CO)$_3$(η-Cp)]$^-$.

The cleavage of [Mn$_2$(CO)$_{10}$] (and also [MnRe(CO)$_{10}$] and [Re$_2$(CO)$_{10}$]) by halogens in polar solvents[69] produces [MnX(CO)$_5$] and [Mn(CO)$_5$(NCMe)]$^+$. A two-step mechanism is proposed. The reaction is initiated by the addition of X$^+$ to the metal–metal bond; solvent then displaces [MnX(CO)$_5$] from the halide-bridged intermediate. Not surprisingly, reaction rates are dependent on the solvent. The order of reaction rates is [Re$_2$(CO)$_{10}$] > [MnRe(CO)$_{10}$] > [Mn$_2$(CO)$_{10}$].

Reactions of [MnRe(CO)$_{10}$] with halogens in MeCN produce [MnX(CO)$_5$] and [Re(CO)$_5$(NCMe)]$^+$. Initial reaction with ICl gave [MnI(CO)$_5$] and [Re(CO)$_5$(NCMe)]$^+$, but the result is complicated by halogen exchange between [MnI(CO)$_5$] and ICl to give [MnCl(CO)$_5$] and I$_2$.

The redox chemistry of the manganese carbonyl system was mentioned earlier in the context of the metal–metal bond energy;[46] we return to this area again. The electrochemistry of the manganese carbonyl system is summarized in a Lattimer diagram (Scheme 2). In MeCN, [Mn(CO)$_5$] is unstable with respect to disproportionation, giving [Mn(CO)$_5$(NCMe)]$^+$ and [Mn(CO)$_5$]$^-$. Large overvoltages exist for both oxidation and reduction of [Mn$_2$(CO)$_{10}$] because these processes occur in one-electron steps to or from the electrode or to or from a reagent in solution, and because they occur through thermodynamically unstable intermediates.

$$
\begin{array}{c}
\underset{-1.90}{\phantom{x}} \quad 2\,[\text{Mn(CO)}_5] \quad \underset{-0.08}{\phantom{x}} \\[1em]
2\,[\text{Mn(CO)}_5(\text{L})]^+ \ \overset{-0.30}{\text{---}}\ \text{Mn}_2(\text{CO})_{10} \ \overset{-0.69}{\text{---}}\ 2\,[\text{Mn(CO)}_5]^- \\[1em]
-2.1 \qquad\qquad\qquad\qquad\qquad\qquad\qquad \\[1em]
[\text{Mn}_2(\text{CO})_{10}]^+ \ \overset{1.50}{\text{---}}\ \phantom{xxxxxxxxx}\ [\text{Mn}_2(\text{CO})_{10}]^-
\end{array}
$$

**Scheme 2**

Stop-flow kinetic techniques were used to measure the rates of electron transfer between [Mn$_2$(CO)$_{10}$] and [Re(CO)$_5$]$^-$, a reaction forming [Mn(CO)$_5$]$^-$. The reaction is first order in each reactant.[70]

Other chemical reactions of [Mn$_2$(CO)$_{10}$] have been reported. A dinuclear complex (**3**) is formed[71] with the ynamine MeC≡CNEt$_2$. The reaction of MeC≡CNMe$_2$ with [Mn$_2$(CO)$_9$(NCMe)] had a different outcome, forming (**4**).[72]

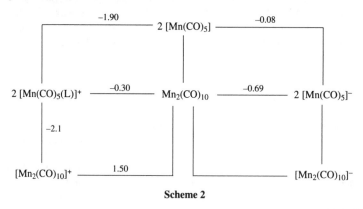

(3)                    (4)

The parent carbonyl [Mn(CO)$_5$]$^-$ and other species are present in samples of [Mn$_2$(CO)$_{10}$] absorbed on alumina and MgO, according to spectroscopic data.[73] Although catalysis involving [Mn$_2$(CO)$_{10}$] is not reviewed extensively here, a few studies deserve mention, including the reduction of PhNO$_2$ to PhNH$_2$,[74] ring chlorination of toluene,[75] and homologation of MeOH in the presence of CO and H$_2$.[76–7]

## 1.2.2 Derivatives of Dimanganese Decacarbonyl [Mn$_2$(CO)$_{10-n}$L$_n$]

Studies done before 1982 produced a great many derivatives of dimanganese decacarbonyl with one or more carbonyl groups replaced by other ligands. A few additional complexes of this type have been characterized in the late 1980s and early 1990s and, in addition, physicochemical methods used to study [Mn$_2$(CO)$_{10}$] have been applied to these substituted complexes.

Several nomenclature schemes are in use to identify structures of these species. Here, *ax* and *eq* are used to signify positions *trans* and *cis* to the metal–metal bond. Primes (as in *ax'* and *eq'*) distinguish between metal centers.

New complexes, prepared by the established routes involving thermal, photochemical, or catalyzed ligand substitution, are listed in Table 1.[78-95] It is interesting to learn of the first thiocarbonyl derivative of [Mn$_2$(CO)$_{10}$]. A small amount of [Mn$_2$(CO)$_9$(CS)], as a mixture of equatorial and axial isomers, is obtained from a reaction between Na[Mn(CO)$_5$] and S=CCl$_2$; major products in this reaction are [Mn$_2$(CO)$_{10}$] and [Mn(CO)$_4$Cl$_2$]$^-$.[78]

**Table 1**   New Mn$_2$(CO)$_{10-n}$(L)$_n$ derivatives.

| Compounds (L) | Synthetic route | Geometry (if known) | Ref. |
|---|---|---|---|
| *Mn$_2$(CO)$_9$(L)* | | | |
| CS | d | *ax + eq* (IR) | 78 |
| 4,5-Me$_2$-1,2-Ph$_2$-1,6-dihydrophosphorin sulfide | a (Δ) | *ax* (IR) | 79 |
| Tetrahydrothiophene | a (ℏv) | *eq* (x ray) | 80 |
| (Me2N)2C=S | a (ℏv) | *ax* (IR) | 81 |
| S̅C̅H̅$_2$CMe$_2$CH$_2$ | d | *eq* (IR) | 82 |
| =C(NMe$_2$)OAl$_2$(NMe$_2$)$_5$ | d | *ax* (x ray) | 83 |
| =C̅C̅H̅$_2$CH$_2$TiCp*$_2$O | d | *ax* (IR) | 84 |
| RCHO (R = Me, *p*-Tol, CH$_2$CH$_2$CH=C(Me)Ph) | c | *eq* (x ray, IR) | 85,86 |
| =C̅(CH$_2$)$_n$CH(Me$_3$)O̅ (n = 3) | b | *eq* (x ray) | 87 |
| =C̅OCH$_2$CH$_2$O̅ | b | *eq* (IR) | 88 |
| pyrazole, imidazole | a (ℏv) | *eq* (eq) | 89 |
| *Mn$_2$(CO)$_8$(L)* | | | |
| Tetrahydrothiophene | | See text (x ray) | 80 |
| S̅C̅H̅$_2$CMe$_2$CH$_2$ | | See text (x ray) | 82 |
| *Mn$_2$(CO)$_8$(L)$_2$* | | | |
| PF$_3$ | a (Δ,ℏv) | *ax, eq* (IR) | 90 |
| CNCy, CNBz, CNxylyl | *a (cat.)* | *eq, eq'* (IR) | 91 |
| *Mn$_2$(CO)$_7$(L)$_3$* | | | |
| PF$_3$ (two isomers) | a | *ax, ax' eq* | 90 |
| | d | *ax, eq, eq* (?) (NMR) | |
| CNR (R = Bz, xylyl) | a (cat.) | | 91 |
| *Mn$_2$(CO)$_6$(L)$_4$* | | | |
| PF$_3$ | a | *ax, ax', eq, eq'* | 90 |
| CNR (R = xylyl) | a | Mixture of isomers | 91 |
| *Mn$_2$(CO)$_8$(L^L)* | | | |
| (EtO)$_2$POP(OEt)$_2$ | a (Δ,ℏv) | *eq, eq'* | 92 |
| *Mn$_2$(CO)$_6$(L^L)$_2$* | | | |
| depm, Cy$_2$PCH$_2$PCy$_2$ | a (Δ) | *eq–eq'* | 93 |
| depe | a (ℏv) | *eq–eq'* | 94 |
| dppm, new isomer | *a (Δ)* | *(mer, fac')* | 95 |
| (EtO)2POP(OEt)2 | *a (Δ)* | *eq, eq'* | 92 |
| *Mn$_2$(CO)$_5$(L^L)$_2$* | | | |
| Cy$_2$PCH$_2$PCy$_2$ | a (Δ) | | 93 |

$^a$ Carbonyl replacement.   $^b$ Reaction at coordinated carbonyl group.   $^c$ [MnH(CO)$_5$] + [Mn(R)(CO)$_5$].   $^d$ Other, see cited reference. Δ thermal; ℏv, photolytic; cat., catalyzed.

In 1983, Suslick and Schubert[96] reported the use of ultrasound in the synthesis of [Mn$_2$(CO)$_8$(L)$_2$] species (L = PPh$_3$, P(OEt)$_3$, P(OPh)$_3$, PBu$_3$) from [Mn$_2$(CO)$_{10}$] and the ligand. Only disubstituted species are formed. Halomanganese carbonyls, [Mn(X)(CO)$_{5-n}$(L)$_n$], are obtained when halocarbon solvents are used. The rates of these processes are independent of the nature of the ligand and its concentration, and there was no evidence of metal–metal bond cleavage in this reaction.

A new preparation of oxygen bonded aldehyde complexes is shown in Equation (3).[85-6] If the products stand in the presence of $[Mn(H)(CO)_5]$, a slow reaction ensues forming $[Mn_3(\mu^3\text{-}CO)_2\text{-}(\mu^2\text{-}CO)(CO)_9]$. The aldehyde ligand in these complexes can be replaced by MeCN. This lability was exploited in a reaction with $[Re(H)(CO)_5]$, giving $[(OC)_5MnMn(CO)_4(\mu\text{-}H)Re(CO)_5]$.[97]

$$[MnH(CO)_5] + [Mn(R)(CO)_5] \longrightarrow eq\text{-}[Mn_2(CO)_9(O{=}CHR)] \tag{3}$$

Among the new complexes are two bimetallic species: $[Mn_2(CO)_9{=}C(NMe_2)OAl_2(NMe_2)_5]$, formed from $[Mn_2(CO)_{10}]$ and $Al(NMe_2)_3$;[83] and $[Mn_2(CO)_9({=}CCH_2CH_2(Ti\text{-}\eta\text{-}Cp*_2)O]$, obtained when $[Ti\text{-}(\eta\text{-}H_2C{=}CH_2)(\eta\text{-}Cp*)_2]$ and $[Mn_2(CO)_{10}]$ are mixed.[84]

Details of $PF_3$ + $[Mn_2(CO)_{10}]$ reactions using thermal and photolytic conditions have been worked out.[90] A mixture of products is amenable to partial separation by preparative gas chromatography. Five derivatives of the formula $[Mn_2(CO)_{10-n}(PF_3)_n]$ are obtained from a thermal reaction and characterized by IR and $^{19}F$ NMR: $ax\text{-}[Mn_2(CO)_9(PF_3)]$, $ax,ax'\text{-}$ and $ax,eq\text{-}[Mn_2(CO)_8(PF_3)_2]$, $ax,eq,ax'\text{-}$ $[Mn_2(CO)_7(PF_3)_3]$, and $ax,eq,ax',eq'\text{-}[Mn_2(CO)_6(PF_3)_4]$. A sixth dinuclear complex was obtained from the reaction of $[Mn_2(CO)_{10}]$ and $[Mn_2(CO)_6(PF_3)_4]$; NMR data suggest that it is $[(OC)_5MnMn(CO)_2(PF_3)_3]$. Four additional compounds were seen when photolytic conditions were employed, but they were not isolated in sufficient purity to allow characterization. Side products include species with bridging hydrido and difluorophosphido groups: $[MnH(CO)_{5-n}(PF_3)_n]$ ($n = 1\text{-}5$), $[Mn_2(\mu\text{-}PF_2)_2(CO)_{8-n}(PF_3)_n]$ ($n = 2\text{-}7$), and $[Mn_2(\mu\text{-}PF_2)(\mu\text{-}H)(CO)_{8-n}(PF_3)_n]$ ($n = 2\text{-}5$). The source of hydrogen in the hydrido complexes was not determined; these compounds were found even when extensive efforts were made to dry the system.

Sulfur ligands played a role in another part of this chemistry. When $[Mn_2(CO)_{10}]$ and tetrahydrothiophene are heated together, two products are obtained. One is the monsubstituted complex $[Mn_2(CO)_9(L)]$ (L = tetrahydrothiophene). The second, $[Mn_2(CO)_8(L)]$, has Structure (**5**) with the bridging sulfur atom of the ligand contributing two electron pairs to bonding.[80] This chemistry was further defined by synthesis of $[Mn_2(CO)_9(\overline{SCH_2CMe_2CH_2})]$ from $[Mn_2(CO)_9(NCMe)]$ and the indicated ligand and its conversion to a ligand-bridged octacarbonyl species using $Me_3NO$ to abstract a carbonyl group.[82]

(**5**)

Photolysis studies on $[Mn_2(CO)_{10}]$ have been extended to $[Mn_2(CO)_{10-n}(L)_n]$ species (L = phosphines). The two photoprocesses observed for the parent species, metal–metal bond homolysis and ligand dissociation, are also seen for the substituted complexes. Flash-photolysis studies[33-4] showed that the rates of the radical recombination process and the readdition of a carbonyl group are inversely related to phosphine size.

Photolysis of $[Mn_2(CO)_{10-n}(L)_n]$ (L = CO, $PR_3$; R = Bu, Et, Ph, OMe, OPh) with alkyl halides (BzCl, $PhCH_2CH_2Br$, PhCl, MeI) gives equal quantities of $[Mn(R)(CO)_4(L)]$ and $[Mn(X)(CO)_4(L)]$,[98] via a radical chain mechanism (Equation (4)). Reactions with HCl[99] and $Bu_3SnH$[35-6] give similar products.

$$[Mn_2(CO)_{10-n}(L)_n] + RX \longrightarrow [Mn(R)(CO)_4(L)] + [Mn(X)(CO)_4(L)] \tag{4}$$

L = CO, $PR_3$; R = Bu, Et, Ph, OMe, OPh; RX = BzCl, $PhCH_2CH_2Br$, PhCl, MeI

Thermal and photolytic reactions of $[Mn_2(CO)_{10-n}(L)_n]$ and alkyl halides give similar products; however, kinetic data indicate different mechanisms.[100] The kinetics of $[Mn_2(CO)_{10-n}(L)_n]$ reactions with $C_2H_2Cl_4$, $C_{16}H_{33}I$, NO, $O_2$, $P(OEt)_3$, and $PBu_3$ indicate two reaction pathways.[101]

The reaction of $[Mn_2(CO)_{10}]$ with depe produces a mixture of $[Mn(CO)_3(depe)]$ and its dimer, the latter in the larger amount.[94] More will be discussed about this work in Section 1.2.3. Formation of the complexes $[Mn_2(CO)_8(L{\wedge}L)]$ and $[Mn_2(CO)_6(L{\wedge}L)_2]$ in which the bidentate ligands bridge equatorial positions on the two metals is well known. Several new complexes of these types have been reported (Table 1, $L{\wedge}L = POP\ \{(EtO)_2POP(OEt)_2\}$, depm, and $Cy_2PCH_2PCy_2$).[92-3] The hexacarbonyl complexes undergo carbonyl loss, giving products with a semibridged carbonyl group (designated as a $\mu,\eta^1\eta^2$-CO ligand). Carbonyl loss is facilitated by a bulky diphosphine ligand.[93]

In dppm + $[Mn_2(CO)_{10}]$ reactions, two isomers with the formula $[Mn_2(CO)_6(dppm)_2]$ are formed.[95] Both contain dppm ligands bridging the two metals, but differ with respect to the geometry at each metal site (*mer,mer'* and *mer,fac'*). The latter isomer is less stable and slowly converts to the former. Both species are electroactive, and can be oxidized to paramagnetic cations.

In earlier work, $[Mn_2(CO)_5(dppm)_2]$ was found to add CO and CNR, converting the carbonyl group from a semibridging to a terminal position. Turney[102] showed that $SO_2$, $CS_2$, and $PhN_2^+$ also add to this complex, giving isolable products. In contrast, no reaction was seen with $P(OMe)_3$, MeCN, or py. A 1:1 adduct was isolated from the reaction with $CH_2N_2$; its structure was later shown to be (6).[103]

$$\left[\begin{array}{c} \underset{OC}{\overset{Ph_2P}{\underset{OC}{\bigg|}}} \underset{Mn}{\overset{O}{\underset{C}{\diagdown}}} \underset{N}{\overset{PPh_2}{\underset{Mn}{\bigg|}}} \underset{Ph_2P}{\overset{CO}{\underset{N}{\diagup}}} \underset{PPh_2}{\overset{CO}{\diagdown}} \end{array}\right]$$

(6)

Protonation of the metal–metal bond in these species with $HBF_4$ in MeCN[104] gives $[Mn_2(\mu\text{-}H)(CO)_5(NCMe)(dppm)_2]^+$. Acetonitrile can be replaced in this species by $CN^-$ and the product, $[Mn_2(\mu\text{-}H)(CO)_5(CN)(dppm)_2]$, can be decarbonylated with $Me_3NO$ to give a species with a semibridged cyanide ligand. NMR spectra indicate fluxionality; at low temperature the $^{13}C$ NMR pattern for the cyanide carbon is a 1:2:1 triplet, while at room temperature a 1:4:6:4:1 pentet is seen.

Further study[105] demonstrated that reactions of $HBF_4$ or $HSO_3F$ in $CH_2Cl_2$ with $[Mn_2(CO)_5(dppm)_2]$ form $[Mn_2(\mu\text{-}H)(CO)_6(dppm)_2]^+$. A pentacarbonyl intermediate was assumed, but not detected; the additional carbonyl group presumably comes from partial decomposition, because the yield improves if CO is added to the system. Reactions of HCl and $CF_3CO_2H$ with the indicated starting complex produce $[Mn_2(\mu\text{-}H)(CO)_5(X)(dppm)_2]$ (X = Cl, $CF_3CO_2$) (Equation (5)).

$$[Mn_2(CO)_5(dppm)_2] + HX \longrightarrow [Mn_2(\mu\text{-}H)(CO)_5(X)(dppm)_2] \tag{5}$$

$$X = Cl, CF_3CO_2$$

Structural studies on various compounds of this type have been carried out. Several (indicated in Table 1) were done during characterization. Other compounds whose structures were defined by x-ray diffraction include *eq,eq'*-$[Mn_2(dpam)(CO)_8]$[106] and $[Mn_2(CO)_8\{P(NMe_2)_3\}_2]$.[107]

### 1.2.3 17-Electron Manganese Carbonyl Complexes $[Mn(CO)_{5-n}L_n]$

Radical species $[Mn(CO)_{5-n}(L)_n]$ as intermediates in various reactions of manganese carbonyls will not be revisited. This section directs attention to species that have a stable existence, and to reactions of transient species not mentioned earlier.

The procedure used to prepare $[Mn(CO)_3(L)_2]$ (L = $PBu_3$, $PBu^i_3$, $P(OPr^i)_3$) is described; these species have a lifetime of weeks in hydrocarbon solvents at room temperature.[108] Several instances in which there is an equilibrium between the monomer and the dimer are reported for complexes with smaller ligands. Complexes with phosphite ligands, $[Mn(CO)_3(L)_2]$ (L = $P(OEt)_3$, $P(OPr^i)_3$), behave in this fashion. Solutions of these complexes are green at room temperature but change to red–orange at 0 °C as the amount of dimer increases; $v(CO)$ for the monomer lie at 1893 $cm^{-1}$ and 1875 $cm^{-1}$, while for the dimer values of 1957 $cm^{-1}$ and 1947 $cm^{-1}$ are recorded. A monomer–dimer equilibrium is also seen with $[Mn(CO)_3(depe)]$,[109] the dimer being the predominant species at room temperature (Equation (6)).

$$[Mn_2(CO)_6(L)_4] \rightleftharpoons 2\,[Mn(CO)_3(L)_2] \tag{6}$$

$$L = P(OEt)_3, P(OPr^i)_3; L_2 = depe$$

Photolysis of $[(OC)_5MnMn(CO)_3(bipy)]$ forms $[Mn(CO)_3(bipy)]_2$, which is apparently a dimer in solution because no ESR signal can be detected.[110-11] However, when $[(OC)_5MnMn(CO)_3(bipy)]$ is photolyzed in the presence of the spin-trap reagent $Bu^tNO$, $[Mn(CO)_3(bipy)(Bu^tNO)]$ is formed.[110]

Various $[Mn(CO)_{5-n}(L)_n]$ radicals have been prepared by one-electron oxidation of the corresponding anions using tropylium tetrafluoroborate. Disubstituted (L = $PR_3$; R = Ph, $OC_6H_4Cl$-*p*, $OC_6H_4OMe$-*p*) and trisubstituted species (L = $P(OMe)_3$) are thermally stable to dimerization, while $[Mn(CO)_5]$ and

[Mn(CO)$_4$(PPh$_3$)] dimerize as expected.[112] The formation of [Mn(CO)$_5$] by electrochemical oxidation of [Mn(CO)$_5$]$^-$ has also been studied.[113]

Kinetic studies on manganese radical species have been carried out. The reaction of [Mn(CO)$_3$(PBu$^i_3$)$_2$] with CO (initially giving [Mn(CO)$_4$(PBu$^i_3$)] which dimerizes) is judged to be first order in each reagent, supporting an associative mechanism.[114] Analysis of the kinetic data on reactions of [Mn(CO)$_5$] with PPh$_3$ and AsPh$_3$ led to a similar conclusion.[115] Flash photolysis of [Mn$_2$(CO)$_8$(L)$_2$] (L = CO, PBu$_3$, PBu$^i_3$, PPr$^i_3$) in the presence of RX (RX = CX$_4$, CHX$_3$; X = Cl, Br) gives [Mn(X)(CO)$_4$(L)]. Halogen abstraction occurs when stable radicals [Mn(CO)$_3$(L)$_2$] are mixed with these halocarbons. Rate constants for the reactions of halocarbons extend over a wide range, with rates enhanced by a higher degree of substitution and by an electron-donor ability of the ligand.[116] Reactions of [Mn(CO)$_3$(L)$_2$] (L = PBu$_3$, PBu$^i_3$, P(OPr$^i$)$_3$) with halocarbons (CX$_4$, CHX$_3$) give primarily *mer,trans*-[Mn(X)(CO)$_3$(L)$_2$]. Kinetic parameters for the reactions of these metal species with Bu$_3$SnH have been obtained. These reactions give [Mn(H)(CO)$_3$(L)$_2$] primarily, via bimolecular hydrogen transfer.[108]

A stable compound, [Mn(CO)$_3$(tppo)] (**7**), is obtained by photolysis of [Mn$_2$(CO)$_{10}$] with this spin-trap reagent (tppo = 2,2,6,6-tetramethylpiperidyl-*N*-oxide). An x-ray crystal structure of this 16-electron species was carried out.[117]

(**7**)

The ESR spectrum of [Mn(CO)$_5$] in an argon or CO matrix was obtained. This species is generated by photolysis of [MnH(CO)$_5$]; in argon, H• and HCO• were also detected, while in CO only the latter was seen. This reference corrects early work; ESR data purportedly for [Mn(CO)$_5$] were actually for the dioxygen adduct.[118] Other studies on [Mn(CO)$_5$] appeared in 1984[119] and 1992.[120] In the latter, [Mn(CO)$_5$] was generated by gas-phase thermolysis and co-condensed with adamantane at 77 K. When dioxygen is intentionally added, [Mn(CO)$_5$(O$_2$)] forms. ESR data on the persistent radical species [Mn(CO)$_3$(L)$_2$] (L = PR$_3$; R = Bu, Bu$^i$, Pr$^i$, OPr$^i$) at 77 K in hexane (glass), indicate that the unpaired electron is primarily metal centered.[121]

ESR was also used in the characterization of the paramagnetic addition products of [Mn(CO)$_5$] with *o*- and *p*-quinones,[122-4] with bis(ethoxythiocarbonyl),[125] and with diketones.[126]

## 1.3 THE PENTACARBONYLMANGANATE ION [Mn(CO)$_5$]$^-$ AND OTHER ANIONIC MANGANESE CARBONYLS

### 1.3.1 Synthesis and Physical Properties

One of the first chemical reactions of [Mn$_2$(CO)$_{10}$] to be reported was its reduction to form [Mn(CO)$_5$]$^-$. The nucleophilic character of this anion soon defined its use in the syntheses of various organometallic complexes. In the 1980s and early 1990s, attention has been directed primarily to substituted derivatives, to the redox properties of this group of compounds, to mechanistic studies, and to the further uses of anionic species as reagents.

Among the recently studied synthetic methods for the formation of anionic complexes is the reduction of cationic metal carbonyl species. These reactions are believed to proceed with initial formation of a 19-electron species. Ligand loss giving a 17-electron radical and addition of a second electron forms the anionic product. Then, in many instances, a further coupling reaction between the cationic starting material and the anion occurs.

The first reported example of this type was the electrochemical reduction of [Mn(CO)$_4$(PPh$_3$)$_2$]$^+$ which produced [Mn(CO)$_4$(PPh$_3$)]$^-$ (Equation (7)); the product was identified by its IR spectrum (v(CO) at 1940, 1850, and 1810 cm$^{-1}$).[127] Subsequently, this work was extended to [Mn(CO)$_5$(L)]$^+$ complexes.[128] Formation of [Mn(CO)$_5$]$^-$ is noted with L = CO, while with L = py or MeCN the final product is [Mn$_2$(CO)$_{10}$]. When L is a phosphine, varying amounts of [Mn$_2$(CO)$_8$(L)$_2$], Mn(H)(CO)$_4$(L), and [Mn(H)(CO)$_3$(L)$_2$] are formed. A third paper[129] reports the reduction of [Mn(CO)$_2$(dppe)$_2$]$^+$ to give [Mn(CO)$_2$(dppe)$_2$]$^-$, a species having both bi- and monodentate diphosphine ligands, and reduction of [Mn(CO)$_3${P(OMe)$_3$}$_3$]$^+$ to give [Mn(CO)$_2${P(OMe)$_3$}$_3$]$^-$.

$$[Mn(CO)_4(PPh_3)_2]^+ + 2\,e^- \longrightarrow [Mn(CO)_4(PPh_3)]^- + PPh_3 \qquad (7)$$

Chemical reduction of $[Mn(CO)_6]^+$ and $[Mn(CO)_{6-n}(L)_n]^+$ (L = PPh$_3$, PPh$_2$Me, PEt$_3$) occurs with the reagent $[Fe(CO)_4]^{2-}$ to form $[Mn(CO)_4(L)]^-$, the overall result being viewed as the transfer of $[CO]^{2+}$ from manganese to iron.[130] A suggested mechanism involves initial nucleophilic addition of the iron species to a terminal carbonyl group on manganese. A similar reaction occurs between $[Mn(CO)_6]^+$ and $[Re(CO)_5]^-$; however, the initially formed products $[Re(CO)_6]^+$ and $[Mn(CO)_5]^-$ can react further to give $[M_2(CO)_{10}]$ (M = Mn, Re) along with MnRe(CO)$_{10}$ (Equation (8)).[131] A labeling experiment shows a similar reaction and mechanism for the reaction of $[Mn(CO)_6]^+$ and $[Mn(^{13}CO)_5]^-$. Reduction of $[Mn_2(CO)_{10}]$ by $[Fe(CO)_4]^{2-}$ [130] and the kinetics of the reduction of this species by $[Re(CO)_5]^-$ [70] are also noted.

$$[Mn(CO)_6]^+ + [Re(CO)_5]^- \longrightarrow [(OC)_5Mn(\mu\text{-}CO)Re(CO)_5] \longrightarrow [Mn(CO)_5]^- + [Re(CO)_6]^+ \qquad (8)$$

Coupling reactions between anionic and cationic manganese complexes are closely related to this work. The reactions between $[Mn(CO)_5]^-$ and $[Mn(CO)_5(L)]^+$ (L = CO, MeCN, py) give $[Mn_2(CO)_{10}]$ as the sole product. In other cases there is scrambling of the noncarbonyl ligands, essentially in a statistical fashion. Thus, the products of reactions between $[Mn(CO)_4(PPh_3)]^-$ and $[Mn(CO)_6]^+$ are $[Mn_2(CO)_{10-n}(PPh_3)_n]$ ($n = 0$, 1, 2), while from the reaction of $[Mn(CO)_4\{P(OPh)_3\}]^-$ and $[Mn(CO)_5(PPh_3)]^+$ three products, $[Mn_2(CO)_8(PPh_3)_{2-n}\{P(OPh)_3\}_n\}]$ ($n = 0$, 1, 2) arise. A mechanism involving initial electron transfer from the anion to the cation, giving 17-electron and 19-electron radicals, has been suggested.[132]

Reduction of manganese carbonyl halides is used to prepare $[Mn(CO)_2\{P(OMe)_3\}_3]^-$ from $[Mn(Br)(CO)_2\{P(OMe)_3\}_3]$.[112] Kochi *et al.*[133] used both chemical (Na/Hg) and electrochemical reductions of $[Mn(Br)(CO)_3(L)_2]$ to prepare a series of $[Mn(CO)_3(L)_2]^-$ complexes (L = PR$_3$; R = Ph, OPh, Bu, Pr$^i$, OPr$^i$) (Equation (9)). Here, the anions were oxidized *in situ* to radical species.

$$[Mn(Br)(CO)_3(L)_2] + 2\,e^- \longrightarrow [Mn(CO)_3(L)_2]^- + Br^- \qquad (9)$$

$$L = PR_3;\ R = Ph,\ OPh,\ Bu,\ Pr^i,\ OPr^i$$

Sodium amalgam reduction of $[Mn(Br)(CO)_3(bipy)]$ is reported to proceed via $[Mn(CO)_3(bipy)]_2$ to the deep-blue colored anion $[Mn(CO)_3(bipy)]^-$ which can be converted to either a hydride or a methyl compound with H$^+$ or MeI.[134]

Phosphines[135] (PPh$_3$, PEt$_3$) and CO displace $[Mn(CO)_4(L)]^-$ (L = CO, PPh$_3$, PEt$_3$) from $[(OC)_5MMn(CO)_4(L)]^-$ (M = Cr, W), as shown in Equation (10).

$$[(OC)_5MMn(CO)_4(L^1)]^- + L^2 \longrightarrow [Mn(CO)_4(L^1)]^- + M(CO)_5(L^2) \qquad (10)$$

$$M = Cr,\ W;\ L^1 = L^2 = CO,\ PPh_3,\ PEt_3$$

Several new anionic carbonyl complexes have been prepared from $[Mn(CO)_5]^-$ via carbonyl substitution. One carbonyl group is replaced by dimethyl maleate (DM) or dimethyl fumarate (DF), giving isolable complexes with these ligands (Structures (**8**) and (**9**)). The alkene ligands are quite labile. Interestingly, the complexes, as sodium and lithium salts (but not the PPN salt), catalyze DM–DF isomerizations.[136]

(8)

(9)

The complex $[Mn(CO)_3(DBCat)]^-$ is formed when $[Mn(CO)_5]^-$, as the Bu$_4$N$^+$ or PPN$^+$ salt, is mixed with DBCat (DBCat = 3,5-di-*t*-butyl-1,2-catechol). The isolated PPN$^+$ salt shows IR absorptions at 1990, 1886, and 1870 cm$^{-1}$, and a single-crystal x-ray diffraction study confirms the identity of this 16-electron species.[137] An electrochemical study[138] identified four different electroactive states, $[Mn(CO)_3(DBCat)]^{2-,-,0,+}$, and a theoretical paper described the bonding in this system.[139]

Brookhart *et al.*[140] characterized PPN[Mn(CO)$_3$($\eta^4$-diene)] (diene = 1,3-cyclooctadiene, cyclohepta-triene, cyclooctatetraene) as products of hydride addition to the appropriate cyclohexadienyl–metal carbonyl precursor.

The paramagnetic complexes [Mn(L)($\eta$-C$_4$H$_6$)$_2$] (L = CO, PR$_3$; R = Me, Et, OMe) can be reduced to monoanions using Na/Hg. The trimethylphosphite complex was isolated as the Na(18-crown-6)$^+$ salt.[141]

Solvent-free Na$_3$[Mn(CO)$_4$] is obtained when HMPA solutions of this salt are dissolved and precipitated from liquid ammonia.[142] When dissolved in HMPA, some [Mn(H)(CO)$_4$]$^{2-}$ is present. A dihydride, [Mn(H)$_2$(CO)$_4$]$^-$, formed on the addition of ethanol, was isolated as a Ph$_4$As$^+$ salt.

Three trimetallic anions have been structurally characterized by single-crystal x-ray diffraction.[143] The first, K$_3$[Mn$_3$($\mu$-CO)$_2$(CO)$_{10}$], is obtained when [Mn$_2$(CO)$_{10}$] is treated with aqueous KOH. Water protonates this anion giving a monohydride salt; two structures, (PPN)$_2$[Mn$_3$($\mu$-H)(CO)$_{12}$] and (Ph$_4$As)$_2$[Mn$_3$($\mu$-H)($\mu$-CO)$_2$(CO)$_{10}$], occur with different cations.

Other structural studies have been carried out on anionic complexes: Na(15-crown-5)[Mn(CO)$_5$],[144] PPN[Mn(CO)$_5$],[145] and PPN[Mn(CO)$_4$(PEt$_3$)].[145]

## 1.3.2  Reactions

Some reactions of [Mn(CO)$_5$]$^-$ and related species have already been mentioned in this chapter, including their oxidation forming stable or transient radical species (Section 1.2.3).[112–13,133]

Salts of [Mn(CO)$_5$]$^-$ with the cations [Co($\eta$-Cp)$_2$]$^+$, PPh$_4$$^+$, Ph$_3$S$^+$, Ph$_3$PCH$_2$PPh$_2$$^+$ have an intense yellow color, resulting from a 400 nm absorption due to charge transfer from anion to cation. Photolysis into this band leads to interesting chemical transformations. The cobaltocenium salt is converted to [Co($\eta$-Cp)$_2$] and [Mn$_2$(CO)$_{10}$][146–7] (Equation (11)), while the PPh$_4$$^+$ salt degrades to [Mn(Ph)(CO)$_4$(PPh$_3$)] (60%) and [Mn$_2$(CO)$_{10}$] (20%).[148]

$$2\,[Co(\eta\text{-Cp})_2]^+ + 2\,[Mn(CO)_5]^- \longrightarrow 2\,[Co(\eta\text{-Cp})_2] + [Mn_2(CO)_{10}] \qquad (11)$$

Oxidation of [Mn(CO)$_5$]$^-$ by [V(CO)$_6$][149] produces [Mn$_2$(CO)$_{10}$], while oxidation using [Co(CO)$_3$(PPh$_3$)$_2$]$^+$ produces [Mn$_2$(CO)$_{10}$] and [Mn$_2$(CO)$_9$(PPh$_3$)], the latter in the greater quantity.[150] A kinetic study has been carried out on the oxidation of [Mn(CO)$_4$(L)]$^-$ (L = CO, PR$_3$; R = Ph, Et, Bu) using [M($o$-phen)$_3$]$^{3+}$ (M = Fe, Co).[145]

Selective $^{13}$CO enrichment of [Mn(CO)$_5$]$^-$ is accomplished by the reaction of this species with [Re(CO)$_6$]$^+$ in a $^{13}$CO atmosphere.[151] A mechanism is proposed with [(OC)$_5$Re($\mu$-CO)Mn(CO)$_5$] as a key intermediate. Work related to this result was cited earlier.[49,130–1]

Use of [Mn(CO)$_5$]$^-$ to prepare several bimetallic complexes is noted; [Mn(CO)$_5${Au(PPh$_3$)}] and [Mn(CO)$_4${Au(PPh$_3$)}$_4$]BF$_4$ (**10**) are formed in reactions with [Au(Cl)(PPh$_3$)] and [{Au(PPh$_3$)}$_3$O]BF$_4$, respectively.[152] Reactions of [Mn(CO)$_5$]$^-$ with [Pd$_2$(Cl)$_2$($\mu$-dppm)$_2$][153] and [Rh(Cl)(dppm)$_2$][154–5] are the starting point in the preparation of a series of manganese–palladium and manganese–rhodium complexes with dppm ligands bridging the two metals.

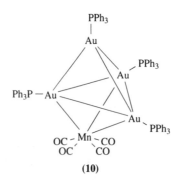

(**10**)

Elementary reactions of anionic metal complexes with hydrocarbons in the gas phase have contributed to the further understanding of catalytic processes including C–H bond activation. The reactions of [Mn(CO)$_3$]$^-$ (g.) with ethane[156] and cyclopropane,[157] and reactions of this anion and [Mn(CO)$_4$]$^-$ (g.) with monoalkenes, dialkenes, and benzene[158] and with silanes and germanes[159] are noted here.

We conclude this section with reference to a study of the reaction between $[Mn(CO)_5]^-$ with $MeNR_3^+$, relative to the catalytic homologation of methanol.[160]

## 1.4 HEXACARBONYLMANGANESE(I) $[Mn(CO)_6]^+$ AND OTHER CATIONIC MANGANESE CARBONYLS

### 1.4.1 Synthesis and Physical Properties

Studies done before 1982 gave rise to a number of methods for the preparation of cationic metal carbonyl complexes, the most common being displacement of a halide ion from the metal coordination sphere by a ligand. Halide acceptors are sometimes required in these reactions, and this method can be extended to complexes with other readily lost anionic ligands such as perchlorate. A second method of formation involves disproportionation in reactions of $[Mn_2(CO)_{10}]$ with ligands. Most of this work was discussed earlier in this chapter.[3,46,63–7] Other general routes include ligand displacement from cationic species, reactions at coordinated ligands, and isomerizations. A brief survey of the applications of these methods to the synthesis of new complexes follows.

#### 1.4.1.1 Displacement of a halide ion or other anionic ligands

(i) Nitrile complexes, $[Mn(CO)_3(NCR)_3]ClO_4$ (R = Et, allyl, Ph, Bz) result when $[Mn(OClO_3)(CO)_5]$ and the nitrile are heated in chloroform; in $CH_2Cl_2$, a mixture of the mono and tris(nitrile) complexes are formed.[161]

(ii) Diphenylphosphine complexes, $[Mn(CO)_{6-n}(PHPh_2)_n]ClO_4$ (n = 1–4) and $[Mn(P^\wedge P)(CO)_{4-n}(PHPh_2)_n]ClO_4$ (P^P = dppm, dppe; n = 1, 2) are produced in reactions of the appropriate manganese carbonyl halide, $AgClO_4$, and $PHPh_2$.[162]

(iii) The *mer* isomers $[Mn(CO)_3(TMEDA)\{P(OR)_3\}]ClO_4$ (R = Me, Et, Pr$^i$) are obtained in a reaction of $[Mn(Br)(CO)_3(TMEDA)]$, $AgClO_4$, and the phosphite at room temperature; the *fac* isomers are formed at 0 °C.[163]

(iv) Formation of $[Mn(CO)_3(ttn)]X$ (ttn = 1,4,7-trithiacyclononane; X = Cl, Br, I) occurs upon mixing the tridentate ligand and $[Mn(X)(CO)_5]$.[164]

(v) The reaction of $[Mn(OClO_3)(CO)_5]$ and tmtu (tmtu = tetramethylthiourea) gives *cis*-$[Mn(CO)_4(tmtu)_2]ClO_4$; additional tmtu displaces another carbonyl ligand, giving *fac*-$[Mn(CO)_3(tmtu)_3]ClO_4$. Other ligands (py, *o*-phen, dppm) react with *cis*-$[Mn(CO)_4(tmtu)_2]ClO_4$ to form *fac*-$[Mn(CO)_3(tmtu)(L^\wedge L)]ClO_4$.[82]

(vi) Triflate anion in $[Mn(O_3SCF_3)(CO)_5]$ is displaced by THF, acetone, $Et_2O$, and $PBu_3$, giving $[Mn(CO)_5(L)]O_3SCF_3$; on standing in the presence of additional ligand, formation of *fac*-$[Mn(CO)_3(L)_3]O_3SCF_3$ occurs. In the presence of strong acid and CO, $[Mn(CO)_6]O_3SCF_3$ is obtained (Equation (12)).[165]

$$[Mn(O_3SCF_3)(CO)_5] + H^+ + CO \longrightarrow [Mn(CO)_6]^+ + CF_3SO_3H \qquad (12)$$

(vii) Using $TlPF_6$ or $TlClO_4$ to assist halide loss, $[Mn(Br)(CO)(dppe)_2]$ converts to $[Mn(L)(CO)(dppe)_2]X$ (L = RNC, RCN; R = Me, Et, Ph, Bz; X = $PF_6$ or $ClO_4$).[166]

(viii) With $AgClO_4$, the ligands ($P(OR)_3$ (R = Me, Ph); $CNBu^t$) displace halide ion from $[Mn(Br)(S_2CPCy_3)(CO)_3]$, and the complexes $[Mn(CO)_3(S_2CPCy_3)(L)]ClO_4$ are isolated. With bidentate ligands (L^L = dppm, dppe), $[Mn(CO)_3(S_2CPCy_3)(L^\wedge L)]ClO_4$, the initial product, contains a monodentate sulfur ligand; this complex is converted to $[Mn(CO)_2(S_2CPCy_3)(L^\wedge L)]ClO_4$ when heated.[167]

(ix) Reaction of $[Mn(Br)(CO)_2(dppm)_2]$ with $TlPF_6$ in the presence of CO gives $[Mn(CO)_3(dppm)_2]PF_6$, a compound having one bi- and one monodentate dppm ligand. If CO is absent, *cis*-$[Mn(CO)_2(dppm)_2]PF_6$ is produced.[168]

(x) Stirring a mixture of $[Mn(Br)(CO)_5]$, $AgClO_4$, and a nitrile gives *fac*-$[Mn(CO)_3(NCR)_3]ClO_4$ (R = Et, Pr, Bz). The nitriles are easily replaced by $NH_3$ or by dien.[169]

(xi) Reactions between $[Mn(H)(CO)_{5-n}(L)_n]$ (L = $P(OR)_3$; R = Me, Et, Pr$^i$, Ph; n = 2–4) and $Ph_3CPF_6$ form, *in situ*, a species with a covalently bonded $PF_6$ group that is easily displaced by additional L, giving $[Mn(CO)_{5-n}(L)_{n+1}]PF_6$.[170]

### *1.4.1.2 Other syntheses*

Displacement of labile $SO_2$ in $[Mn(CO)_5(SO_2)]AsF_6$ by HNSO gives $[Mn(CO)_5(HNSO)]AsF_6$.[171] A simple preparation of the precursor from $[Mn_2(CO)_{10}]$ and $AsF_5$ in liquid $SO_2$ has been described.[172] Alkylation of a cyanide ligand by an allyl–iron complex was applied to the synthesis of *trans*-$[Mn(CO)(CNCH_2CMe=CH_2)(dppm)_2]PF_6$.[173] Trimethylamine-*N*-oxide effects decarbonylation of $[Mn(CO)_3(bipy)(dppm)]X$ (X = $PF_6$, $ClO_4$; monodentate dppm), forming $[Mn(CO)_2(bipy)(dppm)]X$.[174] The same reagent was used to effect substitution of CO in *fac*-$[Mn(CO)_3(CNBu^t)$ (L^L)]ClO_4 (L^L = bipy, *o*-phen, Bu$^t$dab) by either CNPh or $P(OMe)_3$.[175] Reduction (Na/Hg; MeCN) followed by oxidation by $O_2$ converts *cis,cis*-$[Mn(CO)_2(bipy)$ $(CNBu^t)_2]ClO_4$ to the *cis,trans* isomer. A similar procedure was used to carry out *fac* to *mer* isomerization of $[Mn(CNBu^t)_3(CO)(bipy)]ClO_4$.[176]

Structural studies on *cis*-$[Mn(CO)_2(dppm)_2]Br\cdot1/3H_2O$, *trans*-$[Mn(CO)_2(dppm)_2]ClO_4\cdot1/2CH_2$,[177] *fac*-$[Mn(CO)_3(o$-phen$)(CNBu^t)]ClO_4$,[176] and $[Mn(CO)(CNCH_2C(Me)=CH_2)(dppm)_2]PF_6$[173] have been completed.

### 1.4.2 Reactions

Some reactions of cationic complexes have been mentioned in this chapter, including electrochemical reduction and chemical reduction of these species using carbonylmetallate anions and other reagents (Section 1.3.1),[127–32] and ligand substitution (Section 1.4.1).[82,167–9,171,174–5] 

Photolysis converts *cis*-$[Mn(CO)_2(dppm)_2]PF_6$ to the *trans* isomer which reverts to the *cis* isomer on heating (Equation (13)).[168]

$$cis\text{-}[Mn(CO)_2(dppm)_2]PF_6 \quad \underset{\text{ii, heat}}{\overset{\text{i, } h\nu}{\rightleftharpoons}} \quad trans\text{-}[Mn(CO)_2(dppm)_2]PF_6 \qquad (13)$$

With the proper choice of ligand it is possible to oxidize cationic complexes, forming isolable low-spin manganese(II) species. For example, *trans*-$[Mn(CO)(L)(dppm)_2](PF_6)_2$ (L = MeCN, MeNC, Bu$^t$NC, CO) complexes are isolated when the manganese(I) precursor is oxidized.[178] Often, isomerization accompanies oxidation reactions. Connelly *et al.*[179] were able to oxidize *cis,cis*-$[Mn(L)(CO)_2\{P(OPh)_3\}(dppm)]^+$ (L = py, NCMe, CNMe, PPh$_3$), *cis,cis*-$[Mn(L)(L')(CO)_2(dppe)]^+$, and *cis*-$[Mn(CO)_2(dppm)_2]^+$ (all $PF_6^-$ salts) electrochemically, forming the corresponding manganese(II) complexes with *trans* carbonyl groups.

Borohydride reduction of $[Mn(CO)_5(L)]^+$ (L = PR$_3$, R = Ph, OPh) and $[Mn(CO)_4(L)_2]^+$ (L = PPh$_3$, $P(OPh)_3$) results in the formation of formyl compounds, $[Mn(CHO)(CO)_4(L)]$ and $[Mn(CHO)(CO)_3(L)_2]$.[180] *fac*-$[Mn(CO)_3(PPh_3)_2(CMeOMe)]^+$ can be reduced by the borohydride ion, giving $[Mn(CO)_3(PPh_3)_2(CH_2OMe)]$.[181]

Potassium hydroxide deprotonates the methylene group of a dppm ligand in *fac*-$[Mn(CO)_3(CNBu^t)(dppm)]ClO_4$ and $[Mn(CO)_4(dppm)]ClO_4$. Addition of $[Au(Cl)(PPh_3)]$ or $[Ag(Cl)(PPh_3)]$ forms species having either one or two Ag(PPh$_3$) or Au(PPh$_3$) groups covalently attached to the carbon in this ligand; single-crystal x-ray diffraction studies have confirmed several structures.[182]

## 1.5 MANGANESE CARBONYL HALIDES

### 1.5.1 Derivatives of $[MnX(CO)_5]$ (X = Cl, Br, I, and Pseudohalides)

Cleavage of the metal–metal bond in $[Mn_2(CO)_{10-n}(L)_n]$ creates radical species that react with halocarbons to form $[Mn(X)(CO)_{5-n}(L)_n]$ complexes. Similarly, the stable radical species $[Mn(CO)_3(L)_2]$ react with halogen donors to give $[Mn(X)(CO)_3(L)_2]$.[98–102,108–9,116] The mechanism of halogen cleavage of $[Mn_2(CO)_{10}]$ producing $[Mn(X)(CO)_5]$ has been defined.[69] The use of metal carbonyl halides as starting materials in the syntheses of metal carbonyl cations has also been detailed in earlier sections of this chapter.[161–70]

### 1.5.1.1 Syntheses of manganese carbonyl halide complexes

The reaction of $[Mn_2(CO)_{10}]$ with $SO_2Cl_2$ (Equation (14)) is reported to be a useful preparation of $[Mn(Cl)(CO)_5]$.[183] Several procedures for the synthesis of both $[Mn(X)(CO)_5]$ and $[Mn_2(\mu\text{-}X)_2(CO)_8]$ species (X = Cl, Br, I) have been reported.[22,24] The dimer, $[Mn_2(\mu\text{-}I)_2(CO)_8]$, is one of two products of a reaction between $Na[Mn(CO)_5]$ and diiodoacetylene.[184] The complex $[Mn(Cl)_2(CO)_4]^-$ is the major product in the reaction of $Na[Mn(CO)_5]$ and $Cl_2C{=}S$; it was isolated as $K(18\text{-crown-}6)^+$, and its solid-state structure was determined.[78]

$$[Mn_2(CO)_{10}] + SO_2Cl_2 \longrightarrow 2\,[Mn(Cl)(CO)_5] + SO_2 \qquad (14)$$

Palladium oxide catalyzes reactions of $[Mn(Br)(CO)_5]$ with isonitriles (R = Me, Bz, Cy, Pr$^i$, Bu$^t$, xylyl). By controlling the stoichiometry and reaction conditions it is possible to prepare all possible species $[Mn(Br)(CO)_{5-n}(L)_n]$ ($n = 1$–5), as well as the cationic complexes $[Mn(CO)(L)_5]Br$ and $[Mn(L)_6]Br$.[185]

New $[Mn(X)(CO)_{5-n}(L)_n]$ complexes formed by carbonyl substitution in the parent species have been reported, including:

(i) $[Mn(X)(CO)_4(L)]$ (L = phosphines immobilized on a polystyrene backbone,[186] $P(Ph)(Me)(CO_2CH_{3-n}Cl_n)$,[187] $P(Tol)_2OTMS$,[188] $Ph_2PCH_2SiMe_2CH{=}CH_2)$;[189]
(ii) $[Mn(X)(CO)_3(L)_2]$ (L = $PPh_{3-n}(TMS)_n$; $n = 1$–3);[190]
(iii) $[Mn(X)(CO)_3(L^\wedge L)]$ (L$^\wedge$L = $(Tol)_2POSiMe_2OP(Tol)_2$,[190] $Ph_2PCH_2SiMe_2CH{=}CH_2$,[191] chelating dinitriles and isonitriles,[191] $S_2CPCy_3$,[192] $S_2CCMe_2PPh_3$,[193] $Ph_2P({=}S)$ $(N(Me)R)$ (thioformamides) and $Ph_2PC({=}NR)SMe$ (thioformimido esters),[194] M(salen) and several other complexes (M(salen) = $N,N'$-ethylenebis(salicylideneiminate) complexes of Co, Cu, Zn, Pd, Ni, and Sn),[195] $Ph_2PCH_2P({=}E)Ph_2$ (E = S, Se),[196] R-dab (R-dab = $RN{=}CHCH{=}NR$);[65-7,197]
(iv) $[Mn(X)(CO)_2(dppm)(L)]$ (L = monodentate dppm,[168] $P(OPh)_3)$.[198]

In addition, preparation of carbene complexes $[Mn(X)(CO)_4(\overline{COCH_2CH_2}A)]$ (A = O, NH) by the reaction of ethene oxide and aziridine at a coordinated carbonyl group is noted.[199]

There has been considerable interest in the complexes $[Mn_2(\mu\text{-}X)_2(CO)_{8-2n}(\mu\text{-}L^\wedge L)_n]$, which have a chelating ligand bridging the two metal atoms. Syntheses of such compounds have been reported: (i) X = Br, $n = 2$, L$^\wedge$L = PhSeSePh,[200] POP (POP = $(EtO)_2POP(OEt)_2$);[201] (ii) X = Cl, $n = 1$, 2, L$^\wedge$L = POP.[202] Reduction of the species with POP ligands followed by protonation or reaction with $[Au(PPh_3)Cl]$ results in the formation of $[Mn_2(\mu\text{-}X)(\mu\text{-}Y)(CO)_6(POP)]$ (Y = H, Au(PPh$_3$)) (11).[201-2]

$$
\begin{array}{c}
\text{O} \\
(EtO)_2P \diagdown \diagup P(OEt)_2 \\
OC \diagdown \mid Y \mid \diagup CO \\
Mn \diagdown \diagup Mn \\
OC \diagup \mid X \mid \diagdown CO \\
CO \quad CO
\end{array}
$$

(11)

The reaction between $[I(py)_2]^+$ and $[Mn_2(CO)_8(dppm)]$ forms $[Mn_2(\mu\text{-}I)(CO)_8(dppm)]^+$, which has been isolated and structurally characterized as the $BF_4^-$ salt.[203] Formation of $[Mn_2\text{-}(\mu\text{-}Cl)_2(CO)_6(\overline{SCH_2CMe_2CH_2})_2]$ can be accomplished by the reaction of $[Mn_2(\overline{SCH_2CMe_2CH_2})(CO)_8]$ with HCl; a structural study of this compound has been carried out.[81]

Other structures determined by single-crystal x-ray diffraction include: $[Mn_2(\mu\text{-}Br)_2(\mu\text{-}PhSeSePh)(CO)_6]$,[200] $[Mn(Br)(CO)_3(S_2CCMe_2PPh_3)]$,[193] $[Mn_2(\mu\text{-}Br)(\mu\text{-}Y)(CO)_6(POP)]$ (Y = H, Br,[201] Y = Au(PPh$_3$)[202]), $[Mn(I)(CO)_3(PPh_2TMS)_2]$,[190] $[Mn(Br)(CO)_3(S{=}CP(NHPh)PPh_2)]$,[204] and $[Mn(Cl)(CO)_3(bipy)]$.[205]

Several research groups have described syntheses of manganese(II) complexes. Carbon monoxide adds to $[Mn(X)_2(L)]$ to give the pseudotetrahedral and paramagnetic $[Mn(X)_2(CO)(L)]$ (L = PPhMe$_2$, X = Cl, Br, I; L = PPhEt$_2$, PPr$_3$, X = Cl, Br), and is lost from this complex upon heating or when it is subject to low pressure.[206] The reaction of NaCp, MnCl$_2$, and TMEDA in THF gives $[Mn(Cl)(TMEDA)(\eta\text{-}Cp)]$.[207] Reaction of MnCl$_2$ and $[Mn(\eta\text{-}MeCp)_2]$ in THF gives $[MnCl(THF)(\eta\text{-}MeCp)]_2$; a second compound in this series is obtained when PEt$_3$ displaces THF. Structural determinations of the TMEDA and PEt$_3$ compounds have been carried out.[208]

### 1.5.1.2 Reactions of manganese carbonyl halide complexes

Stereochemical consequences of carbonyl group loss are dealt with in two theoretical papers. Davy and Hall[209] calculated that *cis* carbonyl loss is favored over *trans* carbonyl loss in $[Mn(Cl)(CO)_5]$ by 12–38 kJ mol$^{-1}$, a result which is in agreement with experimental data. Pierloot *et al.*[210] examined photochemical pathways for CO loss in this species, concluding that a *trans* carbonyl might dissociate under such conditions.

Although certain reactions of metal carbonyl halides result in ionic products, Tyler *et al.*[211] concluded that for $[Mn(X)(CO)_5]$ heterolytic halide dissociation is not a viable mechanistic pathway. Instead, the formation of ionic products in reactions of this species occurs by initial disproportionation to give $MnX_2$ and $[Mn_2(CO)_{10}]$; the latter species then reacts with the ligand to give ionic products.[3,63-4]

In several instances, metal carbonyl anions have been used as reducing agents, giving mixed-metal carbonyls. Thus, $[MnFe(CO)_9]^-$ is formed[130] by the reaction between $[Mn(Br)(CO)_5]$ and $[Fe(CO)_4]^{2-}$, and $[MnFe_2(CO)_{12}]^-$ is the product obtained in reactions of $[Mn(Br)(CO)_5]$ with either $[Fe_3(CO)_{11}]^{2-}$ or $[Fe_4(CO)_{13}]^{2-}$.[212]

As described earlier, reduction of $[Mn(X)(CO)_{5-n}(L)_n]$ presents an important route to anionic carbonyl species.[112-13]

Reduction of $[Mn(Br)(CO)_4(PPh_2H)]$ with sodium metal gives $Na_2[Mn(CO)_4(PPh_2)]$ *in situ*,[213] which reacts with alkyl halides to give various alkyl metal species. Reactions of $[Mn(Br)(CO)_4(PR_2H)]$ (R = Ph, Cy, Bu$^t$) with $KBH(Bu^s)_3$ giving $K[Mn(CO)_4(PR_2H)]$,[214] and between $[Mn(Br)(CO)_4(PPhMeCO_2CH_2Cl)]$ and magnesium giving $[\overline{Mn(CO)_4(PPhMeCO_2CH_2})]$,[186] are described.

One-electron oxidation of $[Mn(Br)(CO)_2(dppm)\{P(OPh)_3\}]$ yields the corresponding cation; in this process isomerization converts carbonyl groups from the *cis* to the *trans* positions.[179]

The formation of a mixed-metal compound $[MnRe(\mu\text{-}Br)(\mu\text{-}PPh_2)(CO)_8]$ (**12**) occurs in a reaction between $[Mn(Br)(CO)_5]$ and $[Re(X)(CO)_4(PPh_2TMS)]$.[215]

(**12**)

### 1.5.1.3 Pseudohalide complexes

The syntheses of a number of cyanide and thiocyanate complexes have been accomplished by ligand substitution reactions. Cyanide and thiocyanide replace a carbonyl group in $[Mn(CO)_2(dppm)_2]^+$ giving $[Mn(X)(CO)(dppm)_2]$,[177] and $[Mn(X)(CO)_2(bipy)\{P(OPh)_3\}]$ species are formed[216] when CN$^-$ or SCN$^-$ replace one of the phosphite ligands in the cationic species *cis,trans*-$[Mn(CO)_2(bipy)\{P(OPh)_3\}_2]^+$. Cyanide and thiocyanate ions also replace bromide ions in $[Mn(Br)(CO)_3(dppm)]$, giving *fac*-$[Mn(X)(CO)_3(dppm)]$; a crystal-structure study on the thiocyanate compound confirmed that this ligand was bonded to manganese via nitrogen.[217]

One important aspect of this work is the formation of bimetallic complexes having bridging cyanide ligands. For example, reactions between $[Mn(CN)(CO)_2(L^\wedge L)(L)]$ and $[Mn(Br)(CO)_2(L^\wedge L)(L)]$ ($L^\wedge L$ = dppm, dppe, L = P(OPh)$_3$, P(OEt)$_3$) in the presence of TlPF$_6$ form $[Mn_2(\mu\text{-}CN)(CO)_4(L^\wedge L)_2(L)_2]PF_6$.[218] Similar chemistry has been applied to the syntheses of compounds containing manganese and ruthenium.[219] The oxidative chemistry of these compounds is a part of most of these studies.

## 1.6 REFERENCES

1.  N. G. Connelly and W. E. Geiger, *Adv. Organomet. Chem.*, 1984, **23**, 2.
2.  T. J. Meyer and J. V. Caspar, *Chem. Rev.*, 1985, **85**, 187.
3.  A. E. Stiegman and D. R. Tyler, *Acc. Chem. Res.*, 1984, **17**, 61.
4.  A. E. Stiegman and D. R. Tyler, *Coord. Chem. Rev.*, 1985, **63**, 217.
5.  D. R. Tyler and F. Mao, *Coord. Chem. Rev.*, 1990, **97**, 119.
6.  J. K. Kochi, *J. Organomet. Chem.*, 1986, **300**, 139.
7.  J. K. Kochi and T. M. Bockman, *Adv. Organomet. Chem.*, 1991, **33**, 51.
8.  M. Y. Darensbourg, *Prog. Inorg. Chem.*, 1985, **33**, 221.

9. J. E. Ellis, *Adv. Organomet. Chem.*, 1990, **31**, 1.
10. G. A. Carriedo and V. Riera, *J. Organomet. Chem.*, 1990, **394**, 275.
11. K. Yasufuku, N. Hiraga, K. Ichimura and T. Kobayashi, *Coord. Chem. Rev.*, 1990, **97**, 167.
12. M. O. Albers and N. J. Coville, *Coord. Chem. Rev.*, 1984, **53**, 227.
13. R. J. Puddephatt, *Chem. Soc. Rev.*, 1983, **12**, 99.
14. I. Tabushi, *Coord. Chem. Rev.*, 1988, **86**, 1.
15. D. J. Stufkens, *Coord. Chem. Rev.*, 1990, **104**, 39.
16. D. J. Stufkens, T. van der Graaf, G. J. Stor and A. Oskam, *Coord. Chem. Rev.*, 1991, **111**, 331.
17. N. G. Connelly, *Chem. Soc. Rev.*, 1989, **18**, 153.
18. W. C. Trogler, (ed.), 'Organometallic Radical Processes', Elsevier, Amsterdam, 1990.
19. C. G. Kreiter, *Adv. Organomet. Chem.*, 1986, **26**, 297.
20. M. H. Quick and R. J. Angelici, *Inorg. Synth.*, 1979, **19**, 160, 161; K. J. Reimer and A. Shaver, *ibid.*, **159**, 162.
21. S. P. Schmidt, J. Nitschke and W. C. Trogler, *Inorg. Synth.*, 1989, **26**, 113.
22. F. Calderazzo, R. Poli and D. Vitali, *Inorg. Synth.*, 1985, **23**, 32.
23. J. E. Ellis and G. F. P. Warnock, *Organomet. Synth.*, 1988, **4**, 100.
24. M. Martin, B. Rees and A. Mitschler, *Acta Crystallogr., Sect. B*, 1982, **38**, 6.
25. J. A. Connor *et al.*, *Organometallics*, 1982, **1**, 1166.
26. T. A. Seder, S. P. Church and E. Weitz, *J. Am. Chem. Soc.*, 1986, **108**, 7518.
27. D. A. Prinslow and V. Vaida, *J. Am. Chem. Soc.*, 1987, **109**, 5097.
28. A. Fox and A. Poë, *J. Am. Chem. Soc.*, 1980, **102**, 2497.
29. R. W. Wegman, R. J. Olsen, D. R. Gard, L. R. Faulkner and T. L. Brown, *J. Am. Chem. Soc.*, 1981, **103**, 6089.
30. L. J. Rothberg, N. J. Cooper, K. S. Peters and V. Vaida, *J. Am. Chem. Soc.*, 1982, **104**, 3536.
31. H. Yesaka, T. Kobayashi, K. Yasufuku and S. Nagakura, *J. Am. Chem. Soc.*, 1983, **105**, 6249.
32. A. F. Hepp and M. S. Wrighton, *J. Am. Chem. Soc.*, 1983, **105**, 5934.
33. H. W. Walker, R. S. Herrick, R. J. Olsen and T. L. Brown, *Inorg. Chem.*, 1984, **23**, 3748.
34. R. S. Herrick and T. L. Brown, *Inorg. Chem.*, 1984, **23**, 4550.
35. R. J. Sullivan and T. L. Brown, *J. Am. Chem. Soc.*, 1991, **113**, 9155.
36. R. J. Sullivan and T. L. Brown, *J. Am. Chem. Soc.*, 1991, **113**, 9162.
37. J. Z. Zhang and C. B. Harris, *J. Chem. Phys.*, 1991, **95**, 4024.
38. S. Zhang, H.-T. Zhang and T. L. Brown, *Organometallics*, 1992, **11**, 3929.
39. R. E. Tecklenberg, Jr. and D. H. Russell, *J. Am. Chem. Soc.*, 1987, **109**, 7654.
40. W. K. Meckstroth, R. B. Freas, W. D. Reents, Jr. and D. P. Ridge, *Inorg. Chem.*, 1985, **24**, 3139.
41. W. K. Meckstroth and D. P. Ridge, *J. Am. Chem. Soc.*, 1985 **107**, 2281.
42. F. A. Cotton and R. R. Monchamp, *J. Chem. Soc.*, 1960, 533.
43. J. L. Hughey, C. P. Anderson and T. J. Meyer, *J. Organomet. Chem.*, 1977, **125**, C49.
44. J. A. M. Simões, J. C. Schultz and J. L. Beauchamp, *Organometallics*, 1985, **4**, 1238.
45. J. L. Goodman, K. L. Peters and V. Vaida, *Organometallics*, 1986, **5**, 815.
46. J. R. Pugh and T. J. Meyer, *J. Am. Chem. Soc.*, 1992, **114**, 3784.
47. H. Wawersik and F. Basolo, *Inorg. Chim. Acta*, 1969, **3**, 113.
48. N. J. Coville, A. M. Stolzenberg and E. L. Muetterties, *J. Am. Chem. Soc.*, 1983, **105**, 2499.
49. S. P. Schmidt, W. C. Trogler and F. Basolo, *Inorg. Chem.*, 1982, **21**, 1698.
50. A. Marcomini and A. Poë, *J. Am. Chem. Soc.*, 1983, **105**, 6952.
51. A. Marcomini and A. Poë, *J. Chem. Soc., Dalton Trans.*, 1984, 95.
52. F. Calderazzo, A. Juris, R. Poli and F. Ungari, *Inorg. Chem.*, 1991, **30**, 1274.
53. S. P. Schmidt, F. Basolo, C. M. Jensen and W. C. Trogler, *J. Am. Chem. Soc.*, 1986, **108**, 1894.
54. P. Steil, W. Sacher, P. M. Fritz and W. Beck, *J. Organomet. Chem.*, 1989, **362**, 363.
55. N. J. Coville and P. Johnston, *J. Organomet. Chem.*, 1989, **363**, 343.
56. S. Firth, P. M. Hodges, M. Poliakoff, J. J. Turner and M. J. Therien, *J. Organomet. Chem.*, 1987, **331**, 347.
57. T. J. Oyer and M. S. Wrighton, *Inorg. Chem.*, 1988, **27**, 3689.
58. D. J. Robinson, E. A. Darling and N. J. Coville, *J. Organomet. Chem.*, 1986, **310**, 203.
59. P. Johnston, G. J. Hutchings and N. J. Coville, *J. Am. Chem. Soc.*, 1989, **111**, 1902.
60. A. L. Rheingold, W. K. Meckstroth, and D. P. Ridge, *Inorg. Chem.*, 1986, **25**, 3706.
61. A. E. Leins, D. G. Billing, D. C. Levendis, J. du Toit and N. J. Coville, *Inorg. Chem.*, 1992, **31**, 4756.
62. R. J. Klinger and J. W. Rathke, *Inorg. Chem.*, 1992, **31**, 804.
63. A. E. Stiegman and D. R. Tyler, *Inorg. Chem.*, 1984, **23**, 527.
64. A. E. Stiegman, A. S. Goldman, C. E. Philbin and D. R. Tyler, *Inorg. Chem.*, 1986, **25**, 2976.
65. M. W. Kokkes, W. G. J. de Lange, D. J. Stufkens and A. Oskam, *J. Organomet. Chem.*, 1985, **294**, 59.
66. T. van der Graaf, R. M. J. Hofstra, P. G. M. Schilder, M. Rijkhoff, D. J. Stufkens and J. G. M. van der Linden, *Organometallics*, 1991, **10**, 3668.
67. H. K. van Dijk, J. van der Haar, D. J. Stufkens and A. Oskam, *Inorg. Chem.*, 1989, **28**, 75.
68. A. E. Stiegman, A. S. Goldman, D. B. Leslie and D. R. Tyler, *J. Chem. Soc., Chem. Commun.*, 1984, 632.
69. S. P. Schmidt, W. C. Trogler and F. Basolo, *J. Am. Chem. Soc.*, 1984, **106**, 1308.
70. M. S. Corraine and J. D. Atwood, *Inorg. Chem.*, 1989, **28**, 3781.
71. R. D. Adams, G. Chen and Y. Chi, *Organometallics*, 1992, **11**, 1473.
72. R. D. Adams, G. Chen, L. Chen, M. P. Pompeo and J. Yin, *Organometallics*, 1991, **10**, 2541.
73. M. P. Keyes, L. U. Gron and K. L. Watters, *Inorg. Chem.*, 1989, **28**, 1236.
74. H. Alper and L. C. Damude, *Organometallics*, 1982, **1**, 579.
75. R. Davis, J. L. A. Durrant and C. C. Rowland, *J. Organomet. Chem.*, 1986, **315**, 119.
76. M. J. Chen and J. W. Rathke, *Organometallics*, 1987, **6**, 1833.
77. M. J. Chen and J. W. Rathke, *Organometallics*, 1989, **8**, 515.
78. W. Petz and D. Rehder, *Organometallics*, 1990, **9**, 856.
79. E. Deschamps, F. Mathey, C. Knobler and Y. Jeannin, *Organometallics*, 1984, **3**, 1144.
80. E. Guggolz, K. Layer, F. Oberdorfer and M. Ziegler, *Z. Naturforsch., Teil B*, 1985, **40**, 77.

81. R. D. Adams, J. Belinski and L. Chen, *Organometallics*, 1992, **11**, 4104.
82. C. Carriedo, M. V. Sanchez, G. A. Carriedo, V. Riera, X. Solans and M. L. Valin, *J. Organomet. Chem.*, 1987, **331**, 53.
83. J. F. Janik, E. N. Duesler and R. T. Paine, *J. Organomet. Chem.*, 1987, **323**, 149.
84. K. Mashima, K. Jyodoi, A. Ohyoshi and H. Takaya, *J. Chem. Soc., Chem. Commun.*, 1986, 1145.
85. R. M. Bullock, B. J. Rappoli, E. G. Samsel and A. L. Rheingold, *J. Chem. Soc., Chem. Commun.*, 1989, 261.
86. R. M. Bullock and B. J. Rappoli, *J. Am. Chem. Soc.*, 1991, **113**, 1659.
87. J. M. Garner, A. Irving and J. R. Moss, *Organometallics*, 1990, **9**, 2836.
88. M. M. Singh and R. J. Angelici, *Inorg. Chim. Acta*, 1985, **100**, 57.
89. V. V. Gumenyuk, V. N. Babin, Yu. A. Belousov, N. S. Kochetkova and I. V. Dobryakova, *Polyhedron*, 1984, **3**, 707.
90. C. C. Grimm, P. E. Brotman and R. J. Clark, *Organometallics*, 1990, **9**, 1119.
91. M. O. Albers and N. J. Coville, *S. Afr. J. Chem.*, 1983, **35**, 139.
92. V. Riera and M. A. Ruiz, *J. Chem. Soc., Dalton Trans.*, 1986, 2617.
93. T. E. Wolff and L. P. Klemann, *Organometallics*, 1982, **1**, 1667.
94. D. R. Tyler and A. S. Goldman, *J. Organomet. Chem.*, 1986, **311**, 349.
95. F. R. Lemke and C. P. Kubiak, *Inorg. Chim. Acta*, 1986, **113**, 125.
96. K. S. Suslick and P. F. Schubert, *J. Am. Chem. Soc.*, 1983, **105**, 6042.
97. A. Albinati, R. M. Bullock, B. J. Rappoli and T. F. Koetzle, *Inorg. Chem.*, 1991, **30**, 1414.
98. M. A. Biddulph, R. Davis and F. I. C. Wilson, *J. Organomet. Chem.*, 1990, **387**, 277.
99. B. H. Byers, T. P. Curran, M. J. Thompson and L. J. Sauer, *Organometallics*, 1983, **2**, 459.
100. R. Davis and F. I. C. Wilson, *J. Organomet. Chem.*, 1990, **396**, 55.
101. C. V. Sekhar and A. J. Poë, *J. Am. Chem. Soc.*, 1985, **107**, 4874.
102. T. W. Turney, *Inorg. Chim. Acta*, 1982, **64**, L141.
103. G. Ferguson, W. J. Laws, M. Parvez and R. J. Puddephatt, *Organometallics*, 1983, **2**, 276.
104. H. C. Aspinall, A. J. Deeming and S. Donovan-Mtunzi, *J. Chem. Soc., Dalton Trans.*, 1983, 2669.
105. H. C. Aspinall and A. J. Deeming, *J. Chem. Soc., Dalton Trans.*, 1985, 743.
106. B. F. Hoskins and R. J. Steen, *Aust. J. Chem.*, 1983, **36**, 683.
107. J. J. Hunt, E. N. Duesler and R. T. Paine, *J. Organomet. Chem.*, 1987, **320**, 307.
108. S. B. McCullen and T. L. Brown, *J. Am. Chem. Soc.*, 1982, **104**, 7496.
109. A. S. Goldman and D. R. Tyler, *J. Organomet. Chem.*, 1986, **311**, 349.
110. R. R. Andréa, W. G. J. de Lange, T. van der Graaf, M. Rijkhoff, D. J. Stufkens and A. Oskam, *Organometallics*, 1988, **7**, 1100.
111. M. W. Kokkes, D. J. Stufkens and A. Oskam, *Inorg. Chem.*, 1985, **24**, 2934.
112. J. A. Armstead, D. J. Cox and R. Davis, *J. Organomet. Chem.*, 1982, **236**, 213.
113. D. J. Kuchynka and J. K. Kochi, *Inorg. Chem.*, 1989, **28**, 855.
114. S. B. McCullen, H. W. Walker and T. L. Brown, *J. Am. Chem. Soc.*, 1982, **104**, 4007.
115. T. R. Herrinton and T. L. Brown, *J. Am. Chem. Soc.*, 1985, **107**, 5700.
116. R. S. Herrick, T. R. Herrinton, H. W. Walker and T. L. Brown, *Organometallics*, 1985, **4**, 42.
117. P. Jaitner, W. Huber, G. Huttner and O. Scheidsteger, *J. Organomet. Chem.*, 1982, **259**, C1.
118. M. C. R. Symons and R. L. Sweany, *Organometallics*, 1982, **1**, 834.
119. S. A. Fairhurst, J. R. Morton, R. N. Perutz and K. F. Preston, *Organometallics*, 1984, **3**, 1389.
120. K. Mach, J. Novakova and J. B. Raynor, *J. Organomet. Chem.*, 1992, **439**, 341.
121. G. B. Rattinger, R. Linn Belford, H. Walker and T. L. Brown, *Inorg. Chem.*, 1989, **28**, 1059.
122. K. A. M. Creber, T.-I. Ho, M. C. Depew, D. Weir and J. K. S. Wan, *Can. J. Chem.*, 1982, **60**, 1504.
123. A. Vlcek, Jr., *J. Organomet. Chem.*, 1986, **306**, 63.
124. T. van der Graaf, D. J. Stufkens, J. Vichova and A. Vlcek, *J. Organomet. Chem.*, 1991, **401**, 305.
125. W. G. McGimpsey, M. C. Depew and J. K. S. Wan, *Organometallics*, 1984, **3**, 1684.
126. G. A. Abakumov, V. K. Cherkasov, K. G. Shalnova, I. A. Teplova and G. A. Razuvaev, *J. Organomet. Chem.*, 1982, **236**, 333.
127. B. A. Narayanan and J. K. Kochi, *J. Organomet. Chem.*, 1984, **272**, C49.
128. D. J. Kuchynka, C. Amatore and J. K. Kochi, *Inorg. Chem.*, 1986, **25**, 4087.
129. D. J. Kuchynka and J. K. Kochi, *Inorg. Chem.*, 1988, **27**, 2574.
130. Y. Zhen and J. D. Atwood, *Organometallics*, 1991, **10**, 2778.
131. Y. Chen and J. D. Atwood, *J. Am. Chem. Soc.*, 1989, **111**, 1506.
132. K. Y. Lee, D. J. Kuchynka and J. K. Kochi, *Organometallics*, 1987, **6**, 1886.
133. D. J. Kuchynka, C. A. Amatore and J. K. Kochi, *J. Organomet. Chem.*, 1987, **328**, 133.
134. F. J. G. Alonso, A. Llamazares, V. Riera, M. Vivanco, S. G. Granda and M. R. Diaz, *Organometallics*, 1992, **11**, 2826.
135. Y. K. Park, S. J. Kim, J. H. Kim, I. S. Han, C. H. Lee and H. S. Choi, *J. Organomet. Chem.*, 1991, **408**, 193.
136. L. L. Padolik, F. Ungvary and A. Wojcicki, *J. Organomet. Chem.*, 1992, **424**, 319.
137. F. Hartl, A. Vlcek, L. A. deLearie and C. G. Pierpont, *Inorg. Chem.*, 1990, **29**, 1073.
138. F. Hartl and A. Vlcek, *Inorg. Chem.*, 1991, **30**, 3048.
139. F. Hartl, D. J. Stufkens and A. Vlcek, *Inorg. Chem.*, 1992, **31**, 1687.
140. M. Brookhart, S. K. Noh, F. J. Timmers and Y. H. Hong, *Organometallics*, 1988, **7**, 2458.
141. R. L. Harlow, P. J. Krusic, R. J. McKinney and S. S. Wreford, *Organometallics*, 1982, **1**, 1506.
142. G. F. P. Warnock, L. Cammarano Moodie and J. E. Ellis, *J. Am. Chem. Soc.*, 1989, **111**, 2131.
143. W. Schatz, H.-P. Neumann, B. Nuber, B. Kanellakopulos and M. L. Ziegler, *Chem. Ber.*, 1991, **124**, 453.
144. A. Alvanipour, H. Zhang and J. L. Atwood, *J. Organomet. Chem.*, 1988, **358**, 295.
145. M. S. Corraine *et al.*, *Organometallics*, 1992, **11**, 35.
146. T. M. Bockman and J. K. Kochi, *J. Am. Chem. Soc.*, 1989, **111**, 4669.
147. H. Kunkely and A. Vogler, *J. Organomet. Chem.*, 1989, **372**, C29.
148. C. H. Wei, T. M. Bockman and J. K. Kochi, *J. Organomet. Chem.*, 1992, **428**, 85.
149. F. Calderazzo and G. Pampaloni, *J. Chem. Soc., Chem. Commun.*, 1984, 1249.
150. J. D. Atwood, *Inorg. Chem.*, 1987, **26**, 2918.
151. Y. Zhen, W. G. Feighery and J. D. Atwood, *J. Am. Chem. Soc.*, 1991, **113**, 3616.

152. B. K. Nicholson, M. I. Bruce, O. bin Shawkataly and E. R. T. Tiekink, *J. Organomet. Chem.*, 1992, **440**, 411.
153. P. Braunstein, C. de Meric de Bellefon, M. Ries and J. Fischer, *Organometallics*, 1988, **7**, 332.
154. D. M. Antonelli and M. Cowie, *Organometallics*, 1990, **9**, 1818.
155. D. M. Antonelli and M. Cowie, *Organometallics*, 1991, **10**, 2173.
156. R. N. McDonald, M. T. Jones and A. K. Chowdhury, *J. Am. Chem. Soc.*, 1991, **113**, 476.
157. R. N. McDonald, M. T. Jones and A. K. Chowdhury, *J. Am. Chem. Soc.*, 1992, **114**, 71.
158. R. N. McDonald, M. T. Jones and A. K. Chowdhury, *Organometallics*, 1992, **11**, 392.
159. R. N. McDonald, M. T. Jones and A. K. Chowdhury, *Organometallics*, 1992, **11**, 356.
160. S. A. Roth, G. D. Stucky, H. M. Feder, M. J. Chen and J. W. Rathke, *Organometallics*, 1984, **3**, 708.
161. F. J. Garcia Alonso and V. Riera, *Polyhedron*, 1983, **2**, 1103.
162. G. A. Carriedo, V. Riera, M. L. Rodriguez and J. J. Sainz-Velicia, *Polyhedron*, 1987, **6**, 1879.
163. F. J. Garcia Alonso, V. Riera and M. Vivanco, *J. Organomet. Chem.*, 1987, **321**, C30.
164. H. Elias *et al.*, *Inorg. Chem.*, 1989, **28**, 3021.
165. J. Nitschke, S. P. Schmidt and W. C. Trogler, *Inorg. Chem.*, 1985, **24**, 1972.
166. F. J. Garcia Alonso, V. Riera and M. J. Misas, *Transition Met. Chem.*, 1985, **10**, 19.
167. D. Miguel, V. Riera and J. A. Miguel, *J. Organomet. Chem.*, 1991, **412**, 127.
168. G. A. Carriedo, V. Riera and J. Santamaria, *J. Organomet. Chem.*, 1982, **234**, 175.
169. D. A. Edwards and J. Marshalsea, *Polyhedron*, 1984, **3**, 353.
170. H. Berke and G. Weiler, *Z. Naturforsch., Teil B*, 1984, **39**, 431.
171. G. Hartmann, R. Hoppenheit and R. Mews, *Inorg. Chim. Acta*, 1983, **76**, L201.
172. G. Hartmann and R. Mews, *Chem. Ber.*, 1986, **119**, 374.
173. N. G. Connelly, A. G. Orpen, G. M. Rosair and G. H. Worth, *J. Chem. Soc., Dalton Trans.*, 1991, 1851.
174. G. A. Carriedo, M. C. Crespo, C. Diaz and V. Riera, *J. Organomet. Chem.*, 1990, **397**, 309.
175. F. J. Garcia Alonso, V. Riera and M. Vivanco, *J. Organomet. Chem.*, 1990, **398**, 275.
176. F. J. Garcia Alonso, V. Riera, M. L. Valin, D. Moreiras, M. Vivanco and X. Solans, *J. Organomet. Chem.*, 1987, **326**, C71.
177. G. A. Carriedo, J. B. Parra Sota, V. Riera, M. L. Valin, D. Moreiras and X. Solans, *J. Organomet. Chem.*, 1987, **326**, 201.
178. G. A. Carriedo, V. Riera, N. G. Connelly and S. J. Raven, *J. Chem. Soc., Dalton Trans.*, 1987, 1769.
179. N. G. Connelly *et al.*, *J. Chem. Soc., Dalton Trans.*, 1988, 1623.
180. D. H. Gibson, K. Owens, S. K. Mandal, W. E. Sattich and J. O. Franco, *Organometallics*, 1989, **8**, 498.
181. D. H. Gibson, K. Owens, S. K. Mandal, W. E. Sattich and J. O. Franco, *Organometallics*, 1991, **10**, 1203.
182. J. Ruiz, V. Riera, M. Vivanco, S. Garcia Grande and A. Garcia Fernandez, *Organometallics*, 1992, **11**, 4077.
183. A. R. Manning, G. McNally, R. Davis and C. C. Rowland, *J. Organomet. Chem.*, 1983, **259**, C15.
184. J. A. Davies, M. El-Ghanam, A. A. Pinkerton and D. A. Smith, *J. Organomet. Chem.*, 1991, **409**, 367.
185. N. J. Coville, P. Johnston, A. E. Leins and A. J. Markwell, *J. Organomet. Chem.*, 1989, **378**, 401.
186. H. Menzel, W. P. Fehlhammer and W. Beck, *Z. Naturforsch., Teil B*, 1982, **37**, 201.
187. E. Lindner, D. Merkle, W. Hiller and R. Fawze, *Chem. Ber.*, 1986, **119**, 659.
188. K. M. Cooke, T. P. Kee, A. L. Langton and M. Thornton-Pett, *J. Organomet. Chem.*, 1991, **419**, 171.
189. E. C. Alyea, R. P. Shakya and A. E. Vougioukas, *Transition Met. Chem.*, 1985, **10**, 435.
190. G. Effinger, W. Hiller and I.-P. Lorenz, *Z. Naturforsch., Teil B*, 1987, **42**, 1315.
191. R. J. Angelici, M. H. Quick, G. A. Kraus and D. T. Plummer, *Inorg. Chem.*, 1982, **21**, 2178.
192. D. Miguel, V. Riera, J. A. Miguel, C. Bois, M. Philoche-Levisalles and Y. Jeannin, *J. Chem. Soc., Dalton Trans.*, 1987, 2875.
193. W. Winter, R. Merkel and U. Kunze, *Z. Naturforsch., Teil B*, 1983, **38**, 747.
194. U. Kunze, A. Bruns and D. Rehder, *J. Organomet. Chem.*, 1984, **268**, 213.
195. M. Mason *et al.*, *J. Chem. Soc., Dalton Trans.*, 1987, 2599.
196. A. M. Bond, R. Colton and P. Panagiotidou, *Organometallics*, 1988, **7**, 1767.
197. G. Schmidt, H. Paulus, R. van Eldik and H. Elias, *Inorg. Chem.*, 1988, **27**, 3211.
198. N. G. Connelly, S. J. Raven, G. A. Carriedo and V. Riera, *J. Chem. Soc., Chem. Commun.*, 1986, 992.
199. M. M. Singh and R. J. Angelici, *Inorg. Chem.*, 1984, **23**, 2699.
200. J. L. Atwood *et al.*, *Inorg. Chem.*, 1983, **22**, 1797.
201. J. Gimeno, V. Riera, M. A. Ruiz, A. M. M. Lanfredi and A. Tiripicchio, *J. Organomet. Chem.*, 1984, **268**, C13.
202. V. Riera, M. A. Ruiz, A. Tiripicchio and M. Tiripicchio Camellini, *J. Chem. Soc., Dalton Trans.*, 1987, 1551.
203. V. Riera, M. A. Ruiz, A. Tiripicchio and M. Tiripicchio Camellini, *J. Organomet. Chem.*, 1986, **308**, C19.
204. B. Just, W. Klein, J. Kopf, K. G. Steinhauser and R. Kramolowsky, *J. Organomet. Chem.*, 1982, **229**, 49.
205. E. Horn, M. R. Snow and E. R. T. Tiekink, *Acta Crystallogr., Sect. C*, 1987, **43**, 792.
206. C. A. McAuliffe, D. S. Barratt, C. G. Benson, A. Hosseiny, M. G. Little and K. Minten, *J. Organomet. Chem.*, 1983, **258**, 35.
207. J. Heck, W. Massa and P. Weinig, *Angew. Chem., Int. Ed. Engl.*, 1984, **23**, 722.
208. F. H. Kohler, N. Hebendanz, U. Thewalt, B. Kanellakopulos and R. Klenze, *Angew. Chem., Int. Ed. Engl.*, 1984, **23**, 721.
209. R. D. Davy and M. B. Hall, *Inorg. Chem.*, 1989, **28**, 3524.
210. K. Pierloot, P. Hoet and L. G. Vanquickenborne, *J. Chem. Soc., Dalton Trans.*, 1991, 2363.
211. X. Pan, C. E. Philbin, M. P. Castellani and D. R. Tyler, *Inorg. Chem.*, 1988, **27**, 671.
212. W. Deck, A. K. Powell and H. Vahrenkamp, *J. Organomet. Chem.*, 1991, **411**, 431.
213. F. M. Ashmawy, C. A. McAuliffe, K. L. Minten, R. V. Parish and J. Tames, *J. Chem. Soc., Chem. Commun.*, 1983, 436.
214. E. Lindner and D. Goth, *J. Organomet. Chem.*, 1987, **319**, 149.
215. P. J. Manning, L. K. Peterson, F. Wada and R. S. Dhami, *Inorg. Chim. Acta*, 1986, **114**, 15.
216. G. A. Carriedo, M. C. Crespo, V. Riera, M. L. Valin, D. Moreiras and X. Solans, *Inorg. Chim. Acta*, 1986, **121**, 191.
217. G. A. Carriedo *et al.*, *J. Organomet. Chem.*, 1986, **302**, 47.
218. G. A. Carriedo, N. G. Connelly, M. C. Crespo, I. C. Quarmby, V. Riera and G. H. Worth, *J. Chem. Soc., Dalton Trans.*, 1991, 315.
219. A. Christofides *et al.*, *J. Chem. Soc., Dalton Trans.*, 1991, 1595.

# 2

# Manganese Alkyls and Hydrides

THOMAS C. FLOOD

*University of Southern California, Los Angeles, CA, USA*

## 2.1 INTRODUCTION

Since the review of the organometallic chemistry of manganese in _COMC-I_,[1] several hundred new publications have appeared. These cover a wide range of structural types, most of which are discussed here, including manganese $\eta^1$-alkyl, aryl, acyl, alkenyl, alkynyl, carbene, carbyne, and vinylidene ligands, including chelated groups, but not those $\eta^3$ or more at the local site. Locally, $\eta^2$ groups are covered if they involve a formal $\sigma$ bond (e.g., $\eta^2$ acyls and vinyls) but pure $\eta^2$ $\pi$-complexes (e.g., $\pi$-ethene) are not. $\eta$-Allyl, $\eta$-aryl, Cp, and so on are covered in Chapters 4, and 5, this volume, except when they are ancillary ligands and their $\eta^1$ carbon chemistry otherwise is the focus of the chemistry in the molecule. Manganese hydrocarbon complexes of formal oxidation state 2 or 3 are covered if they are in a "conventional" ligand environment, such as in $MnCp(CO)_2(SiR_3)R$. Homoleptic alkyls and those of oxidation state 2 or higher with no strongly $\pi$-acid ligands are discussed in Chapter 7, this volume.

## 2.2 MANGANESE ACYL COMPLEXES

The chemistry of manganese alkyls is intimately interwoven with that of manganese acyls. One of the two most common routes to the alkyls is via the acyls, and the CO insertion reaction, one of the cornerstones of organometallic chemistry, has found a useful paradigm in the chemistry of $MnR(CO)_5$ and its derivatives. We accordingly begin our discussion with manganese acyls.

### 2.2.1 Synthesis of Manganese Acyls

There are two generally employed methods for preparing manganese acyl complexes: acylation of anionic manganese complexes, and "migratory insertion" of CO into manganese–carbon bonds. Historically, the most common method for preparation of both Mn–acyl and Mn–alkyl groups has revolved around the nucleophilicity of the manganese centers in anionic complexes, and this has remained true since the 1980s. The most widely used anion has been the pentacarbonylmanganate anion, $[Mn(CO)_5]^-$, which is readily available from manganese carbonyl, $Mn_2(CO)_{10}$, by reduction, typically with sodium amalgam in THF. This anion is sufficiently nucleophilic to attack acyl halides to afford manganese acyls in variable but generally high yields. The preparation of $Mn(COMe)(CO)_5$ by this route has been described in _Inorganic Syntheses_.[2]

Anion acylation is quite general and has the advantage, unlike anion alkylation, that it is quite insensitive to the organic residue of the acid chloride or acid anhydride acylating agent; aryl, vinyl, perfluorocarbon, acyl chlorides, and so on can be used. Then, combined with thermal or photochemical decarbonylation, one can synthesize a wide variety of alkyl, vinyl, and aryl manganese species that could be very difficult to make in any other way. For example, attempts to prepare $(OC)_5Mn–(CH_2)_4–Mn(CO)_5$ by direct alkylation with $[Mn(CO)_5]^-$ and $Br–(CH_2)_4–Br$ yielded the binuclear, cyclic carbene complex (**1**) shown in Scheme 1.[3] However, use of adipoyl chloride followed by decarbonylation in refluxing hexane gave the desired $(OC)_5Mn–(CH_2)_4–Mn(CO)_5$ in good yield.[4] The acyl groups could be readily regenerated in the usual fashion by adding a ligand, in this case $PMe_3$, to the manganese alkyl in solution. The substituted manganese acyl was in the _cis_ position, as is generally the case when the new ligand is not too large. This same technique was used to prepare the penta- and hexamethylene bridged dimanganese complexes.[4]

It is occasionally observed that the product of an attempted alkylation is actually the CO-inserted acyl product instead. This is the case, for example, with attempted alkylation of $K[Mn(CO)_5]$ by $MeC(CH_2I)_3$ (Equation (1)) where the acyl product (**2**) is obtained.[5] In other cases, the acyl product cannot be successfully decarbonylated to the desired alkyl product, as in Equation (2) where acyl (**3**) formed well but decomposed on heating in solution.[6]

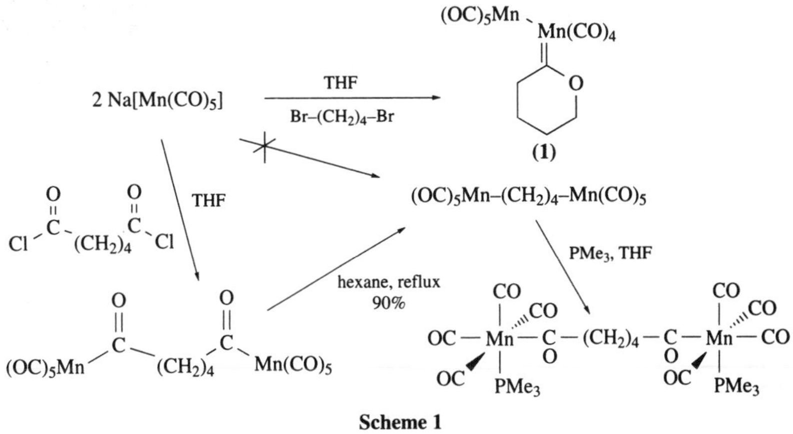

$$(OC)_5Mn—Mn(CO)_4$$

**Scheme 1**

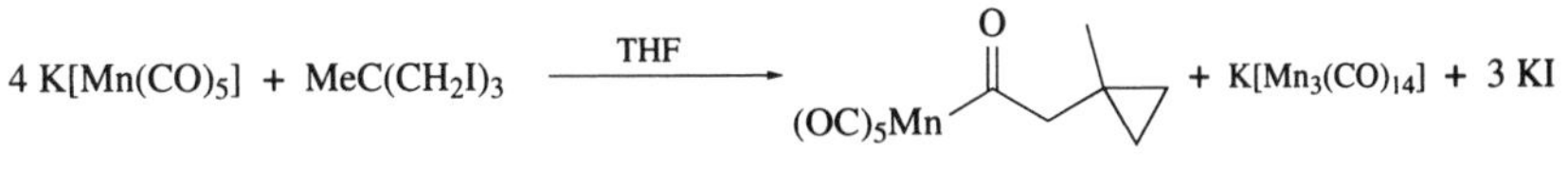

$$2\ Na[Mn(CO)_5] \xrightarrow[\text{Br–(CH}_2)_4\text{–Br}]{\text{THF}}$$

**(1)**

$$(OC)_5Mn–(CH_2)_4–Mn(CO)_5$$

$$\text{Cl}\overset{O}{\overset{\|}{C}}\text{(CH}_2)_4\overset{O}{\overset{\|}{C}}\text{Cl} \quad \text{THF}$$

$$(OC)_5Mn\overset{O}{\overset{\|}{C}}(CH_2)_4\overset{O}{\overset{\|}{C}}Mn(CO)_5$$

hexane, reflux
90%

PMe$_3$, THF

**Scheme 1**

$$4\ K[Mn(CO)_5] + MeC(CH_2I)_3 \xrightarrow{\text{THF}} (OC)_5Mn + K[Mn_3(CO)_{14}] + 3\ KI \tag{1}$$

**(2)**

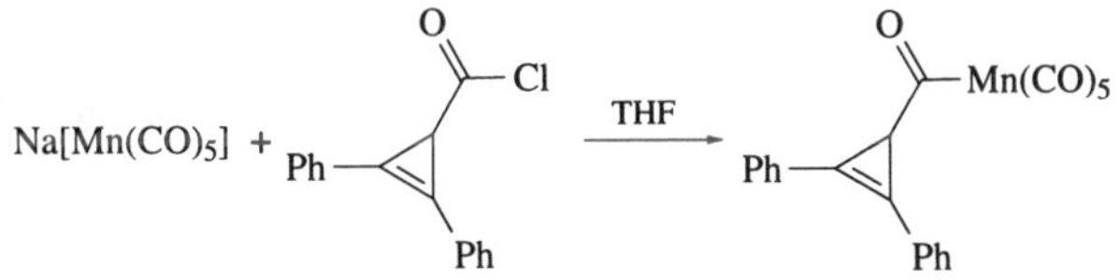

$$Na[Mn(CO)_5] + \underset{Ph}{\underset{Ph}{\overset{O}{\overset{\|}{C}}}-Cl} \xrightarrow{\text{THF}} \underset{Ph}{\underset{Ph}{\overset{O}{\overset{\|}{C}}-Mn(CO)_5}} \tag{2}$$

**(3)**

Another route to manganese acyls is the addition of nucleophiles to the CO group of neutral and cationic complexes. A recent example involves the addition of MeLi or PhLi to cyclohexadienyl manganese complexes (**4**) (Equation (3)).[7] Anions (**5**) are isolated in up to 90% yield as orange solids which are pyrophoric as the Li$^+$ salts, but somewhat less air sensitive with the [N(PPh$_3$)$_2$]$^+$ counterion. Subsequent rearrangement chemistry of anions (**5**) induced by acids was reported. Nucleophilic addition to [MnCp(CO)$_2$(NO)]$^+$ has been known for some time, and a new example of this route to acyl species is shown in Equation (4). In this case, deprotonation of the initial adduct by a second equivalent of strongly basic ylide gives an acyl stabilized ylide (**7**) as product.[8] Since strong nucleophiles will also add to MnCp(CO)$_3$, it was found that CH$_2$=PMe$_3$ and homologous ylides add to Mn(RC$_5$H$_4$)(CO)$_3$ (R = H or Me) to afford the analogous ylide salts Me$_4$P[Mn(RC$_5$H$_4$)(CO)$_2$C(O)CH=PMe$_3$].[9]

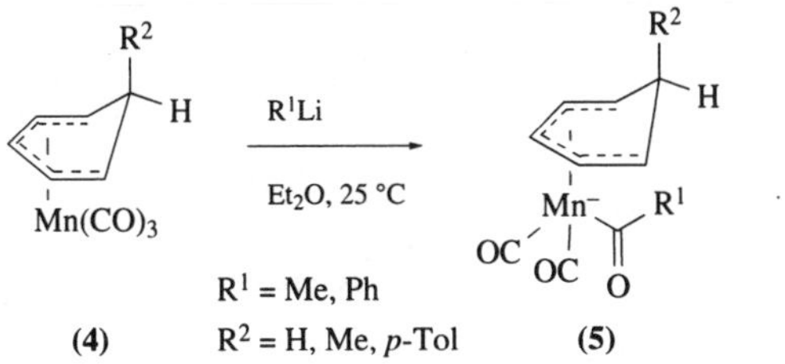

$$\xrightarrow[\text{Et}_2\text{O, 25 °C}]{R^1Li} \tag{3}$$

R$^1$ = Me, Ph
R$^2$ = H, Me, *p*-Tol

**(4)**      **(5)**

$$\xrightarrow[\substack{\text{THF, –40 °C} \\ 89\%}]{2\ CH_2=PMe_3} + PMe_4^+ \tag{4}$$

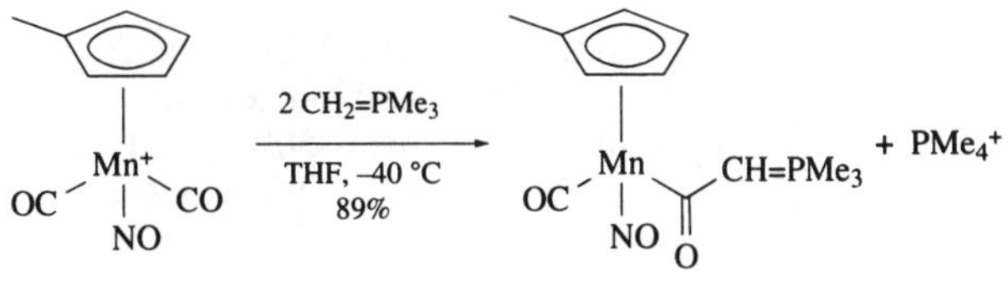

**(6)**      **(7)**

Addition of the lithium enolate (**8**) to Mn(OTf)(CO)$_5$ might be expected to form the manganese alkyl (manganese carbon enolate), but instead affords the manganese acyl (**9**) (Equation (5)).[10] Acyl (**9**) could be formed by direct kinetic attack by the enolate anion on a radial CO carbon atom, or it might first form the manganese–carbon bond with subsequent CO insertion being driven by the internal rhenium acyl nucleophile. Exchange of $D_2O$ with the enolizable hydrogens in the manganese heterocycle of (9) occurred with 11:1 diastereoselectivity.

$$(5)$$

Another example of formation of an acyl by net nucleophilic attack on a Mn–CO group is afforded by the reaction wherein diazomethane adds to the manganese dimer (10) to give the novel cyclic acyl complex (11) (Equation (6)).[11] Structure (11) is consistent with attack by the $\delta^-$ $CH_2$ end of the $CH_2N_2$ molecule on a coordinated CO, either before or after coordination of the nitrogen end to manganese.

$$(6)$$

Oxidation of the carbyne complex (12) by $NO_2^-$ forms the double CO insertion product α-ketoacyl (13) (Equation (7)). Complex (13) is also generated by one-electron oxidation of the known anion $[Mn(Cp')(CO)_2(C(O)Tol)]^-$ (where $Cp' = MeC_5H_4$) to form the 17-electron neutral species, followed by its treatment with NO gas.[12] Although α-ketoacyl ligands had been known for some time, including examples such as $Mn(CO)_5(COCOR)$,[13] the NO-induced formation of (13) was the first demonstration of CO insertion into a metal acyl bond. Both photochemical and thermal decarbonylation attempts showed (13) to be very resistant. However, heating (13) at 65 °C in toluene or photolysis in $CH_2Cl_2$ in the presence of $PPh_3$ or $Bu^tNC$ readily formed the substitution product $Mn(Cp')(L)(NO)(COCOTol)$.[12] Due to this apparent discrepancy, a low-temperature study of (13) employing continuous irradiation at 436 nm and monitoring by FTIR (Fourier transform IR) was carried out.[14] It was concluded that the photodissociated intermediate is present as the internally coordinated species $\overline{Mn(Cp')(NO)(COCOTol)}$ (14), which would persist long enough to be associatively trapped by ligands such as $PPh_3$, but would not tend to undergo unimolecular CO extrusion during its lifetime.

$$(7)$$

With a view to studying metal-bridged, coupled CO ligands, $Na[Mn(CO)_5]$ was treated with 0.5 equiv. of oxalyl chloride in $Et_2O$ at −78 °C. The binuclear species $Mn(CO)_5(\mu\text{-}CO\text{–}CO)Mn(CO)_5$ was isolated as a deep-purple solid in greater than 50% yield. Unfortunately, under all conditions investigated, only decomposition to CO and $Mn_2(CO)_{10}$ was seen.[15]

### 2.2.2 Reactions of Manganese Acyls

Aside from decarbonylation, since the 1980s the main focus of reactivity studies of manganese acyl complexes has been in attempts at reduction and addition. Reaction of $Mn(^{13}C(O)Tol)(CO)_5$ (Tol = $p\text{-}C_6H_4Me$) with $H_2/CO$ at 160 atm in hexane led to exclusive production of $Mn(C(O)O^{13}CH_2\text{-}Tol)(CO)_5$.[16] There was quantitative retention of the carbon isotope label in the site shown, so that no reversible CO extrusion–insertion occurred. In sulfolane solvent, only tolualdehyde ($p$-MeC$_6$H$_4$CHO) was formed. Control experiments ruled out that formation of Mn(C(O)OCH$_2$R)(CO)$_5$ might occur in hexane via reaction of MnH(CO)$_5$ with an aldehyde intermediate, and suggested that an intermediate aldehyde–manganese complex is also unlikely to be involved in that solvent.

Manganese acyl complexes exhibit a variety of interesting reactions with hydridic reagents. For example, reaction of Mn(C(O)Me)(CO)$_5$ with Si(H)Et$_3$ or Sn(H)Bu$^n_3$ in THF is reported to afford acetaldehyde and Mn(ER$_3$)(CO)$_5$ (E = Si or Sn).[17] The rate is first order in manganese, shows saturation behavior toward HSiEt$_3$, and is inhibited by CO. These data fit a classic scheme of reversible CO dissociation from Mn(C(O)Me)(CO)$_5$ followed by uptake of HSiEt$_3$. Apparently, Sn(H)Bu$^n_3$ is more reactive than the silane toward the unsaturated intermediate, as in this case CO dissociation is clearly rate limiting. In contrast, when Mn(C(O)Me)(CO)$_5$ was treated with SiHMe$_2$Ph in C$_6$D$_6$ at room temperature, the hydrosilation product Mn(CH(OSiMe$_2$Ph)Me)(CO)$_5$ was isolated after chromatography in 67% yield. The NMR spectrum revealed no traces of MeCHO, Mn(SiMe$_2$Ph)(CO)$_5$, or EtOSiMe$_2$Ph.[18] At the same time, Mn(C(O)Ph)(CO)$_5$ reacted with SiHMe$_2$Ph under the same conditions to form PhCHO plus Mn(SiMe$_2$Ph)(CO)$_5$ in 75–85% yield. Reaction of SiH$_2$Ph$_2$ with either Mn(C(O)Me)(CO)$_5$ or Mn(C(O)Ph)(CO)$_5$ afforded high yields of Mn(CH(OSiHPh$_2$)R)(CO)$_5$.[18] In the latter cases, the products were too unstable for isolation. Both Mn(C(O)Me)(CO)$_5$ and Mn(C(O)Ph)(CO)$_5$ were found to be extremely efficient hydrosilation catalysts for aldehydes and ketones.

The chemistry of the unusual α-ketoacyl (13), the preparation of which was discussed earlier,[12] and its PPh$_3$-substituted analogue have been studied (Scheme 2).[19] Treatment of (13) (L = PPh$_3$) with strong acid formed tolualdehyde and [Mn(Cp')(CO)(NO)(PPh$_3$)]$^+$. At low temperature the α-hydroxycarbene species (15), presumably an intermediate, was observed by NMR spectroscopy. Although (13) (L = PPh$_3$) was inert to UV irradiation, (13) (L = CO) showed well-defined photochemistry. Photolysis in the presence of HNEt$_2$ afforded a moderate yield of α-ketoamide (16), while in the presence of 2-butyne metallafuran (17) was isolated in low yield. The latter reaction also occurs for other electron-rich alkynes, and the organic ligand could be removed by treatment with HCl in MeCN.

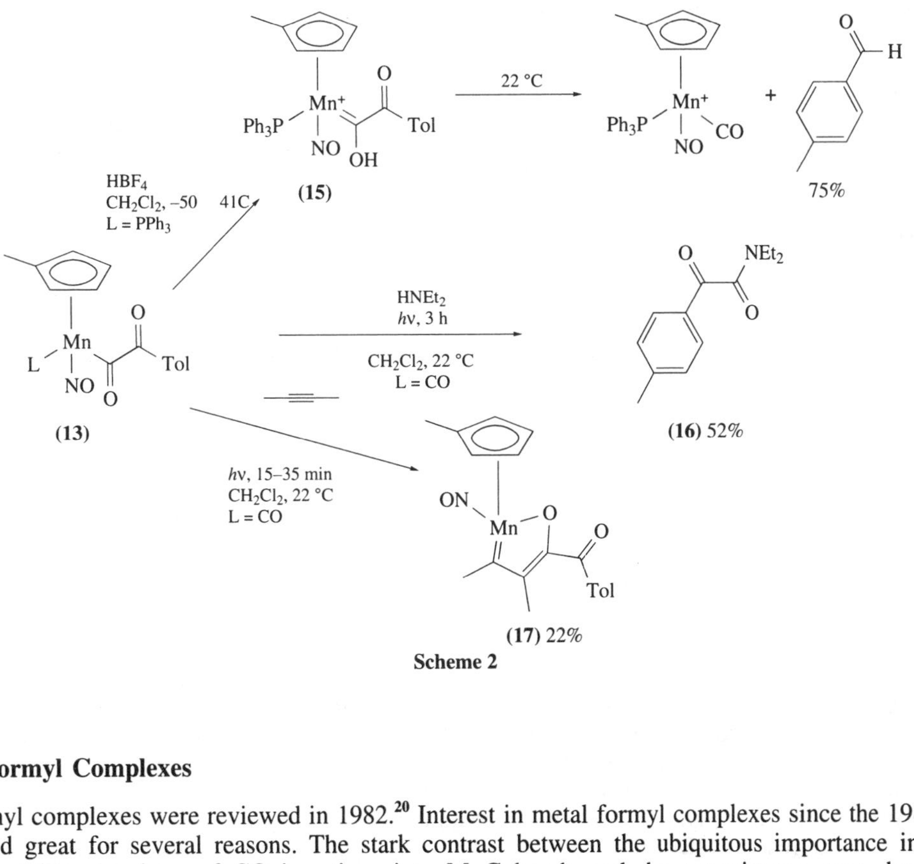

**Scheme 2**

### 2.2.3 Formyl Complexes

Formyl complexes were reviewed in 1982.[20] Interest in metal formyl complexes since the 1980s has remained great for several reasons. The stark contrast between the ubiquitous importance in all of organometallic chemistry of CO insertions into M–C bonds and the conspicuous near absence of examples of CO insertion into M–H bonds has made learning about the chemistry of the formyl group an intriguing intellectual goal. In addition, syngas chemistry continues to be an important preoccupation as we face the need for new feedstock and energy sources in the twenty-first century. Whether or not metal-bound formyl groups are important in heterogeneous Fischer–Tropsch catalysis, as was commonly assumed at first, there is a finite probability that OC–M–H insertion chemistry will become important in new catalytic technology of syngas.

A number of hydridic preparations of formyl complexes of several metals exist. One of the more effective preparations for manganese is shown in Equation (8). Species such as $Mn(CHO)(CO)_5$ are unstable, but substitutions by phosphites or phosphines render them more stable. For example, *mer,trans*-$Mn(CHO)L_2(CO)_3$ (**18**) and *trans*-$Mn(CHO)L_3(CO)_2$ (**19**) (L = $P(OPh)_3$, Equation (8)) can be prepared by the addition of $NaBH_4$ in MeOH or $NaBH(OMe)_3$ in THF to $[MnL_2(CO)_4]^+$ and $[MnL_3(CO)_3]^+$ at −80 °C and isolated in 80–90% yields as pale-yellow crystals.[21] These preparations have been expanded to include *mer,trans*-$Mn(CHO)(PPh_3)_2(CO)_3$ (**20**) (Scheme 4), which is isolable, and *cis*-$Mn(CHO)L(CO)_4$ (L = $PPh_3$ (**21**) (Scheme 5), or $P(OPh)_3$, (**22**)): the former has been isolated and the latter has not.[22] A binuclear formyl, $Li[(OC)_5Mn–Mn(CO)_4(CHO)]$, prepared from $Mn_2(CO)_{10}$ and $LiBHEt_3$ in THF was reported to be stable for hours at −20 °C, but decomposed rapidly at 22 °C.[23]

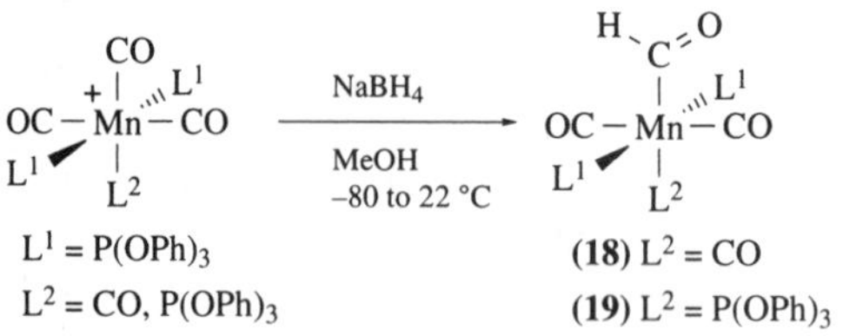

$$(8)$$

$L^1 = P(OPh)_3$
$L^2 = CO, P(OPh)_3$

(**18**) $L^2 = CO$
(**19**) $L^2 = P(OPh)_3$

As mentioned earlier, the thermal stability of formyl complexes is not high, although it was found that manganese formyl (**18**) is slower to lose CO to form the hydride (**23**) than the corresponding acetyl complex (**24**) forms the methyl complex (**25**) (Scheme 3). Complex (**18**) was found to decompose in a kinetically first-order reaction which is not inhibited by added CO and which exhibits a kinetic isotope effect $k(Mn(CHO))/k(Mn(CDO))$ of 3.24. The results were interpreted according to a radical chain mechanism.[21] There is also evidence from an electrochemical study for a radical chain mechanism in the decarbonylation of $Mn(CHO)(PPh_3)_2(CO)_3$ (**20**) to the hydride $Mn(H)(PPh_3)_2(CO)_3$, and the decarbonylation was found to be inhibited in THF solution by the addition of $SnHBu_3$.[24] A separate report observed that crystallized formyl underwent decarbonylation very slowly in pure benzene solution to form $Mn(H)(PPh_3)(CO)_4$ and free $PPh_3$.[22]

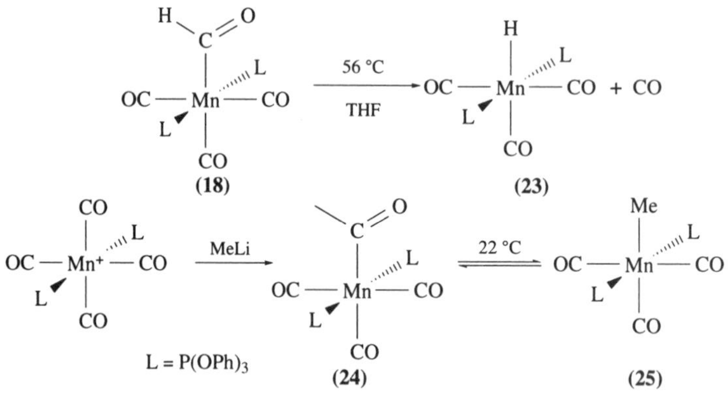

**Scheme 3**

In addition to simple decarbonylation, metal formyls undergo interesting reactions with electrophiles and nucleophiles. The formyl group has resonance structures similar to those of acyl metal species (Equation (9)), such that the oxygen is more electron rich than in normal organic carbonyl compounds. Electrophiles attack the acyl oxygen as demonstrated by Fischer in his syntheses of metal–carbene complexes. The stable formyls of manganese are similarly oxygen protonated by strong acids. As shown in Scheme 4, anhydrous $HBF_4$ protonates the formyl oxygen of (**20**) to form the cation (**26**), a rare example of an isolable α-hydroxymethylidine complex. When dissolved in methanol, (**26**) reacts to form a methoxymethylidine species (**27**), which can also be prepared by alkylation of (**20**) with methyl triflate.[25] When (**27**) is treated with excess (**20**), the manganese cation itself is sufficiently electrophilic to accept a hydride from the formyl complex, leading to the methoxymethyl ether (**28**).[26]

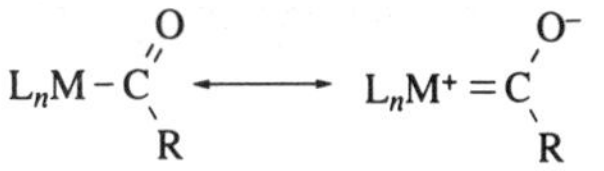

$$L_nM-\overset{\overset{O}{\|}}{C}\diagdown_R \quad\longleftrightarrow\quad L_nM^+=C\diagup^{O^-}_{\diagdown R} \tag{9}$$

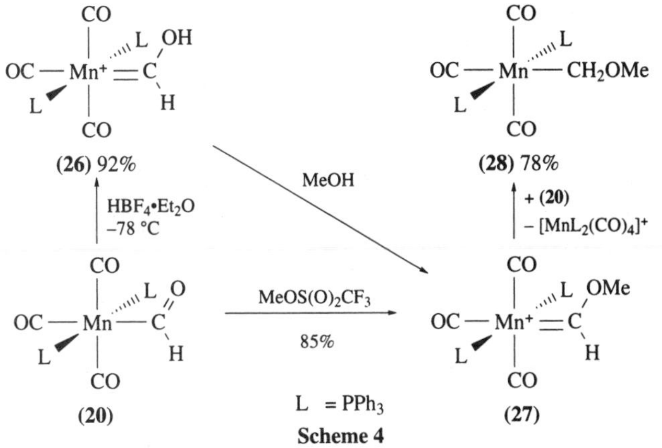

**Scheme 4**

L = PPh$_3$

It appears that during the formation of cations (**26**) and (**27**), these species are not electrophilic enough to exhibit cyclometallation with their own PPh$_3$ or P(OPh)$_3$ arene rings. However, when reaction times are long enough or conditions are otherwise appropriate, cyclometallation does occur, specifically with formyl complexes (**18**), (**21**), and (**22**). For example, this reaction is shown for (**21**) in Scheme 5, along with an outline of the suggested mechanism.[27] In addition, specific procedures for converting the methoxymethyl manganese species (e.g., (**28**)) to their halomethyl derivatives (e.g. Mn(PPh$_3$)$_2$(CO)$_3$(CH$_2$Cl)) have been reported.[28]

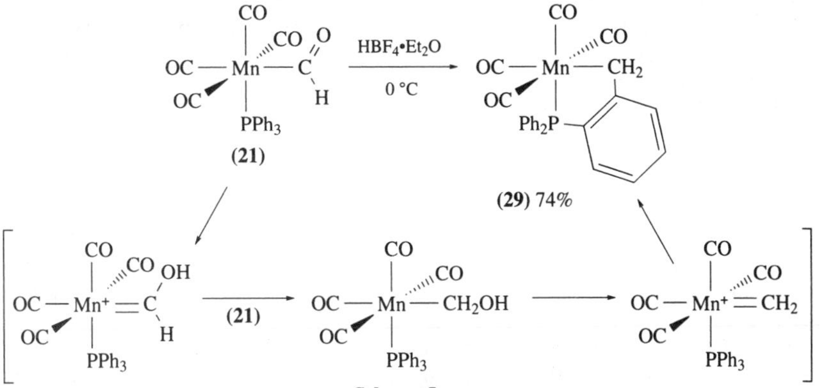

**Scheme 5**

## 2.3 MANGANESE ALKYL COMPLEXES

### 2.3.1 Synthesis

There are a wide variety of methods for the preparation of Mn–C($sp^3$) bonds, most of which have been known for some time. Methods used since the 1980s are described here according to the type of reaction employed. Note that this classification is not meant to imply that the mechanism of any given preparation has been clearly established, or that all examples grouped under a given reaction type necessarily proceed by a common mechanism.

#### 2.3.1.1 Alkylation of manganese anions

Historically, the most common synthetic method for manganese alkyl synthesis has revolved around the nucleophilicity of manganese centers in anionic complexes, and since the 1980s this has remained true. The most used anion has been [Mn(CO)$_5$]$^-$, with alkylations of [CpMn(CO)$_2$(SiR$_3$)]$^-$ and [(η-

$C_6H_6)Mn(CO)_2]^-$ and some other anions having also been reported. Table 1 shows examples of alkylations of manganese anions, which are generally carried out in ether solvents, frequently at reduced temperatures to minimize side reactions. A number of modifications, such as variations of the cation, solvent, and so on, were discussed in *COMC-I*,[1] but few have been used in the intervening years. In addition, some examples of alkylations or acylations of substituted manganese anions (e.g., $[Mn(PR_3)(CO)_4]^-$) have been reviewed,[1] but these too have been virtually unused recently.

Other preparations that were carried out under reducing conditions are mechanistically less clear than the examples given in Table 1, but probably proceed by mechanisms involving or related to metal anions. These are gathered in Table 2. Generally, these are effected by the action of sodium amalgam or other activated metal on an α,ω-dihalide, where one end is a manganese halide functional group.

In the course of studying the homologation of methanol catalyzed by $Mn_2(CO)_{10}$, the anion $[Mn(CO)_5]^-$ was inferred to be methylated by $MeO_2CH$ and the *N*,*N*-dimethylpiperidinium ion, $[Me_2NC_5H_{10}]^+$, at similar rates at 200 °C in MeOH.[50]

### 2.3.1.2 Decarbonylation of manganese acyl complexes

A particularly important method is the extrusion of carbon monoxide from the manganese acyl functional group. This subject was considered earlier in the context of acyl manganese chemistry (Section 2.2), and is considered further below when discussing the mechanism of the CO insertion reaction (Section 2.3.2.3 (i)). It is one of the most common ways of making manganese alkyls, and it is probably the most general. For example, as reviewed before,[1] manganese anions form manganese halides rather than (perfluoroalkyl)manganese species in reaction with perfluoroalkyl halides. Therefore, the best route to perfluoroalkyl manganese species is the acylation–decarbonylation sequence. *COMC-I*[1] should be consulted for examples of this type. A recent publication gives details of the preparation of nine different $MnR(CO)_5$ or $MnAr(CO)_5$ by the decarbonylation of the corresponding acyls.[51]

The acyls prepared from $Na[Mn(CO)_5]$ and long-chain acid chlorides in THF were decarbonylated in refluxing hexane forming $Mn(CO)_5(R)$, where R is a normal alkyl of length 4–18 carbon atoms. The substantially increased stability of the long-chain manganese alkyls towards air and thermal decomposition, in comparison with short-chain alkyls, led the authors to question the conventional interpretation that the diminished stability of the latter molecules comes simply from the facility of β-hydrogen elimination.[52]

### 2.3.1.3 Additions of manganese anions to metal-coordinated π-systems

The manganese anion $[Mn(CO)_5]^-$ is strong enough to add to assorted π-complexed unsaturates, and recent examples include additions to carbon π-systems of two, three, and four carbon atoms. For example, $[Mn(CO)_5]^-$ adds to $[M(CO)_5(CH_2=CH_2)]^+$ where M = $Mn^{53}$ and M = $Re^{54}$ to give dimethylene bridged $(CO)_5M–CH_2CH_2–Mn(CO)_5$. The anion $[Mn(CO)_5]^-$ adds stereospecifically to the allyl terminus *cis* to NO in the *exo* rotamer of π-allyl cation (30) (Equation (10)).[55] In a similar manner, complexes (32)[56] and (33)[57] are formed by additions of this anion to π-complexed hydrocarbons. In some cases, $[Mn(CO)_5]^-$ reacts with cationic π-complexes via apparent electron transfer, resulting in C–C coupling of the two reduced cations without the incorporation of manganese.[56]

$$\qquad\qquad\qquad (10)$$

### 2.3.1.4 Addition of manganese hydrides to organic unsaturation

The addition of manganese hydrides to alkynes to form vinyl manganese species is well established (see Section 2.6), but additions to alkenes to afford manganese alkyls are much less common. The latter are more generally encountered in catalytic processes such as alkene isomerization, or in stoichiometric reduction of alkenes by manganese hydride. However, one recent example arose from the study of

**Table 1** Manganese alkyls prepared from anionic complexes.

| Entry | Anion | Conditions | Product | Yield (%) | Ref. |
|---|---|---|---|---|---|
| 1 | $NaMn(CO)_5$ | RBr, THF, –78 °C | (OC)$_5$Mn | 75 | 29 |
| 2 | $NaMn(CO)_5$ | i, ClCH$_2$CO$_2$TMS, THF, –78 °C, ii, Silica gel | (CO)$_5$MnCH$_2$CO$_2$H | 56 | 30 |
| 3 | $NaMn(CO)_5$ | TfO–(CH$_2$)$_n$–OTf, Et$_2$O | (CO)$_5$Mn–(CH$_2$)$_n$–Mn(CO)$_5$ | | |
| | | $n$ = 2, –38 °C | | 90 | 31 |
| | | $n$ = 3, –60 °C | | 81 | 32 |
| | | $n$ = 5, –40 °C | | 58 | 33 |
| | | $n$ = 10, –30 °C | | 79 | 33 |
| 4 | $NaMn(CO)_5$ | p-(BrCH$_2$)$_2$C$_6$H$_4$, THF | (CO)$_5$MnCH$_2$ ⟨C$_6$H$_4$⟩ CH$_2$Mn(CO)$_5$ | 68 | 34 |
| 5 | $NaMn(CO)_5$ | RBr, THF, –78 °C | 4 similar examples | 75 | 35 |
| 6 | $NaMn(CO)_5$ | RCl, THF, –20 °C | N–C–Mn(CO)$_5$ H$_2$ | 90 | 36 |
| 7 | [Mn(CO)$_5$]$^-$ | CH$_2$ICl, THF, –20 °C | (OC)$_5$Mn–CH$_2$–Cl | 45 | 37 |
| 8 | [Mn(CO)$_4$(PPh$_3$)]$^-$ | CH$_2$ICl, THF, –78 °C | (OC)$_4$(PPh$_3$)Mn–CH$_2$–Cl | 60 | 37 |
| 9 | [Mn(CO)$_4$(PMe$_2$O)]$^{2-}$ | Me$_2$C(CH$_2$OTf)$_2$, DME, 40 °C | (OC)$_4$Mn—PMe$_2$ | 54 | 38 |

**Table 1** (continued)

| Entry | Anion | Conditions | Product | Yield (%) | Ref. |
|-------|-------|------------|---------|-----------|------|
| 10 | $[Mn(CO)_4(PR_2E)]^{2-}$ | O=C(CH$_2$Cl)$_2$, DME, 70 °C | (OC)$_4$Mn, P—E, R$_2$, O <br> E = O; R = Me, Ph <br> E = S, R = Me | 41–50 | 39 |
| 11 | $[Mn(CO)_4(PPh_2O)]^{2-}$ | Cl$_2$CHC$_6$H$_4$X, DME–heptane–ether, 20 °C | (OC)$_4$Mn—CH, P—O, Ph$_2$, X <br> X = H, Cl, Br | 30–34 | 39 |
| 12 | $[Mn(CO)_4(PMe_2S)]^{2-}$ | Cl$_2$CHPh, DME–heptane–ether, 20 °C | (OC)$_4$Mn—CH, P—S, Me$_2$ | 45 | 39 |
| 13 | $[Mn(CO)_4(PPh_2)]^{2-}$ | Br(CH$_2$)$_3$Br, THF, 22 °C | (OC)$_4$Mn——PPh$_2$ | 12 | 40 |

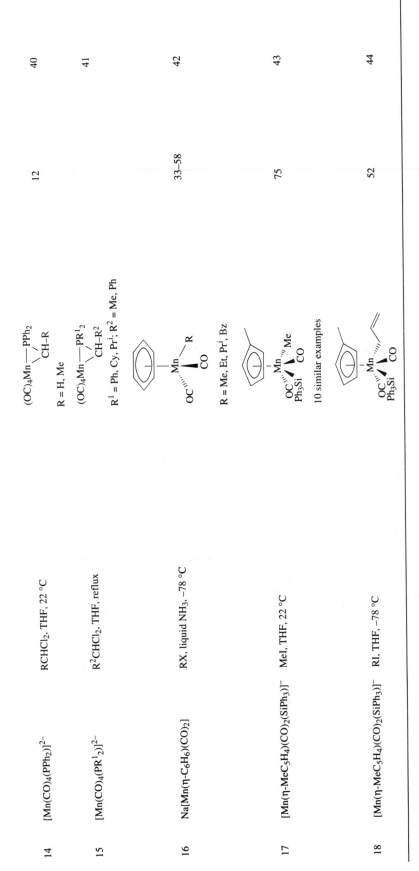

| 14 | $[Mn(CO)_4(PPh_2)]^{2-}$ | $RCHCl_2$, THF, 22 °C | (OC)$_4$Mn—PPh$_2$ / CH-R  R = H, Me | 12 | 40 |
| 15 | $[Mn(CO)_4(PR^1_2)]^{2-}$ | $R^2CHCl_2$, THF, reflux | (OC)$_4$Mn—PR$^1_2$ / CH-R$^2$  R$^1$ = Ph, Cy, Pr$^i$; R$^2$ = Me, Ph | | 41 |
| 16 | $Na[Mn(\eta-C_6H_6)(CO)_2]$ | RX, liquid NH$_3$, −78 °C | R = Me, Et, Pr$^i$, Bz | 33–58 | 42 |
| 17 | $[Mn(\eta-MeC_5H_4)(CO)_2(SiPh_3)]^-$ | MeI, THF, 22 °C | 10 similar examples | 75 | 43 |
| 18 | $[Mn(\eta-MeC_5H_4)(CO)_2(SiPh_3)]^-$ | RI, THF, −78 °C | | 52 | 44 |

**Table 2** Manganese alkyls prepared by reduction.

| Entry | Substrate | Conditions | Product | Yield (%) | Ref. |
|---|---|---|---|---|---|
| 1 | $MnBr(CO)_4[PPh_2OCHR^2CHR^1Cl]$ | Na/Hg, ether, 0 °C | $R^1, R^2 = H, Me$ | 35–40 | 45 |
| 2 | $MnBr(PPh_3)(CO)_3[PPh_2OCH_2CH_2Cl]$ | Na/Hg, ether, –5 °C | | 24 | 46 |
| 3 | $MnBr(CO)_4[PPh_2(CH_2)_5Cl]$ | Na/Hg, THF, 15 °C | | 15 | 47 |
| 4 | $cis$-$MnBr(CO)_2(dppm)[PPh_2CH_2CH_2Cl]$ | Na/Hg, THF, 22 °C | | 50 | 48 |
| 5 | $MnBr(CO)_4[PPh_2C(O)CH_2Cl]$ | Mg, THF, 20 °C | | 54 | 49 |

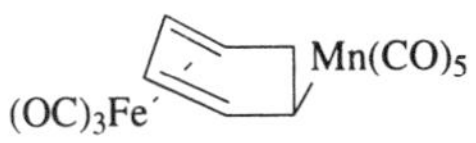

(32) 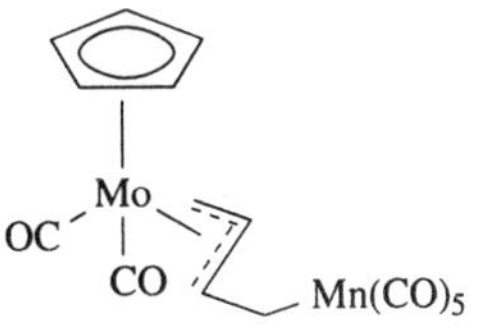  (33)

reduction of cyclopropylstyrene by Mn(H)(CO)$_5$ (Equation (11)).[58] It was reported that up to 72% of this styrene was converted by the hydride to alkenyl species (34). At longer times under these conditions, (34) consumed a second equivalent of Mn(H)(CO)$_5$ with formation of the aldehyde complex (35). This study was complicated by the fact that the course of the reaction was critically dependent on the way in which Mn(H)(CO)$_5$ had been purified. The rearrangement was thought to occur via a radical pair mechanism, because the cyclopropylcarbinyl to 3-butenyl radical ring opening is known to occur very rapidly.

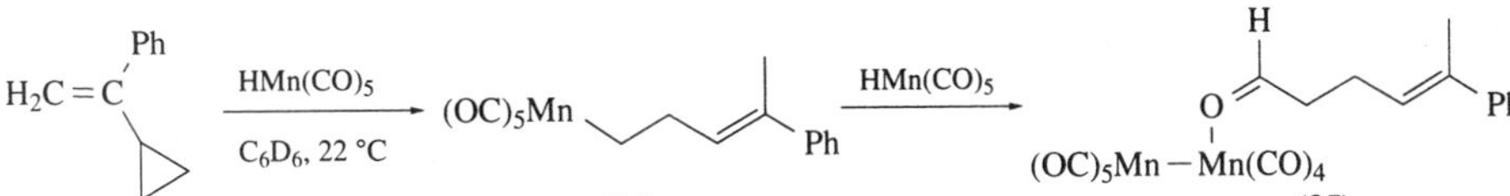

(34)  (35)  (11)

It is well known that metal hydrides can function as catalysts for the hydrosilylation of aldehydes and ketones, perhaps through the intermediacy of metal alkoxides such as (36) (Equation (12)). In principle, the addition could occur with the opposite regiochemistry to give α-hydroxyalkyl metal species such as (37). Attempts to prepare α-hydroxyalkyl manganese compounds such as (38) via the path shown in Scheme 6 resulted in the formation of MnH(CO)$_5$ and benzaldehyde.[59] Thus, at least in this case and probably generally, the metal hydride to α-hydroxyalkyl metal equilibrium is observed to lie in the direction of the metal hydride plus aldehyde.

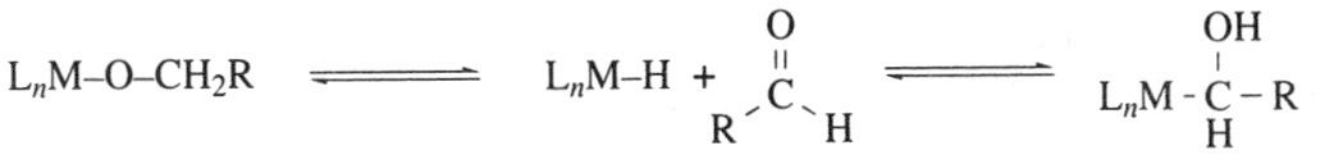

(36)  (37)  (12)

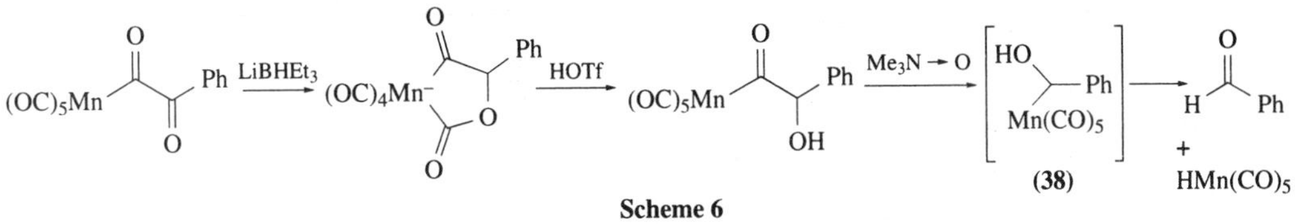

**Scheme 6**

However, when the aldehyde was appended in a phosphine ligand, chelation stabilized the α-hydroxyalkyl complex (39) (Scheme 7). Attempts to carbonylate (39) failed, so the carbonylated material (40) was prepared by other means (Scheme 7) and was stable to decarbonylation under conditions of workup and isolation.[60] In contrast with (39), Mn(H)(CO)$_4$[PPh$_2$CH$_2$CH$_2$CHO] did not cyclize, and attempts to prepare the cyclized material via indirect routes led only to the acyclic aldehyde.[61] Preparation of (41), however, was readily accomplished (Equation (13)), so the factors that control the equilibrium between the open chain hydride–aldehyde and the α-hydroxyalkyl complex are subtle.[61]

In related chemistry, Mn(TMS)(CO)$_5$ was reported to effect conversion of methyl ketals to methyl enol ethers, as shown in Equation (14), with more than 10 examples being provided. Under 1400 kPa of CO, species of the type Mn(CO)$_5$[C(O)CR$^1$(OMe)CH$_2$R$^2$], inferred to be CO-insertion-trapped α-methoxyalkyl intermediates, were isolated from two of these reactions.[62] Reactions of Mn(TMS)(CO)$_5$ with aldehydes and cyclic ethers, with and without CO, have also been reported to afford manganese acyls and alkyls.[63]

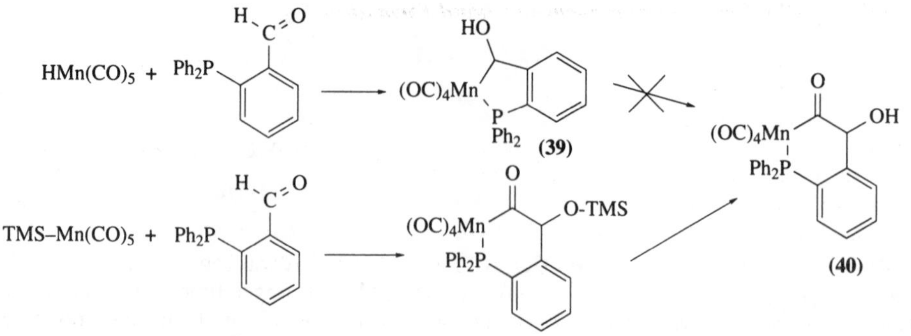

**Scheme 7**

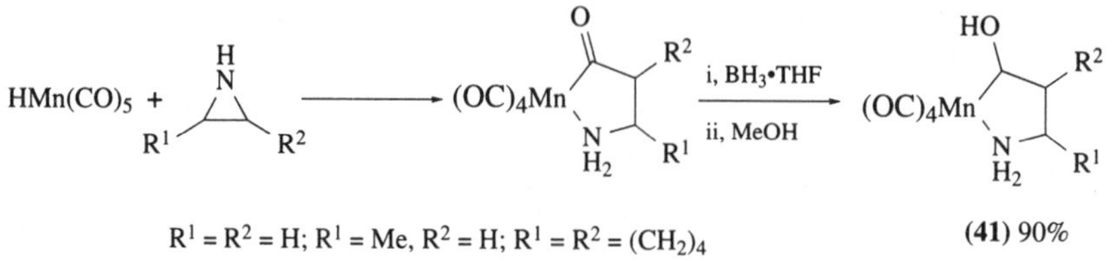

$$R^1 = R^2 = H; R^1 = Me, R^2 = H; R^1 = R^2 = (CH_2)_4 \qquad \textbf{(41)} \ 90\%$$

(13)

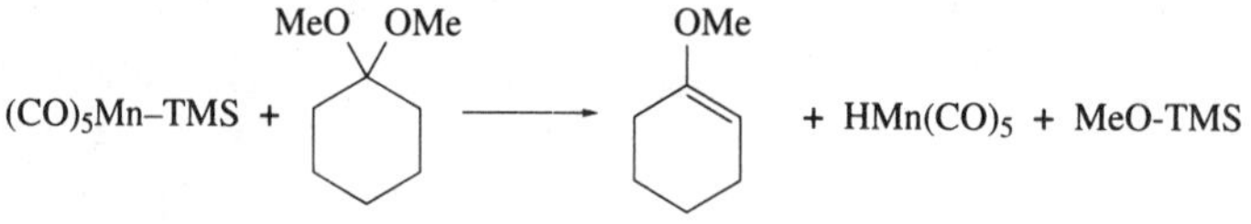

(14)

### 2.3.1.5 Transmetallation between Mn–X and M–R

Another means of making metal–carbon bonds is transmetallation between main-group organometallics such as lithium, manganese, zinc, or aluminum alkyls, and transition metals bearing halides or other good leaving groups. While common for transition metals in general, this approach is much less used for manganese, because $[Mn(CO)_5]^-$ is as readily available as $Mn^I$ halides, and nucleophilic alkylations of the metal anion tend to work well. One example, however, is the reaction of $(\eta\text{-}C_6R_6)Mn(CO)_2Cl$ (R = H or Me) and MeLi in THF at −78 °C which affords $(\eta\text{-}C_6R_6)Mn(CO)_2Me$ in 75% isolated yield.[64]

### 2.3.1.6 Additions of nucleophiles to manganese π-bound hydrocarbons

Nucleophiles generally add to extended π-coordinated complexes to give new π-coordinated systems. However, in the case of (pentadienyl)Mn(CO)$_3$, addition of alkyl lithium reagents at low temperature followed by protonation under CO yields (**42**). The σ-alkyl intermediate (**43**) was detected by $^1$H and $^{13}$C NMR.[65] Addition of nucleophiles to $[(\eta\text{-Ar})Mn(CO)_2(R^1CH=CHR^2)]^+$ complexes, where the alkene is ethene, propene, cyclopentene, and so on, gives neutral alkyl species $(\eta\text{-Ar})Mn(CO)_2(CHR^1CHR^2Nu)$. Nucleophiles used in this way include phosphines, phosphites, hydride, and cyanide.[66,67] In most cases the addition is clean and rapid in THF, $CH_2Cl_2$, or pentane, although in some of the more sterically congested systems alkene displacement is a competing reaction.

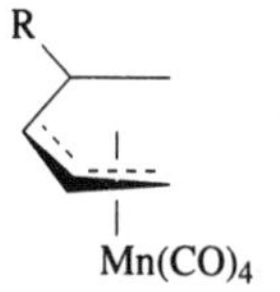

(42)

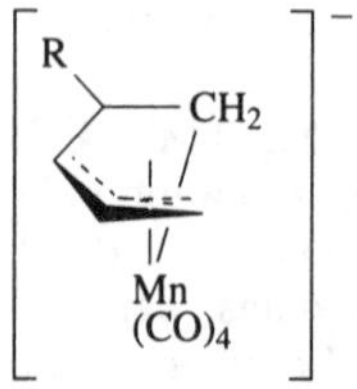

(43)

### 2.3.1.7 Radical paths for manganese–carbon bond formation

Over the years chemists have appreciated the fact that radical mechanisms play a role, sometimes a dominant one, in various organometallic reactions. Since the 1980s several studies have focused on this aspect of manganese chemistry. The heterogeneous reactions of $Na[Mn(CO)_5]$ suspended in benzene or alkane solutions containing various organic halides have been investigated.[68] For example, this anion was allowed to react with 4-bromo-2-pentene over a period of several hours between −30 °C and 5 °C. Ultimately, 2-pentene, 1,3-pentadiene, 4,5-dimethyl-2,6-octadiene and its regioisomers, $MnBr(CO)_5$ and $Mn_2(CO)_{10}$, were formed. IR and NMR data were used to infer the presence of intermediate $Mn(\eta^1\text{-}CHMeCH=CHMe)(CO)_5$. The likelihood of radical mechanisms was discussed.

Photolysis of $Mn_2(CO)_8L_2$ (L = CO, $PBu_3$, $PEt_3$, $P(OMe)_3$, etc.) in the presence of an excess of certain alkyl halides gave $MnX(CO)_4L$ and $MnR(CO)_4L$, in some cases, such as $PhCH_2Cl$, in high yields.[69] Phenethyl bromide ($PhCH_2CH_2Br$) gave significant quantities of $Mn(CH_2CH_2Ph)(CO)_4(PBu_3)$, but $PhCH_2Br$ formed only $MnBr(CO)_4(PBu_3)$. A radical chain mechanism was postulated for the formation of $MnR(CO)_4L$. In the case of $Mn_2(CO)_8(PBu_3)_2$ it was found that simply heating in hexane at 40 °C in the presence of $PhCH_2Cl$ afforded equal quantities of $Mn(CH_2Ph)(CO)_4(PBu_3)$ and $MnCl(CO)_4(PBu_3)$. Based on mechanistic investigations, an outer-sphere electron-transfer path was postulated.[70]

### 2.3.2 Reactivity of Manganese Alkyls

### 2.3.2.1 Thermolytic reactions of manganese alkyls

Flash vacuum pyrolysis of $Mn(benzyl)(CO)_5$ yielded a material whose low-temperature NMR spectrum and mass spectrum were consistent with (**44**). Treatment of this with acid at low temperature formed (**45**) isolable as the $PF_6^-$ salt (Equation (15)). Similarly, pyrolysis of dimetallated (**46**) (Equation (16)) deposited a material which when warmed to room temperature afforded poly(*p*-xylylene) in 40% yield.[71]

$$(15)$$

$$(16)$$

Thermolysis of $MnMe(CO)_5$ in the presence of $SiMe_2(CH_2PPh_2)_2$ led to disubstitution at manganese and activation of one of the methyl groups on silicon, eventually forming (**47**) (Equation (17)). The fate of the methyl group (e.g., formation of methane or acetaldehyde, etc.) was not determined.[72]

$$(17)$$

The heterocycle (**48**) eliminates ethene or propene at temperatures of 0–40 °C and gives dimer (**49**) (Equation (18)). The stability of (**48**) follows the order ($R^1$ = Me, $R^2$ = H) < ($R^1 = R^2$ = H) < ($R^1$ = H, $R^2$ = Me).[45] The $PPh_3$ substituted analogue of (**48**) ($R^1 = R^2$ = H, one CO replaced by $PPh_3$) reacts analogously but much more rapidly.[46]

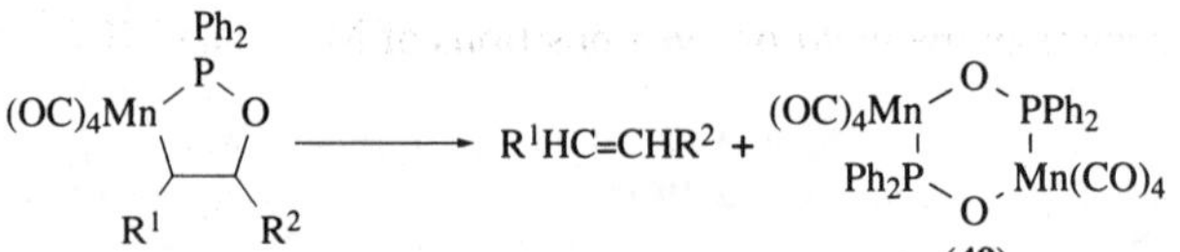

$$(OC)_4Mn \overset{\underset{\displaystyle P}{Ph_2}}{\diagdown} O \atop R^1 \quad R^2 \qquad \longrightarrow \qquad R^1HC{=}CHR^2 \; + \; \begin{array}{c} (OC)_4Mn \diagup\!\!\!\diagdown\!\!\overset{O}{\diagdown} PPh_2 \\ \quad | \qquad\qquad | \\ Ph_2P\diagdown\!\!\!\diagup\!\!\overset{}{\diagup} Mn(CO)_4 \\ O \end{array} \tag{18}$$

$$(48)$$

$$(49)$$

$$R^1, R^2 = H, Me$$

### 2.3.2.2 Reaction of manganese alkyls with metal hydrides

The reductive cleavage of $M(C(O)OEt)(CO)_n$ and $M(CH_2C(O)OEt)(CO)_n$ (M = Co, $n = 4$; M = Mn, $n = 5$) by $MnH(CO)_5$ and $CoH(CO)_4$ was examined in all permutations under CO atmosphere in heptane solvent at 25 °C. In all cases, the kinetics were consistent with nearly rate-determining loss of CO from the alkyl or acyl complex, followed by fast attack by the metal hydride. The metal-containing products were $Mn_2(CO)_{10}$, $Co_2(CO)_8$, or $MnCo(CO)_9$, with no homonuclear dimer formed in any mixed-metal experiment. The partitioning ratio for each unsaturated intermediate was calculated from kinetic plots, and it was found for both $Mn(C(O)OEt)(CO)_4$ and $Mn(CH_2C(O)OEt)(CO)_4$ that $CoH(CO)_4$ was 4–5 times faster at trapping the intermediate than $MnH(CO)_5$.[73]

Reactions between $MnR(CO)_4L$ (R = $p$-MeOC$_6$H$_4$CH$_2$; L = CO or ($p$-MeOC$_6$H$_4$)$_3$P) and $Mn(H)(CO)_4L$ exhibit diverse stoichiometries and mechanistic pathways, depending on the solvent and CO concentration and on whether L is CO or ($p$-MeOC$_6$H$_4$)$_3$P (Scheme 8).[74] Other similar studies led to very similar conclusions.[75]

**In benzene**

$R{-}Mn(CO)_5 \;\rightleftharpoons\; R{-}Mn(CO)_4 + CO$

$R{-}Mn(CO)_4 + H{-}Mn(CO)_5 \;\longrightarrow\; R{-}H + Mn_2(CO)_{10}$

**In acetone or acetonitrile**

$R{-}Mn(CO)_5 \;\rightleftharpoons\; R{-}CO{-}Mn(CO)_4(S)$

$R{-}CO{-}Mn(CO)_4(S) + H{-}Mn(CO)_5 \;\longrightarrow\; R{-}CO{-}H + Mn_2(CO)_9(S)$

**In benzene** (PAr$_3$ = P(C$_6$H$_4$-$p$-OMe)$_3$)

$R{-}Mn(CO)_4(PAr_3) \;\longrightarrow\; R{\bullet} + {\bullet}Mn(CO)_4(PAr_3)$

$R{\bullet} + HMn(CO)_4(PAr_3) \;\longrightarrow\; R{-}H + Mn_2(CO)_8(PAr_3)_2$

**In benzene** (PAr$_3$ = P(C$_6$H$_4$-$p$-OMe)$_3$)

$R{-}Mn(CO)_4(PAr_3) + CO \;\rightleftharpoons\; R{-}CO{-}Mn(CO)_4(PAr_3)$

$R{-}CO{-}Mn(CO)_4(PAr_3) + HMn(CO)_4(PAr_3) \;\longrightarrow\; R{-}CO{-}H + Mn_2(CO)_8(PAr_3)_2$

**Scheme 8**

Reaction of $MnR(CO)_5$ (R = Me, C$_6$H$_4$-$p$-Me, or CH$_2$CH$_2$CH=C(Ph)Me) with $MnH(CO)_5$ was found to lead to formation of the $\eta^1$-aldehyde complex $Mn_2(CO)_9(\eta^1$-OCHR) when carried out in nondonor solvents.[76]

### 2.3.2.3 Insertion reactions

#### (i) CO and CO$_2$ insertion

The CO insertion reaction is central to organomanganese chemistry, and examples of CO insertion or extrusion chemistry permeate this review as well as that in *COMC-I*. In particular, $MnR(CO)_5$ has served as a paradigm for this reaction, presumably due to its experimental tractability, its nearly thermoneutral equilibrium, and its high symmetry and simplicity. There has been much activity since the 1980s concerning the mechanism of the reaction.

Electron-donating groups in the *meta* or *para* positions of Mn(benzyl)(CO)$_5$ lead to rate accelerations in the CO insertions under conditions where the insertion step itself is rate limiting.[77]

Many years ago, Noack and Calderazzo[78] carried out elegant experiments using $^{13}$C-labeled CO and IR spectroscopy to monitor the stereochemistry at manganese of the CO-induced CO insertion. Their results were consistent with methyl migration to one of the *cis* CO sites. An issue that was not addressed explicitly in this classic study was the geometry of the intermediate. In a later study, the stereochemistry of the intermediate was inferred by trapping with the small phosphite (P(OCH$_2$)$_3$CMe) which gives only *cis* product (**51**) (Scheme 9).[79] Any intramolecular mechanism would yield 25% of (**51α**), but the ratio of (**51**) isotopomers **β/γ/δ** would reflect the stereochemical attributes of the intermediate(s). These could have structures ranging from (**52a**) to (**52e**) and could be either static or stereochemically nonrigid during their short lifetime. The label ratio observed in acetone/THF or in hexamethylphosphoramide (HMPA) was 2.0/0/1.0, consistent only with a rigid structure of geometry (**52a**), (**52a(S)**), or (**52c**), or any combination of these.

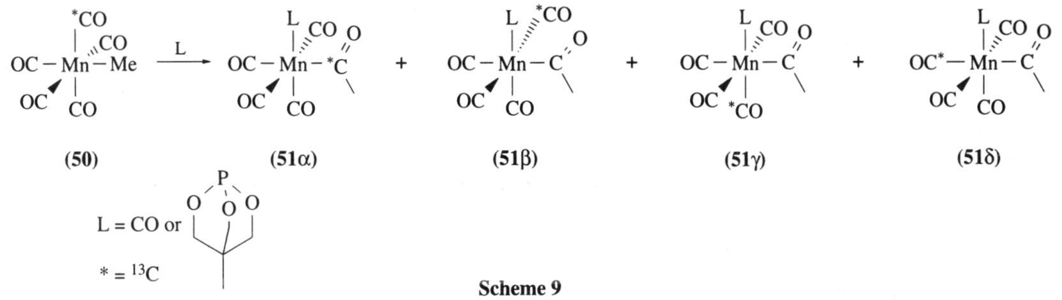

**Scheme 9**

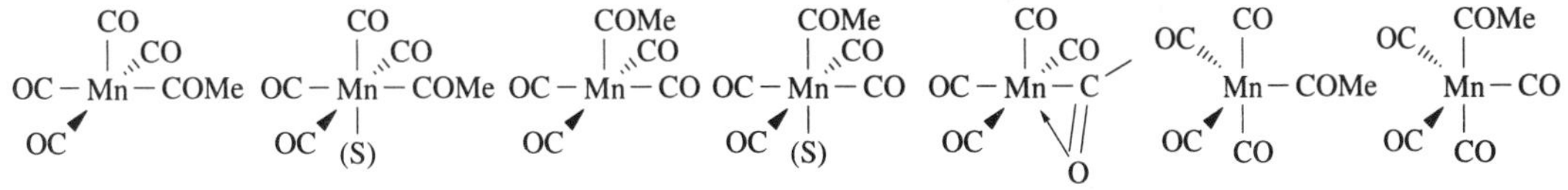

The above result is complicated by the issue of solvent coordination, and the possibility of an η$^2$-coordinated acyl group. There are many examples of rates of CO insertion being accelerated by donating solvents, and in a number of cases solvates of the intermediates have been detected spectroscopically. For example, MnR(CO)$_5$ equilibrates with MnC(O)R(CO)$_4$(solvent), where the extent of solvate formation decreases with R = Et ≫ Me > 2,3,4,5,6-pentamethylbenzyl > 2,4,6-trimethylbenzyl > 4-methylbenzyl ~ phenyl. Solvent-coordinating ability decreases in the order DMSO > DMF > pyridine > acetonitrile.[80] It has frequently been suggested that the acceleration of the rate of CO insertion by nucleophilic solvents arises from one or both of two effects: (a) bimolecular nucleophilic assistance by the solvent in the transition state; and (b) an increase in the equilibrium concentration of solvated acyl intermediate, which would increase the overall rate by its concentration effect on the rate of the second step, ligand uptake to form product. The second assertion implies that the reaction has been carried out under conditions where displacement of solvent from the intermediate by the final ligand is at least partially rate determining, and that this displacement must be associative in nature. If solvent displacement by the final ligand were dissociative, the steady-state concentration of the unsaturated intermediate (e.g., (**52a**)) would remain essentially unchanged by increases in the amount of solvated intermediate present (e.g., (**52a(S)**)), and increases in concentration of (**52a(S)**) should have no effect on the reaction rate.

In this context, Halpern and co-workers[81] have conducted an important experiment. In benzene at 45 °C, Mn(CH$_2$C$_6$H$_4$-*p*-OMe)(CO)$_5$ reacts with MnH(CO)$_4$(PMe$_2$Ph) to form axial Mn$_2$(CO)$_9$(PMe$_2$Ph) and *p*-MeOC$_6$H$_4$CH$_2$CHO (Equation (19)). The insertion step is accelerated by donor solvents, but even more by nucleophilic catalysts such as OPPh$_3$. An important feature of this reaction is that MnH(CO)$_4$(PMe$_2$Ph) traps intermediate (**53**) or (**54**) sufficiently slowly that rates of product formation and starting material reformation are comparable. With most other traps, such as PPh$_3$ or MnH(CO)$_5$, product formation is much too fast compared with reversion of the intermediate to starting material for the kinetic analysis to work. Their kinetic results clearly show that it is intermediate (**53**), not (**54**), which is trapped by the hydride. Thus, to the extent that this result is general, catalysts such as polar solvents and OPPh$_3$ exercise their influence not by increasing the concentration of solvate (**54**) (or (**52a(S)**)), but by catalyzing the formation of the unsaturated intermediates (**53**) (or (**52a**)). If this result is general, the stereochemistry of the CO insertion at the metal is a reflection of the geometry of intermediate (**52a**) or (**53**) and not (**52a**(S)), (**54**), or (**52c**). In addition, the lifetime of the intermediate actually being trapped would generally be less than a nanosecond rather than of the order of microseconds, as is usually inferred. In less coordinating solvents $\eta^2$-coordinated acyl intermediate (**52c**) is probably important (see Section 2.4.4.1), but kinetically would simply fill the role of the solvated species (**52a**(S)).

The stereochemistry at the metal summarized above is not the result of a robust energetic preference. Rather, that path is easily altered when ligands with large differences in *trans* influence are present in the coordination sphere. For example, *fac*-[Mn(CO)₃(bipy)Me] (**55**) undergoes CO insertion induced by a variety of ligands, all of which are shown to have entered kinetically *trans* to the new acetyl group in [Mn(CO)₂(L)(bipy)COMe] (**56**).[82]

Other aspects of the geometry and state of coordination of unsaturated alkyl or acyl manganese species, as studied by matrix isolation and flash photolysis techniques, are discussed in Section 2.4.4.1.

It has been reported that the chelated complex (**57**) underwent carbonylation under the conditions shown in Equation (20). IR spectra of the product obtained from (**57**) (R = H) were consistent with hydrogen bonding of the hydroxyl group to the acyl oxygen. On this basis, it was suggested that the 16-fold greater rate when R = H over when R = TMS might be due in part to electrophilic assistance by intramolecular hydrogen bonding to the developing acyl group.[61] Enough thermodynamic driving force for the insertion of CO is apparently generated by the strong Si–O bond in (**58**) to form its three-membered ring, as shown in Equation (21).[83] Subsequent removal of the silyl group by chromatography on wet silica gel reveals that the equilibrium between (**59**) and (**60**) lies towards the open-chain acyl.

Only a single example of the reaction of a Mn–C bond with $CO_2$ has been reported since the 1980s. Insertion of $CO_2$ into the Mn–C bond of a strained four-membered ring compound produced (**61**) in 40% yield (Equation (22)), but no insertion into the corresponding five-membered manganacycle was detected.[84]

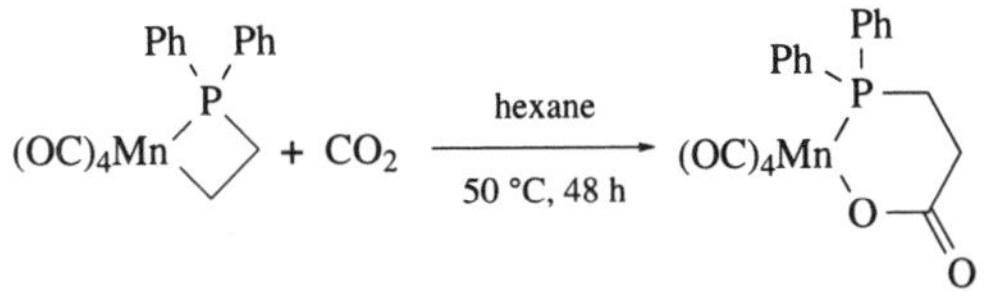

$$(OC)_4Mn \overset{\overset{\displaystyle Ph \diagdown \underset{\displaystyle P}{\,} \diagup Ph}{}}{\diamondsuit} + CO_2 \xrightarrow[50\ ^\circ C,\ 48\ h]{hexane} (OC)_4Mn \begin{matrix} \end{matrix} \qquad (22)$$

**(61)**

*(ii) Alkene insertion*

Additions of manganese alkyl or acyl groups across organic unsaturation are apparently more difficult than for Mn–H, and examples are not common. It has been reported that the sequential insertion of CO followed by alkenes and alkynes can be induced, often with high yields, by the application of high pressures $(2–10 \times 10^5\ kPa)$.[85] As shown in Scheme 10, benzyl- or methylpentacarbonylmanganese(I) will undergo insertions with a reasonable variety of alkynes and mono-, di-, and trisubstituted alkenes. These ketone-chelated manganese alkenyls (**62**) and alkyls (**63**) can be protolytically cleaved to liberate the organic ligands. Chelated alkyls (**63**) could be demetallated under photochemical conditions.

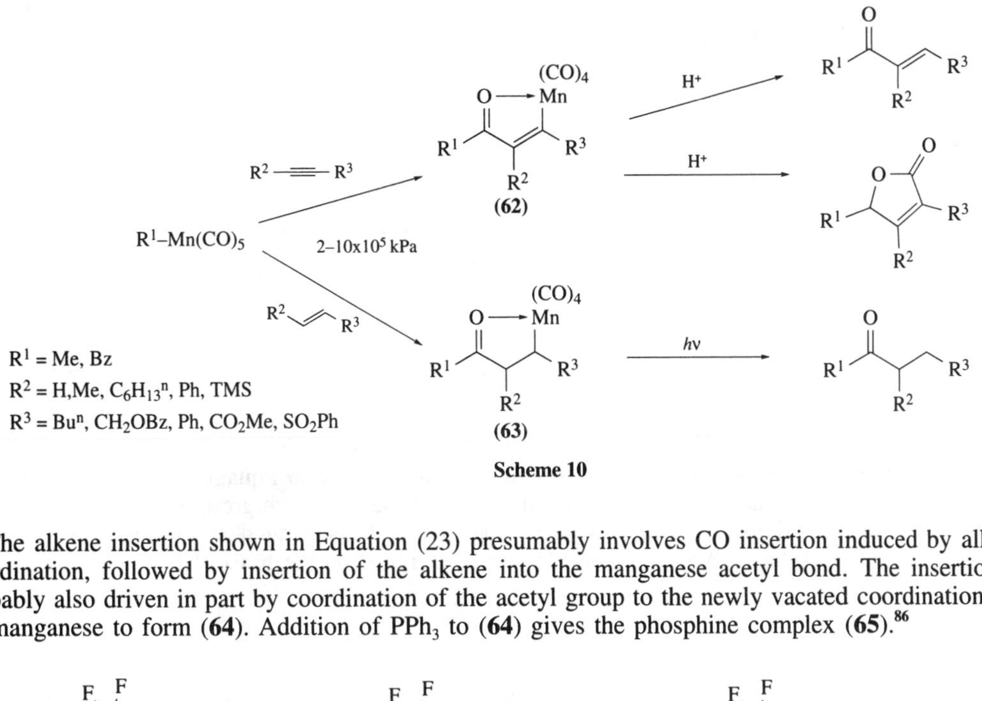

$R^1 = Me, Bz$
$R^2 = H, Me, C_6H_{13}^n, Ph, TMS$
$R^3 = Bu^n, CH_2OBz, Ph, CO_2Me, SO_2Ph$

**Scheme 10**

The alkene insertion shown in Equation (23) presumably involves CO insertion induced by alkene coordination, followed by insertion of the alkene into the manganese acetyl bond. The insertion is probably also driven in part by coordination of the acetyl group to the newly vacated coordination site on manganese to form (**64**). Addition of $PPh_3$ to (**64**) gives the phosphine complex (**65**).[86]

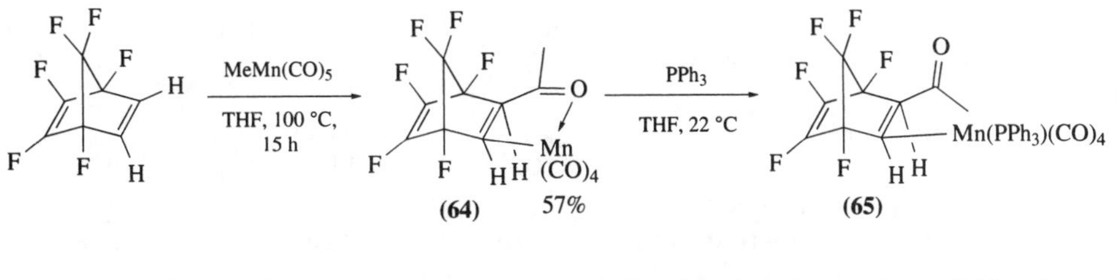

(23)

Nucleophilic substitution of glycosyl bromides by $Na[Mn(CO)_5]$ to form glycosyl Mn–C bonds is shown in Table 1. These molecules, in turn, undergo several synthetically interesting transformations as shown in Scheme 11,[35] including carbomethoxylation and conjugate addition to polarized alkenes and alkynes. Four other examples of carbohydrates were used, two of which demonstrated that the chemistry works for furanoses as well as pyranoses.

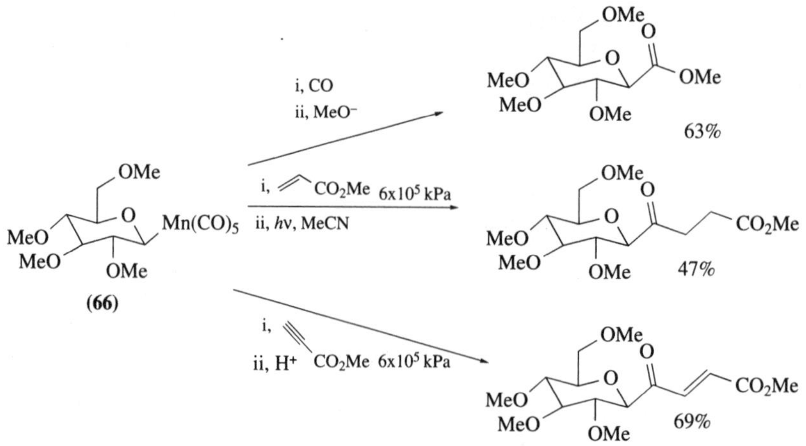

**Scheme 11**

### (iii) Insertions of η-polyenyl ligands into manganese alkyls

Preparations of Mn(η-arene)(CO)$_2$R (**67**) are mentioned in Table 1. On treatment with PPh$_3$ at 76 °C, a steady-state mixture of acyl (**68**) and alkyl (**67**) is established, while the mixture is converted to the ring alkylated pentadienyl complexes (**69**), as shown in Equation (24).[42] Excess PPh$_3$ generates a very low alkyl:acyl ratio and yet inhibits Pr$^i$ migration to the ring, so it was inferred that ring alkylation does not occur directly from (**68**).

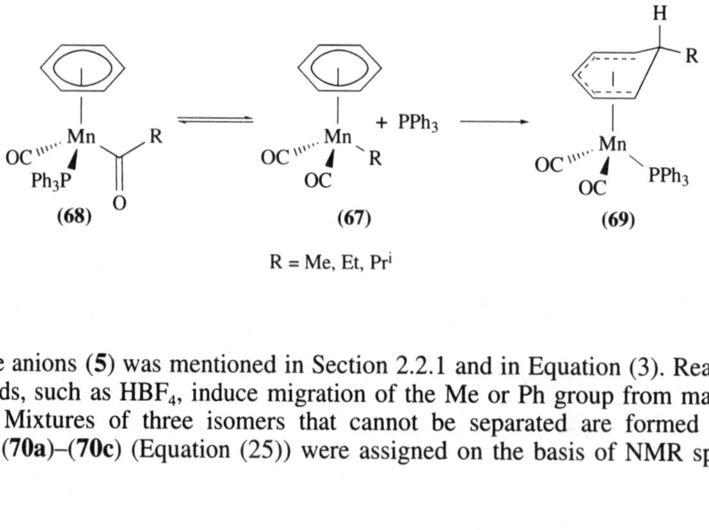

$$(24)$$

R = Me, Et, Pr$^i$

Preparation of the anions (**5**) was mentioned in Section 2.2.1 and in Equation (3). Reactions of these acyl anions with acids, such as HBF$_4$, induce migration of the Me or Ph group from manganese to the pentadienyl ligand. Mixtures of three isomers that cannot be separated are formed for which the "agostic" structures (**70a**)–(**70c**) (Equation (25)) were assigned on the basis of NMR spectra.[7]

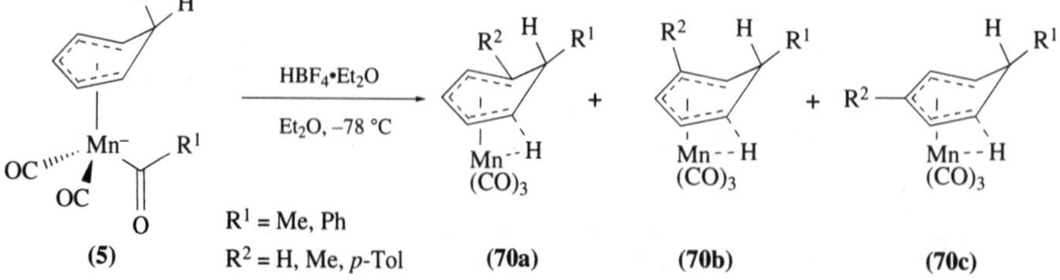

$$(25)$$

Reaction of Mn(benzyl)(CO)$_5$ with butadiene and pentadiene (Equation (26)) gives generally inseparable mixtures of isomers. However, thermal substitution of one CO by PMe$_3$ generates isomers that can be separated and characterized.[87]

$$\text{Bz-Mn(CO)}_5 \ + \quad \overset{}{\underset{}{\diagup\!\!\diagdown}} \quad \xrightarrow[\substack{\text{hexane} \\ -20\,°C}]{h\nu} \quad \underset{\text{Mn(CO)}_4}{\overset{\text{CO}}{\underset{\text{OC}}{\text{Mn}}}} \quad \xrightarrow{69\,°C} \quad \text{Ph}\diagup\!\!\diagdown\!\!\diagup\!\!\diagdown \quad + \text{ isomers} \qquad (26)$$

### (iv) SO₂ and other insertions into manganese alkyls

Although no notable work has been carried out since the 1980s on $SO_2$ insertion *per se*, this reaction continues to be widely used to prepare characterizable derivatives of manganese alkyls.[29,33] Isoelectronic analogues of $SO_2$ will insert into Mn–C bonds in a similar manner, and as with $SO_2$ when the organic ligand is allylic, cyclizations occur. For example, $Mn(CH_2CH=CH_2)(CO)_5$ and $MeS(O)_2N=S=O$ give (**71**), $Mn(CH_2C\equiv CPh)(CO)_5$ and $MeS(O)_2N=S=NS(O)_2Me$ give (**72**), and $Mn(CH_2CH=CH_2)(CO)_5$ and $MeS(O)_2N=S=NS(O)_2Me$ give (**73**). All of these were conducted in $CH_2Cl_2$ at 22 °C for 30 min.[88]

(71)   (72)   (73)

When delivered as the triethylphosphine telluride, tellurium can be inserted into the Mn–C bond of $Mn(benzyl)(CO)_5$.[89] Excess $PEt_3$ must be used to prevent the dissociation of $TePEt_3$, so concurrent substitution for CO by L also occurs. The stronger Mn–C bond of $MnMe(CO)_5$ affords a mixture of $Mn(TeMe)(CO)_3L_2$ and $MnMe(CO)_3L_2$.

### 2.3.2.4 Electrophilic cleavage of manganese alkyls

Cleavage of metal–carbon bonds by anhydrous strong acids, such as HCl in ether or HBr in $CH_2Cl_2$, has been known for many years as a way of removing organic ligands from their metal complexes. This technique has seen increasing use in recent years to generate highly reactive metal centers of the type M–X where X⁻ is a very weakly bound anion. Manganese is no exception. $HSO_3C_6F_{13}$ has been used in $CH_2Cl_2$ at room temperature to cleave the methyl group in $MnMe(CO)_5$ to form $Mn[OS(O)_2C_6F_{13}](CO)_5$, although in poor yield (22%).[90] Cleavage of a series of $MnR(CO)_5$ by $HSO_3CF_3$ and $HBF_4$ was carried out in $CH_2Cl_2$ at room temperature.[51] The qualitative rate of cleavage was found to decrease across the series: H, Me, Ph, *p*-Tol, *p*-BrPh, *p*-CF₃Ph > Bz > *p*-ClBz ~ *p*-MeOBz ≫ PhCH₂CH₂. In all cases cleavage with $HSO_3CF_3$ was much faster than with $HBF_4$. On treatment with $HSO_3CF_3$ or $HBF_4$, the corresponding acyl complexes formed no aldehydes. Instead, acid cleavage was much slower than with the alkyls, and the sole organic products were decarbonylated RH. Acid cleavage of $MnMe(CO)_5$ was also carried out with $HOTeF_5$ in $CH_2Cl_2$ at room temperature, affording orange crystalline $Mn(OTeF_5)(CO)_5$ in 67% (recrystallized) yield.[91] In contrast with $Mn[OS(O)_2CF_3](CO)_5$, the teflate does not form $[Mn(CO)_6]^+$ in solution under CO atmosphere even after many hours. It is readily converted by chloride ion to $MnCl(CO)_5$, but this is apparently via facile CO dissociation.

The path of cleavage of $(OC)_5Mn(CH_2)_4Mn(CO)_5$ (**4**) by halogens showed a profound solvent dependence.[4] In THF, treatment of (**4**) by $Br_2$ or $I_2$ followed by aqueous workup yielded adipic acid $(HO_2C(CH_2)_4CO_2H)$. In $CH_2Cl_2$ solvent, though, the usual dihalide $X(CH_2)_4X$ was obtained.

### 2.3.2.5 Photochemical reactions of manganese alkyls

For some organometallic systems, such as $FeCp(COR)(CO)L$ (L = CO or $PR_3$), photolysis tends to be the most effective means for substitution and/or decarbonylation. In the case of manganese carbonyl complexes, ligand dissociation is readily accomplished thermally, so photochemical procedures are not commonly used for these purposes. Photochemical ligand dissociations have been used in low-temperature and matrix-isolation investigations; these are discussed in Section 2.4.4.1. Photolysis of

$MnR(CO)_5$ (R = Me, Et, $CH_2Ph$, or Ph) has been reported to yield both radical and (when R = Et) $\beta$-hydride elimination products.[92]

As shown in Scheme 10, irradiation of the chelated alkyl manganese species (63) proved to be an effective means of removing the organic ligand from the metal. For example, irradiation of (74) for 90 min at 350 nm under reduced pressure afforded the keto ester (75) in 75% isolated yield (Equation (27)).[85] The added hydrogen in the product came from water during workup and not from photolysis solvents $CD_3CN$ or $CDCl_3$. Photodemetallation was inhibited completely by 1 atm of CO.

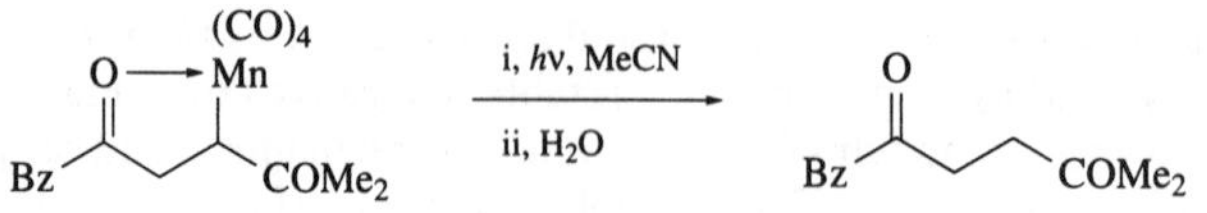

$$\text{(27)}$$

### 2.3.2.6 $\eta^1$–$\eta^n$-conversions of allyl, pentadienyl, etc.

Both thermal and photochemical procedures have been described for conversion of a variety of substituted $\eta^1$-allylmanganese complexes, $Mn(\eta^1\text{-}CH_2R^1C=CHR^2)(CO)_4L$ (L = CO, $PPh_3$, or $PMe_2Ph$; $R^1,R^2$ = H, Me, Ph, Cl, or $Bu^t$) into the $\eta^3$ form.[34] The $\eta^1$-pentadienyl group of (76) can be selectively converted into either the $\eta^5$ form (77) or the $\eta^3$ form (78), depending on conditions (Scheme 12).[29]

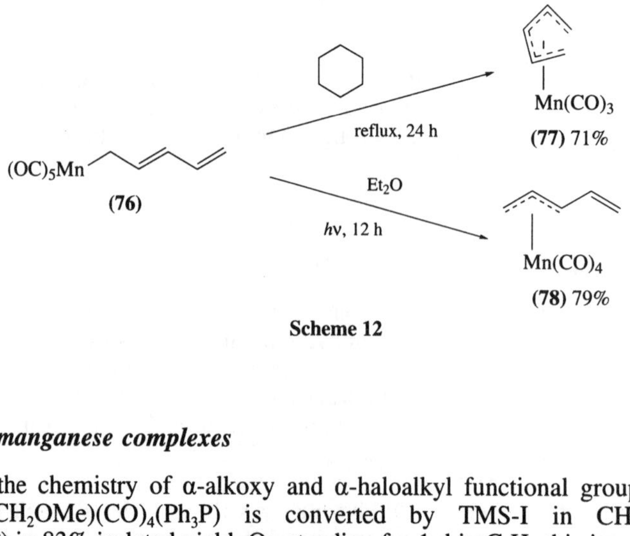

**Scheme 12**

### 2.3.2.7 α-Haloalkyl manganese complexes

Investigations of the chemistry of α-alkoxy and α-haloalkyl functional groups have continued in recent years. $Mn(CH_2OMe)(CO)_4(Ph_3P)$ is converted by TMS-I in $CH_2Cl_2$ at $-78\,^\circ C$ to $Mn(CH_2I)(CO)_4(Ph_3P)$ in 83% isolated yield. On standing for 1 d in $C_6H_6$ this is converted in high yield to zwitterionic $Mn(I)(CH_2PPh_3)(CO)_4$. In addition, treatment of $Mn(CHO)(CO)_5$ in toluene at $-78\,^\circ C$ with a 57% aqueous solution of HI, followed by warming to room temperature afforded the ether $O[CH_2Mn(CO)_5]_2$ in 23% isolated yield.[93]

Facile halogen exchange takes place on reaction of $Mn(CF_3)(CO)_5$ with 1 equiv. of $BX_3$ (X = Cl, Br, or I) in noncoordinating solvents.[94] For example, $Mn(CCl_3)(CO)_5$ was produced in 91% yield after purification by sublimation. $Mn(CBr_3)(CO)_5$ is a white solid which is stable under nitrogen for several days, but decomposes in hexane solution within a day. $Mn(CI_3)(CO)_5$ was stable in solution for less than 1 h. Analogous complexes of $CHCl_2$, $CH_2Cl$, $CHBr_2$, and $CH_2Br$ were prepared by the same $BX_3$ exchange. Formation of only $Mn(CCl_2CF_3)(CO)_5$ from the corresponding $C_2F_5$ complex confirmed that the exchange was associated with C–F bonds located α to the metal.[94] Several reactions of these trihalomethyl manganese species have been investigated. IR spectral data suggest that an equilibrium is established in $CH_2Cl_2$ between $Mn(CF_3)(CO)_5$ and $BF_3$ on the one hand, and the carbene complex $[Mn(CF_2)(CO)_5]BF_4$ on the other. Addition of water to this mixture leads to formation of $[Mn(CO)_6]^+$ by hydrolysis of the $[Mn=CF_2]^+$ group.[95] Radical reactions of these trihalomethyl complexes were also clearly evident. For example, decomposition of $Mn(CBr_3)(CO)_5$ in benzene solution at room temperature with or without alkenes present gave a complex set of radical-derived products with no significant cyclopropanation. In addition, stepwise reduction of $Mn(CBr_3)(CO)_5$ all the way to $MnMe(CO)_5$ was readily accomplished with $SnHBu^n_3$.[95]

### 2.3.2.8 Reactions of unsaturation in alkenyl and alkynyl complexes

The Mn–C bond in $Mn(R)(CO)_5$ complexes is robust enough to allow some types of chemistry to go on with alkene or alkyne functional groups present in the R fragment. For example, reaction of $Mn(CH_2C{\equiv}CPh)(CO)_5$ with $Co_2(CO)_8$ at 22 °C in pentane generated the standard dicobalt–alkyne cluster $[Co_2(CO)_6(PhC{\equiv}CCH_2Mn(CO)_5)]$ (**79**) without disruption of the propargylic Mn–C bond.[96]

Complexes such as (**80**) containing a 1,3-diene moiety successfully undergo cycloaddition reactions (Scheme 13). When R = Me, the tetracyanoethene addition product (**81**) contains stereochemical information on the addition, and an x-ray structure determination of (**81**) revealed the reaction to be stereospecific. Stereospecific cycloaddition of maleic anhydride to (**80**) readily occurs between 0 °C and 5 °C to generate (**82**). On standing in solution at 23 °C for 36 h, (**82**) underwent CO insertion induced by intramolecular acyl coordination, and crystals of (**83**) formed in high yield. Increased steric hindrance to *s-cis* diene arrangement as in (**84**) in Equation (28) favors formation of (**85**), presumably resulting from a two-step electrophilic–nucleophilic addition sequence.[97]

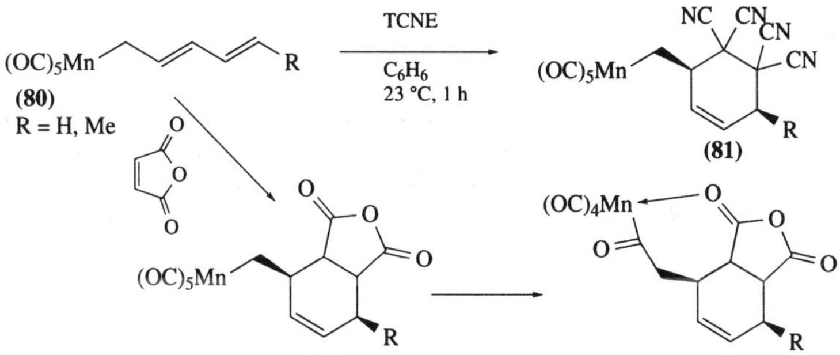

**Scheme 13**

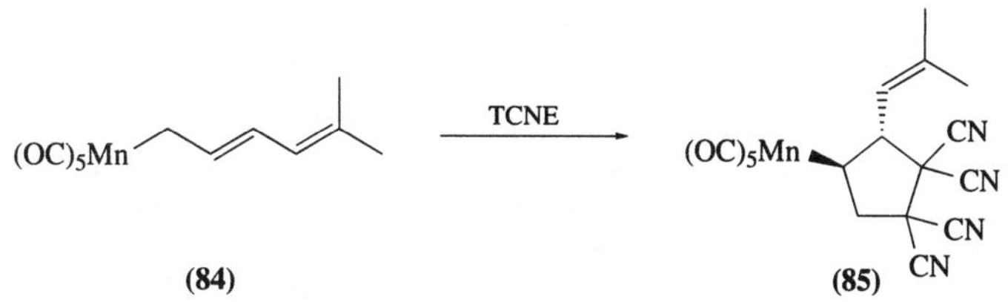

(28)

### 2.3.2.9 Nucleophilic alkyl transfer to other metals

Nucleophilic transalkylation between metal centers is not a common reaction, but there are several well-documented examples. One of these is afforded by manganese. When $MnR(CO)_5$ was treated with $[FeCp(CO)_2]^-$ in THF at 22 °C, the organic ligand was transferred from manganese to iron to give $FeCp(CO)_2R$ and $[Mn(CO)_5]^-$. The rates followed the order: R = H > $CH_2Ph$ > Me > Ph. Rate constants were ~100 $s^{-1}$ for methyl and ~900 $s^{-1}$ for benzyl. Methyl transfer was clean and quantitative, while phenyl transfer was messy and showed properties that suggest the involvement of radical processes.[98]

### 2.3.2.10 Reductive elimination

As shown in Table 1 (entries (17) and (18)), the anions $[Mn(MeCp)(CO)_2(EPh_3)]^-$ (E = Si or Ge) are alkylated by various alkyl halides affording formally $Mn^{III}$ species of type (**86**) in Equation (29) and type (**88**) in Scheme 14. When allyl complex (**86**) is heated in THF at reflux, reductive elimination of 3-(triphenylsilyl)propene occurs so that the incipient vacant site at $Mn^I$ is taken up by the propene π-bond itself (**87**).[44] With butenyl complex (**88**) (R = $CH_2CH_2CH{=}CH_2$, E = $SiPh_3$), "standing" in THF results in migration of the butenyl group to the MeCp ring and, ultimately, alkene chelate (**91**) and $SiHPh_3$ are obtained. With a simple alkyl like methyl in (**88**) (R = Me), it is necessary to add a suitable ligand for the new vacant site at manganese in order to isolate well-defined organomanganese product in addition to the organosilicon or organogermanium product. As shown in Scheme 14, $PPh_3$ works well for this purpose.[43]

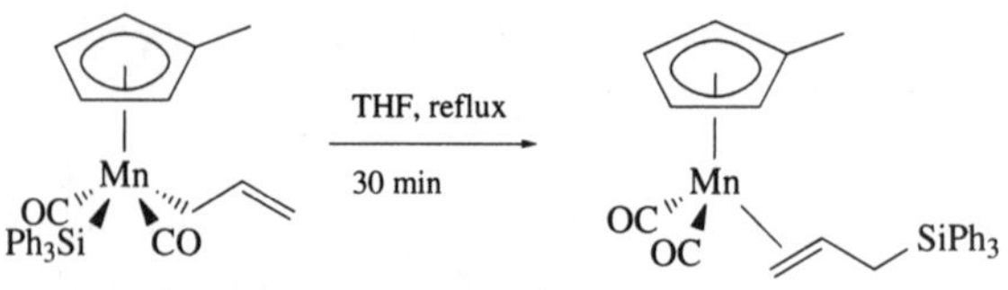

**Scheme 14**

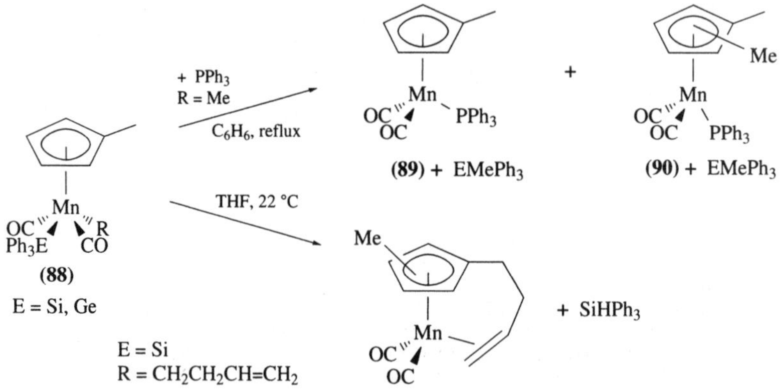

## 2.4 PHYSICAL PROPERTIES

### 2.4.1 Spectroscopy

Solid-state, cross-polarization, magic-angle spinning (CP/MAS) $^{31}$P NMR spectra of Mn[$\eta^2$-PPh$_2$(CH$_2$)$_n$](CO)$_4$ ($n$ = 3 or 4) have been used to determine the $^{31}$P shielding tensors and $^{55}$Mn–$^{31}$P coupling constants. Estimates were made of the quadrupolar coupling constant and the asymmetry parameter of the electric field gradient at manganese.[99] Ring-size effects were noted on both the $^{31}$P and $^{13}$C NMR chemical shifts in these same molecules where $n$ = 3, 4, or 5.[100] The $^{55}$Mn nuclear magnetic shielding constants in the complexes MnR(CO)$_5$ (R = H, Me, CN, or Cl) were calculated using the *ab initio* finite perturbation SCF-MO method. It was concluded that the chemical shifts are determined predominantly by $3d$ contributions of the paramagnetic term.[101]

Several detailed IR spectroscopic studies of MnMe(CO)$_5$ have been conducted in recent years, where most of the interest has focused on Mn–C and C–H vibrations. One such study included Raman spectra, incoherent inelastic neutron-scattering data, and an attempted x-ray crystal structure determination.[102] The x-ray work was thwarted by severe disorder of the methyl group position. Deuterium labeling of the methyl group in MnMe(CO)$_5$ has been used to help assess rotational barriers.[103] Additional assignments were made with the use of deuterium and $^{13}$C labeling in various combinations from IR and Raman spectra in the gas and solid phases in conjunction with energy-factored and $A_1$ force-field calculations.[104]

It has been concluded from x-ray PES measurements of the core binding energies of MnR(CO)$_5$ (R = Me, Pr$^n$, and $\eta^1$-allyl) and Mn($\eta^3$-allyl)(CO)$_4$ that the R groups of the first three are negatively charged. Both carbon $1s$ binding energies and theoretical considerations indicate an unusually large relaxation energy associated with the ionization of the terminal carbon atom of the $\sigma$-allyl group. The $\pi$-allyl group has a relatively low electron density, and it is concluded that the $\pi$-Cp group of Mn($\eta$-Cp)(CO)$_3$ is probably positively charged.[105]

### 2.4.2 Thermochemistry

Microcalorimetry has been used to measure the standard enthalpy of formation of a series of MnR(CO)$_5$ derivatives, from which a set of internally consistent Mn–R bond dissociation enthalpies (BDEs) have been calculated.[106] Attempts to assign absolute BDEs have rested on determinations of the Mn–Mn BDE in Mn$_2$(CO)$_{10}$, but unfortunately, there is not universal agreement on the strength of this bond. Suggested values span a range from 94 kJ mol$^{-1}$ [106] to >176 kJ mol$^{-1}$.[107] The BDEs in Table 3 are reported relative to the Mn–Mn bond in $Mn_2(CO)_{10}$ having a value of 94 kJ mol$^{-1}$. For example, this value of the Mn–Mn bond places the Mn–Me BDE in $MnMe(CO)_5$ at 153 kJ mol$^{-1}$, and that of Mn–H in $MnH(CO)_5$ at 213 kJ mol$^{-1}$. In Section 2.8.3.2, independent inferences of the Mn–H BDE are discussed which place it from 264 kJ mol$^{-1}$ to 284 kJ mol$^{-1}$ (implying a Mn–Mn BDE as large as 165 kJ mol$^{-1}$), and thus further adjustments in the absolute numbers in Table 3 may be anticipated.

**Table 3** Relative bond dissociation enthalpies[106] of R–Mn(CO)$_5$ with respect to the Mn–Mn bond in $Mn_2(CO)_{10}$ which was assigned a reference value of 94.$^{a}$

| Disrupted bond | BDE$^{b}$ (kJ mol$^{-1}$) | Disrupted bond | BDE$^{b}$ (kJ mol$^{-1}$) |
|---|---|---|---|
| Mn–Mn$^{a}$ | 94 | Mn–H | 213 |
| Mn–Me | 153 | Mn–Cl | 294 |
| Mn–CH$_2$Ph | 87 | Mn–Br | 242 |
| Mn–C(O)Me | 129 | Mn–I | 195 |
| Mn–C(O)Ph | 89 | MnBr(CO)$_4$–CO | 102 |
| Mn–Ph | 170 | Mn$_2$(CO)$_9$–CO | 78 |
| Mn–CF$_3$ | 172 | | |

$^{a}$ Estimates for the actual Mn–Mn BDE in $Mn_2(CO)_{10}$ range from 94 kJ mol$^{-1}$ [106] to >176 kJ mol$^{-1}$.[107]    $^{b}$ Uncertainties in the tabulated values tend to be ≤±12 kJ mol$^{-1}$.

Good linear correlations have been found between standard enthalpies of formation of crystalline organomanganese compounds of the type MnR(CO)$_5$ and the enthalpy of formation of ligands RH in their standard reference state.[108] R groups include halides, hydride, and various hydrocarbon and fluorocarbon alkyls and acyls. These plots are intended for interpolation of enthalpies of formation of other similar manganese complexes.

In the 1970s it was reported that treatment of MnMe(CO)$_5$ with aluminum halides would induce rapid CO insertion, resulting in isolation of complexes containing the aluminum salt, such as in Scheme 15.[109] More recently, thermochemical measurements on these processes (outlined in Scheme 15) have been published. Additional measurements of the enthalpy of coordination of 1/2Al$_2$Br$_6$ with FeCp(CO)$_2$(COMe), Ph–CO–Ph, and Ph–CO–NPh$_2$ (−129, −104, and −113 kJ mol$^{-1}$, respectively), led the authors to conclude that the presence of the transition metal does not have much effect on the strength of the Al–acyl bond.[110]

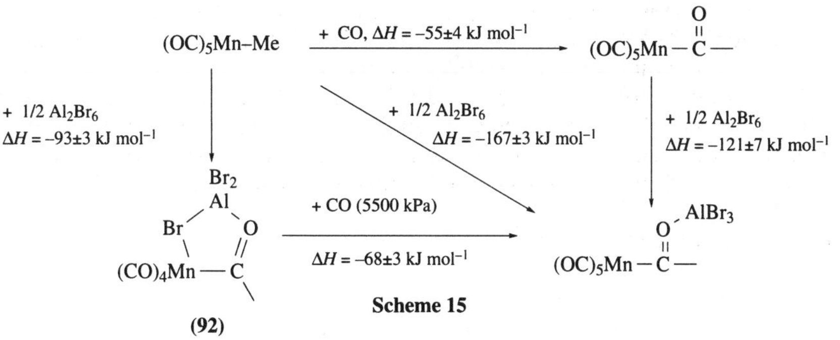

**Scheme 15**

## 2.4.3 Theoretical Calculations

Quantum-mechanical calculations have increasingly been used to address a variety of issues in structural and mechanistic organomanganese chemistry, the methods used covering a wide range of sophistication. Examination of the structural data for MnR(CO)$_5$ reported in the literature shows a rough correlation between the group electronegativity of R and the R–Mn–CO$_{cis}$ angle; the lower the electronegativity of R, the more the equatorial CO groups bend toward R, although the largest deformation found was 7°. Extended Hückel calculations reproduce these distortions and reproduce the trend as a function of an increasingly electropositive "H" group. It is suggested that the same electronic factors that cause the distortions also account for the rate dependence of the CO insertion reaction on the nature of R.[111] A Hartree–Fock–Slater method was used to calculate the energy of the rearrangement of $MnR(CO)_5$ (R = H or Me) to $Mn(\eta^1\text{-}COR)(CO)_4$ (R = Me (**52a**)) and to $Mn(\eta^2\text{-}COR)(CO)_4$ (R = Me (**52c**)), and the energy of migrations of R to *cis*-Mn–C≡S and *cis*-Mn=CH$_2$ (Table 4).[112] Since there is no reasonable direct path from $MnR(CO)_5$ to an $\eta^2$-acyl, the reticence of $MnH(CO)_5$ to undergo CO insertion was concluded to arise from the thermodynamic inaccessibility of $Mn(\eta^1\text{-}CHO)(CO)_4$, even though in principle $Mn(\eta^2\text{-}CHO)(CO)_4$ is energetically accessible on the global reaction surface.

**Table 4**   Calculated enthalpies of rearrangement of R–MnL(CO)$_4$.[112]

|  |  | $\Delta H$ | |
| --- | --- | --- | --- |
| *Inserting group  L* | *Migrating group R* | $Mn(\eta^1\text{-}LR)(CO)_4$ (kJ mol$^{-1}$) | $Mn(\eta^2\text{-}LR)(CO)_4$ (kJ mol$^{-1}$) |
| Mn–C≡O | H | 159 | 91 |
|  | Me | 75 | −4 |
| Mn–C≡S | H | 71 | −59 |
|  | Me | 20 | −116 |
| Mn=CH$_2$ | H | −113 |  |
|  | Me | −71 |  |

The CO insertion reaction was also studied by *ab initio* Hartree–Fock (HF) calculations using a basis set for $Mn(COMe)(CO)_5$ of 178 independent functions, including 4–31 G bases for all carbon, oxygen, and hydrogen atoms, and six *d*-type polarization functions for the reacting carbon and oxygen atoms. The method of partial retention of diatomic differential overlap (PRDDO), which is said to mimic well the corresponding minimum-basis-set *ab initio* calculations in a fraction of the time, was used to preliminarily probe the reaction surface.[113] A $\Delta H$ of 42 kJ mol$^{-1}$ and an $E_a$ of 71 kJ mol$^{-1}$ were calculated at the HF level for the isomerization from $MnMe(CO)_5$ to $Mn(COMe)(CO)_4$ in the form of (**52a**). The transition state was calculated to be very late with a Me–C(O) C–C bond length of 0.157 nm, and a Me–Mn distance of 0.241 nm. In this event, the basal metal orbital in (**52a**) would be almost completely exposed, and so it was concluded that solvent participation in the transition state would be virtually certain. Thus, any real $E_a$ would probably be less than 70 kJ mol$^{-1}$, and experimental values do range from 58 kJ mol$^{-1}$ to 70 kJ mol$^{-1}$ for $MnMe(CO)_5$.[113] The overall energy for the conversion of $MnMe(CO)_5$ and CO to $Mn(COMe)(CO)_5$ was calculated to be −33 kJ mol$^{-1}$, reported experimental values for which range from −33 kJ mol$^{-1}$ to −54 kJ mol$^{-1}$. Isomerization of the intermediate (**52a**) to $Mn(\eta^2\text{-}COMe)(CO)_4$ (**52c**) was predicted to be exothermic with $\Delta H$ = −46 kJ mol$^{-1}$ and an $E_a$ value of only 13 kJ mol$^{-1}$, which is consistent with the results of spectroscopic investigations (see Section 2.4.4.1). The results of the calculations were also discussed with respect to the stereochemical findings outlined in Section 2.3.2.3 (i).[113] Further calculations of the same type were used to examine substituent effects in R on the $\Delta H$ and $E_a$ values of the initial insertion–migration step.[114]

A study of the chelated complexes $Mn(\kappa^2\text{-}CR^2=CR^3COR^1)(CO)_4$ (**62**) and $Mn(\kappa^2\text{-}CHR^2CHR^3COR^1)(CO)_4$ (**63**), shown in Scheme 10 (see Section 2.3.2.3 (ii)) was conducted employing $^{55}$Mn NMR and Fenske–Hall calculations. It was suggested that (**62**) possesses some aromatic character.[115] Fenske–Hall calculations have also been carried out on $(\mu\text{-}CH_2)[MnCp(CO)_2]_2$ in the context of comparisons to analogous methylene-bridged dimers of other metals and mixed metals.[116,117]

## 2.4.4 Study of Organomanganese Transients

### 2.4.4.1 Spectroscopy of transients in solution and in matrices

Spectroscopy of species isolated in matrices and time-resolved spectroscopy of flash-generated transients in solution are techniques that have added greatly to our understanding of transient intermediates in organometallic chemistry. This area has continued to develop rapidly since the 1980s. Photolysis of $Mn(COMe)(CO)_5$ at 12 K in a CH$_4$ matrix allows the study of $Mn(COMe)(CO)_4$. Bands appearing at 1634 cm$^{-1}$ and 1613 cm$^{-1}$ were assigned to acetyl stretching modes, $\nu_{CO(acetyl)}$, with the doublet ascribed to the presence of two conformations about the Mn–C(acetyl) bond. The frequencies were concluded to be in the range for an $\eta^1$-acyl group.[118] A reinvestigation of the geometry of photogenerated $MnMe(CO)_4$ was carried out in CH$_4$ and argon matrices at 20 K. It was found that the situation is very similar to that of $MnH(CO)_4$ (see Section 2.8.3.3). There were inferred to be two isomers formed in the photolysis: a $C_s$ structure (**93a**(S)), and $C_{4v}$ structure (**93b**(S)), where (S) is a coordinated solvent molecule. Previously postulated (**93c**) was thought not to be present.[119] More recent data indicate that (**93a**(S)) is formed directly and is itself then photoisomerized to (**93b**(S)).[120]

(**93a**)  (**93a**(S))  (**93b**(S))  (**93c**)

Even more detailed structural information regarding these transients was obtained from a careful analysis of the $\nu_{C-H}$ stretching band in Mn–CD$_2$H derivatives, called $\nu^{is}_{C-H}$ since the C–H stretch is isolated from Fermi resonance and coupling by the two deuteriums. A shift of $-16$ cm$^{-1}$ in $\nu^{is}_{C-H}$ was found upon photosubstitution of Mn(CD$_2$H)(CO)$_5$ in an N$_2$ matrix to form *cis*-Mn(CD$_2$H)(CO)$_4$(N$_2$). The isolated CH vibration was too weak to assign reliably in *cis*-Mn(CD$_2$H)(CO)$_4$(argon). From known $\nu^{is}_{C-H}$ vs. $r_{0_{C-H}}$ (bond length) correlations, it was inferred that the C–H bond was lengthened and weakened, while the Mn–C bond was strengthened on substitution of CO by N$_2$.[120]

Laser flash photolysis of MnMe(CO)$_5$ has been carried out in cyclohexane, isooctane, and THF under argon or CO with time-resolved IR and optical detection.[121] The second-order rate constant for the reassociation of CO with MnMe(CO)$_4$(S) (**93a**(S)) was $2.1 \times 10^6$ M$^{-1}$ s$^{-1}$ in C$_6$H$_{12}$, and $1.4 \times 10^2$ M$^{-1}$ s$^{-1}$ in THF. An unexpected observation was that under CO at 1 atm in C$_6$H$_{12}$ solvent, the yield of MnMe(CO)$_4$(C$_6$H$_{12}$) was reproducibly reduced fivefold compared with photolyses carried out under argon. This corresponds to a CO/C$_6$H$_{12}$ selectivity of $\sim 5 \times 10^3$ for coordination to the initial photolysis intermediate. In THF under CO, no reduction in initial yield of MnMe(CO)$_4$(THF) was seen, and a similar yield reduction has not been seen on photolysis of Cr(CO)$_6$ in C$_6$H$_{12}$ under CO. This result suggests that the initial intermediate may not have a simple $C_s$ singlet structure like (**93a**). One possibility mentioned was that a very short-lived triplet might show more selectivity toward incoming ligands.[121] Decarbonylation of Mn(COMe)(CO)$_5$ by laser flash photolysis exhibits a long-lived intermediate when monitored by time-resolved IR spectroscopy. The intermediate, which has an acyl band at 1607 cm$^{-1}$, is formed in less than 100 ns. At 200 K the intermediate is stable enough to be monitored by standard FTIR. Only small frequency shifts are seen on changing solvents; methylcyclohexane, THF, 2-methyl-THF, and 2,5-dimethyl-THF, and a feature at $\sim$384 nm in the UV–visible spectra is essentially invariant in these solvents. These data strongly suggest that the intermediate is much too stable to be unsolvated (**52a**) or (**52d**), and is much too solvent insensitive to be (**52a**(S)). The postulated structure is therefore $\eta^2$-acetyl species (**52c**).[122] MO calculations confirm the reasonableness of this structure (see Section 2.4.3; see also the related discussion of the CO insertion mechanism in Section 2.3.2.3(i)).

Photolysis of Mn($\eta^1$-allyl)(CO)$_5$ at 12 K in various gas matrices affords Mn($\eta^3$-allyl)(CO)$_4$. Continued irradiation did not reverse this reaction, even in CO matrix, but instead gave dissociative substitution of one additional CO.[123] IR spectroscopic evidence, including $^{13}$CO labeling, has been reported to show that a CO ligand is photolytically ejected from Mn[C(O)C$_7$H$_7$](CO)$_5$ in argon, CH$_4$, N$_2$, and CO matrices at $\sim$12 K to form Mn($\sigma$-C$_7$H$_7$)(CO)$_5$. Subsequent photolyses led to sequential formation of Mn($\eta^3$-C$_7$H$_7$)(CO)$_4$, Mn($\eta^5$-C$_7$H$_7$)(CO)$_3$, and Mn($\eta^7$-C$_7$H$_7$)(CO)$_2$ without detectable 16-electron intermediates, even at 12 K.[124]

### 2.4.4.2 Gas-phase studies

Activity has continued since the 1980s in the area of reactions of gas-phase naked metal ions with hydrocarbons. It has been noted that Mn$^+$ is much less reactive toward hydrocarbons than are other $3d$ transition metal ions (M$^+$), and several rationalizations have been given.[125] However, generation of Mn$^+$ in the presence of 4-octyne led to extensive reaction with formation of H$_2$ (23), CH$_4$ (7), C$_2$H$_4$ (40), C$_3$H$_6$ (19), and C$_8$H$_{14}$ (11), where the number in parentheses is the "branching ratio." 3-Heptyne also formed methane, but 2-hexyne and 1-pentyne did not. This observation and the results from variously deuterium-labeled 4-octynes led to the conclusion that the formation of methane is a 1,6-elimination, most probably proceeding through intermediate (**94**). The extant data did not permit conclusions to be drawn regarding the paths of formation of the other products. An upper limit for the ring strain in (**94**) was estimated to be 75 kJ mol$^{-1}$.[125]

There has been increased activity with more complex, ligand-coordinated metal ions in the gas phase. In particular considerable work on [Mn(CO)$_3$]$^-$ has appeared. In contrast with the naked Mn$^+$ ion,

$$\text{(94)}$$

[Mn(CO)$_3$]$^-$ has high reactivity toward hydrocarbons, and is isoelectronic with solution phase, 14-valence-electron, three-coordinate transition metal species that have been postulated as reactive intermediates which activate hydrocarbon C–H bonds in solution.[126] A major difference between [Mn(CO)$_3$]$^-$ and solution phase 14-electron ML$_3$, is that the relative reactivity toward CH bonds is $3° > 2° > 1° \gg CH_4$ (no detectable adduct), while in solution it is $CH_4 > 1° > 2° > 3°$ (no adducts detected). The ions [Mn(CO)$_4$]$^-$ and [Mn(CO)$_5$]$^-$ were shown to be unreactive.[127] With methanol in the gas phase, [Mn(CO)$_3$]$^-$ shows two ions; deuterium labeling shows a more intense ion which comes from C–H activation, and a less intense ion from O–H activation.[128] Cyclopropane reacts with [Mn(CO)$_3$]$^-$ to give two principal ions of composition [Mn(CO)$_3$(C$_3$H$_6$)]$^-$ and [Mn(CO)$_3$(C$_3$H$_4$)]$^-$ which are inferred from their subsequent reactivity with other standard reagents to be saturated (18-electron) complexes and are believed to be better represented as [Mn(CO)$_3$(H)$_2$(C$_3$H$_4$)]$^-$ and [Mn(CO)$_3$(H)(C$_3$H$_3$)]$^-$, respectively.[129]

Dissociative electron attachment has been used to generate [Mn(CF$_3$)(CO)$_n$]$^-$ ($n = 2$–5) from Mn(CF$_3$)(CO)$_5$ and Mn(COCF$_3$)(CO)$_5$ which were studied by FT ion cyclotron resonance spectroscopy.[130] In a similar manner, the IR multiphoton dissociation spectrum of [Mn(CF$_3$)(CO)$_3$(NO)]$^-$ has been investigated.[130]

## 2.5 MANGANESE ARYL COMPLEXES

The chemistry of aryl complexes of transition metals is distinctly different from that of metal alkyls. The π-system of the arene affords mechanistic paths related to the organic chemist's aromatic substitution, electrophilic on one end and nucleophilic on the other end of the polarization scale. The availability of this wide range of facile mechanisms means that aryl metal complexes tend to be labile in their formation and in their reactions. At the same time, metal–aryl M–C bond strengths tend to be high because of the greater overlap and the greater electronegativity of $sp^2$ carbon orbitals compared with $sp^3$ orbitals. The strong resistance of metal aryls to β-hydride elimination means that their reactions are cleaner and generally more controllable than those of metal alkyls. Altogether, these factors suggest that the chemistry of metal aryls should be quite rich, and this is true for manganese.

### 2.5.1 Synthesis of Manganese Aryls

Nucleophilic aromatic substitution generally requires the presence of electron-withdrawing groups on the arene. Thus, although it is one of the standard methods of preparation of manganese alkyls, nucleophilic substitution of arenes by manganese anions only occurs for activated arenes such as pentafluoropyridine.[131]

Occasionally, arylation is accomplished by a formally carbanionic route, although the yields are not generally high. Examples are shown in Equations (30),[132] (31),[133] and (32).[134] In the case of Equation (32), treatment of the crude reaction mixture, while still at low temperature, with PPh$_3$, P(OMe)$_3$, dppe, or TMEDA resulted in isolation of the arylmanganese products with the corresponding ligands substituted for CO on manganese in yields of 37–69%.[134]

$$\text{Br–Mn(CO)}_5 + \text{Li}\!\!-\!\!\underset{F\ \ \ \ F}{\overset{F\ \ \ \ F}{\bigodot}}\!\!-\!\!\text{Li} \xrightarrow[22\%]{\text{Et}_2\text{O, 22 °C}} \text{(OC)}_5\text{Mn}\!\!-\!\!\underset{F\ \ \ \ F}{\overset{F\ \ \ \ F}{\bigodot}}\!\!-\!\!\text{Mn(CO)}_5 \tag{30}$$

One of the two most common ways of preparing arylmanganese species is through the acylation–decarbonylation sequence described in Section 2.3.1.2. For example, this method was used to prepare the bismanganese aryl complex (**96**)(Equation (33)),[135] and (**97**) and (**98**) were synthesized in analogous fashion with overall yields of 37% and 35%, respectively.[132]

Br–Mn(CO)$_5$ + [ferrocenyl-Li] $\xrightarrow[\text{75\%}]{\text{THF, –78°C}}$ [ferrocenyl-Mn(CO)$_5$] (95)  (31)

Br–Mn(CO)$_5$ + [arene-Li with Cr(CO)$_3$] $\xrightarrow{\text{THF, –50°C}}$ [product with Mn(CO)$_5$, Cr(CO)$_3$]  X = H, 31%; X = F, 52%  (32)

NaMn(CO)$_5$ + [terephthaloyl chloride] $\xrightarrow{\text{THF, 22 °C}}$ [bis-acyl Mn(CO)$_5$] $\xrightarrow[\text{74\%}]{\text{toluene,}}$ (OC)$_5$Mn—C$_6$H$_4$—Mn(CO)$_5$ (96)  (33)

(OC)$_5$Mn—C$_6$H$_4$—Mn(CO)$_5$ (97)

(OC)$_5$Mn—C$_6$H$_3$(Mn(CO)$_5$)—Mn(CO)$_5$ (98)

The other common way to prepare arylmanganese species is via a cyclometallation method originally developed in the 1970s,[1] where the arene is heated in the presence of a manganese alkyl, most commonly MnMe(CO)$_5$ or Mn(CH$_2$Ph)(CO)$_5$. Many recent examples of this method are gathered in Table 5, and a typical example is presented in Equation (34).[145] In the latter case, an isotope effect of 3.0 was observed by intramolecular competition, indicating that site selection is determined by a step involving hydrogen migration. An example of a cyclometallation of a different kind is shown in Equation (35).[146]

[trimethoxy acetophenone, R = H,D] $\xrightarrow[\text{heptane,}]{\text{Bz–Mn(CO)}_5}$ [cyclomanganated product with Mn(CO)$_4$] (99) 90%  $k_H/k_D = 3.0$  (34)

Mn$_2$(CO)$_{10}$ + [phosphorine sulfide] $\xrightarrow[\text{3.5 h}]{\text{xylene, 140 °C}}$ [Mn$_2$(CO)$_9$ product] 38% $\xrightarrow[\text{5 h}]{\text{xylene, 140 °C}}$ [Mn(CO)$_4$ cyclometallated product]  (35)

Many examples of cyclometallation were known prior to 1983,[1] particularly those induced by nitrogen. Table 5 also illustrates the wide range of molecules that can be cyclomanganated through the directing effect of carbonyl oxygen. It has become apparent that while many metals engage in cyclometallations induced by phosphorus and nitrogen, manganese reagents (and those of rhenium) are particularly effective in the direct cyclometallation of aromatic substrates where the guiding donor atom is carbonyl oxygen.[136] Entries 11–15 in Table 5 involve molecules where double cyclometallations are possible. Compounds (101) and (103) are both molecules where the two metals are substituents on the

**Table 5**  Manganese alkyls prepared by cyclometallation.

| Entry | Product | | | | Yield (%) | Ref. |
|-------|---------|--|--|--|-----------|------|
| 1 | | | | | | 136[a] |

| $R^1$ | $R^2$ | $R^3$ | $R^4$ | | | |
|-----|-----|-----|-----|--|--|--|
| H | H | H | H | | 85 | |
| H | H | OMe | H | | 86 | |
| OMe | OMe | OMe | H | | 80 | |
| H | OMe | OMe | OMe | | 86 | |
| H | OMe | OMe | H | | 60 | |
| H | H | OMe | OMe | | 20 | |
| OBz | OMe | OMe | H | | 90 | |
| O-TBDMS | H | H | H | | 86 | |
| O-TBDMS | OMe | OMe | H | | 80 | |
| Me | H | Me | H | | 56 | |
| H | H | Br | H | | 96 | |

| 2 | | | | | | 136[a] |

| X = S | | | | | 75 | |
| X = NMe | | | | | 28 | |
| X = O | | | | | 18 | |

| 3 | | | | | 70 | 136[a] |

| 4 | | | | | 90 | 136[a] |

| 5 | | | | | 21 | 137[a] |

| 6 | | | | | | 138[a] |

| R = Me | | | | | 36 | |
| R = Ph | | | | | 59 | |

**Table 5**  (continued)

| Entry | Product | Yield (%) | Ref. |
|-------|---------|-----------|------|
| 7 | | | 138[a] |
| | X = OMe | 33 | |
| | X = NMe₂ | 61 | |
| 8 | | | 138[a] |
| | R = Me | 100 | |
| | R = Et | 33 | |
| | R₂ = (CH₂)₄ | 77 | |
| 9 | | | 138[a] |
| | R = Me | 68 | |
| | R = Ph | | |
| 10 | | 18 | 139[b] |
| 11 | **(100)** | 46 | 140[a] |
| 12 | **(101)** | 69 | 140[a] |
| 13 | | | 141[c] |
| | **(102a)** R = H | 50 | |
| | **(102b)** R = NMe₂ | 50 | |

**Table 5** (continued)

| Entry | Product | Yield (%) | Ref. |
|-------|---------|-----------|------|
| 14 | | | 141[c] |
| | (103a) R = H | 45 | |
| | (103b) R = NMe$_2$ | 60 | |
| 15 | | 52 | 142[c] |
| 16 | (104) | 95 | 143[a] |
| 17 | (105) | 100 | 143[a] |
| 18 | (106) | 71 | 143[a] |
| 19 | | 83 | 144[d] |

[a] MnBz(CO)$_5$ in heptane at reflux.  [b] MnBz(CO)$_5$ in CH$_2$Cl$_2$ at reflux.  [c] MnMe(CO)$_5$ in octane at reflux.  [d] MnMe(CO)$_5$ in benzene at reflux.

same aromatic ring, and in these cases the incorporation of one or two metals was determined by simple manipulation of the reaction stoichiometry. The implication is that, as a ring substituent, the first Mn(CO)$_4$ is neither particularly activating nor deactivating toward incorporation of the second metal.

Woodgate and co-workers have examined preparations and reactions of many cyclomanganated diterpenoids and related molecules. For example, one paper[143] presents details on the synthesis of more than 20 manganese aryls; five derivatives of 13-acylpodocarpic acid (e.g., (**104**), entry 16 in Table 5), seven derivatives related to 7-oxopodocarpic acid (e.g., (**105**), entry 17 in Table 5), and 7-oxodehydroabietic acid (e.g., (**106**), entry 18 in Table 5), among others.

### 2.5.2 Reactions of Manganese Aryls

Arylmanganese species exhibit the types of reactivity that one expects for this kind of transition metal–carbon bond, including several illustrated for $1,4\text{-}C_6H_4[Mn(CO)_5]_2$ (**96**) in Scheme 16. Reactions with halogens are typical of electrophilic cleavage. Halogen cleavage of the simpler cyclometallated arenes tends to proceed in high yield, as in the case of (**96**) with $Br_2$ or $I_2$ in THF which afforded $p\text{-}C_6H_4X_2$ (**107**) as the only identifiable organic product.[132] Similarly, reaction of (**99**) (Equation (34)) with $Br_2$ in $CCl_4$ quantitatively affords (**112**).[145] Attempts to cleave the diterpene derivatives (**104**) and (**105**) (Table 5) with halogen sources, however, gave generally messy results. The most selective cleavages were effected with *N*-bromosuccinimide (NBS) in $CCl_4$ forming, for example, (**113**) from (**104**) in 55% yield, along with 30% of the unhalogenated, demetallated ketone. With NBS/$CCl_4$, (**105**) gave 50% of (**114**) along with 30% of the free ketone.[147]

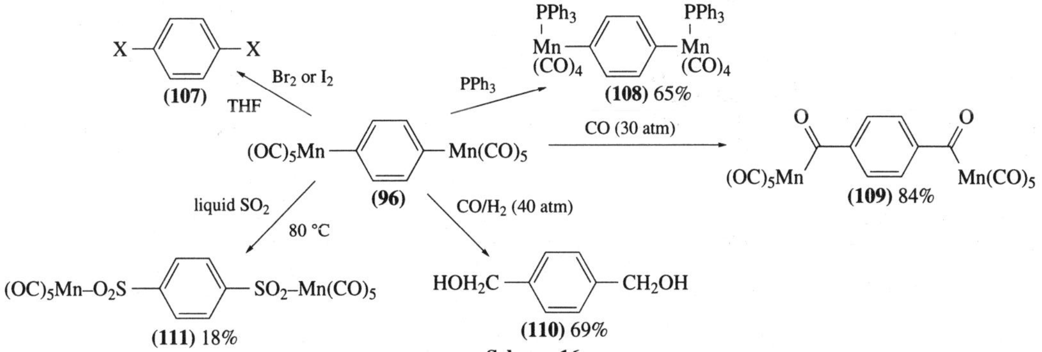

**Scheme 16**

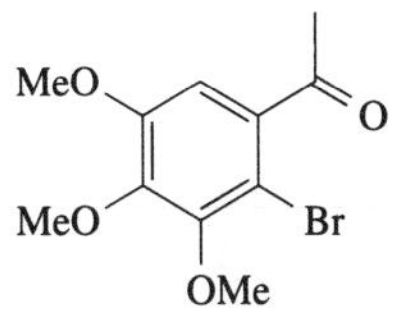

(**112**) 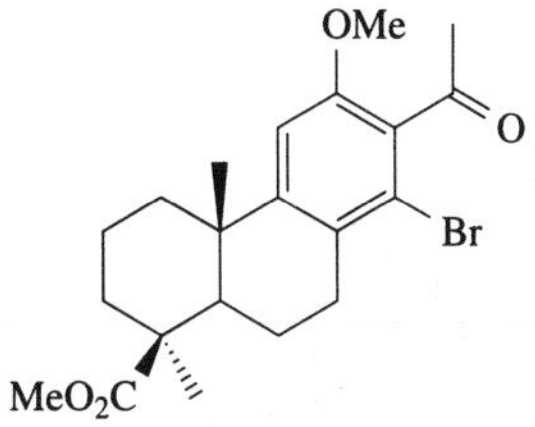    (**113**) 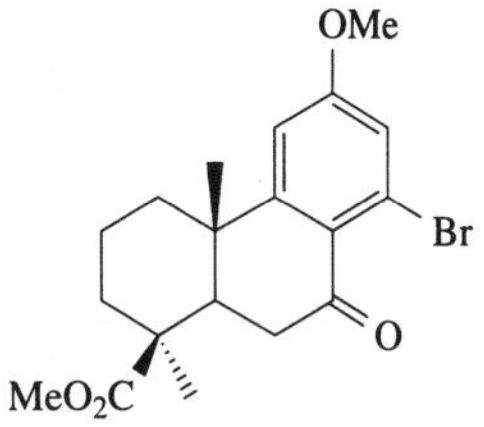    (**114**)

Organomercurials are useful reagents for organic synthesis, and these too can be generated by electrophilic cleavage of arylmanganese bonds. The arylmercuric chlorides (**115**)–(**117**) were generated in good to excellent yields from their corresponding cyclometallated tetracarbonylmanganese aryl species on treatment with $HgCl_2$ in methanol at reflux.[148] Oxidative cleavage of (**99**) (Equation (34)) by $Ce^{IV}$ in $CD_3CO_2D$ solvent gave a 50% yield of isotopically pure 2-deutero-3,4,5-trimethoxyacetophenone.[145]

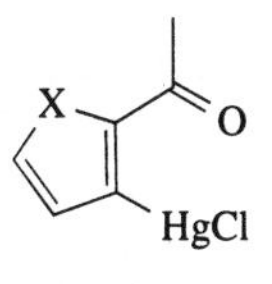

(**115**) 76% 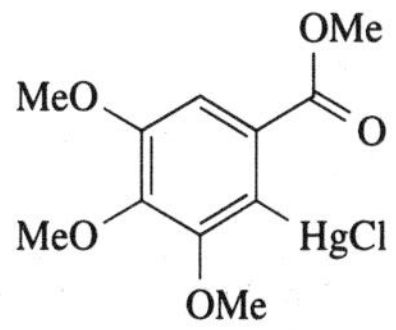    (**116**) 90%

Treatment of manganese aryl complexes with nucleophilic ligands tends to result more in substitution of CO than CO insertion, in contrast with alkyl complexes, consistent with the well-known reticence of

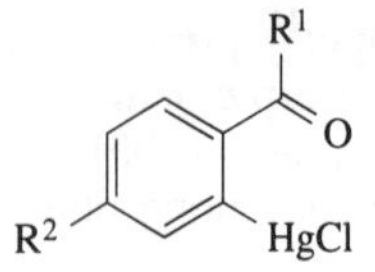

| | R¹ | R² | Yield (%) |
|---|---|---|---|
| (a) | Me | H | 48 |
| (b) | Me | OMe | 76 |
| (c) | Ph | H | 60 |

**(117)**

CO to insert into M–Ar bonds. For example, treatment of (**96**) (Scheme 16) with PPh$_3$ (or PMePh$_2$) slowly yields the bis-substituted species (**108**) without any detectable acyl (CO insertion) product.[135] This same process can be made much more efficient by treatment of (**96**) with Me$_3$NO followed by phosphine. Another example of the preference for substitution is the reaction of (**95**) (Scheme 17) with PPh$_3$ which effects only substitution to form (**120**).[133]

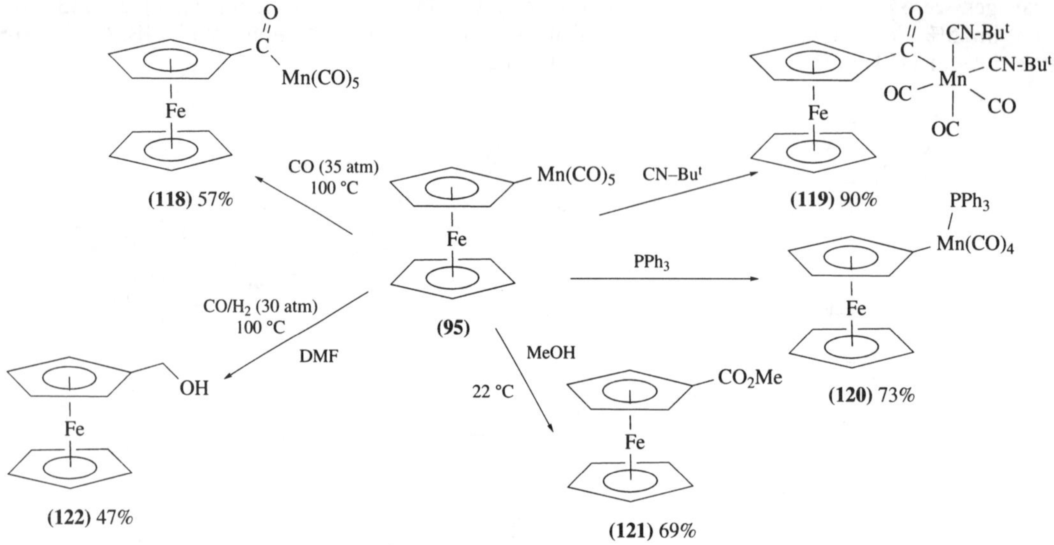

**Scheme 17**

Insertion reactions comprise an important area of reactivity for arylmanganese complexes. In spite of the mentioned reticence of manganese aryls to undergo CO insertions, under more vigorous conditions such reactions are very much in evidence. Again using (**95**) and (**96**) as examples, under 30–35 atm of CO, good yields of acylmanganese complexes (**118**) (Scheme 17)[133] and (**109**) (Scheme 16),[135] respectively, are obtained. In the case of (**95**), $t$-butylisonitrile induces both CO insertion and substitution at manganese, forming (**119**) in excellent yield (Scheme 17).[133] Under reducing conditions with CO and H$_2$ at 30–40 atm both (**95**) and (**96**) undergo CO insertion with subsequent reduction and hydrogenolysis to alcohols (**122**) and (**110**), respectively. In methanol solution (**95**) exhibits CO insertion and then solvolysis to form ester (**121**).[133]

The insertion of SO$_2$ is occasionally used as a method of routine characterization of arylmanganese species. The double insertion into (**96**) to form (**111**) in low yield (Scheme 16) provides one example.[135]

From the standpoint of organic synthesis, the insertions of alkenes and alkynes into manganese aryl bonds are particularly interesting reactions. A variety of cyclometallated species undergo such reactions. On simple thermolysis, complex (**123**) (Equation (36)) inserts diphenylacetylene into the Mn–C bond and then cyclizes to the indenol derivative (**124**).[149] If the original chelating acyl group bears a substituent that will eliminate with manganese, such as H or NMe$_2$, then substituted indenone (**125**) forms directly. The reaction was shown to proceed as well with HC≡CH, HC≡CPh, and TMS-C≡C-TMS, although the last gave only an 8% yield of the corresponding indenol.

Activated alkenes were shown[150] to insert into the Mn–C bond of species like (**123**) in addition to more functionalized substrates like thiophene (**126**) (Equation (37)). These reactions were effected through the intervention of an equivalent of PdCl$_2$. Some cyclization–dehydration occurred spontaneously (**127**), but most of the product was the noncyclized result of manganese hydride elimination (**128**).

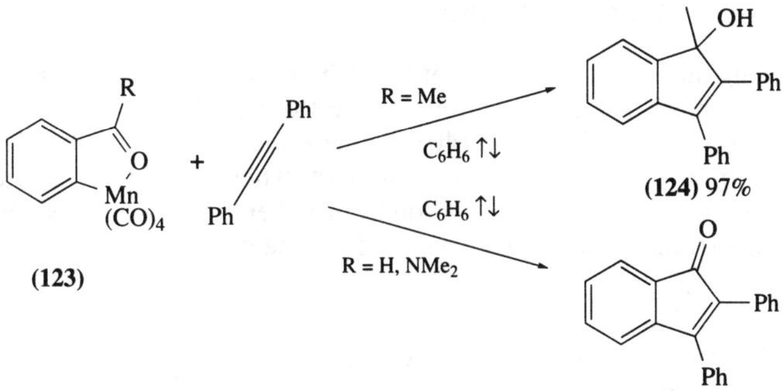


(123) (124) 97% (125) 56% (36)

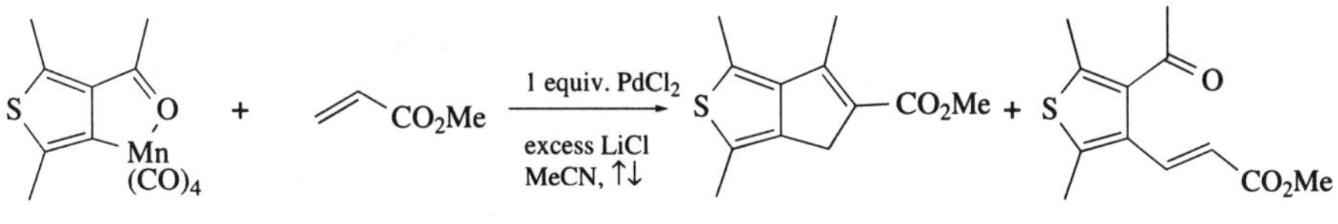


(126) (127) 34% (128) 56% (37)

It was subsequently shown that 10 mol.% of $Pd(OAc)_2(PPh_3)_2$ would catalytically accomplish the same reaction on similar cyclometallated substrates (Equation (38)).[151] The example of (**129**) illustrates the further point that with two neighboring carbonyl groups, the regioselectivity of *in situ* cyclization can be a concern. In this case, two regioisomers (**130**) and (**131**) are formed.

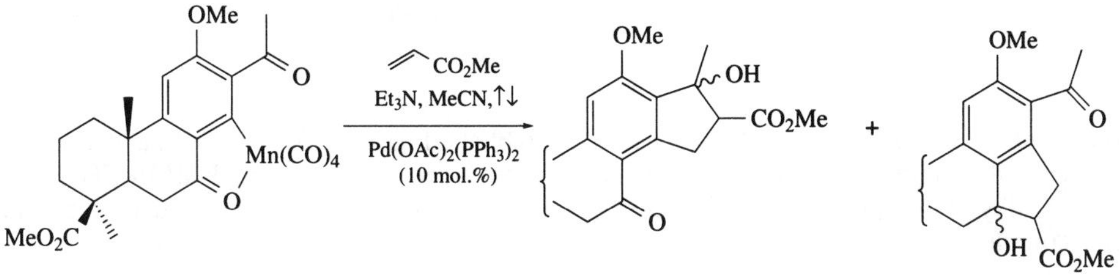


(129) (130) 32% (131) 9% (38)

The effectiveness of both alkene and alkyne insertions can be enhanced substantially through the use of $Me_3NO$, which initially generates a site at the manganese center that can be easily substituted. Scheme 18 shows several such insertions of the 13-acylpodocarpic acid derivative (**104**) which proceed through the weakly solvated intermediate (**132**). Numerous examples using various alkenes, alkynes, and cyclometallated substrates have been reported.[152,153]

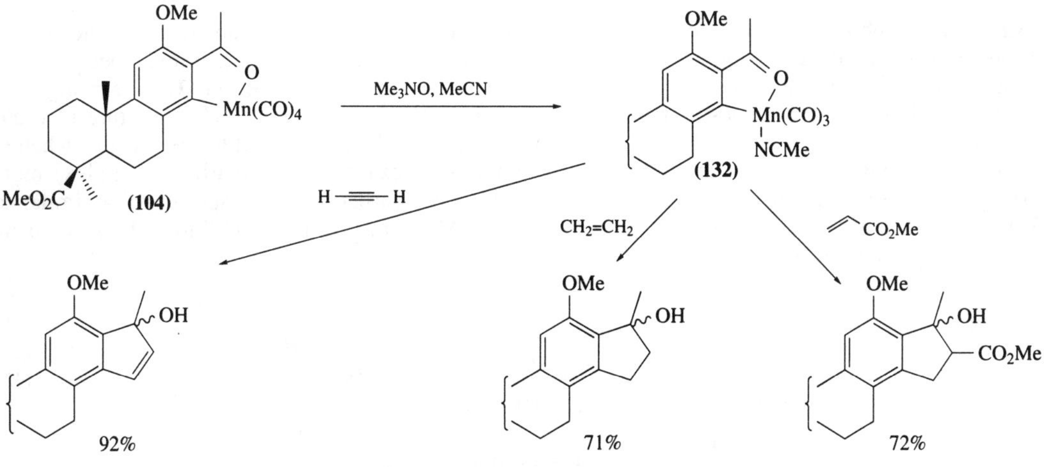


**Scheme 18**

## 2.6 MANGANESE VINYL AND ALKYNE COMPLEXES

As with the preparation of manganese aryls, reaction of vinyl halides with manganese anions is not expected to be a general preparation of manganese vinyls. Nevertheless, there are a few specialized examples such as the reactions in Equations (39)[86] and (40).[154] These reactions probably proceed via an addition–elimination mechanism, particularly in the second case where conjugate addition to the unsaturated carbonyl should be facile. In addition, perfluoronorbornadiene reacts with $Na[Mn(CO)_5]$ in THF at ambient temperature to afford the vinyl manganese product analogous to (133) in 80% yield.[155]

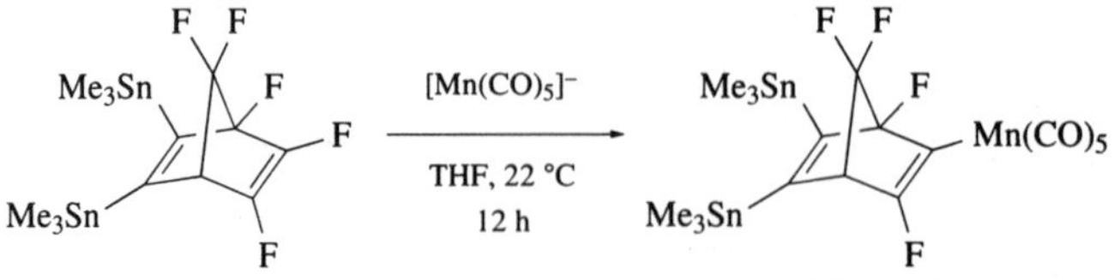

(39)

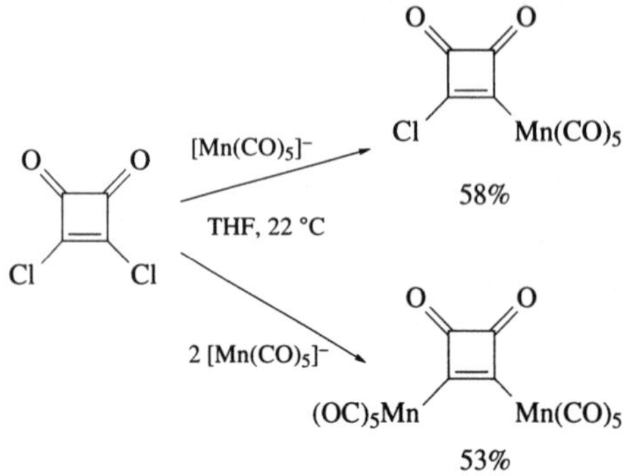

(40)

Nucleophilic addition to π-alkyne complexes is a well-established method for vinyl metal production in general, and several examples have been reported in recent years. Addition of $Na[Mn(CO)_5]$ to a ruthenium cationic alkyne complex (134) (Equation (41)) led to isolation in 80% yield of the *trans* bimetallic vinyl complex (135).[156]

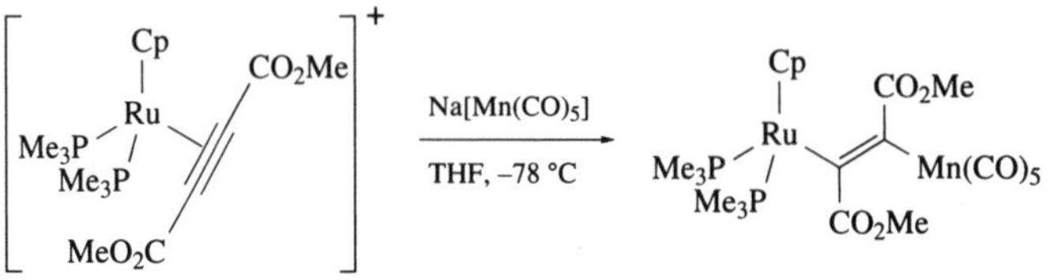

(41)

Additions of phosphines to manganese coordinated alkynes has been studied since the 1980s. Initially, it was reported that $PPh_3$ adds to (136) generating the zwitterionic vinyl species (137) (Equation (42)).[157] It was later found that with smaller phosphines, the zwitterion (139) rearranges on standing in solution to the α-phosphonium zwitterion (140) (Equation (43)).[158] The structure of (139) (with MeCp and $PEt_3$) was studied by x-ray diffraction, particularly with regard to the issue of whether the resonance contributor (139z) (the zwitterionic vinyl complex) or (139c) (carbene complex) more nearly describes the structure.[159] It was concluded to be more carbene-like. A species very similar to (139), assigned the structure $MnCp[=C(OUCp_3)CH=PPhMe_2]$, results from addition of the carbene complex $[UCp_3(=CH–PPhMe_2)]$ to $MnCp(CO)_3$ in THF.[160]

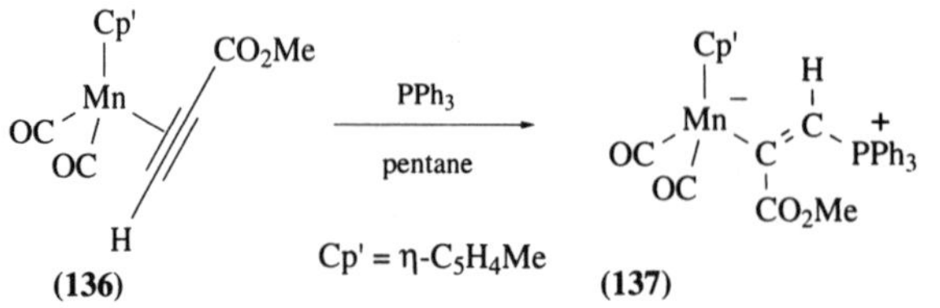

(42)

$$
\text{(138)} \quad \xrightarrow[\text{pentane}]{\text{PMe}_3 \text{ or PEt}_3} \quad \left[ \text{(139z)} \leftrightarrow \text{(139c)} \right] \longrightarrow \text{(140)} \quad (43)
$$

Cp' = η-Cp, η-$C_5H_4$Me, or η-Cp*

In one instance, electrophilic addition to an alkyne π-complex of manganese has resulted in the formation of a vinyl complex. Addition of X–Y ($Br_2$, HCl, or $HO_2CCF_3$) to Mn(Cp*)(CO)$_2$(HC≡CH) generates Mn(Cp*)(CO)$_2$($\eta^2$-*trans*-XCH=CHY). However, addition of Cl–NO leads to vinyl species Mn(Cp*)(CH=CHCl)(CO)(NO).[161]

Some vinyl manganese species have been prepared from alkynyl manganese precursors. For example, formal [2 + 2] cycloaddition of (NC)$_2$C=C(CF$_3$)$_2$ to (141) affords the cyclobutenyl manganese derivative (142) (Equation (44)).[162] The structure and regiochemistry were confirmed by x-ray crystallography. TCNE reacts similarly with (141) to form a cyclobutenyl complex, but it affords the product of formal electrocyclic ring opening of the cyclobutene, butadiene (143).[163]

$$
\text{(141)} \quad \xrightarrow[\text{CH}_2\text{Cl}_2]{(\text{NC})_2\text{C=C(CF}_3)_2} \quad \text{(142)} \quad (44)
$$

(143)

Insertion of alkynes into Mn–X bonds is a source of vinyl complexes. Heterocycle (144) (Equation (45)) inserts the activated alkyne into the Mn–P bond to afford (145).[164]

$$
\text{(144)} \quad + \quad \overset{\text{CO}_2\text{Me}}{\underset{\text{CO}_2\text{Me}}{|||}} \quad \xrightarrow[]{\text{hexane, } \uparrow\downarrow} \quad \text{(145)} \quad (45)
$$

A related example is shown in Scheme 19, where (146) adds an activated alkyne to form regioisomeric heterocycles (148a) and (148b).[165] It is thought that the dimer (146) is in equilibrium with the monomer (147), and that the alkyne inserts into either the Mn–S or the Mn–P bond of the monomer. In the presence of excess alkyne, regioisomer (148a) undergoes a regiospecific [2 + 2 + 2] cycloaddition to form bicyclic (149).

Simple addition of MeC≡CH to bridged binuclear hydride (150) (Equation (46)) leads to two regioisomeric bridging manganese vinyls (151a) and (151b). These additions also occur with MeC≡CMe, MeC≡CPh, and EtC≡CH.[166]

Two specialized routes to vinyl manganese species have been reported. Buchner's mercurial diazo compound Hg[C(N$_2$)CO$_2$Et]$_2$ reacts with MnBr(CO)$_5$ in ether at reflux to give (152).[167] Also, Mn$_2$(CO)$_9$(NCMe) reacts with HC≡COEt with CO coupling to afford the bicyclic complex (153), which has the unusual feature of a carbonyl group coordinated to two independent metals. The structure of the latter was determined by x-ray diffraction.[168]

The chemistry of alkynylmanganese complexes is not extensive. Early attempts to prepare manganese acetylides by anionic routes, such as from MnBr(CO)$_5$ and LiC≡CPh, were not generally successful. More recently, however, metathesis with tin acetylides has been found to work reasonably well. For example, MnBr(CO)$_5$ and Sn(C≡CPh)Me$_3$ form Mn(C≡CPh)(CO)$_5$ in 42% yield, and (OC)$_5$Mn(C≡C-*p*-C$_6$H$_4$-*p*-C$_6$H$_4$-C≡C)Mn(CO)$_5$ was synthesized from the bis(stannylated) diacetylide.[169] An alternative

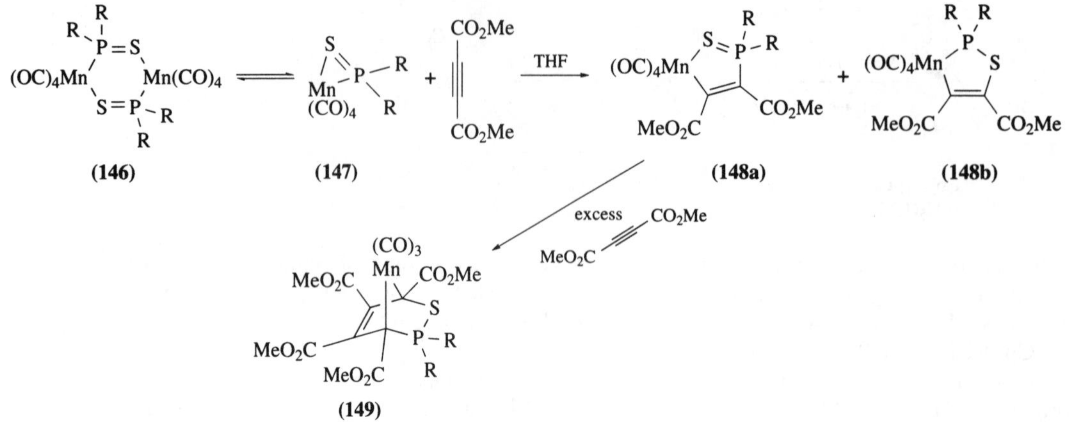

**(146)**     **(147)**     **(148a)**     **(148b)**

**(149)**

**Scheme 19**

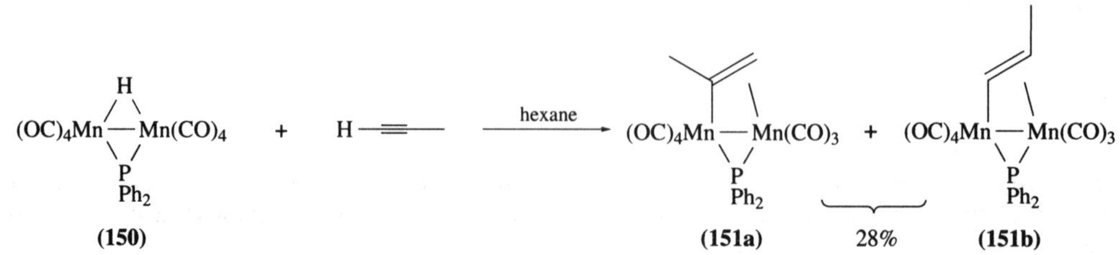

**(150)**     **(151a)**     28%     **(151b)**     (46)

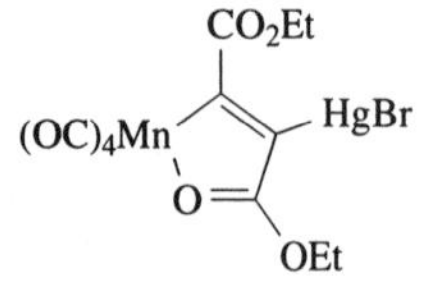

**(152)** 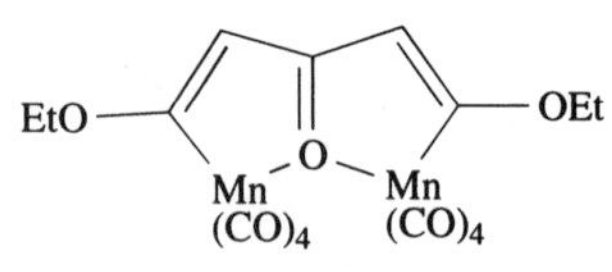     **(153)**

method involves the use of terminal alkynes and silver ion with $MnBr(CO)_5$, as shown in Equation (47).[170] Yields were 45–80% and were quite sensitive to the combination of ligands (L) and their geometry. These products could be adorned with various permutations of ancillary ligands by thermal and photochemical substitution of the acetylide complexes.

$$Mn(CO)_n(L)_{5-n}Br + AgBF_4 + H\!\!-\!\!\equiv\!\!-R \xrightarrow{CH_2Cl_2/H_2O} Mn(CO)_n(L)_{5-n}(\equiv\!\!-R) + AgBr + HBF_4 \quad (47)$$

R = Me, Ph; n = 1–3

L = $PEt_3$, $PCy_3$, dppe, dppm, $P(OMe)_3$, $P(OPh)_3$, phen, bipy, $CNBu^t$

Anionic acetylide complex $[Cp'(CO)(PPh_3)MnC\equiv CMe]^-$ (**155**) has been generated *in situ* by treatment of the carbene complex $[Cp'(CO)(PPh_3)Mn=C(OMe)Et]$ (**154**) with *n*-butyllithium.[171] The chemistry of this anion has been examined extensively, almost all in the context of carbene complexes, and is discussed in Section 2.7.1.

Another acetylide preparation, shown in Scheme 20, begins with nucleophilic addition to a CO ligand of $Mn_2(CO)_{10}$ to form the acyl anion (**156**). Photochemical decarbonylation generates the alkyl anion (**157**) which is unstable to loss of $[Mn(CO)_5]^-$ and leads to the alkoxide of manganese acetylide (**158**).[172] The alkoxide function is subject to manipulation as shown in Scheme 20.

Few reactions of vinyl manganese complexes have been reported. One example involving a binuclear vinyl species (**159**) is shown in Equation (48).[173] In this case CO insertion was induced by the addition of hydride. Additional chemistry of (**159**) induced by CO, $CNBu^t$, and $PEt_3$ was examined.

Alkynes σ-bonded to manganese show typical π-complexation chemistry toward other metals. The alkyne group of *fac*-$Mn(C\equiv CR)(CO)_3$(dppe) (R = $CH_2OMe$, $Bu^t$, or Ph) is a conventional π-ligand toward $Cu^I$, $Ag^I$, and $Au^I$ with a variety of ancillary ligands.[174] Coordination of *cis*-$Mn(C\equiv CPh)(CO)_4(PCy_3)$ by $Co_2(CO)_8$ yields $[Co_2(CO)_6(PhC\equiv CMn(CO)_4PCy_3)]$ (**161**) where the configuration is *trans* at manganese. Oxidation of (**161**) by $[I(py)_2]BF_4$ in $CH_2Cl_2$ regenerates $Mn(C\equiv CPh)(CO)_4(PCy_3)$ except in its *trans* form. Heating in hexane causes isomerization back to the *cis* material.[175]

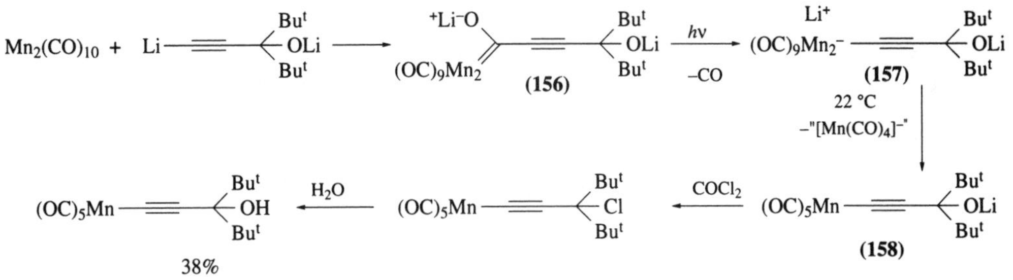

Scheme 20 shows reaction schemes with structures (156), (157), (158).

**Scheme 20**

Equation (48) with structures (159) and (160):

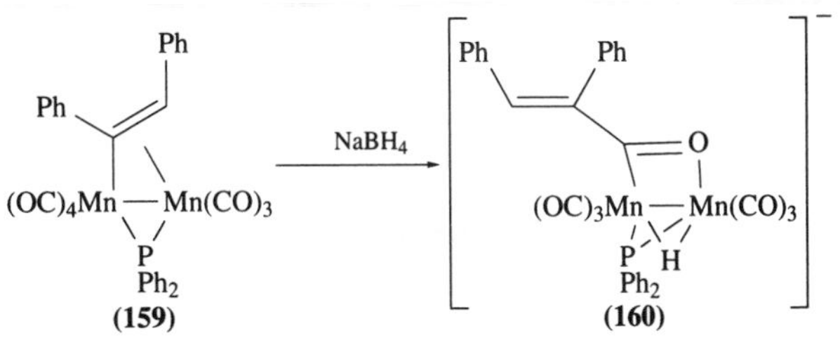

$$(48)$$

The chemistry of vinyl and alkyne manganese species is intimately interwoven with that of manganese alkylidene and alkylidyne complexes. Several reactions of manganese acetylides are included as preparations of alkylidene manganese complexes (see Section 2.7).

## 2.7 MANGANESE CARBENE AND CARBYNE COMPLEXES

### 2.7.1 Synthesis of Manganese Carbenes and Carbynes

The area of metal–carbon multiple bonding has advanced since the 1970s for many transition metals including manganese. In 1967, Fischer and Maasböl[176] described the preparation of Mn(Cp')(CO)$_2$[=C(R)OMe] (Cp' = $\eta^5$-C$_5$H$_4$Me; R = Me or Ph). Since that time, many carbene complexes of manganese in both Cp-based and conventional octahedral environments have been prepared. The Fischer preparation is general and is still relied on heavily. Some examples of this synthesis form the basis of the chemistry shown in Equations (49)[177] and (50),[178] and the preparation of complex (166).[179] Equation (49) also illustrates a general means of exchanging the stabilizing heteroatom group of "Fischer carbenes," that is, substitution by nucleophilic reagents at the carbene carbon, in this case interconverting acetyl species (162) and thioether (163). Equation (50) shows that the acyl anion intermediate formed by anion addition to a neutral Mn–CO complex can be silylated instead of alkylated, in this case yielding the series of compounds Mn(Cp')(CO)$_2$(=C(Ph)OSiMe$_2$R) (R = Me, vinyl, or allyl) (164).[178] Photolysis of (164) (R = vinyl) generated the chelated carbene–alkene complex (165). Structures of this last type are of current interest as possible models for intermediates implicated in the catalysis of the celebrated alkene metathesis reaction. Protonation or alkylation of neutral acylmetal groups is known to generate cationic carbene complexes, and syntheses of the neutral formyl species Mn[C(O)H](CO)$_3$(PPh$_3$)$_2$ has given access to substituted methylene complexes (167a) by protonation, (167b) by methylation, and (167c) by group exchange of the methoxy derivative with benzylamine, all in high yield.[180] X-ray structure determinations were reported for (167b) and (167c). Protonation of acetyl derivative Mn[C(O)Me](CO)$_3$(dppp) affords the hydroxy carbene complex (168), with yields depending on the acid used (e.g., HOTf, 87%).[181] A crystal structure was also reported for (168) as the triflate.

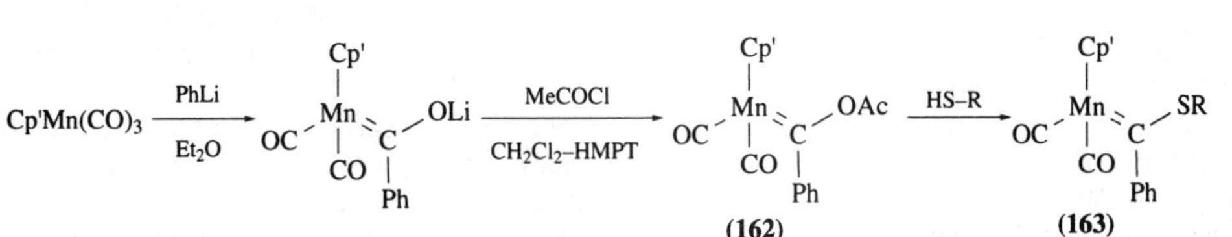

Equation (49) reaction scheme:

Cp'Mn(CO)$_3$ $\xrightarrow[\text{Et}_2\text{O}]{\text{PhLi}}$ ... $\xrightarrow[\text{CH}_2\text{Cl}_2\text{–HMPT}]{\text{MeCOCl}}$ (162) $\xrightarrow{\text{HS–R}}$ (163)

Cp' = $\eta$-C$_5$H$_4$Me

R = Ph (80%), allyl (90%), cyclohexyl (66%)

$$(49)$$

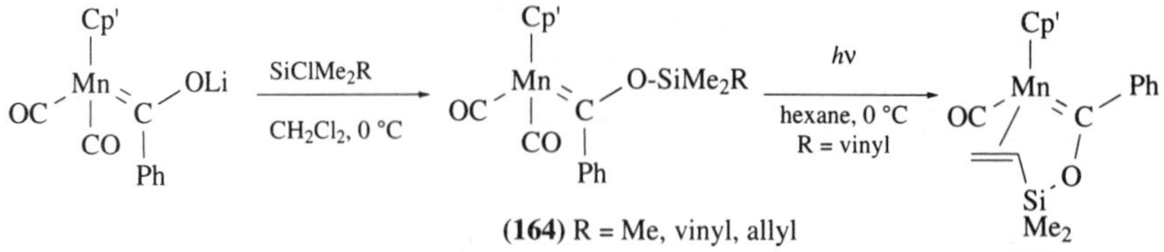

(50)

(164) R = Me, vinyl, allyl

(165)

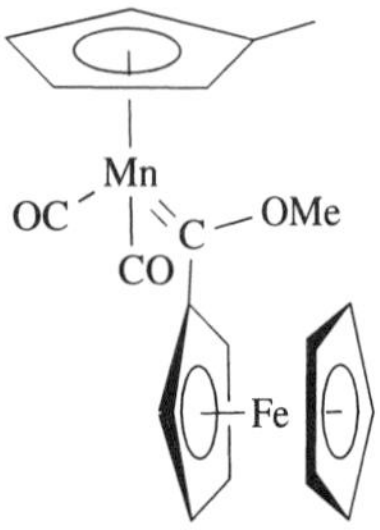

(166)

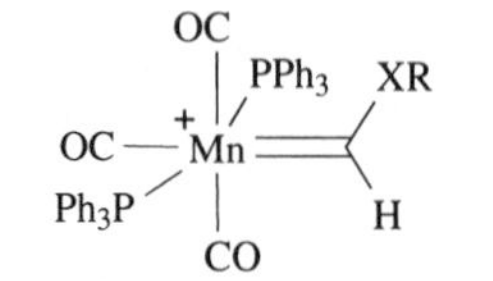

(167a) XR = OH, 94%
(167b) XR = OMe, 98%
(167c) XR = NHBz, 80%

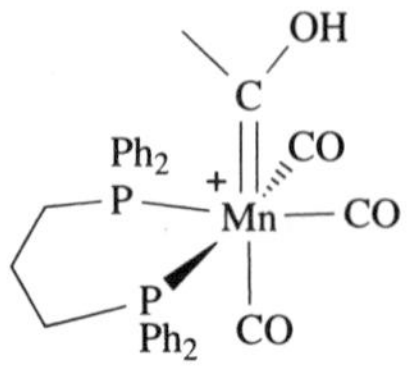

(168)

Another example of a nucleophilic addition to a coordinated CO ligand is shown in Scheme 21.[9] The adding reagent is the phosphoranyl ylide $Me_3P=CH_2$, a second equivalent of which deprotonates a zwitterion to form the salt (169). Double alkylation of (169) leads to stable carbene complex (170). Further deprotonation of (170) generates (171) which is a hybrid of vinyl zwitterion and carbene–ylide forms.

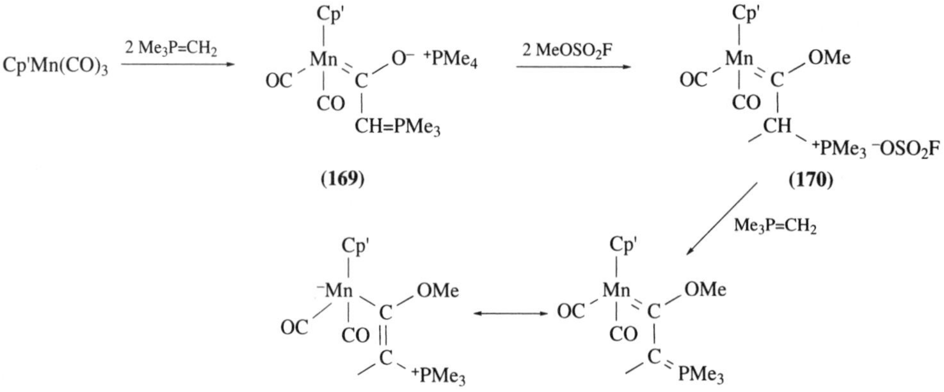

(169)

(170)

(171)

**Scheme 21**

The anion $[Mn(MeCp)(EPh_3)(CO)_2]^-$ (172) is prepared by deprotonation of the corresponding hydride with NaH in THF, and reacts with geminal dichlorides or iminoacyl chlorides, designated R–Cl in Equation (51). Intermediates are presumably generated in this reaction which probably resemble (173) and wherein the $ER_3$ group should be efficiently removed by chloride to form carbene complexes (174).[182] Various R–Cl and the corresponding carbene product are shown below Equation (51). Yields are not high, generally in the range 24–43%.

Octahedral acyl anions can be prepared through the use of $^-GeR_3$, as shown in Equation (52), where $GeR_3$ is $GePh_3$ or $Ge(\alpha$-naphthyl$)MePh$, and the salt $[NEt_4][(175)](GeR_3 = GePh_3)$ was isolated in 57% yield. The germanium anion apparently induces CO insertion into the Mn–Me bond rather than adding to a CO ligand. Subsequent alkylation of the acyl oxygen affords the neutral carbene complex, for example, (176), in moderate yield.[183]

Dibromoalkanes have been employed to prepare cyclic Fischer carbenes of manganese carbonyl using $Na[Mn(CO)_5]$ (Equation (53)). With a twofold excess of $Na[Mn(CO)_5]$, the second equivalent apparently functions as a nucleophilic ligand to induce the CO insertion reaction and form the acyl anion (177). The anion is then efficiently trapped by the proximate alkyl bromide to form the cyclic carbene (178).[184] Several 1,3- and 1,4-dibromides have been used, as shown in Equation (53).

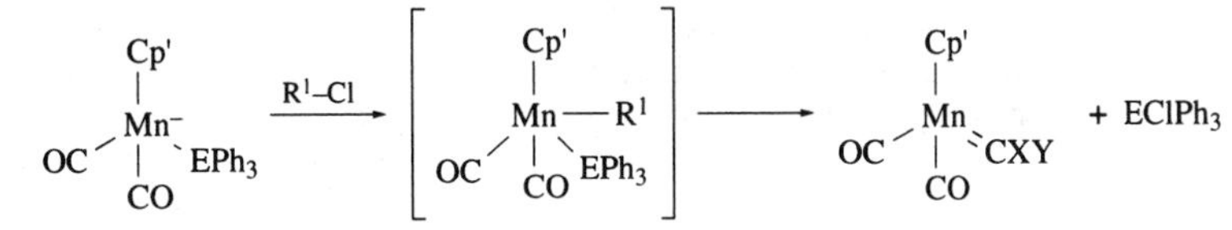

$$\text{(172)} \xrightarrow{\text{R}^1\text{-Cl}} \left[ \text{(173)} \right] \longrightarrow \text{(174)} + \text{EClPh}_3 \tag{51}$$

(172)   (173)   (174)

E = Si, Ge, Sn

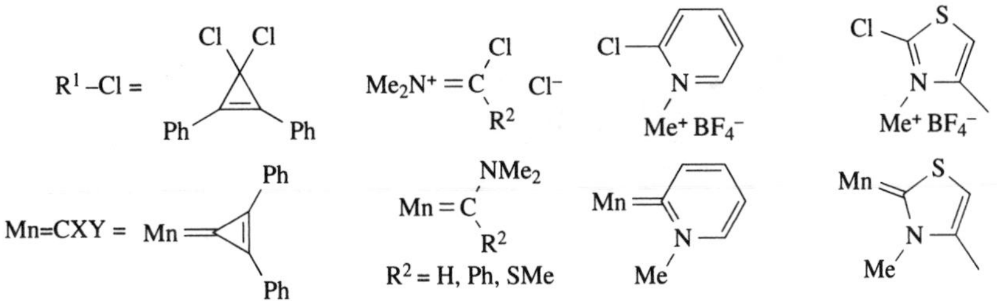

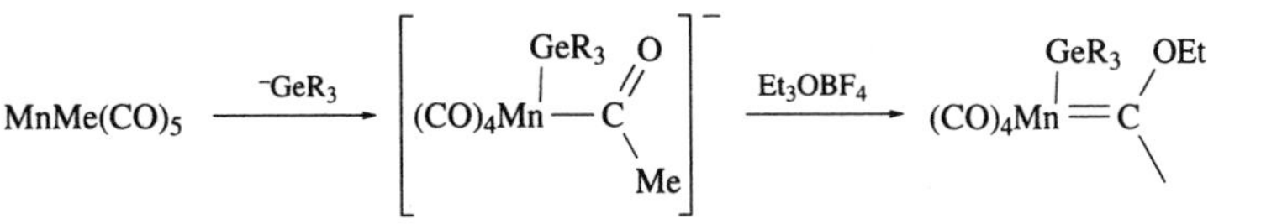

$$\text{R}^1\text{-Cl} =$$

$$\text{Mn=CXY} = \text{Mn}$$

$$\text{R}^2 = \text{H, Ph, SMe}$$

$$\text{MnMe(CO)}_5 \xrightarrow{\text{-GeR}_3} \left[ (\text{CO})_4\text{Mn}-\text{C} \begin{array}{c} \text{GeR}_3 \\ \text{O} \\ \text{Me} \end{array} \right]^- \xrightarrow{\text{Et}_3\text{OBF}_4} (\text{CO})_4\text{Mn}=\text{C} \begin{array}{c} \text{GeR}_3 \quad \text{OEt} \end{array} \tag{52}$$

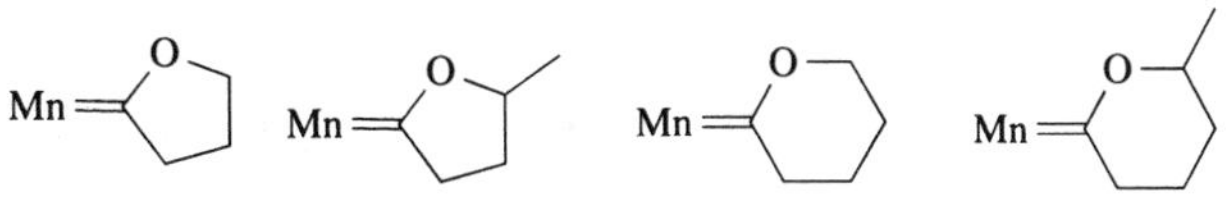

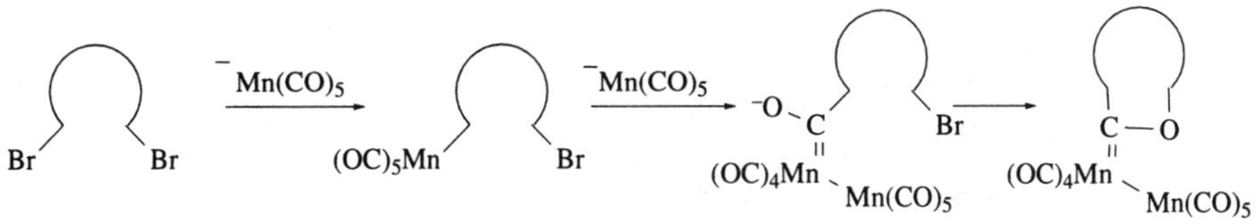

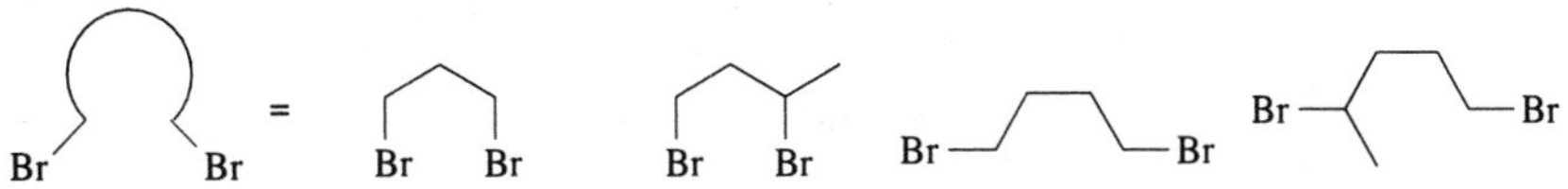

(175)   (176) R = Ph, 63%

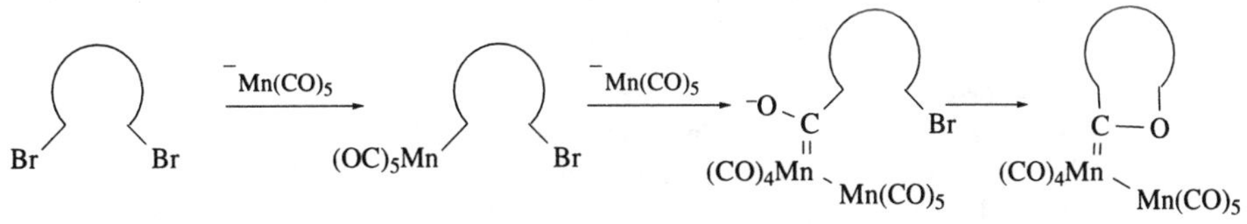

$$\xrightarrow{^-\text{Mn(CO)}_5} (\text{OC})_5\text{Mn} \cdots \text{Br} \xrightarrow{^-\text{Mn(CO)}_5} \underset{(\text{CO})_4\text{Mn}\cdots\text{Mn(CO)}_5}{^-\text{O}-\text{C}} \cdots \text{Br} \longrightarrow \underset{(\text{CO})_4\text{Mn}\cdots\text{Mn(CO)}_5}{\text{C}-\text{O}} \tag{53}$$

(177)   (178)

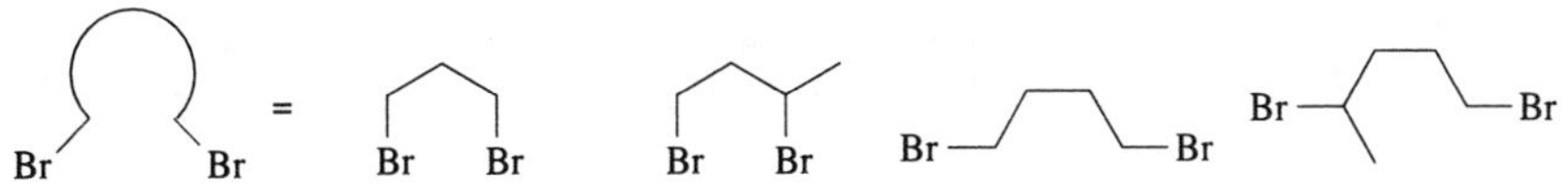

(178) = Mn= ...

$$\xrightarrow{^-\text{Mn(CO)}_5} (\text{OC})_5\text{Mn} \cdots \text{Br} \xrightarrow{^-\text{Mn(CO)}_5} \underset{(\text{OC})_4\text{Mn}\cdots\text{Mn(CO)}_5}{^-\text{O}-\text{C}} \cdots \text{Br} \longrightarrow \underset{(\text{OC})_4\text{Mn}\cdots\text{Mn(CO)}_5}{\text{C}-\text{O}} \tag{53}$$

178 = Mn= ...

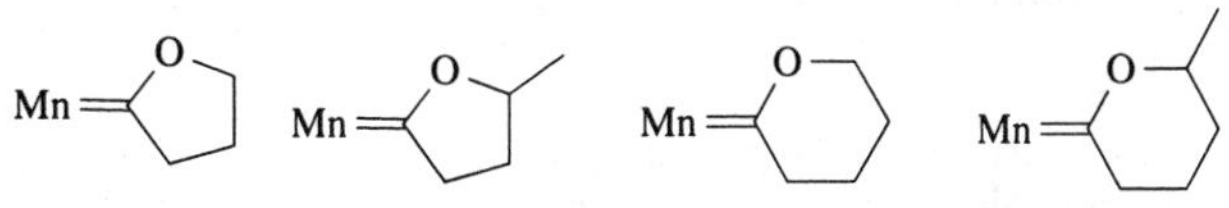

A sequence closely related to that of Equation (53) has been used to prepare bisheterosubstituted cyclic carbene complexes of $[\text{MnCp(NO)(CO)}_2]^+$ (179) from aziridine and oxirane as shown in Scheme 22. Catalytic bromide is used to open the strained rings to form $\text{BrCH}_2\text{CH}_2\text{NH}_2$ and $\text{BrCH}_2\text{CH}_2\text{O}^-$, respectively, which are strong enough nucleophiles to add to a CO ligand of the active cation (179).

Internal alkylation of the alkyl bromide by the acyl oxygen anion then completes the cyclization to (**180**) and (**181**).[185] In a completely analogous fashion, bromide ion catalyzes formation of (**182**) from MnBr(CO)$_5$ and either aziridine or oxirane.[186]

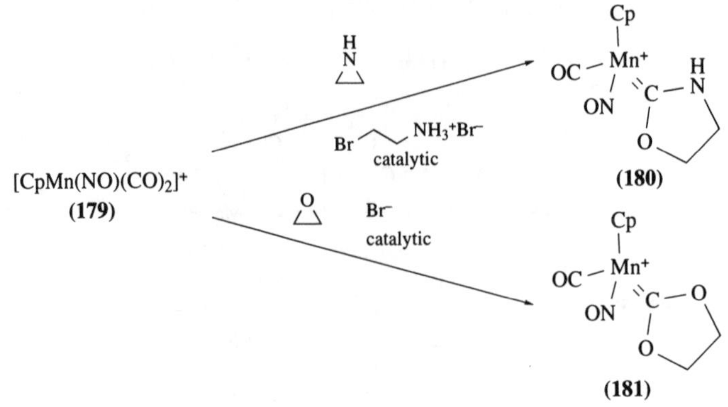

**Scheme 22**

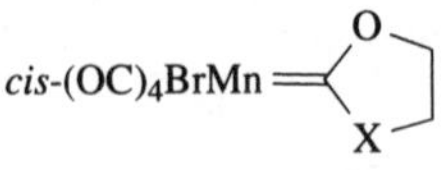

(**182**)

X = O, NH

Another synthetic approach to manganese carbenes is via nucleophilic additions to coordinated alkynes. Simple aliphatic amines, hydrazine, and phenylhydrazine add to the alkyne complex [Mn(Cp')(CO)$_2$(HC≡CH)] (**183**) with rearrangement to give aminocarbene complexes [(Cp')(CO)$_2$Mn=C(NR$_2$)Me] (NR$_2$ = NH$_2$, NHMe, NMe$_2$, NHNH$_2$, or NHNHPh) (**184**).[187] Conditions are mild and yields are generally 70–86%. The chemistry is insensitive to the presence of Cp, Cp', or Cp* ligands. The hydrazine derivatives were found to fragment to MnCp(CO)$_2$(NCMe) and NH$_2$R (R = H or Ph) on heating at 30 °C for 20 h. The attacking nucleophile can be internal as the cyclization reaction shown in Equation (54) illustrates. Carbene complex (**185**) was isolated in 51% yield.[188]

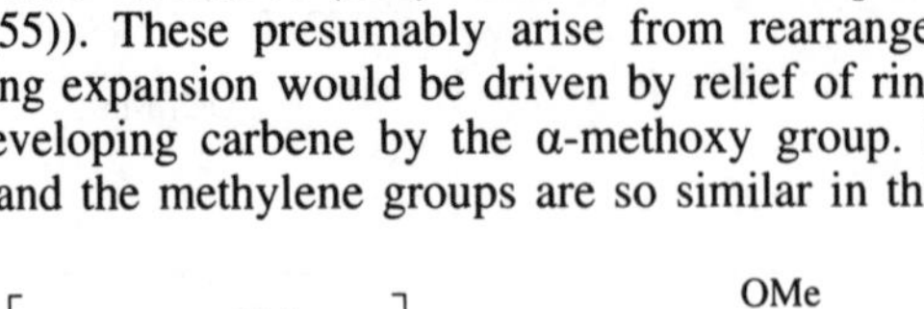

(54)

Several other miscellaneous methods have been used to prepare manganese carbene complexes. Thermolysis of the benzocyclobutene derivative (**186**) forms two isomeric pentacyclic Fischer carbenes (**188a**) and (**188b**) (Equation (55)). These presumably arise from rearrangement of decarbonylated intermediate (**187**), where the ring expansion would be driven by relief of ring strain and by the extra stabilization provided to the developing carbene by the α-methoxy group. It is interesting that the migratory aptitudes of the aryl and the methylene groups are so similar in this case.[189]

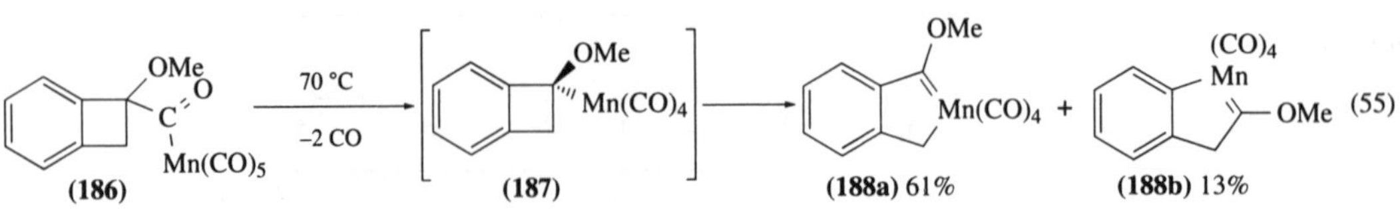

(55)

One might logically suppose that carbene complexes might be accessible from diazo compounds in reaction with a suitable coordinatively unsaturated manganese intermediate. Compounds (**189**)–(**192**) have been prepared from the corresponding N$_2$=CAr$_2$ and MnCp(CO)$_2$(THF) in THF at 20 °C in yields of 52–84%.[190]

The Fischer carbene (**193**) can be twice deprotonated to form anionic acetylide (**194**) as shown in Scheme 23.[171] This in turn reacts with strained-ring ethers in the presence of a strong Lewis acid to generate cyclic carbene compounds (**195**) (89%), (**196**) (68%), and (**197**) (68%) as shown. All three

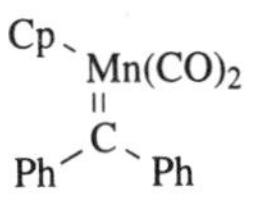

**(189)** 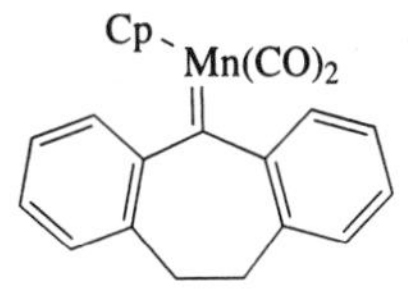  **(190)**

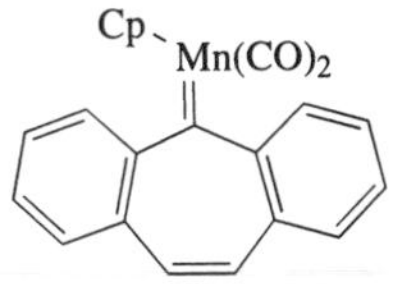

**(191)** 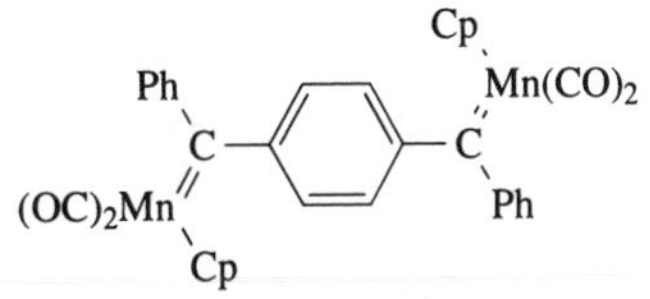  **(192)**

carbene moieties are chiral, and the metal center is chiral. Overall, the products are formed with high diastereoselectivity; only a single diastereomer of (195) and (196) was detected by NMR. Two x-ray structure determinations established the relative configurations of the carbon group and the metal center for (195) and (196). Two diastereomers of (197) were formed in a 7:3 ratio, and NMR was used to establish the stereochemistry of the carbene ligand of the major isomer as that of (197a). The relative carbon configurations of (197b) could not be established.

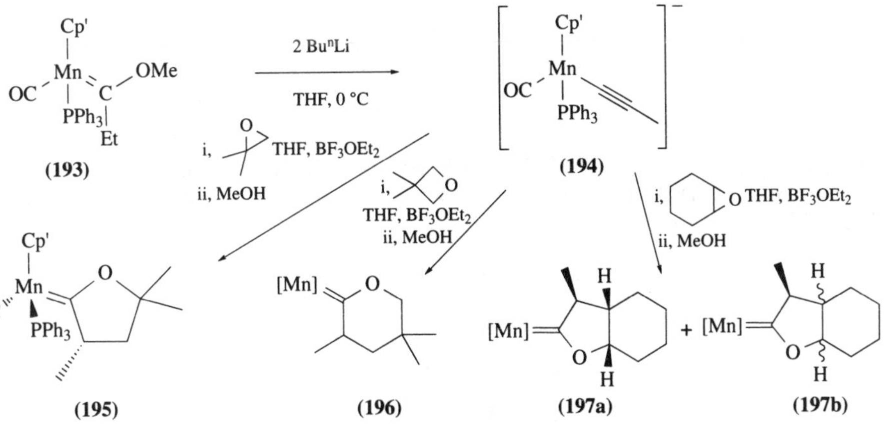

**(193)**  **(194)**

**(195)**  **(196)**  **(197a)**  **(197b)**

**Scheme 23**

As elaborated below, neutral carbene compounds of the type $MnCp(CO)_2[=C(OR^1)R^2]$ and cationic carbyne complexes $[MnCp(CO)_2(\equiv CR^2)]^+$ are mutually interconvertible. Thus, addition of $Et_4NX$ salts to $[MnCp(CO)_2(\equiv CPh)]SbCl_6$ in $CH_2Cl_2$ at $-78\,°C$ forms the halocarbene complexes $MnCp(CO)_2[=C(X)Ph]$: X = Cl (63%), Br (43%), and I (37%).[191] All of these decompose rapidly at room temperature.

Preparations of vinylidene manganese complexes from anionic alkynyl intermediate (194) have been explored extensively (Scheme 24).[171] The β-carbon of anion (194) is alkylated and acylated by a variety of reagents to afford neutral vinylidine compounds (198). A surprising result is that *t*-butyl iodide alkylates rather than protonates the anion, forming (198) (R = $Bu^t$) in 78% yield. This result was confirmed by an x-ray crystal structure of the product. Anion (194) conjugately adds to an α,β-unsaturated carbonyl (199) and adds to polar cumulenes, such as $CO_2$ (200) and ketenes and isocyanates (201).

π-Coordinated alkynes can undergo base-induced rearrangement to vinylidine complexes. Photolysis of $Mn(Cp)(CO)_3$ in ether generates the solvate $Mn(Cp)(CO)_2(OEt_2)$ which on treatment with $HC\equiv CCH(OR)_2$ (R = Me, Et, etc.) affords $Mn(Cp)(CO)_2[\eta^2\text{-}HC\equiv CCH(OR)_2]$. Addition of butyllithium to this π-complex presumably forms an alkynic anion similar to (194), which upon protonation gives the desired vinylidine $Mn(Cp)(CO)_2[=C=CHCH(OR)_2]$.[192] In contrast, with a good anionic leaving group on the alkyne, rearrangement can occur as shown from (202) to (203) in Equation (56), presumably formally via a cationic manganese species.[193]

A different route to vinylidine complexes is shown in Equation (57).[194] Photochemical ejection of CO from $MnCp'(CO)_2L$ should allow oxidative addition of the vinyl chloride to form presumed intermediate

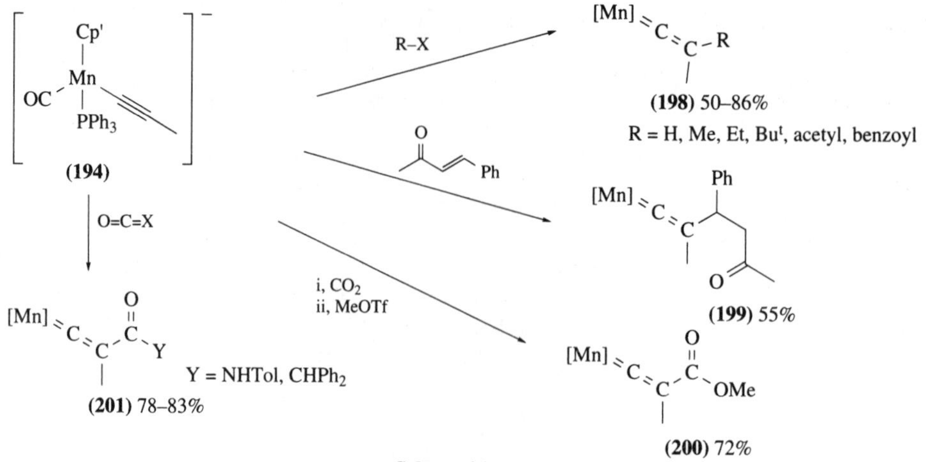

**Scheme 24**

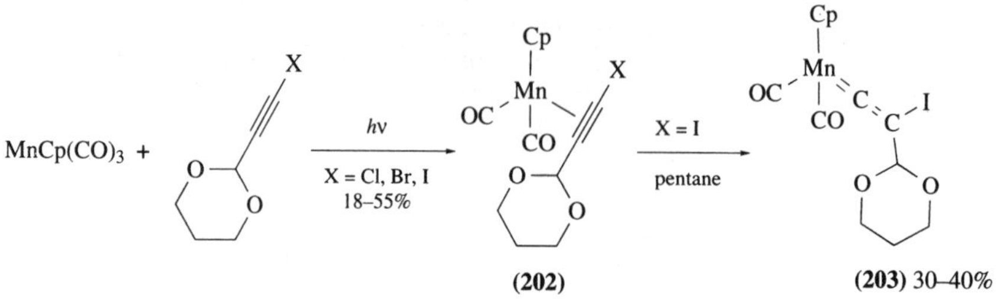

(56)

(204) from which β-elimination of TMS-Cl would produce vinylidine (205). Yields were not reported. With excess (chlorovinyl)silane, a coupling reaction occurred with (205) and $MnCp'(CO)(L)(\eta^2\text{-}Me_2C=C=C=CMe_2)$ was isolated.

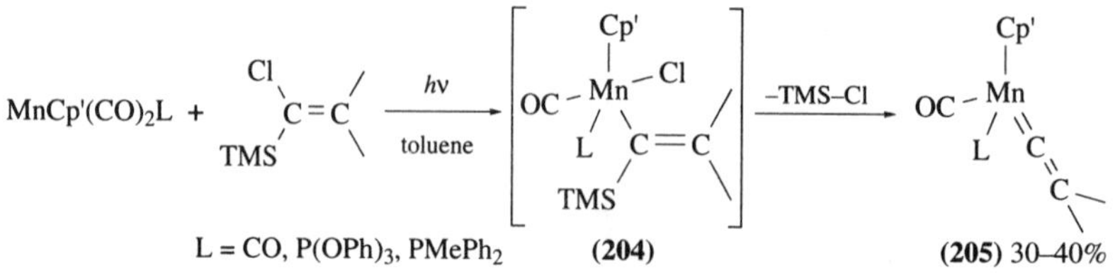

(57)

$L = CO, P(OPh)_3, PMePh_2$     (204)                         (205) 30–40%

Very few preparations of manganese alkylidynes have been reported. One synthesis is based on a general method developed by Fischer. Treatment of (206) with boron halides, as shown in Equation (58) affords cationic ferrocenylcarbyne complex (207) in very good yield.[179] The carbyne is electrophilic, and reverts to a carbene on addition of a nucleophile. In the example shown, phenyl chalcogenides were used.

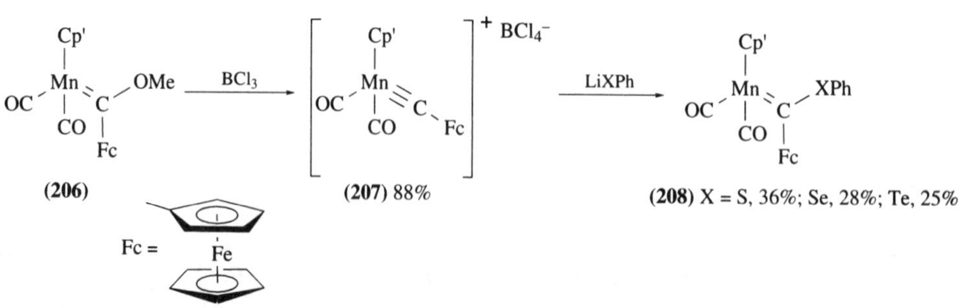

(58)

(206)                         (207) 88%                    (208) X = S, 36%; Se, 28%; Te, 25%

$Fc =$

Protonation of cumulated carbene complexes also affords alkylidyne cations.[195] For example, treatment of $MnCp(CO)_2(=C=C=CR_2)$ (R = Ph or $Bu^t$) with HX (HCl, HBF$_4$, or HO$_2$CCF$_3$) affords $[MnCp(CO)_2(\equiv C\text{-}CH=CR_2)]X$ in 33–93% yields. An x-ray crystal structure where R = Ph and $X^- = BF_4^-$ confirmed the structure.

### 2.7.2 Reactions of Manganese Carbenes and Carbynes

As with chromium carbene complexes, manganese alkylidene complexes have found use in organic synthesis through addition chemistry with alkenes and alkynes, although the chemistry of manganese is far less developed than that of chromium. Methyl or phenyl methoxycarbene complexes (**209**) undergo reaction with enyne substrates (**210**) at elevated temperature to afford bicyclic structures (**211**) (Equation (59)).[196] No evidence of cyclobutanones, furans, or metathesis products was found, as is typically found in the analogous chemistry of $Cr(CO)_5[=C(OMe)Me]$. Enol ether products (**211**) were hydrolyzed to the ketones in wet $CHCl_3$ and isolated in 63–71% overall yields. If acyl anion (**212**) (Scheme 25) is used in reaction with enyne (**213**), ketone (**214**) is isolated, in higher yield than from the same reaction with (**209**). The anion (**212**) also acts as an acyl anion equivalent in reaction with the Michael acceptor $\alpha,\beta$-unsaturated esters, as shown in Scheme 25, to form $\beta$-ketoesters (**215**). 1-Hexyne couples with (**212**) and CO also inserts, ultimately forming lactone (**216**).

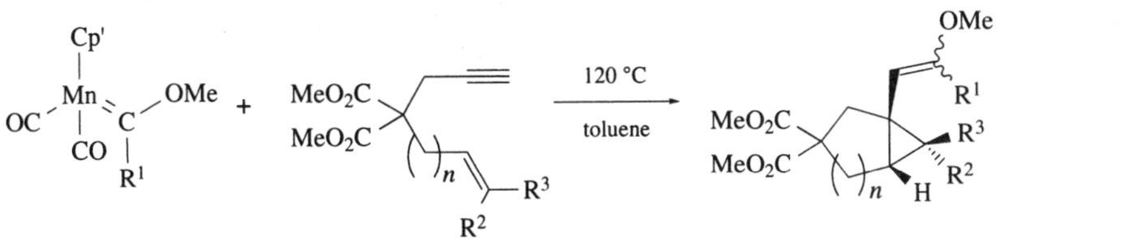

(59)

(**209**) $R^1$ = Me, Ph          (**210**)          (**211**) $n$ = 1,2; $R^2$, $R^3$ = H, Me, Ph

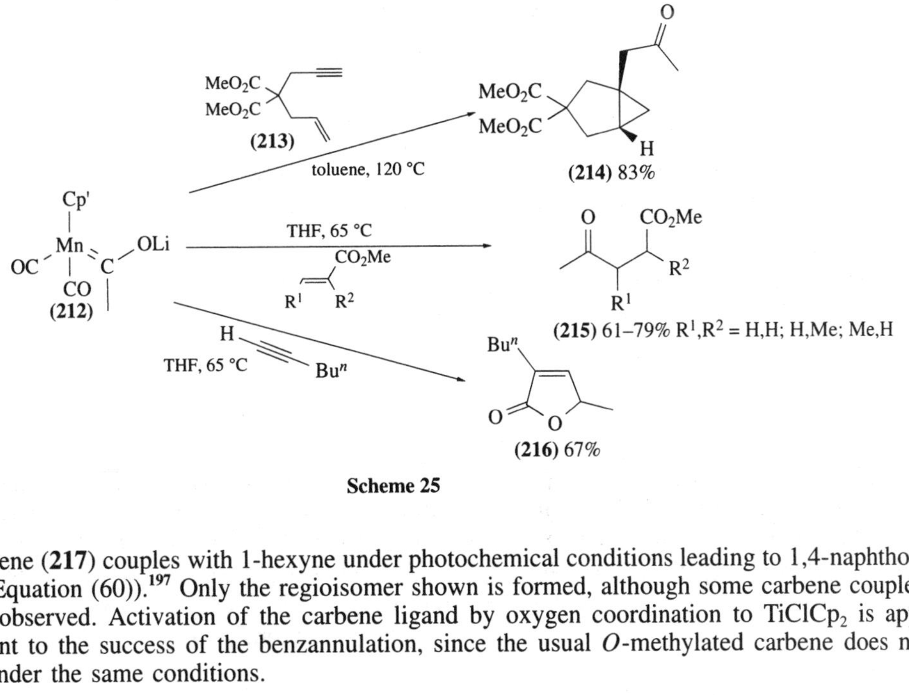

**Scheme 25**

Carbene (**217**) couples with 1-hexyne under photochemical conditions leading to 1,4-naphthoquinone (**218**) (Equation (60)).[197] Only the regioisomer shown is formed, although some carbene coupled (**219**) is also observed. Activation of the carbene ligand by oxygen coordination to $TiClCp_2$ is apparently important to the success of the benzannulation, since the usual *O*-methylated carbene does not yield (**218**) under the same conditions.

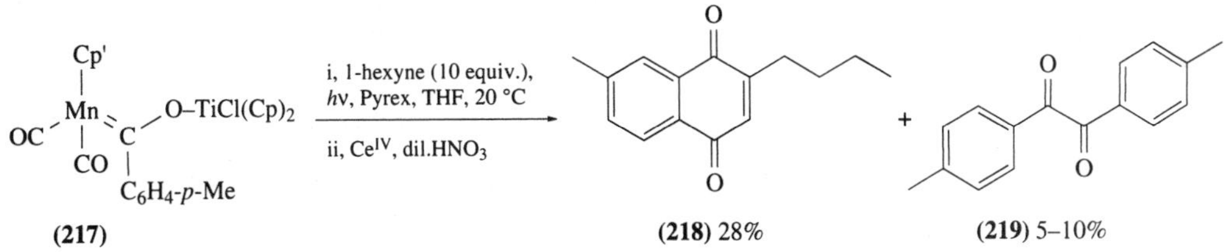

(60)

(**217**)          (**218**) 28%          (**219**) 5–10%

The benzannulation has been carried out intramolecularly by tethering the alkyne within carbene complex (**220**) via the siloxy bridge (Equation (61)).[198] As is general for such carbene annulations, yields are low when $R^3$ = H, but otherwise the intramolecular nature of the reaction greatly enhances the yield over the intermolecular case (Equation (60)).

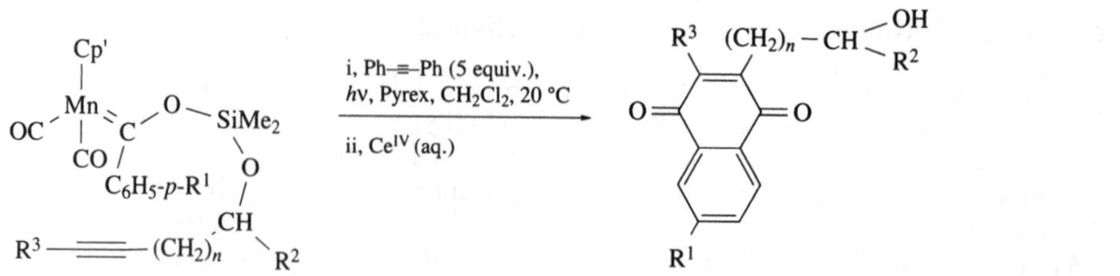
Cp'
Mn
OC
CO
O
SiMe2
C6H5-p-R1
CH
R3
(CH2)n
R2
i, Ph-≡-Ph (5 equiv.),
hv, Pyrex, CH2Cl2, 20 °C
ii, CeIV (aq.)
R3 (CH2)n − CH
OH
R1

(220) $R^1$, $R^2$, $R^3$ = H, Me, Et, etc. $n$ = 0–2

(221) 71% when $R^1$ = Me, $R^2$ = H, $R^3$ = Et, $n$ = 0

(61)

Isonitriles react readily with typical manganese carbenes (**222**) (Scheme 26). Ketene imine complexes (**223**) are initially generated which are stable in solution, but the precise nature of the bonding of the ketene imine to the metal has not been settled. In the presence of excess CNR (**223**) reacts further, but the direction of the reaction depends largely on the nature of the heterosubstituent X. With X = oxygen, cycloaddition of the isonitrile to the carbene is observed, with both possible regioisomers being formed ((**224a**) and (**224b**)).[199] With X = sulfur, simple displacement of the ketene imine ligand (**225**) is found.[200] With aldehydes, ketones, isocyanates, and so on, cycloaddition to the ends of the coordinated ketene imine occurs forming cyclic carbene complex (**226**).[201] Methanol is observed to add, also at each end, and carbene complex (**227**) is generated.[202]

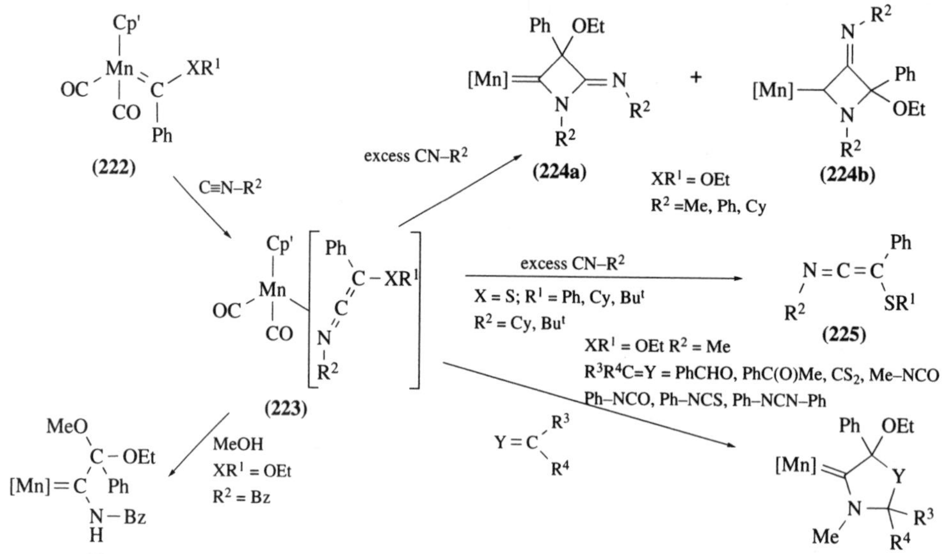
Cp'
Mn
XR1
OC
CO
Ph
(222)
C≡N–R2
excess CN–R2
Ph OEt
[Mn]
N
R2
(224a)
+
(224b)
XR1 = OEt
R2 =Me, Ph, Cy
(223)
excess CN–R2
X = S; R1 = Ph, Cy, But
R2 = Cy, But
N=C=C
SR1
(225)
XR1 = OEt R2 = Me
R3R4C=Y = PhCHO, PhC(O)Me, CS2, Me-NCO
Ph-NCO, Ph-NCS, Ph-NCN-Ph
Y=C
R3
R4
MeOH
XR1 = OEt
R2 = Bz
MeO
C−OEt
[Mn]=C Ph
N−Bz
H
(227)
Me
(226)

**Scheme 26**

Similarly, isonitriles add to vinylidine complexes to generate coordinated cumulenes. For example, $Mn(Cp)(CO)_2(=C=CHPh)$ and CN–R (R = Bu$^t$, Cy, or $CH_2Ph$) react to form $Mn(Cp)(CO)_2(RN=C=C=CHPh)$. This was not isolated, but was characterized by addition of various nucleophiles in overall yields of 50–80%.[203]

Vinylidine (**228**) undergoes a high-yield addition reaction with phosphites under very mild conditions to form an alkene π-complex (**229**) (Equation (62)). That addition occurs at the carbenoid carbon was confirmed, where R = ethyl, by x-ray diffraction.[204] The mechanism was surmised to be related to the Arbuzov reaction.

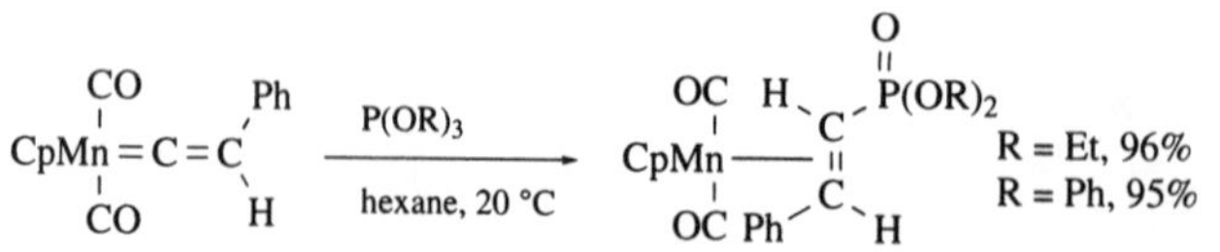
CO
CpMn=C=C
Ph
H
CO
P(OR)3
hexane, 20 °C
O
OC H
P(OR)2
CpMn
OC Ph
H
R = Et, 96%
R = Ph, 95%

(228) (229) (62)

The cumulated carbene complex $[Cp(CO)_2Mn=C=C=CR_2]$ (R = Bu$^t$, Ph) (**230**) undergoes nucleophilic addition at the carbenoid carbon to give the zwitterionic species

[Cp(CO)$_2$MnC(PPh$_3$)=C=CR$_2$] (**231**) which was isolable since no Arbuzov-type reactivity is possible.[205] In addition, MnCp(CO)$_2$(=C=CHR$^1$) (R$^1$ = Ph, CO$_2$Me, or C(Bu$^t$)$_2$OH) gives analogous products on reaction with PPh$_2$R$^2$ (R$^2$ = Ph or Me) in 61–72% yield.

The aminocarbene complex (**232**) has been shown to react with the sequence of reagents shown in Equation (63) to form a benzonitrile complex (**233**) in 43% yield.[206] No mechanism for the transformation was proposed, but an α-elimination of a dithiocarbamate ester seems possible. Compound (**232**) was prepared from Mn(CO)$_5$Mn[=C(OEt)Ph](CO)$_4$ and NH$_3$ in ether in 95% yield.

$$(OC)_5Mn—Mn=C \overset{NH_2}{\underset{Ph}{}} \quad \xrightarrow[\substack{\text{ii, CS}_2 \\ \text{iii, Et}_3O^+BF_4^-}]{\text{i, Bu}^n\text{Li}} \quad (OC)_5Mn—Mn—(NC–Ph) \qquad (63)$$

(**232**)         (**233**)

Alkylidyne cations exhibit as their most characteristic reactivity the addition of nucleophiles to the carbyne carbon. One example of this is shown in Equation (58),[195] and another is shown in Equation (64).[207] In the latter example, carbene complex (**235**) was observed by NMR to have two C(carbene)–N rotational isomers with a barrier to interconversion of greater than 88 kJ mol$^{-1}$. Photolysis expels a CO ligand allowing coordination of the allyl group. Such structures are of interest in the investigation of alkene metathesis mechanisms, and the conformation of (**236**) was carefully examined by x-ray diffraction.

(**234**)      (**235**)      (**236**)     (64)

Advantage can be taken of the electrophilic nature of alkylidyne cations, as shown in Equation (65).[208] The nucleophilic nitrogen of *N*-methylbenzaldimine could add to the electrophilic carbyne to form iminium ion (**237**), which presumably could effect an electrophilic aromatic substitution at the Cp ring to form (**238**). Similarly, dimethyl cyanamide is attacked electrophilically by the cation (**234**) (Equation (66)).[209] In this case, if methyllithium is added at low temperature, the presumed intermediate (**239**) is trapped as carbene complex (**240**). Untrapped, the cation rearranges and attacks the Cp ring forming (**241**).

(**234**)      (**237**)      (**238**) 50%     (65)

(**234**)      (**239**)      (**240**) NMe$_2$      (**241**) 50%     (66)

Oxidation of the tolyl carbyne cation analogue of (**234**), namely $[Mn(Cp)(CO)_2(\equiv C–C_6H_4\text{-}p\text{-}Me)]^+$, by $[N(PPh_3)_2]^+$ $NO_2^-$ leads to $Mn(Cp)(CO)(NO)[C(O)–C(O)–C_6H_4\text{-}p\text{-}Me)]$.[210] Presumably, $NO_2^-$ transfers oxygen to the carbyne carbon and incorporates the NO into the coordination sphere of the manganese. At some point, a highly unusual CO insertion is induced into the acyl–manganese bond.

Photochemically generated $MnCp'(CO)_2(THF)$ reacts thermally with $SiHClBu^t_2$ to form the bis(carbyne)-bridged dimer (Equation (67)).[211] An x-ray crystal structure reveals that the Cp' rings are *cis*, and all four Mn–C(carbyne) bonds are equivalent. Thermogravimetric analysis of solid (**243**) shows weight loss over the temperature range 78–200 °C, and mass spectroscopy identifies the carbyne dimer (**244**) which is formed in low yield.

$$MnCp'(CO)_2(THF) \xrightarrow[\text{THF}]{SiHClBu^t_2} \text{(243)} \xrightarrow{80\text{–}200\,°C} Bu^t_2HSi-O\!\equiv\!\!\equiv\!\!\equiv\!O-SiHBu^t_2 \quad \textbf{(244)} \tag{67}$$

Several x-ray crystal structure determinations of manganese–carbene complexes have found that the dihedral angle between the plane of the carbene and its two substituents and the plane of the Cp ring is generally 80–90°, although there are exceptions.[212] Fenske–Hall SCF MO calculations reproduce this experimental observation with both "Fischer carbenes" (e.g., $MnCp(CO)_2[=C(OMe)Ph]$) and "Schrock carbenes" (e.g., $MnCp(CO)_2(=CMe_2)$). The conformational preference, however, had no simple explanation, but depended on the cumulative effect of several molecular orbitals.[213] In addition, Hartree–Fock–Roothaan, Fenske–Hall SCF MO calculations suggest that the Mn–C carbene bond order in $MnCp(CO)_2(=C=CH_2)$ is between two and three. The high bond order implies that the rotational barrier should be small, and again, this reproduces the experimental observation for manganese vinylidenes.[214] Extended Hückel MO theory is consistent with the above results for $MnCp(CO)_2(=C=CH_2)$ and further suggests that the terminal methylene carbon should be the preferred site for attack by electrophiles, as has been found experimentally.[215]

## 2.8 MANGANESE HYDRIDE COMPLEXES

### 2.8.1 Synthesis of Manganese Hydrides

The synthetic methodology for the preparation of manganese hydrides has not changed significantly since the publication of *COMC-I*.[1] The principal method is protonation of various anionic manganese complexes. In recent years, this method has been extended to dianions and even trianions. $MnCp(CO)_2(py)$ was reduced with sodium naphthalene in THF at −78 °C to form $[MnCp(CO)_2]^{2-}$ which was isolated in 55% yield as the tetraethylammonium salt. The dianion was then treated with water to give $[MnHCp(CO)_2]^-$. The monohydride anion, in turn, could be protonated again with $NH_4PF_6$ affording $MnH_2Cp(CO)_2$, or could be treated with $PbClPh_3$ giving $MnHCp(CO)_2(PbPh_3)$.[216] Ellis and co-workers[217] have prepared the remarkable trianion of manganese $Na_3[Mn(CO)_4]$ by reduction of $Na[Mn(CO)_5]$ with sodium in HMPA solvent. It is noteworthy that direct reduction of $Mn_2(CO)_{10}$ under the same conditions did not give a clean reaction. Addition of a large excess of liquid ammonia to the HMPA solution caused precipitation of tan, unsolvated, analytically pure $Na_3[Mn(CO)_4]$ in 98% yield. This solid salt is remarkably stable, decomposing above 320 °C. By various protonation and coupling reactions, solutions of $Na_2[MnH(CO)_4]$ and $Na[MnH_2(CO)_4]$, solid $(AsPh_4)[MnH_2(CO)_4]$ (75% yield, slow decomposition at room temperature), and solutions of $(NEt_4)[MnH(CO)_4(SnPh_3)]$ were all prepared.

Transient generation of unsaturated, formally zero-valent manganese species can lead to hydride formation. For example, reaction of manganese vapor with a mixture of benzene and $PMe_3$ gives $MnH(\eta^6\text{-}C_6H_6)(PMe_3)_2$.[218] In addition, thermolysis of the presumably rigid and crowded chelating diphosphine (**245**) with $Mn_2(CO)_{10}$ in toluene leads to hydridic material (**246**) (Equation (68)), for which an x-ray structure has been reported.[219]

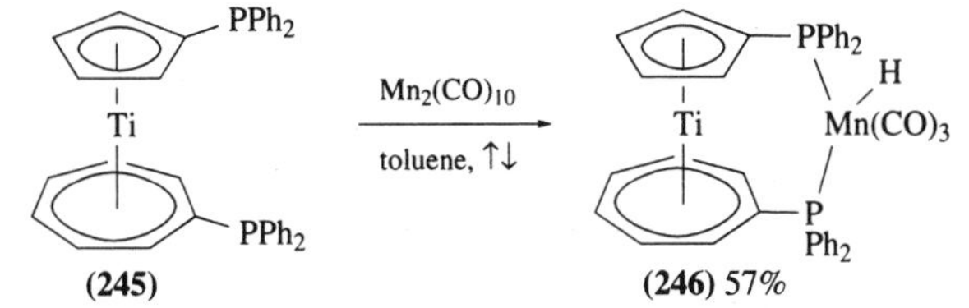

(68)

(245)     (246) 57%

Protonation of neutral complexes generates cationic species which are formally regarded as hydrides. A recent example is protonation of (**247**) as shown in Scheme 27.[218] In addition, it has been reported that (**247**) could be prepared from metal vapor (as shown), in quantities of over 8 g in 4 h. Cationic hydride (**248**) could be deprotonated back to (**247**), as shown. In addition, the Mn[III] dihydride (**249**) could be prepared by simple addition of $H_2$ to (**247**).

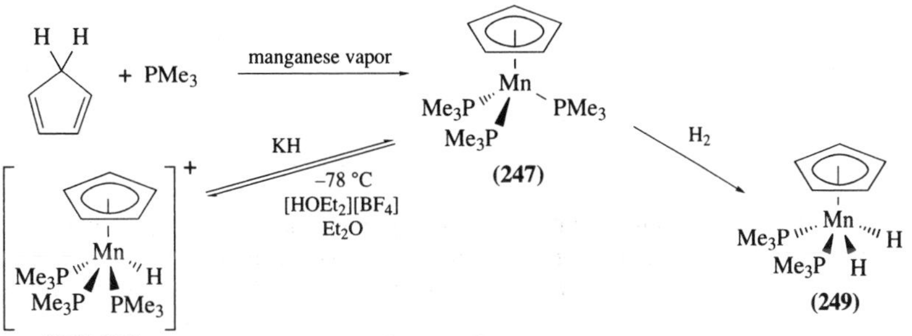

(248) 82%     **Scheme 27**

Assorted reductions of manganese cations and dimers by electrochemical means have afforded manganese hydrides. When $[Mn(S)_3(CO)_3]^+$ (S = MeCN) is electrochemically reduced in MeCN solvent in the presence of $PPh_3$ or $PMe_2Ph$, the disubstituted hydride $MnH(CO)_3L_2$ is formed exclusively. Cyclic voltammetry and coulometry showed that reduction first forms a catalytic amount of the very substitution-labile, 19-electron intermediate $[Mn(S)_3(CO)_3]\bullet$. This chain-carrying intermediate then undergoes rapid ligand exchange to form $[Mn(S)(CO)_3L_2]\bullet$, which in turn transfers an electron to $[Mn(S)_3(CO)_3]^+$ to produce $[Mn(S)(CO)_3L_2]^+$ and to regenerate $[Mn(S)_3(CO)_3]\bullet$. After complete conversion to $[Mn(S)(CO)_3L_2]^+$ has occurred, stoichiometric reduction to $[Mn(S)(CO)_3L_2]\bullet$ followed by hydrogen abstraction from the solvent produces $MnH(CO)_3L_2$.[220] At −30 °C the cyclic voltammagram of formyl complex $Mn(S)(CO)_2(PPh_3)_2(CHO)$ grew in.[221] Raising the temperature to ambient led to rapid electrocatalytic conversion of the formyl to $MnH(CO)_3L_2$. In a related study, electrochemical reduction of $[Mn(CO)_5L]^+$ in THF generated $MnH(CO)_4L$.[222] It is believed that the hydrogen abstracting species is 19-electron $[Mn(CO)_5L]\bullet$ and not 17-electron $[Mn(CO)_4L]\bullet$, as electrooxidation of $[Mn(CO)_4L]^-$ led only to dimer $Mn_2(CO)_8L_2$ and no hydride.

Use of main-group hydridic reagents to prepare manganese hydrides is still uncommon. One example is the $NaBH_4$ reduction of manganese iodide $[Mn(C_6Me_6)(CO)_2I]$ (**250**) to hydride $[Mn(C_6Me_6)(CO)_2H]$ (**251**).[223]

Tin hydrides reduce manganese complexes to hydrides. These reductions can proceed by way of radical mechanisms (Equation (69)),[224] or by what are believed to be oxidative-addition and reductive-elimination sequences. An example of the latter process is shown in Scheme 28. Photolysis of $Mn_2(CO)_8L_2$ (L = CO, $PMe_3$, $PBu^n_3$, $PPr^i_3$, or $PCy_3$) in hexane solution under CO atmosphere in the presence of $SnHBu^n_3$ affords a mixture of $MnH(CO)_4L$ and $Mn(SnR_3)(CO)_4L$ (**252**). At low CO pressures or under argon atmosphere, the tin-containing product is $MnH(SnR_3)_2(CO)_3L$ (**253**). The mechanism shown in Scheme 28 was deduced from continuous and flash-photolysis studies.[225]

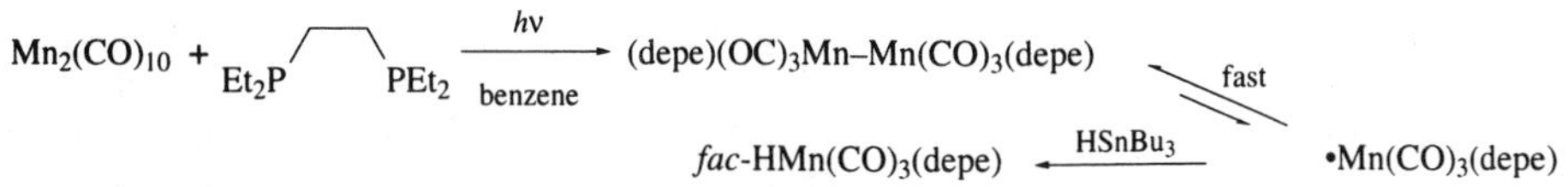

(69)

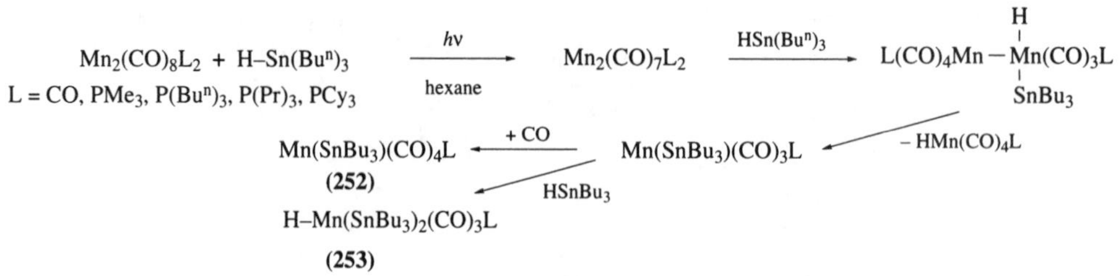

**Scheme 28**

## 2.8.2 Reactions of Manganese Hydrides

Although many manganese hydrides are prepared by protonation of metal anions, occasionally the hydride is more directly available than the anion, and so deprotonation of the hydride can provide the most convenient route to the anion. Thus, hydride (251) can be deprotonated by methyllithium in THF to afford $Li[Mn(\eta^6\text{-}C_6Me_6)(CO)_2]$ (254) (Scheme 29).[64,226] In the example shown, formic acetic anhydride was used to generate a formyl group on manganese, but formyl (255) was not observed. Instead, the reduced arene *exo*-(256) was found. This hydrogen-atom transfer is thought to occur directly from the formyl group. In contrast, hydride (251) itself under CO atmosphere exhibits hydrogen-atom transfer to the arene, but to the *exo* face to afford *endo*-(256). The mechanism was not explored.[226]

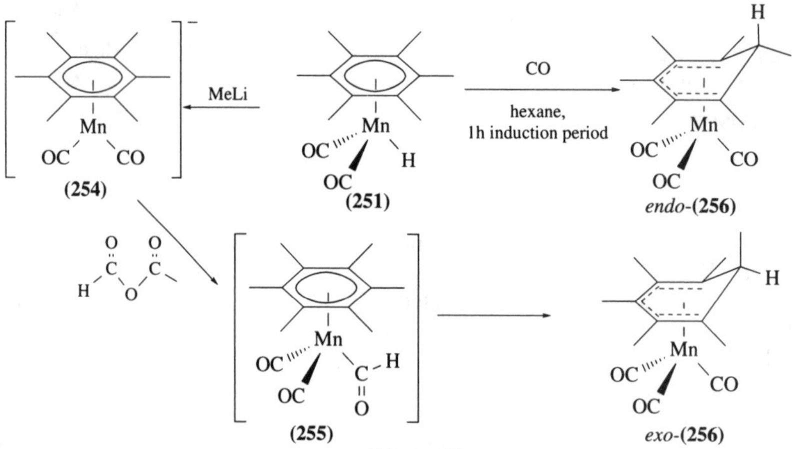

**Scheme 29**

It has previously been observed that manganese hydrides can be used to reduce conjugated alkenes and polynuclear arenes.[227] These reactions are believed to proceed via hydrogen-atom transfer to form solvent-caged radical pairs which may combine or diffuse apart (Scheme 30).[228] In this case, internal return of cage radical pair (257) is believed to lead to aldehyde (258), and separation of the radicals by diffusion leads to reduction to alkane (259). On the other hand, reactions of conjugated dienes with $MnH(CO)_5$ proceed as in Equation (70).[229] The reaction is second order overall, and shows no inhibition by added CO. A radical path essentially the same as that shown in Scheme 30 was proposed.

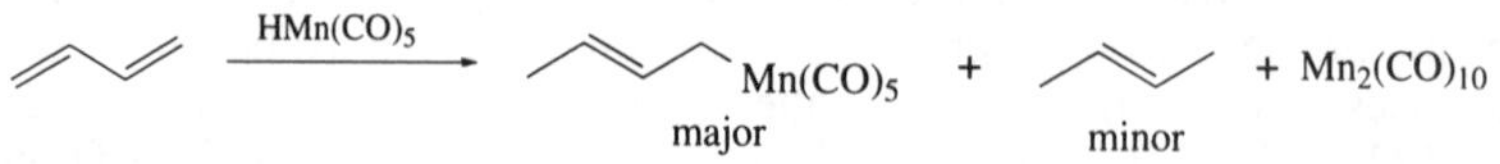

The substrate that is most reactive toward hydrogen-atom transfer from $MnH(CO)_5$ found to date is tetramethylallene (260) (Scheme 31).[230] The organic radical in the radical pair (261) is the same as that from H· transfer to pentadiene (262) except that the former addition is not reversible and the latter is. Thus, at low $MnH(CO)_5$ concentration, allene (260) is irreversibly isomerized to the diene.

Various other reactions of manganese hydride have been reported. $MnH(CO)_5$ undergoes rapid exchange with other metal carbonyl dimers under mild conditions, as in the case of $Co_2(CO)_8$ and $MnH(CO)_5$ reacting to form $CoH(CO)_4$ and $MnCo(CO)_9$ at 25 °C.[231] Reduction of metal alkyls and acyls is a common reaction of metal hydrides. For examples of manganese hydrides trapping metal alkyls and acyls, see Sections 2.3.2.2, 2.3.2.3(i), and 2.8.3.1.[232] Manganese hydrides can serve as catalysts for the

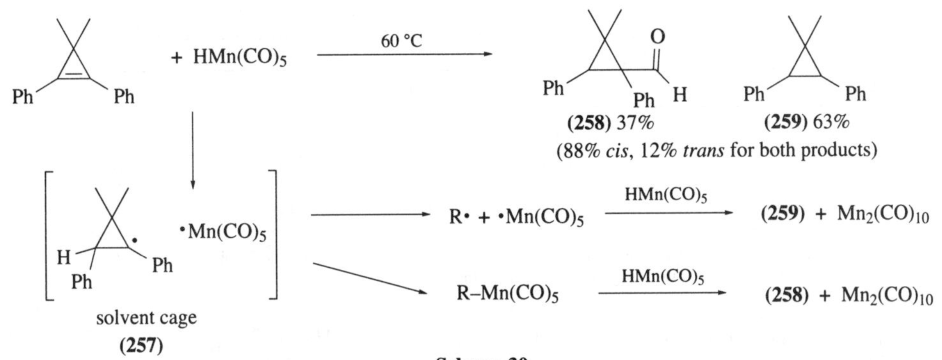

**Scheme 30**

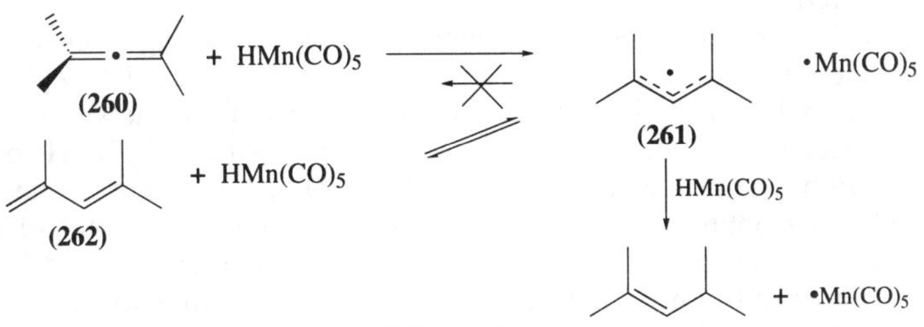

**Scheme 31**

isomerization of α-alkenes, and in the presence of an $H_2$ atmosphere, a species such as *cis*-$MnH(CO)_4(PPh_3)$ catalyzes competitive isomerization and hydrogenation of such alkenes.[233] Intramolecular addition of MnH to aldehyde groups was described in Section 2.3.1.4.[234]

Finally, it is noted that hydrides can be abstracted by strong Lewis acids. Equation (71) shows abstraction from (263) by $Ph_3C^+$.[235] Intramolecular coordination at the newly vacated site affords bis-chelated *cis*-(264), which isomerizes to the *trans* isomer over a matter of hours.

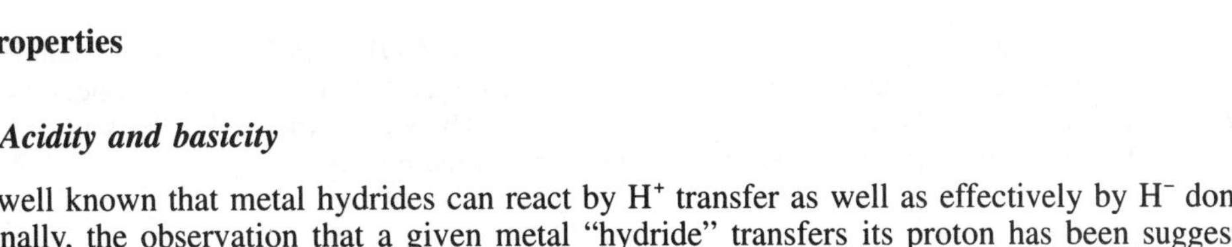

$$(71)$$

### 2.8.3 Properties

#### 2.8.3.1 Acidity and basicity

It is well known that metal hydrides can react by $H^+$ transfer as well as effectively by $H^-$ donation. Occasionally, the observation that a given metal "hydride" transfers its proton has been suggested to imply that the bond is polarized $M^{(\delta-)}-H^{(\delta+)}$. However, acidic behavior does not require a specific direction for the M–H dipole. This is supported by the results of energy-factored force-field calculations in which $MnH(CO)_5$ and $[Mn(CO)_5]^-$ were compared. It was concluded that the manganese hydride bond has the polarization $Mn^{(\delta+)}-H^{(\delta-)}$.[236] Incidentally, the same conclusion was drawn for $CoH(CO)_4$, $FeH_2(CO)_4$, and $ReH(CO)_5$.

The $pK_a$ values of a number of transition metal hydrides in acetonitrile solvent at 25 °C have been measured by Norton and co-workers.[237,238] Values for several manganese complexes, and a few others for comparison, are given in Table 6. The $pK_a$ values in aqueous solution have been measured or estimated for several of the hydrides considered, and in those cases, the $pK_a$ values were 7 or 8 units more positive in MeCN than in water.

Ion cyclotron resonance (ICR) studies of gas-phase acidity show proton affinities (of the conjugate base of the acid shown) as follows: $FeH_2(CO)_4$, $1335 \pm 21$; $MnH(CO)_5$, $1330 \pm 17$; and $CoH(CO)_4$, $\leq 1314 \pm 4$ kJ mol$^{-1}$.[239] From these numbers and estimates of the M–H bond dissociation enthalpies, the

**Table 6**  $pK_a$ values of metal hydrides determined in MeCN at 25 °C.[a]

| Metal hydride | $pK_a$ | Metal hydride | $pK_a$ |
|---|---|---|---|
| $CoH(CO)_4$ | 8.3 | $ReH(CO)_5$ | 21.1 |
| $FeH_2(CO)_4$ | 11.4 | $MnH(CO)_4(PEtPh_2)$ | 21.6 |
| $MnH(CO)_5$ | 15.1 | $Mn(\eta^3\text{-}C_6H_9)(CO)_2{}^b$ | 22.2 |
| $MnH(CO)_4(PPh_3)$ | 20.4 | $Mn(\eta\text{-}C_6H_6)H(CO)_2$ | 26.8 |

[a] See Refs. 237 and 238. Uncertainties in the least-significant digit were typically 1 or 2 units.  [b] The acidic hydrogen of $\eta^3\text{-}C_6H_9$ is allylic and agostically bound to the metal.

electron affinities of the corresponding radicals were estimated: $[FeH(CO)_4]\cdot$, $\leq 2.8 \pm 0.8$; $[Mn(CO)_5]\cdot$, $2.4 \pm 0.3$; and $[Co(CO)_4]\cdot$, $\leq 2.3$ eV. Additional studies[240] have expanded the scale of qualitative gas-phase acidities:    $HCl < ReH(CO)_5 < CHF_2CO_2H < HBr < CCl_3CO_2H < MnH(CO)_4 < HI < MnH(CO)_3(PF_3)_2$, $MnH(CO)_2(PF_3)_3 < CF_3SO_3H < CoH(CO)_4$.

Since the 1980s, mechanisms involving metal hydride complexes acting as nucleophiles toward other unsaturated metal species have been proposed. This concept has substantial implications for catalytic processes, such as the hydrogenolysis step in the hydroformylation shown in Equation (72). The unsaturated acyl intermediate may undergo attack by either $H_2$ (path a) or M–H (path b). In the latter case, the M–H bond can be viewed as a Lewis base coordinating to the unsaturated intermediate. One scale of metal hydride nucleophilicity has been established by measuring rates of reaction of various hydrides with solvated rhenium propionyl complex (**266(S)**) (Equation (73)).[232] Reaction of (**266(S)**) with MH occurs at a conveniently measurable rate, but is much faster than the initial insertion–solvation equilibrium between (**265**) and (**266(S)**). Convincing evidence was presented that trapping by MH is via a dissociative mechanism involving an intermediate, presumably (**266**), and that the measured rate of trapping is $k_{MH}$. A partial list of relative reactivities (nucleophilicities) is: $ReH(CO)_5$, 139; $MnH(CO)_5$, 100; $OsH_2(CO)_4$, 58; and $WHCp(CO)_3$, 9 (all with respect to $CrHCp(CO)_3 = 1.0$). These nucleophilicities are very roughly in reverse order with respect to kinetic acidity of the metal hydrides. For additional examples of manganese hydrides trapping metal acyls, see Sections 2.3.2.2 and 2.3.2.3(i).

$$M\text{–}H + R\text{–}\overset{\displaystyle O}{\underset{\displaystyle H}{C}} \xleftarrow[\;\;a\;\;]{H_2} M\text{–}\overset{\displaystyle O}{\underset{\displaystyle R}{C}} \xrightarrow[\;\;b\;\;]{M\text{–}H} R\text{–}\overset{\displaystyle O}{\underset{\displaystyle H}{C}} + M\text{–}M \qquad (72)$$

$$(OC)_5Re\text{–}Et \underset{\text{slow}}{\rightleftharpoons} (OC)_4Re\overset{\displaystyle \overset{O}{\|}}{\underset{\displaystyle (S)}{\overset{C\!\sim\! Et}{}}} \underset{\text{fast}}{\rightleftharpoons} \left[ (OC)_4Re\text{–}\overset{\displaystyle O}{\underset{\displaystyle Et}{C}} \right] \xrightarrow[k_{MH}]{M\text{–}H} Et\text{–}\overset{\displaystyle O}{\underset{\displaystyle H}{C}} + M\text{–}Re(CO)_4(S) \qquad (73)$$

(**265**)                         (**266(S)**)                         (**266**)

(S) = MeCN

SCF-X$\alpha$ scattered wave calculations[241] on $MnH(CO)_5$, $FeH_2(CO)_4$, and $CoH(CO)_4$ conclude that the HOMOs of these complexes are the metal $d$-orbitals and not the M–H $\sigma$-bonds. In connection with the preceding discussion of M–H bonds as bases, naturally the HOMO of the molecule is not required to be the site of greatest basicity, either kinetically or thermodynamically.

### 2.8.3.2  Thermochemistry

The hydrogenolysis equilibrium between $[Mn_2(CO)_{10}]$ and $[Mn(CO)_5H]$ has been studied in supercritical $CO_2$ under 142–300 atm pressure at 165–200 °C.[242] The measurements were made using high-pressure NMR techniques monitoring $^{55}Mn$ and $^1H$ nuclei, and yielded values of $\Delta S° = 36 \pm 3$ J K$^{-1}$ and $\Delta H° = 36 \pm 1$ kJ mol$^{-1}$. The approach to equilibrium at each temperature was substantially accelerated by the presence of catalytic $Co_2(CO)_8$, but the equilibrium concentrations were unaffected.

Ignoring solvation effects, M–H acidity depends on the BDE and on the electron affinity of M·. Good values for either of these quantities are known for relatively few species in organometallic systems. Some relative BDEs are given in Table 3 (Section 2.4.2),[106] but the absolute numbers are uncertain because the reference enthalpy, the BDE of the Mn–Mn bond in $Mn_2(CO)_{10}$, is uncertain. The Mn–H BDE listed in Table 3 is 213 kJ mol$^{-1}$, but there are several suggestions that this value is too low. Using the $pK_a$ values for acetonitrile (Table 6) and electrochemical oxidation potentials of the conjugate bases, the bond dissociation energies of $MnH(CO)_5$ and $MnH(CO)_4(PPh_3)$ were estimated to be 285 and 290 kJ mol$^{-1}$, respectively.[243] In addition, an estimate of 264 kJ mol$^{-1}$ for the BDE of $MnH(CO)_5$ has been made from kinetic measurements on hydrogen transfers between anthracene and dihydroanthacene, compared with transfer from $MnH(CO)_5$ to anthracene.[244] An additional observation can be made as follows. Assuming that heat capacity, pressure, and phase corrections are relatively small for the above-mentioned $\Delta H_{rxn}$ for the equilibrium between $[Mn_2(CO)_{10}]$ and $[Mn(CO)_5H]$,[242] 36 kJ mol$^{-1}$ should be a reasonable approximation of the $\Delta H°_{rxn}$ value for that reaction at 25 °C. If, in addition, it is assumed that the relative BDEs in Table 3 are accurate (i.e., BDE(Mn–H) – BDE(Mn–Mn) = 119 kJ mol$^{-1}$), then it can be calculated that the Mn–H BDE must be near 276 kJ mol$^{-1}$, and the Mn–Mn BDE must be near 155 kJ mol$^{-1}$. Thus, three estimates put the value of the BDE of Mn–H in $MnH(CO)_5$ at 264, 276, and 284 kJ mol$^{-1}$.

### 2.8.3.3 Other physical and theoretical studies

Fenske–Hall calculations[245] on M–H complexes find a correlation between acidity, which is most closely related to M$^-$ stability, and the HOMO–LUMO gap of the conjugate base. Both $MnH(CO)_5$ and $MnH(CO)_4(PH_3)$ were included in these calculations. Inner-shell excitation of $Mn_2(CO)_{10}$, $MnBr(CO)_5$, and $MnH(CO)_5$ has been studied by electron-energy-loss spectroscopy.[246] *Ab initio* valence bond calculations with configuration interaction have been carried out on diatomic cation hydrides MH$^+$ for all the metals from calcium to zinc.[247] For MnH$^+$, the percentages of bonding of the hydride to the metal orbitals were: $s$, 76%; $p$, 13%; and $d$, 11%. This is a sharp break from CrH$^+$ ($s$, 41%; $p$, 13%; $d$, 47%) and earlier metals, because MnH$^+$ is $d^5$.

Photolysis of $MnCp(CO)_3$ in the presence of $H_2$ in supercritical xenon solvent at 25 °C led to observation of a material postulated to be the molecular dihydrogen complex $MnCp(H_2)(CO)_2$.[248] The assignment was made by comparing the IR spectra of the dihydrogen complex with those of $MnCp(N_2)(CO)_2$, *trans*-$ReH_2Cp(CO)_2$, and $ReCp(N_2)(CO)_2$. The primary product from photolysis of $MnH(CO)_5$ in argon or methane matrices at 20 K is the solvent-coordinated molecule of $C_s$ symmetry (**267**(S)).[249] Solvent coordination was deduced from the 2500 cm$^{-1}$ red shift in the UV–visible on moving from methane to argon. Photolysis of (**267**(S)) at 367 nm in methane or at 403 nm in argon yielded the $C_{4v}$ isomer (**268**(S)). These results are parallel to those of the matrix photolysis of $MnMe(CO)_5$ (see Section 2.4.4.1). Prolonged photolysis of $MnH(CO)_5$ at 193 nm in an argon matrix generated some $[Mn(CO)_5]·$, but the quantum yield was very low. Hartree–Fock–Roothaan calculations have been carried out on the saturated species $Mn(X)(CO)_5$ and the corresponding unsaturated species $Mn(X)(CO)_4$ (X = H, Cl, NH$_3$, or PH$_3$).[250] The results of the calculations on the hydrides are in good agreement with the matrix photolysis results just mentioned, with the $C_s$ structure (**267**) being 13 kJ mol$^{-1}$ below the $C_{4v}$ structure (**268**) in energy.

(267)          (267)(S)          (268)          (268)(S)

## 2.8.4 Silane Complexes of Manganese

In 1971, Hart-Davis and Graham[251] first prepared $Mn(H)(SiR_3)Cp(CO)_2$ (R = Cl or Ph) and noted that $SiHCl_3$ is more tightly bound to manganese than $SiHPh_3$. They postulated a resonance hybrid between canonical structures (**270**) and (**271**) (Equation (74)). Since the 1980s, activity regarding "σ-complexes" of alkanes and their group 14 analogues has increased substantially. Manganese shows an affinity for such interaction with Si–H bonds in particular, and has given rise to a rich diversity of structures ranging from σ-complexed structures like (**270**) to species that have undergone complete oxidative addition of the Si–H bond to form a formal Mn$^{III}$ complex like (**271**) at the other extreme. An authoritative review has appeared.[252] Points of particular focus have included the extent of Si–H bond breaking as a function of the substituents on the silane, the ancillary ligands, and the cyclopentadienyl substituents, and, more recently, the energetics of the Mn–(Si–H) interaction as a function of Si–H bond order.

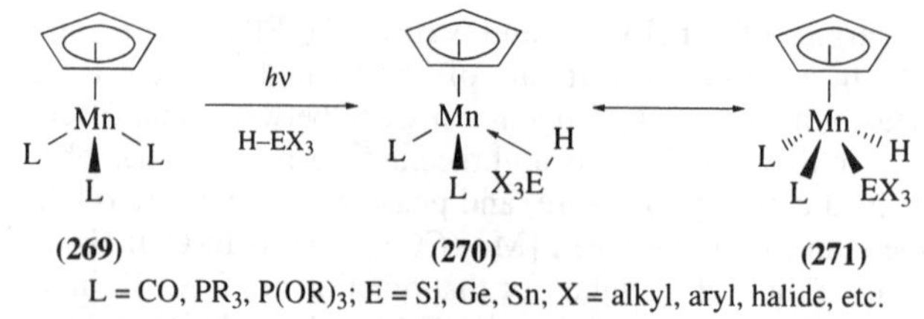

(74)

**(269)**                    **(270)**                    **(271)**

L = CO, PR$_3$, P(OR)$_3$; E = Si, Ge, Sn; X = alkyl, aryl, halide, etc.

Yang and co-workers[253] have used photoacoustic calorimetry and variable-temperature kinetic studies to examine the energy surface for the substitution of seven different silanes into MnCp(CO)$_2$(S) (S = heptane). The $\Delta H_{rxn}$ for formation of Mn(H)(SiR$_3$)Cp(CO)$_2$ ranged from −54 to −92 kJ mol$^{-1}$, with the value depending on the cone angle of the SiR$_3$ group at or above a threshold value of 135°. Rate constants for silane coordination were all about $2.5 \times 10^6$, the $\Delta H^{\ddagger}$ for SiHEt$_3$ was 115 kJ mol$^{-1}$, and $\Delta S^{\ddagger}$ was 48 J mol$^{-1}$ K$^{-1}$. An estimate of 35 kJ mol$^{-1}$ was made for the manganese–heptane interaction.

Valence photoelectron spectra have been recorded for a wide variety of species ((**270**) and (**271**), Equation (74)), including those with Cp, Cp', and Cp* ligands.[254-7] Ionization energies of the metal and ligands were found to correlate with the extent of the breaking of the E–H bond (E = Si or Ge). The findings parallel the trends known from structural and reactivity studies: the more electron donating the ligand L or the Cp ligand (i.e., Cp < Cp' < Cp*), the tighter the ligand-to-metal binding and the greater the extent of E–H bond breaking. This bond breaking was by far most sensitive to the groups X on silicon or germanium, with electronegative groups causing a greater extent of breaking (oxidative addition).

### 2.8.5 Di- and Trimanganese Hydrides

A large number of dimanganese hydrides have been prepared, and these are gathered in Table 7. Most of the recent molecules belong to a group containing one or two bridging phosphorus ligands, usually bis(diphenylphosphino)methane or tetraethyldiphosphite. Preparations are varied, with the source of hydride being most commonly either one of the complex metal hydrides (NaBH$_4$, Li[BHEt$_3$], etc.) or a strong acid. The source of hydride for those species containing phosphido bridges is frequently the P–H group of a starting phosphine. Compounds (**279a**) and (**279b**) (entry 13, Table 7), assigned the side-on bridged CO (**279a**) or isonitrile (**279b**) structures shown, are stable only below −60 °C.[265] Their unprotonated neutral precursors also have a single sideways-bridged ligand, which is fluxional with regard to side-to-side oscillation. It was found by means of NMR that the barrier for this oscillation was reduced by more than 39 kJ mol$^{-1}$ on protonation, consistent with the dominant interaction being donation of metal electrons to the side-on ligand rather than ligand to metal donation.

Doubly proton-bridged dimers such as (**272**) (entry 1, Table 7) and (**273**) (entry 5, Table 7) behave as though they are unsaturated, and so have a rich reaction chemistry. It should be noted that the metal–metal double bond in these species is not to be taken literally. The convention is that the M–M and M–H–M designations together represent one three-center, two-electron bond. Semiempirical extended Hückel calculations and parameter-free Fenske–Hall calculations were both used in an investigation of (**284**), with the result that practically no Mn–Mn bonding was found.[272] Nevertheless, such hydride bridged species do show facile reactivity. For example, (μ-H)$_2$-(**272a**) (entry 1, Table 7) reacts with BH$_3$·THF or SiH$_2$Ph$_2$ to afford electron-deficient hydride-bridged species (**280**) (entry 14) or (**281**) (entry 15), respectively.[266] There are numerous examples of addition of bridging hydrides to various types of organic unsaturation. Alkynes RC≡CR (R = H, CF$_3$, or Ph) readily insert into the Mn–H bond of (**282**) (entry 16, Table 7; L = CO) to form complexes (**285**) in yields of 46–99%.[267] In addition, dihydride (**273**) (entry 5, Table 7) inserts acetaldehyde or methylisonitrile generating μ-ethoxide complex (**274**) (entry 6) or bridged imine (**275**) (entry 7), respectively.[261] Ethyne readily inserts into the Mn–H bond in (**272a**) (entry 1, Table 7) to form the bridged vinyl species (**276**) (entry 8), and Equation (75) illustrates the insertion of *t*-butylacetylene into (**272a**) to form (**277**) and (**286**).[258]

An interesting example of hydride reactivity is shown in Equation (76). Carbon dioxide inserts into a Mn–H group of (**272b**) ((**272a**) shows parallel chemistry) to afford a mixture of CO$_2$ adduct (**287**) and μ-hydroxo complex (**278b**). The mixture could not be separated, so (**287**) could only be characterized spectroscopically. Chromatography gave >50% yield of hydroxide (**278b**), and so conversion of (**287**) to (**278b**) was induced by the column.[264]

Dinuclear hydrides readily undergo substitution of neutral ligands to provide series of analogous complexes. For example, heating phosphido bridged dimer (**283a**) (entry 18, Table 7) in the presence of

**Table 7** Binuclear and trinuclear manganese hydrides.

| Entry | Product | Yield (%) | Ref. |
|---|---|---|---|

1  x ray

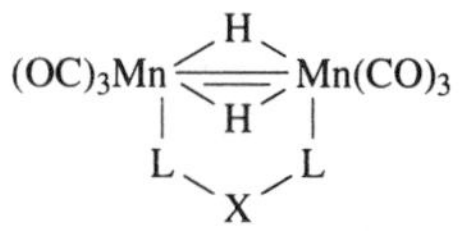

$(OC)_3Mn$ ⟨ H / H ⟩ $Mn(CO)_3$, L–X–L

**(272a)** L–X–L = $Ph_2P–CH_2–PPh_2$ (dppm) — 55 — 258
**(272b)** L–X–L = $(EtO)_2P–O–P(OEt_2)_2$ (pop) — nr[a] — 259

2  x ray

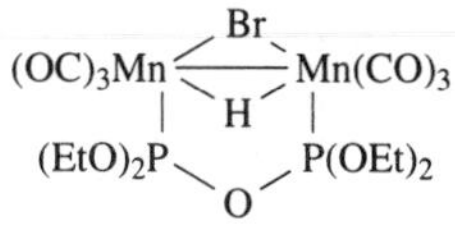

$(OC)_3Mn$ ⟨ Br / H ⟩ $Mn(CO)_3$, $(EtO)_2P$–O–$P(OEt)_2$ — 84 — 260

3  x ray

L–X–L / Br H / L–X–L with $(OC)_2Mn$⟨⟩$Mn(CO)_2$

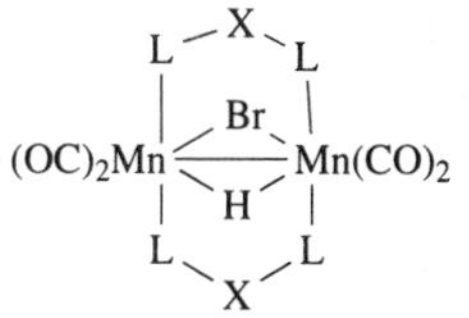

L–X–L = dppm — 62 — 261
L–X–L = pop — 80 — 260

4

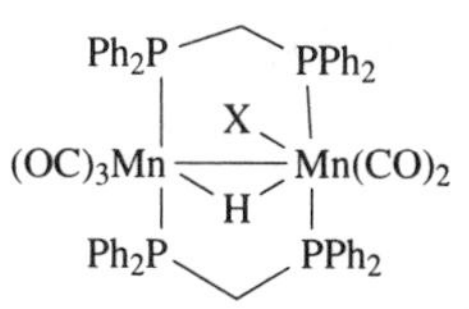

$Ph_2P$...$PPh_2$ / X H / $Ph_2P$...$PPh_2$, $(OC)_3Mn$⟨⟩$Mn(CO)_2$

X = Br — 87 — 261
X = Cl — 54 — 262

5

$Ph_2P$...$PPh_2$ / H H / $Ph_2P$...$PPh_2$, $(OC)_2Mn$⟨⟩$Mn(CO)_2$ — 80–90 — 261

**(273)**

6

$Ph_2P$ Et $PPh_2$ / O H / $Ph_2P$...$PPh_2$, $(OC)_2Mn$⟨⟩$Mn(CO)_2$ — nr — 261

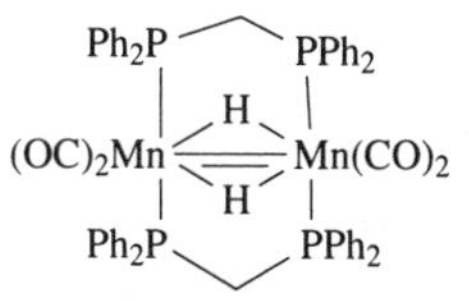

**(274)**

7

H Me / N / $Ph_2P$...$PPh_2$ / H / $Ph_2P$...$PPh_2$, $(OC)_2Mn$⟨⟩$Mn(CO)_2$ — nr — 261

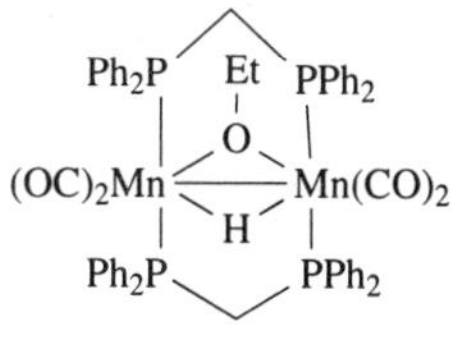

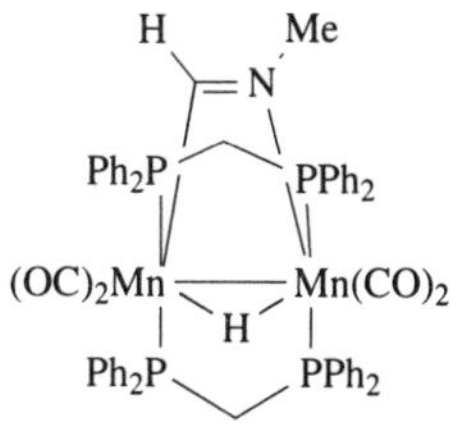

**(275)**

**Table 7**  (continued)

| Entry | Product | Yield (%) | Ref. |
|-------|---------|-----------|------|
| 8 | (276) | nr | 258 |
| 9 | (277) | nr | 258 |
| 10 | $PR_2 = PH_2$, PHCy <br> L–X–L = dppm, pop | 68–76 | 263 |
| 11 | (278a) L–X–L = dppm <br> (278b) L–X–L = pop | >50 | 264 |
| 12 | | 71 | 262 |
| 13 | (279a) E = O <br> (279b) E = $NC_6H_4$-*p*-Me | nr | 265 |
| 14 | (280) | nr | 266 |

**Table 7** (continued)

| Entry | | Product | Yield (%) | Ref. |
|-------|---|---------|-----------|------|
| 15 | x ray | 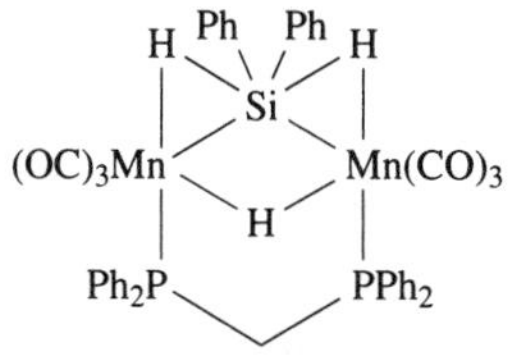 **(281)** | nr | 266 |
| 16 | x ray[268] | 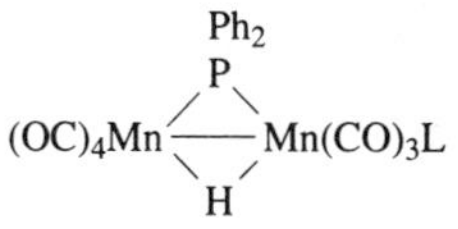 **(282)** L = CO, RCN, RNC, PPh$_3$, P(OR)$_3$ | 16–60 | 267 |
| 17 | x ray | 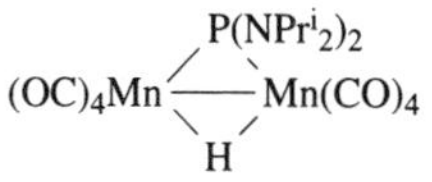 | 51 | 269 |
| 18 | | **(283a)** L$^1$ = CO, L$^2$ = PHCy$_2$<br>**(283b)** L$^1$ = L$^2$ = PMe$_3$ | 60<br>90 | 270<br>270 |
| 19 | | 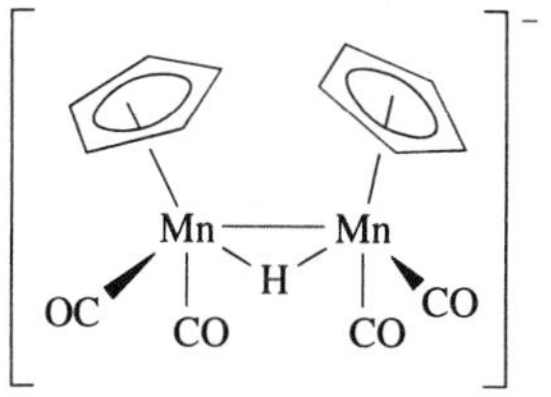 [Na(crypt)]$^+$ | 48 | 271 |

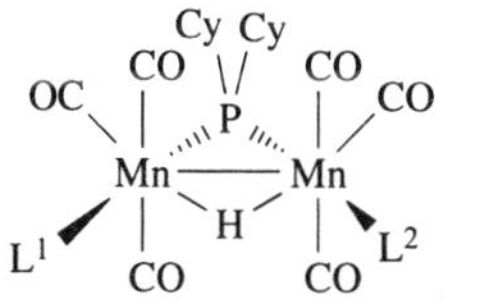

crypt = N(C$_2$H$_4$OC$_2$H$_4$OC$_2$H$_4$)$_3$N

[a] nr, Not reported.

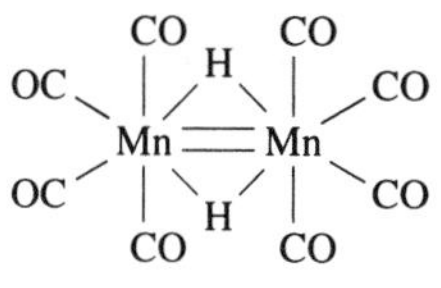

**(284)**

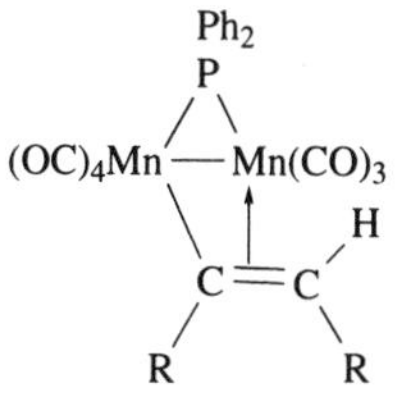

**(285)**
R = H,CF$_3$, Ph

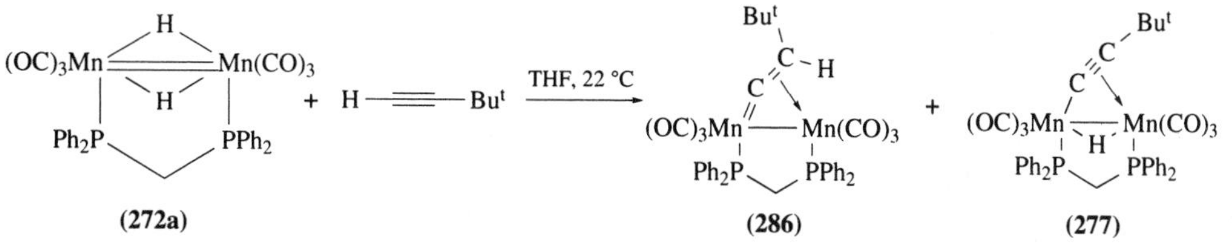

$$(OC)_3Mn \cdots Mn(CO)_3 \text{ (272a)} + H\!-\!\!\equiv\!\!-\!Bu^t \xrightarrow[\text{THF, 22 °C}]{} \textbf{(286)} + \textbf{(277)} \qquad (75)$$

$$(76)$$

PMe$_3$ quite specifically yields symmetrically disubstituted (283b).[270] Photolysis or heating of (282) (entry 16, L = CO) in the presence of potential ligands yields (282) where L is any of a variety of phosphines, phosphites, isonitriles, and so on.[267]

Resonance Raman and IR spectra of various hydrogen–deuterium isotopomers of Mn$_3$H$_3$(CO)$_{12}$ have been reported.[273,274]

## 2.9 REFERENCES

1. P. M. Treichel, in *COMC-I*, vol. 4, chap. 29, p. 7.
2. C. M. Lukehart, G. P. Torrence and J. V. Zeile, *Inorg. Synth.*, 1978, **18**, 57.
3. C. P. Casey, *J. Chem. Soc., Chem. Commun.*, 1970, 1220.
4. S. F. Mapolie and J. R. Moss, *J. Chem. Soc., Dalton Trans.*, 1990, 299.
5. R. Poli, G. Wilkinson, M. Motevalli and M. B. Hursthouse, *J. Chem. Soc., Dalton Trans.*, 1985, 931.
6. D. M. DeSimone, P. J. Desrosiers and R. P. Hughes, *J. Am. Chem. Soc.*, 1982, **104**, 4842.
7. J. B. Sheridan, R. S. Padda, K. Chaffee, C. Wang, Y. Huang and R. Lalancette, *J. Chem. Soc., Dalton Trans.*, 1992, 1539.
8. H. Blau, W. Malisch and P. Weickert, *Chem. Ber.*, 1982, **115**, 1488.
9. W. Malisch, H. Blau and U. Schubert, *Chem. Ber.*, 1983, **116**, 690.
10. J. M. O'Connor, R. Uhrhammer, A. L. Rheingold and D. M. Roddick, *J. Am. Chem. Soc.*, 1991, **113**, 4530.
11. G. Ferguson, W. J. Laws, M. Parvez and R. J. Puddephatt, *Organometallics*, 1983, **2**, 276.
12. J. B. Sheridan, J. R. Johnson, B. M. Handwerker, G. L. Geoffroy and A. L. Rheingold, *Organometallics*, 1988, **7**, 2404.
13. C. P. Casey, C. A. Bunnell and J. C. Calabrese, *J. Am. Chem. Soc.*, 1976, **98**, 1166.
14. D. J. Stufkens, J. B. Sheridan and G. L. Geoffroy, *Inorg. Chem.*, 1990, **29**, 4347.
15. E. J. M. de Boer, J. de With, N. Meijboom and A. G. Orpen, *Organometallics*, 1985, **4**, 259.
16. D. J. Sheeran, J. D. Arenivar and M. Orchin, *J. Organomet. Chem.*, 1986, **316**, 139.
17. R. W. Wegman, *Organometallics*, 1986, **5**, 707.
18. B. T. Gregg, P. K. Hanna, E. J. Crawford and A. R. Cutler, *J. Am. Chem. Soc.*, 1991, **113**, 384.
19. S. L. Bassner, J. B. Sheridan, C. Kelley and G. L. Geoffroy, *Organometallics*, 1989, **8**, 2121.
20. J. A. Gladysz, *Adv. Organomet. Chem.*, 1982, **20**, 1.
21. H. Berke, G. Weiler, G. Huttner and O. Orama, *Chem. Ber.*, 1987, **120**, 297.
22. D. H. Gibson, K. Owens, S. K. Mandal, W. E. Sattich and J. O. Franco, *Organometallics*, 1989, **8**, 498.
23. W. Tam, M. Marsi and J. A. Gladysz, *Inorg. Chem.*, 1983, **22**, 1413.
24. B. A. Narayanan, C. Amatore and J. K. Kochi, *Organometallics*, 1986, **5**, 926.
25. D. H. Gibson, S. K. Mandal, K. Owens and J. F. Richardson, *Organometallics*, 1987, **6**, 2624.
26. D. H. Gibson, K. Owens, S. K. Mandal, W. E. Sattich and J. O. Franco, *Organometallics*, 1991, **10**, 1203.
27. D. H. Gibson, K. Owens, S. K. Mandal, W. E. Sattich and J. F. Richardson, *Organometallics*, 1990, **9**, 424.
28. D. H. Gibson, S. K. Mandal, K. Owens, W. E. Sattich and J. O. Franco, *Organometallics*, 1989, **8**, 1114.
29. T. W. Lee and R.-S. Liu, *J. Organomet. Chem.*, 1987, **320**, 211.
30. M. Akita, N. Kakinuma and Y. Moro-oka, *J. Organomet. Chem.*, 1988, **348**, 91.
31. E. Lindner, M. Pabel and K. Eichele, *J. Organomet. Chem.*, 1990, **386**, 187.
32. E. Lindner and M. Pabel, *J. Organomet. Chem.*, 1991, **414**, C19.
33. E. Lindner, M. Pabel, R. Fawzi, H. A. Mayer and K. Wurst, *J. Organomet. Chem.*, 1992, **435**, 109.
34. Y. K. Chung, D. A. Sweigart, N. G. Connelly and J. B. Sheridan, *J. Am. Chem. Soc.*, 1985, **107**, 2388.
35. P. Deshong, G. A. Slough, V. Elango and G. L. Trainor, *J. Am. Chem. Soc.*, 1985, **107**, 7788.
36. W. Petri and W. Beck, *Chem. Ber.*, 1984, **117**, 3265.
37. J. R. Moss and S. Pelling, *J. Organomet. Chem.*, 1982, **236**, 221.
38. E. Lindner, G. von Au, H.-J. Eberle and S. Hoehne, *Chem. Ber.*, 1982, **115**, 513.
39. E. Lindner, K. A. Starz, N. Pauls and W. Winter, *Chem. Ber.*, 1983, **116**, 1070.
40. E. Lindner, K. A. Starz, H.-J. Eberle and W. Hiller, *Chem. Ber.*, 1983, **116**, 1209.
41. E. Lindner, E. Ossig and M. Darmuth, *J. Organomet. Chem.*, 1989, **379**, 107.
42. P. K. Rush, S. K. Noh and M. Brookhart, *Organometallics*, 1986, **5**, 1745.
43. E. Colomer, R. J. P. Corriu and A. Vioux, *J. Organomet. Chem.*, 1984, **267**, 107.
44. F. Carre, E. Colomer, R. J. P. Corriu and A. Vioux, *Organometallics*, 1984, **3**, 970.
45. E. Lindner and A. Brosamle, *Chem. Ber.*, 1984, **117**, 2730.
46. E. Lindner and A. Brosamle, *Chem. Ber.*, 1985, **118**, 2134.

47. E. Lindner, F. Zinsser, W. Hiller and R. Fawzi, *J. Organomet. Chem.*, 1985, **288**, 317.
48. G. A. Carriedo, J. B. Parra Soto, V. Riera, X. Solans and C. Mieavitlles, *J. Organomet. Chem.*, 1985, **297**, 193.
49. E. Lindner, R. D. Merkle, W. Hiller and R. Fawzi, *Chem. Ber.*, 1986, **119**, 659.
50. M. J. Chen and J. W. Rathke, *Organometallics*, 1989, **8**, 515.
51. P. L. Motz, D. J. Sheeran and M. Orchin, *J. Organomet. Chem.*, 1990, **383**, 201.
52. J. M. Andersen and J. R. Moss, *J. Organomet. Chem.*, 1992, **439**, C25.
53. K. Raab, U. Nagel and W. Beck, *Z. Naturforsch., Teil B*, 1983, **38**, 1466.
54. J. Breimair, B. Niemer, K. Raab and W. Beck, *Chem. Ber.*, 1991, **124**, 1059.
55. H.-J. Müller, U. Nagel and W. Beck, *Organometallics*, 1987, **6**, 193.
56. B. Niemer, J. Breimair, B. Wagner, K. Polborn and W. Beck, *Chem. Ber.*, 1991, **124**, 2227.
57. H.-J. Müller and W. Beck, *J. Organomet. Chem.*, 1987, **330**, C13.
58. R. M. Bullock and B. J. Rappoli, *J. Organomet. Chem.*, 1992, **429**, 345.
59. J. C. Selover, G. D. Vaughn, C. E. Strouse and J. A. Gladysz, *J. Am. Chem. Soc.*, 1986, **108**, 1455.
60. G. D. Vaughn, C. E. Strouse and J. A. Gladysz, *J. Am. Chem. Soc.*, 1986, **108**, 1462.
61. G. D. Vaughn and J. A. Gladysz, *J. Am. Chem. Soc.*, 1986, **108**, 1473.
62. M. Marsi and J. A. Gladysz, *Organometallics*, 1982, **1**, 1467.
63. K. C. Brinkman and J. A. Gladysz, *Organometallics*, 1984, **3**, 147.
64. R. J. Bernhardt, M. A. Wilmoth, J. J. Weers, D. M. LaBrush, D. P. Eyman and J. C. Huffman, *Organometallics*, 1986, **5**, 883.
65. B. C. Roell, Jr. and K. F. McDaniel, *J. Am. Chem. Soc.*, 1990, **112**, 9004.
66. W. A. Halpin, J. C. Williams, Jr., T. Hanna and D. A. Sweigart, *J. Am. Chem. Soc.*, 1989, **111**, 376.
67. T. Hanna, N. S. Lennhoff and D. A. Sweigart, *J. Organomet. Chem.*, 1989, **377**, 133.
68. I. Kovacs, F. Ungvary and J. F. Garst, *Organometallics*, 1993, **12**, 389.
69. M. A. Biddulph, R. Davis and F. I. C. Wilson, *J. Organomet. Chem.*, 1990, **387**, 277.
70. R. Davis and F. I. C. Wilson, *J. Organomet. Chem.*, 1990, **396**, 55.
71. H.-Y. Parker, C. E. Klopfenstein, R. A. Wielesek and T. Koenig, *J. Am. Chem. Soc.*, 1985, **107**, 5276.
72. J. M. Ressner, P. C. Wernett, C. S. Kraihanzel and A. L. Rheingold, *Organometallics*, 1988, **7**, 1661.
73. I. Kovacs, C. D. Hoff, F. Ungvary and L. Marko, *Organometallics*, 1985, **4**, 1347.
74. M. J. Nappa, R. Santi and J. Halpern, *Organometallics*, 1985, **4**, 34.
75. K. E. Warner and J. R. Norton, *Organometallics*, 1985, **4**, 2150.
76. R. M. Bullock and B. J. Rappoli, *J. Am. Chem. Soc.*, 1991, **113**, 1659.
77. J. D. Cotton and R. D. Markwell, *J. Organomet. Chem.*, 1990, **388**, 123.
78. K. Noack and F. Calderazzo, *J. Organomet. Chem.*, 1967, **10**, 101.
79. T. C. Flood, J. E. Jensen and J. A. Statler, *J. Am. Chem. Soc.*, 1981, **103**, 4410.
80. T. L. Bent and J. D. Cotton, *Organometallics*, 1991, **10**, 3156.
81. S. L. Webb, C. M. Giandomenico and J. Halpern, *J. Am. Chem. Soc.*, 1986, **108**, 345.
82. F. J. Garcia Alonso, A. Llamazares, V. Riera, M. Vivanco, M. R. Diaz and S. Garcia Granda, *J. Chem. Soc., Chem. Commun.*, 1991, 1058.
83. G. D. Vaughn, K. A. Krein and J. A. Gladysz, *Angew. Chem., Int. Ed. Engl.*, 1984, **23**, 245.
84. A. Behr, U. Kanne and G. Thelen, *J. Organomet. Chem.*, 1984, **269**, C1.
85. P. DeShong, D. R. Sidler, P. J. Rybczynski, G. A. Slough and A. L. Rheingold, *J. Am. Chem. Soc.*, 1988, **110**, 2575.
86. B. L. Booth, S. Casey, R. P. Critchley and R. N. Haszeldine, *J. Organomet. Chem.*, 1982, **226**, 301.
87. W. Lipps and C. G. Kreiter, *J. Organomet. Chem.*, 1983, **241**, 185.
88. T. W. Leung, G. G. Christoph, J. Gallucci and A. Wojcicki, *Organometallics*, 1986, **5**, 846.
89. K. McGregor, G. B. Deacon, R. S. Dickson, G. D. Fallon, R. S. Rowe and B. O. West, *J. Chem. Soc., Chem. Commun.*, 1990, 1293.
90. M. Appel, K. Schloter, J. Heidrich and W. Beck, *J. Organomet. Chem.*, 1987, **322**, 77.
91. K. D. Abney, K. M. Long, O. P. Anderson and S. H. Strauss, *Inorg. Chem.*, 1987, **26**, 2638.
92. T. E. Gismondi and M. D. Rausch, *J. Organomet. Chem.*, 1985, **284**, 59.
93. S. K. Mandal, D. M. Ho and M. Orchin, *J. Organomet. Chem.*, 1990, **397**, 313.
94. T. G. Richmond and D. F. Shriver, *Organometallics*, 1984, **3**, 305.
95. T. G. Richmond, A. M. Crespi and D. F. Shriver, *Organometallics*, 1984, **3**, 314.
96. T. M. Wido, G. H. Young, A. Wojcicki, M. Calligaris and G. Nardin, *Organometallics*, 1988, **7**, 452.
97. G.-H. Lee, S.-M. Peng, G.-M. Yang, S.-F. Lush and R.-S. Liu, *Organometallics*, 1989, **8**, 1106.
98. P. Wang and J. D. Atwood, *J. Am. Chem. Soc.*, 1992, **114**, 6424.
99. E. Lindner, R. Fawzi, H. A. Mayer, K. Eichele and W. Pohmer, *Inorg. Chem.*, 1991, **30**, 1102.
100. E. Lindner, R. Fawzi, H. A. Mayer, K. Eichele and K. Pohmer, *J. Organomet. Chem.*, 1990, **386**, 63.
101. K. Kanda, H. Nakatsuji and T. Yonezawa, *J. Am. Chem. Soc.*, 1984, **106**, 5888.
102. M. A. Andrews, J. Eckert, J. A. Goldstone, L. Passell and B. Swanson, *J. Am. Chem. Soc.*, 1983, **105**, 2262.
103. C. Long, A. R. Morrisson, D. C. McKean and G. P. McQuillan, *J. Am. Chem. Soc.*, 1984, **106**, 7418.
104. G. P. McQuillan, D. C. McKean, C. Long, A. R. Morrisson and I. Torto, *J. Am. Chem. Soc.*, 1986, **108**, 863.
105. A. J. Ricco, A. A. Bakke and W. L. Jolly, *Organometallics*, 1982, **1**, 94.
106. J. A. Connor *et al.*, *Organometallics*, 1982, **1**, 1166.
107. G. P. Smith, *Polyhedron*, 1988, **7**, 1605.
108. A. R. Dias, J. A. M. Simoes, C. Teixeira, C. Airoldi and A. P. Chagas, *J. Organomet. Chem.*, 1987, **335**, 71.
109. S. B. Butts, S. H. Strauss, E. M. Holt, R. E. Stimson, N. W. Alcock and D. W. Shriver, *J. Am. Chem. Soc.*, 1980, **102**, 5093.
110. S. P. Nolan, R. L. de la Vega and C. D. Hoff, *J. Am. Chem. Soc.*, 1986, **108**, 7852.
111. S. A. Jackson, O. Eisenstein, J. D. Martin, A. C. Albeniz and R. H. Crabtree, *Organometallics*, 1991, **10**, 3062.
112. T. Ziegler, L. Versluis and V. Tschinke, *J. Am. Chem. Soc.*, 1986, **108**, 612.
113. F. U. Axe and D. S. Marynick, *Organometallics*, 1987, **6**, 572.
114. F. U. Axe and D. S. Marynick, *J. Am. Chem. Soc.*, 1988, **110**, 3728.
115. P. DeShong *et al.*, *Organometallics*, 1989, **8**, 1381.
116. B. E. Bursten and R. H. Cayton, *J. Am. Chem. Soc.*, 1987, **109**, 6053.

117. B. E. Bursten and R. H. Cayton, *Polyhedron*, 1988, **7**, 943.
118. R. B. Hitam, R. Narayanaswamy and A. J. Rest, *J. Chem. Soc., Dalton Trans.*, 1983, 615.
119. A. Horton-Mastin, M. Poliakoff and J. J. Turner, *Organometallics*, 1986, **5**, 405.
120. S. Firth *et al.*, *Organometallics*, 1989, **8**, 2876.
121. S. T. Belt, D. W. Ryba and P. C. Ford, *Inorg. Chem.*, 1990, **29**, 3633.
122. W. T. Boese, B. Lee, D. W. Ryba, S. T. Belt and P. C. Ford, *Organometallics*, 1993, **12**, 4739.
123. R. B. Hitam, K. A. Mahmoud and A. J. Rest, *J. Organomet. Chem.*, 1985, **291**, 321.
124. A. K. Campen, R. Narayanaswamy and A. J. Rest, *J. Chem. Soc., Dalton Trans.*, 1990, 823.
125. C. Schulze and H. Schwarz, *J. Am. Chem. Soc.*, 1988, **110**, 67.
126. R. S. Shinomoto, P. J. Desrosiers, T. G. P. Harper, M. A. Deming and T. C. Flood, *J. Am. Chem. Soc.*, 1990, **112**, 704.
127. R. N. McDonald, M. T. Jones and A. K. Chowdhury, *J. Am. Chem. Soc.*, 1991, **113**, 476.
128. R. N. McDonald and M. T. Jones, *Organometallics*, 1987, **6**, 1991.
129. R. N. McDonald, M. T. Jones and A. K. Chowdhury, *J. Am. Chem. Soc.*, 1992, **114**, 71.
130. S. K. Shin and J. L. Beauchamp, *J. Am. Chem. Soc.*, 1990, **112**, 2057, 2066.
131. G. A. Artamkina, A. Y. Mil'chenko, I. P. Beletskaya and O. A. Reutov, *J. Organomet. Chem.*, 1986, **311**, 199.
132. A. D. Hunter and A. B. Szigety, *Organometallics*, 1989, **8**, 2670.
133. M. Herberhold and H. Kniesel, *J. Organomet. Chem.*, 1987, **334**, 347.
134. S. Lotz, M. Schindehutte and P. H. van Rooyen, *Organometallics*, 1992, **11**, 629.
135. S. F. Mapolie and J. R. Moss, *Polyhedron*, 1991, **10**, 717.
136. J. M. Cooney, L. H. P. Gommans, L. Main and B. K. Nicholson, *J. Organomet. Chem.*, 1988, **349**, 197.
137. R. C. Cambie, M. R. Metzler, P. S. Rutledge and P. D. Woodgate, *J. Organomet. Chem.*, 1990, **398**, C22.
138. N. P. Robinson, L. Main and B. K. Nicholson, *J. Organomet. Chem.*, 1988, **349**, 209.
139. M. I. Bruce, M. G. Humphrey and M. J. Liddell, *J. Organomet. Chem.*, 1987, **321**, 91.
140. N. P. Robinson, L. Main and B. K. Nicholson, *J. Organomet. Chem.*, 1992, **430**, 79.
141. R. M. Ceder, J. Sales, X. Solans and M. Font-Altaba, *J. Chem. Soc., Dalton Trans.*, 1986, 1351.
142. J. M. Vila, M. Gayoso, M. T. Pereira, M. Lopez, G. Alonso and J. J. Fernandez, *J. Organomet. Chem.*, 1993, **445**, 287.
143. R. C. Cambie, M. R. Metzler, C. E. F. Rickard, P. S. Rutledge and P. D. Woodgate, *J. Organomet. Chem.*, 1992, **425**, 59.
144. A. Suarez, J. M. Vila, M. T. Pereira, E. Gayoso and M. Gayoso, *J. Organomet. Chem.*, 1987, **335**, 359.
145. L. H. P. Gommans, L. Main and B. K. Nicholson, *J. Chem. Soc., Chem. Commun.*, 1986, 12.
146. E. Deschamps, F. Mathey, C. Knobler and Y. Jeannin, *Organometallics*, 1984, **3**, 1144.
147. R. C. Cambie, M. R. Metzler, C. E. F. Rickard, P. S. Rutledge and P. D. Woodgate, *J. Organomet. Chem.*, 1992, **431**, 177.
148. J. M. Cooney, L. H. P. Gommans, L. Main and B. K. Nicholson, *J. Organomet. Chem.*, 1987, **336**, 293.
149. N. P. Robinson, L. Main and B. K. Nicholson, *J. Organomet. Chem.*, 1989, **364**, C37.
150. L. H. P. Gommans, L. Main and B. K. Nicholson, *J. Chem. Soc., Chem. Commun.*, 1987, 761.
151. R. C. Cambie, M. R. Metzler, P. S. Rutledge and P. D. Woodgate, *J. Organomet. Chem.*, 1992, **429**, 59.
152. R. C. Cambie, M. R. Metzler, P. S. Rutledge and P. D. Woodgate, *J. Organomet. Chem.*, 1992, **429**, 41.
153. R. C. Cambie, M. R. Metzler, C. E. F. Rickard, P. S. Rutledge and P. D. Woodgate, *J. Organomet. Chem.*, 1992, **426**, 213.
154. W. Beck, M. J. Schweiger and G. Muller, *Chem. Ber.*, 1987, **120**, 889.
155. B. L. Booth, S. Casey and R. N. Haszeldine, *J. Organomet. Chem.*, 1982, **226**, 289.
156. J. Breimair, M. Steimann, B. Wagner and W. Beck, *Chem. Ber.*, 1990, **123**, 7.
157. N. E. Kolobova, L. L. Ivanov, O. S. Zhvanko, I. N. Chechulina, A. S. Batsanov and Yu. T. Struchkov, *J. Organomet. Chem.*, 1982, **238**, 223.
158. H. G. Alt, H. E. Engelhardt and E. Steinlein, *J. Organomet. Chem.*, 1988, **344**, 227.
159. R. D. Rogers, H. G. Alt and H. E. Maisel, *J. Organomet. Chem.*, 1990, **381**, 233.
160. R. E. Cramer, K. T. Higa and J. W. Gilje, *J. Am. Chem. Soc.*, 1984, **106**, 7245.
161. H. G. Alt and H. E. Engelhardt, *J. Organomet. Chem.*, 1988, **346**, 211.
162. M. I. Bruce, M. J. Liddell, M. R. Snow and E. R. T. Tiekink, *Organometallics*, 1988, **7**, 343.
163. M. I. Bruce, D. N. Duffy, M. J. Liddell, M. R. Snow and E. R. T. Tiekink, *J. Organomet. Chem.*, 1987, **335**, 365.
164. E. Lindner, M. Darmuth, R. Fawzi and M. Steimann, *Chem. Ber.*, 1992, **125**, 2713.
165. E. Lindner, V. Käss and H. A. Mayer, *Chem. Ber.*, 1990, **123**, 783.
166. A. D. Horton, A. C. Kemball and M. J. Mays, *J. Chem. Soc., Dalton Trans.*, 1988, 2953.
167. W. A. Herrmann, M. L. Ziegler and O. Serhadli, *Organometallics*, 1983, **2**, 958.
168. R. D. Adams, G. Chen, L. Chen, W. Wu and J. Yin, *J. Am. Chem. Soc.*, 1991, **113**, 9406.
169. S. J. Davies, B. F. G. Johnson, J. Lewis and M. S. Khan, *J. Organomet. Chem.*, 1991, **401**, C43.
170. D. Miguel and V. Riera, *J. Organomet. Chem.*, 1985, **293**, 379.
171. C. Kelley, N. Lugan, M. R. Terry, G. L. Geoffroy, B. S. Haggerty and A. L. Rheingold, *J. Am. Chem. Soc.*, 1992, **114**, 6735.
172. H. Berke, P. Harter, G. Huttner and L. Zsolnai, *Chem. Ber.*, 1984, **117**, 3423.
173. K. Henrick, M. McPartlin, J. A. Iggo, A. C. Kemball, M. J. Mays and P. R. Raithby, *J. Chem. Soc., Dalton Trans.*, 1987, 2669.
174. G. A. Carriedo, D. Miguel, V. Riera and X. Solans, *J. Chem. Soc., Dalton Trans.*, 1987, 2867.
175. G. A. Carriedo, D. Miguel and V. Riera, *J. Organomet. Chem.*, 1988, **342**, 373.
176. E. O. Fischer and A. Maasböl, *Chem. Ber.*, 1967, **100**, 2445.
177. R. Aumann and J. Schröder, *J. Organomet. Chem.*, 1989, **378**, 57.
178. M. J. McGeary and J. L. Templeton, *J. Organomet. Chem.*, 1987, **323**, 199.
179. E. O. Fischer and J. K. R. Wanner, *Chem. Ber.*, 1985, **118**, 2489.
180. D. H. Gibson, S. K. Mandal, K. Owens and J. F. Richardson, *Organometallics*, 1990, **9**, 1936.
181. P. L. Motz, D. M. Ho and M. Orchin, *J. Organomet. Chem.*, 1991, **407**, 259.
182. U. Kirchgassner, H. Piana and U. Schubert, *J. Am. Chem. Soc.*, 1991, **113**, 2228.
183. F. Carre, G. Cerveau, E. Colomer and R. J. P. Corriu, *J. Organomet. Chem.*, 1982, **229**, 257.
184. J.-A. M. Garner, A. Irving and J. R. Moss, *Organometallics*, 1990, **9**, 2836.
185. M. M. Singh and R. J. Angelici, *Inorg. Chem.*, 1984, **23**, 2691.
186. M. M. Singh and R. J. Angelici, *Inorg. Chem.*, 1984, **23**, 2699.
187. H. G. Alt, H. E. Engelhardt, E. Steinlein and R. D. Rogers, *J. Organomet. Chem.*, 1987, **344**, 321.

188. K. H. Dötz, W. Sturm and H. G. Alt, *Organometallics*, 1987, **6**, 1424.
189. D. J. Crowther, S. Tivakornpannarai and W. M. Jones, *Organometallics*, 1990, **9**, 739.
190. W. A. Herrmann *et al.*, *J. Organomet. Chem.*, 1984, **264**, 327.
191. E. O. Fischer, J. Chen and K. Scherzer, *J. Organomet. Chem.*, 1983, **253**, 231.
192. C. Löwe, H. U. Hund and H. Berke, *J. Organomet. Chem.*, 1989, **378**, 211.
193. C. Löwe, H. U. Hund and H. Berke, *J. Organomet. Chem.*, 1989, **371**, 311.
194. U. Schubert, U. Kirchgassner, J. Gronen and H. Piana, *Polyhedron*, 1989, **8**, 1589.
195. N. E. Kolobova, L. L. Ivanov, O. S. Zhvanko, O. M. Khitrova, A. S. Batsinov and Yu. T. Struchkov, *J. Organomet. Chem.*, 1984, **262**, 39.
196. T. R. Hoye and G. M. Rehberg, *Organometallics*, 1990, **9**, 3014.
197. B. L. Balzer, M. Cazanoue, M. Sabat and M. G. Finn, *Organometallics*, 1992, **11**, 1759.
198. B. L. Balzer, M. Cazanoue and M. G. Finn, *J. Am. Chem. Soc.*, 1992, **114**, 8735.
199. R. Aumann and H. Heinen, *Chem. Ber.*, 1988, **121**, 1085.
200. R. Aumann, J. Schröder, C. Krüger and R. Goddard, *J. Organomet. Chem.*, 1989, **378**, 185.
201. R. Aumann and H. Heinen, *Chem. Ber.*, 1989, **122**, 77.
202. R. Aumann and H. Heinen, *Chem. Ber.*, 1988, **121**, 1739.
203. V. N. Kalinin, V. V. Derunov, M. A. Lusenkova, P. V. Petrovsky and N. E. Kolobova, *J. Organomet. Chem.*, 1989, **379**, 303.
204. A. B. Antonova *et al.*, *J. Organomet. Chem.*, 1983, **244**, 35.
205. N. E. Kolobova, L. L. Ivanov, O. S. Zhvanko, O. M. Khitrova, A. S. Batsanov and Yu. T. Struchkov, *J. Organomet. Chem.*, 1984, **265**, 271.
206. H. G. Raubenheimer, G. J. Kruger and H. W. Viljoen, *J. Chem. Soc., Dalton Trans.*, 1985, 1963.
207. M. J. McGeary, T. L. Tonker and J. L. Templeton, *Organometallics*, 1985, **4**, 2102.
208. B. M. Handwerker, K. E. Garrett, G. L. Geoffroy and A. L. Rheingold, *J. Am. Chem. Soc.*, 1989, **111**, 369.
209. H. Fischer and C. Troll, *J. Organomet. Chem.*, 1992, **427**, 77.
210. J. B. Sheridan, G. L. Geoffroy and A. L. Rheingold, *J. Am. Chem. Soc.*, 1987, **109**, 1584.
211. H. Handwerker, H. Beruda, M. Kleine and C. Zybill, *Organometallics*, 1992, **11**, 3542.
212. U. Schubert, *Organometallics*, 1982, **1**, 1085.
213. N. M. Kostic and R. F. Fenske, *J. Am. Chem. Soc.*, 1982, **104**, 3879.
214. N. M. Kostic and R. F. Fenske, *Organometallics*, 1982, **1**, 974.
215. F. Delbecq, *J. Organomet. Chem.*, 1991, **406**, 171.
216. V. S. Leong and N. J. Cooper, *Organometallics*, 1988, **7**, 2080.
217. G. F. P. Warnock, L. C. Moodie and J. E. Ellis, *J. Am. Chem. Soc.*, 1989, **111**, 2131.
218. M. L. H. Green, D. S. Joyner and J. M. Wallis, *J. Chem. Soc., Dalton Trans.*, 1987, 2823.
219. M. D. Rausch, M. Ogasa, M. A. Ayers, R. D. Rogers and A. N. Rollins, *Organometallics*, 1991, **10**, 2481.
220. B. A. Narayanan, C. Amatore and J. K. Kochi, *J. Chem. Soc., Chem. Commun.*, 1983, 397.
221. B. A. Narayanan, C. Amatore and J. K. Kochi, *Organometallics*, 1987, **6**, 129.
222. D. J. Kuchynka, C. Amatore and J. K. Kochi, *Inorg. Chem.*, 1986, **25**, 4087.
223. R. J. Bernhardt and D. P. Eyman, *Organometallics*, 1984, **3**, 1445.
224. D. R. Tyler and A. S. Goldman, *J. Organomet. Chem.*, 1986, **311**, 349.
225. R. J. Sullivan and T. L. Brown, *J. Am. Chem. Soc.*, 1991, **113**, 9155, 9162.
226. M. A. Wilmoth, R. J. Bernhardt, D. P. Eyman and J. C. Huffman, *Organometallics*, 1986, **5**, 2559.
227. R. Sweany, S. C. Butler and J. Halpern, *J. Organomet. Chem.*, 1981, **213**, 487.
228. T. E. Nalesnik, J. H. Freudenberger and M. Orchin, *J. Organomet. Chem.*, 1982, **236**, 95.
229. B. Wassink, M. J. Thomas, S. C. Wright, D. J. Gillis and M. C. Baird, *J. Am. Chem. Soc.*, 1987, **109**, 1995.
230. J. F. Garst, T. M. Bockman and R. Batlaw, *J. Am. Chem. Soc.*, 1986, **108**, 1689.
231. I. Kovacs, A. Sisak, F. Ungvary and L. Marko, *Organometallics*, 1989, **8**, 1873.
232. B. D. Martin, K. E. Warner and J. R. Norton, *J. Am. Chem. Soc.*, 1986, **108**, 33.
233. P. L. Bogdan, P. J. Sullivan, T. A. Donovan, Jr. and J. D. Atwood, *J. Organomet. Chem.*, 1984, **269**, C51.
234. G. D. Vaughn, K. A. Krein and J. A. Gladysz, *Organometallics*, 1986, **5**, 936.
235. D. J. Kuchynka and J. K. Kochi, *Organometallics*, 1989, **8**, 677.
236. R. L. Sweany and J. W. Owens, *J. Organomet. Chem.*, 1983, **255**, 327.
237. E. J. Moore, J. M. Sullivan and J. R. Norton, *J. Am. Chem. Soc.*, 1986, **108**, 2257.
238. S. S. Kristjansdottir, A. E. Moody, R. T. Weberg and J. R. Norton, *Organometallics*, 1988, **7**, 1983.
239. A. E. Stevens Miller and J. L. Beauchamp, *J. Am. Chem. Soc.*, 1991, **113**, 8765.
240. A. E. Stevens Miller, A. R. Kawamura and T. M. Miller, *J. Am. Chem. Soc.*, 1990, **112**, 457.
241. C. J. Eyermann and A. Chung-Phillips, *J. Am. Chem. Soc.*, 1984, **106**, 7437.
242. R. J. Klinger and J. W. Rathke, *Inorg. Chem.*, 1992, **31**, 804.
243. M. Tilset and V. D. Parker, *J. Am. Chem. Soc.*, 1989, **111**, 6711; 1990, **112**, 2843.
244. R. Billmers, L. L. Griffith and S. E. Stein, *J. Phys. Chem.*, 1986, **90**, 517.
245. B. E. Bursten and M. G. Gatter, *Organometallics*, 1984, **3**, 895.
246. E. Ruhl and A. P. Hitchcock, *J. Am. Chem. Soc.*, 1989, **111**, 2614.
247. J. B. Schilling, W. A. Goddard III and J. L. Beauchamp, *J. Am. Chem. Soc.*, 1986, **108**, 582.
248. S. M. Howdle and M. Poliakoff, *J. Chem. Soc., Chem. Commun.*, 1989, 1099.
249. S. P. Church, M. Poliakoff, J. A. Timney and J. J. Turner, *Inorg. Chem.*, 1983, **22**, 3259.
250. R. D. Davy and M. B. Hall, *Inorg. Chem.*, 1989, **28**, 3524.
251. A. J. Hart-Davis and W. A. G. Graham, *J. Am. Chem. Soc.*, 1971, **93**, 4388.
252. U. Schubert, *Adv. Organomet. Chem.*, 1990, **30**, 151.
253. D. M. Hester, J. Sun, A. W. Harper and G. K. Yang, *J. Am. Chem. Soc.*, 1992, **114**, 5234.
254. D. L. Lichtenberger and A. Rai-Chaudhuri, *Inorg. Chem.*, 1990, **29**, 975.
255. D. L. Lichtenberger and A. Rai-Chaudhuri, *J. Chem. Soc., Dalton Trans.*, 1990, 2161.
256. D. L. Lichtenberger and A. Rai-Chaudhuri, *Organometallics*, 1990, **9**, 1686.
257. D. L. Lichtenberger and A. Rai-Chaudhuri, *J. Am. Chem. Soc.*, 1990, **112**, 2492.

258. F. J. Garcia Alonso, M. Garcia Sanz, V. Riera, M. A. Ruiz, A. Tiripicchio and M. Tiripicchio Camellini, *Angew. Chem.,
       Int. Ed. Engl.*, 1988, **27**, 1167.
259. V. Riera, M. A. Ruiz, A. Tiripicchio and M. Tiripicchio Camellini, *J. Chem. Soc., Chem. Commun.*, 1985, 1505.
260. J. Gimeno, V. Riera, M. A. Ruiz, A. M. Manotti Lanfredi and A. Tiripicchio, *J. Organomet. Chem.*, 1984, **268**, C13.
261. H. C. Aspinall and A. J. Deeming, *J. Chem. Soc., Chem. Commun.*, 1983, 838.
262. H. C. Aspinall and A. J. Deeming, *J. Chem. Soc., Dalton Trans.*, 1985, 743.
263. R. Carreno, V. Riera and M. A. Ruiz, *J. Organomet. Chem.*, 1991, **419**, 163.
264. F. J. Garcia Alonso, M. Garcia Sanz and V. Riera, *J. Organomet. Chem.*, 1991, **421**, C12.
265. A. J. Deeming and S. Donovan-Mtunzi, *Organometallics*, 1985, **4**, 693.
266. R. Carreno, V. Riera, M. A. Ruiz, Y. Jeannin and M. Philoche-Levisalles, *J. Chem. Soc., Chem. Commun.*, 1990, 15.
267. J. A. Iggo, M. J. Mays, P. R. Raithby and K. Henrick, *J. Chem. Soc., Dalton Trans.*, 1983, 205.
268. J. A. Iggo, M. J. Mays, P. R. Raithby and K. Henrick, *J. Chem. Soc., Dalton Trans.*, 1984, 633.
269. R. B. King, W.-K. Fu and E. M. Holt, *Inorg. Chem.*, 1986, **25**, 2390.
270. A. M. Arif, R. A. Jones and S. T. Schwab, *J. Organomet. Chem.*, 1986, **307**, 219.
271. K. Plößl, G. Huttner and L. Zsolnai, *Angew. Chem., Int. Ed. Engl.*, 1989, **28**, 446.
272. B. Jezowska-Trzebiatowska and B. Nissen-Sobocinska, *J. Organomet. Chem.*, 1988, **342**, 215.
273. M. W. Howard, P. Skinner, R. K. Bhardwaj, U. A. Jayasooriya, D. B. Powell and N. Sheppard, *Inorg. Chem.*, 1986, **25**,
       2846.
274. U. A. Jayasooriya and P. Skinner, *Inorg. Chem.*, 1986, **25**, 2850.

# 3
# Manganese Complexes Containing Nonmetallic Elements

## PAUL M. TREICHEL
*University of Wisconsin–Madison, WI, USA*

## 3.1 INTRODUCTION

### 3.1.1 Scope of this Chapter

Coverage in this chapter includes organometallic complexes of manganese with group 15 and 16 ligands, but not halogen ligands (covered in Chapter 1, this volume, and nitrosyl ligands, covered in Chapter 6, this volume). Discussion of $[\{Mn(CO)_2(\eta\text{-}Cp)\}_2X(R)_n]$ complexes (X = group 15 or 16 element) has been deferred to Chapter 5, this volume, which covers cyclopentadienyl manganese compounds. Specifically exempted from this chapter are compounds in which the group 15 or 16 ligand is a simple two-electron donor since these species are included as derivatives of other classes of compounds.

Review articles describe complexes with $R_2PS$[1] and $\beta$-diketonate ligands.[2] Syntheses of $[Mn(O_3SCF_3)(CO)_5]$,[3] $[Mn_2(\mu\text{-}SPMe_2)_2(CO)_8]$, and $[Mn\{SPMe_2(CR = CR)_n\}(CO)_4]$ ($n = 1$, 2; R = $CO_2Me$),[4] and $[Mn_2(\mu\text{-}X)(\mu\text{-}PPh_2)(CO)_8]$ (X = H, $Au(PPh_3)$, HgCl) and $[PPN][Mn_2(\mu\text{-}PPh_2)(CO)_8]$[5] (PPN = $[Ph_3P)_2N]$) are found in *Inorganic Syntheses*.[3-5]

### 3.1.2 Overview of Earlier Work

Earlier coverage in *COMC-I*[6] was based on ligand type and this organization is continued here.

While a few monometallic complexes with group 15 and 16 ligands are known, in most complexes in this area the nonmetallic element bridges two or more metals. Especially common are species in which $PR_2$ and SR groups bridge two metals. As might be expected, the chemistry of the lower elements in these groups parallels that of phosphorus and sulfur, respectively, while complexes with second period elements are somewhat less common and often unique. An interest in heterobimetallic compounds was emerging in 1982, and a few complexes with nonmetal-containing ligands bridging two different metals were known. In addition, complexes with nonmetal donors in a variety of bidentate ligands had been investigated.

## 3.2 COMPLEXES WITH OXYGEN AND NITROGEN LIGANDS

Among the best-known complexes with oxygen-containing ligands are manganese carbonyls of oxyanions having the general formula $[Mn(OA)(CO)_5]$. Often these complexes are prepared *in situ* for subsequent preparation of cationic complexes since the $OA^-$ group is readily displaced; $[Mn(OClO_3)(CO)_5]$ is particularly well known in this regard. This is documented in Chapter 1, this volume and not surveyed here.

Further study on $[Mn(OA)(CO)_5]$ compounds is described in the literature. Oxygen-bonded sulfonate complexes, $[Mn(O_3SR)(CO)_5]$ (R = F, $CF_3$, $C_3F_7$, $C_6F_{13}$), have been prepared by acid cleavage of the alkyl group in $[Mn(R)(CO)_5]$ (Equation (1)).[7] Several of these complexes were known previously, most often having been prepared from a metal carbonyl halide and a silver salt of the anion. The fluorosulfonic acid complex had been previously prepared by the reaction of $[Mn(Br)(CO)_5]$ and $AgO_3SF$. Heating this compound causes CO loss and formation of $[Mn(O_3SF)(CO)_4]$ which contains a chelating $SO_3F^-$ ligand.[8]

$$[Mn(R)(CO)_5] + RSO_3H \longrightarrow [Mn(O_3SR)(CO)_5] + RH \qquad (1)$$

$$R = F, CF_3, C_3F_7, C_6F_{13}$$

Another synthetic procedure for these species is shown in Equation (2). This reaction takes advantage of the hydridic character of a manganese carbonyl complex in a reaction with acid. The tosylate compound reacts with alcohols to give alkoxide complexes $[Mn(OR)(CO)_3(dppe)]$ (R = Me, Et, Ph) (Equation (3)).[9]

$$[Mn(H)(CO)_3(dppe)] + TolSO_3H \longrightarrow [Mn(O_3STol)(CO)_3(dppe)] + H_2 \qquad (2)$$

$$[Mn(O_3STol)(CO)_3(dppe)] + ROH \longrightarrow [Mn(OR)(CO)_3(dppe)] + HO_3STol \qquad (3)$$

$$R = Me, Et, Ph$$

The complex $[Mn(OTeF_5)(CO)_5]$ was prepared by the reaction of $HOTeF_5$ and $[Mn(Me)(CO)_5]$ and a crystal structure study carried out.[10] This complex, and several other $[Mn(X)(CO)_5]$ species (X = $O_3SCF_3$, $OClO_3$) are reported to react with THF to form $[Mn(X)(CO)_3(THF)_2]$.

Although several complexes having monodentate nitrogen donor ligands have been characterized since the mid-1980s, these species remain rare. The succinimide manganese carbonyl complex, $[Mn(\overline{NCOCH_2CH_2C}O)(CO)_5]$, was prepared by the reaction of $[Mn(Br)(CO)_5]$ and thallium succinimide,[11] and $[PPN][Mn(phth)_2(CO)_4]$ was one of two products of a reaction between $[PPN][Mn(CO)_5]$ and phthSPh (phth is the anion of phthalimide); $[PPN][Mn_2(\mu-SPh)_3(CO)_6]$ was the second product.[12]

Other complexes contain an imido group as part of a chelating ligand. The complexes $[Et_4N][Mn(bibzim)(CO)_4]$ and $[Et_4N][Mn(biim)(CO)_4]$ were prepared by reaction of the dipotassium salts of the bibenzimidazolate and biimidazolate ions, respectively, with $[Mn(Br)_2(CO)_4]^-$. A number of derivatives of these species with other ligands replacing CO were also reported.[13] The preparation of $Li[Mn(RNCONR)(CO)_4]$ from $Li_2(RNCONR)$ (R = Me, Et, $Pr^i$, Cy, Ph, Bz, Np, $p$-$PhC_6H_4$) and $[Mn(Br)(CO)_5]$ is also noted here.[14]

Other complexes contain oxygen donors as part of a chelating ligand. Complexes of the anions of several α-amino acids, $[Mn(O-N)(CO)_4]$ (O–N = histidinate, ornithinate, lysinate anions), coordinating through the carboxylate oxygen and the amino group, are mentioned in this regard; they are prepared by heating the silver salt of the acid and $[Mn(Br)(CO)_5]$.[15] In addition, phosphonate-bridged species (1), $[Mn_2(\mu-OPR_2)_2(CO)_8]$ (R = Ph, OEt), have been made by photolysis of $[Mn_2(CO)_{10}]$ and $Hg[PO(OEt)_2]_2$[16] and by pyrolysis of $[\overline{Mn(PPh_2OCH_2C}H_2)(CO)_4]$.[17]

Two new species, $[Bu_4N]_2[Mn(CO)_3(P_3O_9)]$ (2)[18] and $[Bu_4N]_3[Mn(CO)_3(Nb_2W_4O_{19})]$,[19] are among the more interesting compounds in this category. They are prepared by displacement of the nitriles in $[Mn(CO)_3(NCMe)_3]^+$ by the appropriate oxyanion. A crystallographic study provided structural information for $[Bu_4N]_3[Mn(CO)_3(Nb_2W_4O_{19})]$.[19] Another complex, $[Mn(CO)_3\{(OPR_2)_3Co(\eta-Cp)\}]$ (R = OMe, OEt) was prepared by reaction of $[Co(OPR_2)_3(\eta-Cp)]^-$ with $[Mn(Br)(CO)_5]$. In this species, the three phosphonate oxygen atoms in the cobalt complex coordinate to manganese.[20]

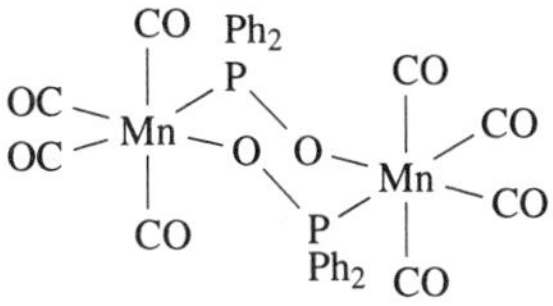

(1)

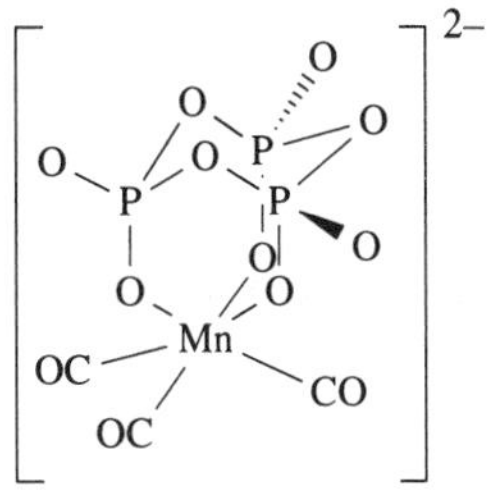

(2)

Two products are obtained from $[Mn_2(\mu\text{-}H)_2(CO)_4(P\text{–}P)_2]$ (P–P = dppm, $(EtO)_2POP(OEt)_2$) and $CO_2$ under 50 atm pressure; $[Mn_2(\mu\text{-}H)(\mu\text{-}OH)(CO)_4(P\text{–}P)_2]$ and $[Mn_2(\mu\text{-}H)(\mu\text{-}O_2CH)(CO)_4(P\text{–}P)_2]$ contain bridging hydroxide and formate ligands, respectively.[21]

Addition of ethylenediamine to $[Mn_4(\mu\text{-}OH)_4(CO)_{12}]$ gives $[Mn_4(\mu\text{-}OH)_4(CO)_{12}]\cdot2en$, a species having an undistorted triple diamondoid network of tetramers linked with en bridges. The authors described this as the "supramolecular" chemistry of the tetrameric species.[22]

The product of the reaction between $[Mn(Br)(CO)_5]$ and $R_2NP(=NR)_2$ (R = TMS) is (3), a 16-electron species. Reversible addition of CO to (3) is observed. The halide on phosphorus is readily substituted by $OPh^-$. Structural evidence for (3) was obtained through a single-crystal x-ray diffraction study of this derivative.[23]

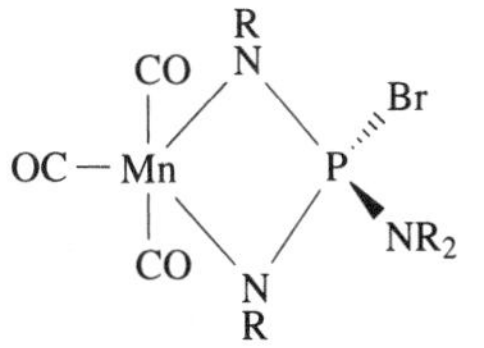

(3)

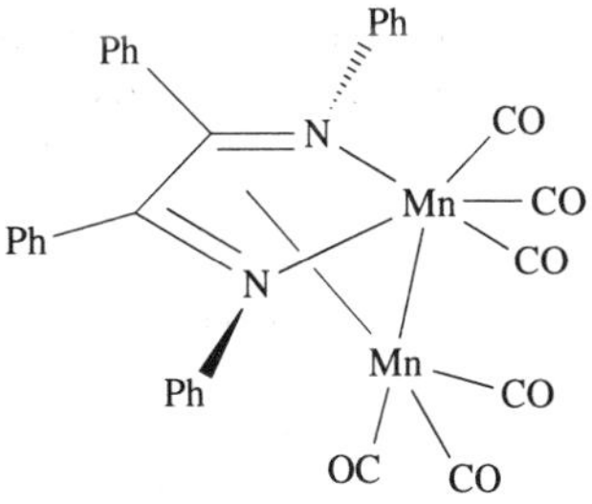

(4)

The reaction of $[Mn_2(CO)_{10}]$ and PhC(=NPh)Cl was found to give (4),[24] while $[(OC)_3Mn\{(p\text{-}Tol)N=C(Bz)C(Bz)=N(p\text{-}Tol)\}Mn(CO)_3]$ is the product in a reaction between $[Mn(COBz)(CO)_4\{CN(p\text{-}Tol)\}]$ and 1,4-diazabutadiene.[25]

Storr and co-workers have studied pyrazolylgallate complexes of manganese. They have prepared and studied a series of compounds with the general formula $[Mn\{pzGa(Me)_2(XCH_2CH_2Y)\}(CO)_3]$ (pz = pyrazolyl); coordination of the $pzGa(Me)_2(XCH_2CH_2Y)$ (X = S, O; Y = $NH_2$, $NMe_2$, SEt, SPh) to the metal occurs via the pyrazolyl nitrogen, and the X and Y groups.[26-9]

### 3.3 COMPLEXES WITH PHOSPHORUS, ARSENIC, ANTIMONY, AND BISMUTH LIGANDS

Several manganese carbonyl complexes with monodentate phosphorus(V) ligands have now been described. Reactions of $Na[Mn(CO)_5]$ with $ClP(O_2C_6H_4)_2$[30-1] and $ClP(NRCONR)_2$ (R = Me, Ph)[32] were used to synthesize $[Mn\{P(O_2C_6H_4)_2\}(CO)_5]$ (5) and $[Mn\{P(RNCONR)_2\}(CO)_5]$. Reactions of (5) with various ligands lead to monosubstituted derivatives[33] while hydrolysis (Equation (4)) results in formation of $[Mn\{PO(O_2C_6H_4)\}(CO)_5]$, (6), apparently via an intermediate, $[Mn\{PO(OC_6H_4OH)_2\}(CO)_5]$.[34-5]

A compound similar to (6), $[Mn\{P(OMe)_3\}_2\{PO(OMe)_2\}(CO)_3]$,[36] forms by $KBHBu^s_3$ reduction of $[Mn\{P(OMe)_3\}_3(CO)_3]^+$.

Two manganese–carbonyl complexes with a terminal $PR_2$ group were encountered. The first, $[Mn\{P(Cp^*)(OR)\}(CO)_4]$ (R = 2,4,6-tri-$t$-butylphenyl group), was obtained from a reaction of $Na[Mn(CO)_5]$ and $ClP(Cp^*)(OR)$. To provide manganese with an 18-electron configuration in this species and account for the short metal–phosphorus distance of 0.2084 nm, the phosphorus ligand is assumed to act as a three-electron donor.[37] The second complex is $[(OC)_3Fe\{\mu\text{-}Pr^iPP\text{-}[Mn(CO)_5]PPr^i\}Fe(CO)_3]$, prepared from $Na[Mn(CO)_5]$ and $[(OC)_3Fe(\mu\text{-}Pr^iPPClPPr^i)Fe(CO)_3]$.[38]

(4)

(5)　　　　　　　　　　　　　　(6)

Two compounds, $[As\{Mn(CO)_5\}\{Cr(CO)_5\}_2]$ and $[As\{Cr_2(\mu\text{-}CO)(CO)_8\}\{Mn(CO)_5\}]$, are reported among a larger group of compounds containing three metals bonded to arsenic.[39] Other complexes with arsenic, antimony, and bismuth ligands include $[As\{Mn(CO)_5\}_3]$ and a decarbonylation product $[(OC)_4Mn=As\{Mn(CO)_5\}_2]$ **(7)**[40], $[Bi\{Mn(CO)_5\}_{3-n}X_n]$ (X = Cl, Br, I; $n$ = 1, 2,[41] 3[42]), and $[Mn(BiPh_2)(CO)_5]$.[43] In addition, $[Mn(SbCl_2)(CO)_5]$ and $[ClSb\{Mn(CO)_5\}_2]$, are reported to be formed in reactions of $Na[Mn(CO)_5]$ and $SbCl_3$; they were not isolated, but isolable derivatives, $[Mn(CO)_5\text{-}(\mu\text{-}SbCl_2)Mn(CO)_2(\eta\text{-}C_5H_4Me)]$ and $[Mn_2(CO)_{10}(\mu\text{-}SbCl)Mn(CO)_2(\eta\text{-}C_5H_4Me)]$, form in a further reaction with $[Mn(CO)_2(THF)(\eta\text{-}C_5H_4Me)]$.[44]

(7)

As mentioned earlier, complexes with $PR_2$ groups bridging two metals are among the most common entries in this section. This area has been extended by further study since the mid-1980s.

The reaction between $[Mn(Cl)(CO)_5]$ and TMS-$PH_2$ yields two products, $[Mn_2(\mu\text{-}PH_2)_2(CO)_8]$ and $[Mn_3(\mu\text{-}PH_2)_3(CO)_{12}]$. From these species, $[Mn_2(\mu\text{-}PX_2)_2(CO)_8]$ and $[Mn_3(\mu\text{-}PX_2)_3(CO)_{12}]$ (X = Cl, Br, I) can be prepared by H–X exchange using $CX_4$. Fluorophosphine analogues can be obtained from the $PCl_2$ complexes by chlorine–fluorine exchange using $AgBF_4$;[45] $[Mn_2(\mu\text{-}PF_2)_2(CO)_8]$ is also a product of the reaction between $[Mn_2(CO)_{10}]$ and $PF_3$.[46] An NMR study on the dimeric $PH_2$ and $PF_2$ compounds is noted.[47]

The preparation of $[Mn_2(\mu\text{-}PHPh)_2(CO)_8]$ by reaction of $PhPH_2$ and $[Mn_2(CO)_{10}]$ is described. *Cis-* and *trans*-phenyl isomers are present according to NMR data. The *trans*-isomer can be isolated by careful crystallization; it isomerizes to the *cis*-isomer on heating. Also, $[Mn_2(\mu\text{-}PHPh)_2(CO)_8]$ is a precursor to $[Mn_2(\mu\text{-}PClPh)_2(CO)_8]$, which is formed by H–Cl exchange, and to $[Mn_2(\mu\text{-}PRPh)_2(CO)_8]$ (R = Me, Et, Pr, COMe, $CH_2COMe$, $CO_2Et$), which are formed in a two-step process involving deprotonation using BuLi and alkylation by the appropriate alkyl halide.[48]

Preparation of $[Mn_2(\mu\text{-}PPhSPh)(\mu\text{-}SPh)(CO)_8]$ is accomplished by a reaction between $Na[Mn(CO)_5]$ and PPh(SPh)Cl; $[Mn(SPh)(CO)_5]$ and $[Mn\{PPh(SPh)\}(CO)_5]$ are identified in the reaction mixture at its early stages.[49] Reactions between $[Mn_2(\mu\text{-}H)_2(CO)_6(P\text{–}P)]$ (P–P = dppm, $(EtO)_2POP(OEt)_2$) and a phosphine ($PH_3$, $PCyH_2$, $PPh_2H$) lead to complexes having bridging phosphido groups, $[Mn_2(\mu\text{-}H)\text{-}(\mu\text{-}PR^1R^2)(CO)_6(P\text{–}P)]$ ($PR^1R^2$ = $PH_2$, PHCy, $PPh_2$).[50]

Photolysis of $[Mn_2(\mu\text{-}PPh_2)_2(CO)_8]$ results in formation of $[Mn_2(\mu\text{-}PPh_2)_2(\mu\text{-}CO)(CO)_6]$.[51] Radiochemical oxidation of $[Mn_2(\mu\text{-}AsPh_2)_2(CO)_8]$ produces the cation radical, $[Mn_2(\mu\text{-}AsPh_2)_2(CO)_8]^+$; ESR studies indicate that the unpaired electron resides in the Mn–Mn $\sigma^*$-orbital.[52] Reactions between $[Mn_2(\mu\text{-}PPh_2)_2(CO)_8]$ and ethyne and allene have been reported;[53] products characterized in these reactions include **(8)** and **(9)**.

(8)　　　　　　　　　　　　　(9)

The compound $[Mn_2(\mu\text{-}H)(\mu\text{-}PPh_2)(CO)_8]$, first reported in 1967 as a product of the reaction between $Na[Mn(CO)_5]$ and $PPh_2Cl$, has been the starting point for considerable further study since the mid-1980s. A much improved preparation, from $[Mn_2(CO)_{10}]$ and $PHPh_2$, has been developed (Equation (5)).[5,54] A similar procedure was also used to prepare two related compounds, $[Mn_2(\mu\text{-}H)(\mu\text{-}PR_2)(CO)_8]$ (R = $NPr^i_2$,[55] $Cy^{56}$).

$$[Mn_2(CO)_{10}] + PHR_2 \longrightarrow [Mn_2(\mu\text{-}H)(\mu\text{-}PR_2)(CO)_8] \tag{5}$$

R = Ph, Cy, $P(NPr^i_2)$

Further chemistry of $[Mn_2(\mu\text{-}H)(\mu\text{-}PPh_2)(CO)_8]$ has been vigorously pursued. Deprotonation of $[Mn_2(\mu\text{-}H)(\mu\text{-}PPh_2)(CO)_8]$ with $NaBH_4$ gave $[Mn_2(\mu\text{-}PPh_2)(CO)_8]^-$, which was characterized as a $PPN^+$ salt.[5,57] This anionic complex reacts with various XCl species to give $[Mn_2(\mu\text{-}X)(\mu\text{-}PPh_2)(CO)_8]$ (X = $Cu(PEt_3)_3$, $Au(PPh_3)$, $Ag(PPh_3)$, HgCl, HgPh, HgCN).[57,58] A second group of reactions of $[Mn_2\text{-}(\mu\text{-}H)(\mu\text{-}PPh_2)(CO)_8]$ involves addition to ethynes, to give complexes with a bridging vinyl group, $[Mn_2\text{-}(\mu\text{-}CR=CHR)(\mu\text{-}PPh_2)(CO)_7]$ (R = H, Ph).[59] Reactions of the $\mu$-vinyl complexes with nucleophiles ($H^-$, $CNBu^t$, CO, $PEt_3$) vary considerably. With hydride ion, products include $[PPN][Mn_2(\mu\text{-}H)(\mu\text{-}COCR=CHR)(\mu\text{-}PPh_2)(CO)_6]$ and $[PPN][Mn_2(\mu\text{-}H)_2(\mu\text{-}PPh_2)(CO)_6]$; with isonitrile and phosphine, carbonyl replacement occurs.[60]

Compounds (**10**) and (**11**) are products in the reactions of $Li_2[Mn(PPh_2)(CO)_4]$ with $R^1CHCl_2$ ($R^1$ = Me, Ph) and $R^2PCl_2$ ($R^2$ = 2,4,6-tri-*t*-butylphenyl), respectively.[61] The organometallic precursor forms by deprotonation of $[Mn(H)(CO)_4(PPh_2H)]$ using BuLi. Compound (**11**) undergoes a reaction with $XC{\equiv}CX$ (X = $CO_2Me$), forming $[Mn(PPh_2PR^2CX{=}CX)(CO)_4]$.[62]

(**10**)　　　　　　　　(**11**)

Compounds with $R_2PS$ ligands, studied extensively by Lindner *et al.*,[63-5] are prepared via reaction of $[Mn(Br)(CO)_5]$ and secondary phosphine sulfides, $R^1_2PH{=}S$, in the presence of a tertiary amine. Mononuclear or dinuclear species may be obtained depending on the size of the organic groups on phosphorus, and these species exist in equilibrium in solution. One mononuclear complex, $[Mn(SPCy_2)(CO)_4]$, adds CO under pressure, forming $[Mn(PCy_2{=}S)(CO)_5]$ (phosphorus bonded to manganese) and the dimeric complexes ($R^1$ = Me, Et, Cy) react with $R^2O_2CC{\equiv}CCO_2R^2$ ($R^2$ = Me, Et, $Pr^i$, Cy) in two steps giving (**12**) and (**13**).

(**12**)　　　　　　　　(**13**)

The reaction of $MeP({=}S)Cl_2$ and $[Mn_2(CO)_{10}]$ in the presence of activated magnesium produces $[Mn\{SPMe[Mn(CO)_5]\}(CO)_4]$. This compound adds two molecules of $RO_2CC{\equiv}CCO_2R$ (R = Me, Et, Cy) sequentially, to give products similar to (**12**) and (**13**).[66-7]

Phosphido-bridged heterobimetallic complexes have gained much attention. Various synthetic routes are available to form such complexes.

A general synthetic route uses the reaction of a metal halide complex and $Na[Mn(CO)_5]$. Examples include reactions of $Na[Mn(CO)_5]$ with $[Ru(Cl)_2(PPh_3)_3]$, giving $[Mn(CO)_3(PPh_3)(\mu\text{-}PPh_2)\text{-}Ru\text{-}(CO)_3(PPh_3)]$ as the product,[68] with $[Fe\{P(H)(Me)Cl\}(CO)_4]$, giving $[Mn(CO)_4\{\mu\text{-}P(H)Me\}Fe(CO)_4]$,[69] and with $[Ru(Cl)(CO)(PPh_2H)(\eta\text{-}Cp)]$, giving $[(\eta\text{-}Cp)Ru(CO)(\mu\text{-}H)(\mu\text{-}PPh_2)Mn(CO)_4]$.[70]

Another general route involves reaction of a bimetallic metal carbonyl and a secondary phosphine. This is the route used to prepare $[Mn(CO)_4(\mu\text{-H})(\mu\text{-PPh}_2)Mo(CO)_2(\eta\text{-Cp})]$ from $PPh_2H$ and $[Mn(CO)_5Mo(CO)_3(\eta\text{-Cp})]$.[71] The procedure can also be used for the synthesis of $[(\eta\text{-Cp})Ru(CO)(\mu\text{-H})(\mu\text{-PPh}_2)Mn(CO)_4]$; this is one of six products obtained in the photolysis of $[(OC)_5MnRu(CO)_2(\eta\text{-Cp})]$ in the presence of $PPh_2H$.[72]

*In situ* formation of $[Pd(NCPh)_2\{Mn(CO)_5\}_2]$ in a reaction between $[Pd(Cl)_2(NCPh)_2]$ and $Na[Mn(CO)_5]$ has been noted; this species reacts with $PCy_2H$ to give $[(Cy_2PH)_2Pd(\mu\text{-PCy}_2)Mn(CO)_4]$.[73] A similar platinum complex, $[(Ph_3P)_2Pt(\mu\text{-PPh}_2)Mn(CO)_4]$ (14), is formed in a reaction of $Na[Mn(CO)_5]$, $PPh_2Cl$, and $[Pt(\eta\text{-C}_2H_4)(PPh_3)_2]$.[74] When $[Pt(cod)_2]$ replaces $[Pt(\eta\text{-C}_2H_4)(PPh_3)_2]$ in this reaction a different platinum–manganese compound, $[(OC)Pt\{(\mu\text{-PPh}_2)Mn(CO)_4\}_2]$, is obtained. The platinum–manganese species (15) is formed by reaction of $[(Ph_3P)_2Pt(\mu\text{-H})(\mu\text{-SiMe}_2)Mn(CO)_4]$ with water.[75]

(structures 14 and 15)

(14)        (15)

Other routes to bimetallic complexes having hydrido and phosphido bridges include photolysis of $[Mn(CO)_4\{\mu\text{-P}(p\text{-Tol})_2\text{-}\eta\text{-C}_5H_4\}Mo(CO)_3]$ with $H_2$, giving $[Mn(CO)_4(\mu\text{-H})\{\mu\text{-P}(p\text{-Tol})_2\}Mo(CO)_2(\eta\text{-Cp})]$,[76] and the reaction between $[Mn(H)(CO)_5]$ and $[Fe(PPh_2)(CO)_2(\eta\text{-Cp})]$, giving $[Mn(CO)_4(\mu\text{-H})(\mu\text{-PPh}_2)Fe(CO)(\eta\text{-Cp})]$, (Equation (6)).[77] When $[Mn(Me)(CO)_5]$ is used in the latter reaction system, $[Mn(CO)_4(\mu\text{-COMe})(\mu\text{-PPh}_2)Fe(CO)(\eta\text{-Cp})]$ is formed as the product.

$$[Mn(H)(CO)_5] + [Fe(PPh_2)(CO)_2(\eta\text{-Cp})] \longrightarrow [Mn(CO)_4(\mu\text{-H})(\mu\text{-PPh}_2)Fe(CO)(\eta\text{-Cp})] \quad (6)$$

Various reactions of these complexes have been studied. Reactions of $[(\eta\text{-Cp})Ru(CO)(\mu\text{-H})(\mu\text{-PPh}_2)Mn(CO)_4]$ with alkynes have been shown to give $[(\eta\text{-Cp})Ru(CO)(\mu\text{-RC=CHR})(\mu\text{-PPh}_2)Mn(CO)_3]$.[70] Photolysis of $[Mn(CO)_4(\mu\text{-H})(\mu\text{-PPh}_2)Mo(CO)_2(\eta\text{-Cp})]$ with ethene forms $[Mn(CO)_4(\mu\text{-PPh}_2)(\mu\text{-COEt})Mo(CO)_2(\eta\text{-Cp})]$,[78] while a related study describes reactions of this precursor with dienes.[79] In a different study, $[Mn(CO)_4(\mu\text{-H})(\mu\text{-PPh}_2)Mo(CO)_2(\eta\text{-Cp})]$ was deprotonated (with KOH/EtOH) and the product, $[Et_4N][Mn(CO)_4(\mu\text{-PPh}_2)Mo(CO)_2(\eta\text{-Cp})]$, used to prepare several other complexes including $[Mn(CO)_4(\mu\text{-I})(\mu\text{-PPh}_2)Mo(CO)_2(\eta\text{-Cp})]$ (from a reaction with $I_2$) and $[Mn(CO)_4\{\mu\text{-M}(PPh_3)\}(\mu\text{-PPh}_2)Mo(CO)_2(\eta\text{-Cp})]$ (M = Ag, Au).[80] Heating $[Mn(CO)_4(\mu\text{-PPh}_2\text{-}\eta\text{-C}_5H_4)Mo(CO)_2(PPh_2H)]$ leads to formation of $[Mn(CO)_3(\mu\text{-H})(\mu\text{-PPh}_2)(\mu\text{-PPh}_2\text{-}\eta\text{-C}_5H_4)Mo(CO)_2]$.[81]

Complexes having manganese and gold atoms with bridging phosphido groups have been prepared by reactions of several gold complexes and cationic complexes such as $[Mn(CO)_5(PPh_2H)]^+$; the simplest of these is $[Mn(CO)_5(\mu\text{-PPh}_2)Au(C_6F_5)]$.[82]

## 3.4 COMPLEXES WITH SULFUR, SELENIUM, AND TELLURIUM LIGANDS

Significant advances in synthetic methodology in this area have been made since the mid-1980s. Previously unknown anionic complexes, $[Mn_2(\mu\text{-SR})_3(CO)_6]^-$ (R = Me, Ph), were prepared in 1985 by two research groups.[83-4] Several synthetic methods were described, including reactions of $SR^-$ with $[Mn(Br)_2(CO)_4]^-$, $[Mn_2(\mu\text{-X})_2(CO)_8]$ (X = Br, SR), or $[Mn_2(\mu\text{-Br})_3(CO)_6]^-$. The following year, other routes to these species were reported using reactions of $Na[Mn(CO)_5]$ with either phthSPh[12] (phth = phthalimide) or RSSR.[85] These reactions apparently occur via the intermediate $[Mn(SR)(CO)_5]$, formed when $[Mn(CO)_5]^-$ displaces either $phth^-$ or $SR^-$ from the organic reagent.

Cyclic voltametric studies on the $[Mn_2(\mu\text{-SR})_3(CO)_6]^-$ complexes reveal a pair of one-electron oxidations. The thiolate groups are labile in acid, and so it is possible to carry out thiolate group exchange (complexed SPh favored over SMe) (Equation (7)). Preparation of $[Mn_2(\mu\text{-SR})_2(CO)_6(L)_2]$ (L = $PR_3$, CO) can be accomplished by the reaction of $[Mn_2(\mu\text{-SR})_3(CO)_6]^-$, $H^+$, and the ligand while, in the absence of added ligand, $[Mn_4(\mu\text{-SR})_4(CO)_{12}]$ forms.[84]

$$[Mn_2(\mu\text{-SR}^1)_3(CO)_6]^- + 3 R^2SH \longrightarrow [Mn_2(\mu\text{-SR}^2)_3(CO)_6]^- + 3 R^1SH \quad (7)$$

Preparations of several complexes with an SH⁻ ligand are reported by two routes. The reaction of propylene sulfide and $[Mn(H)(CO)_5]$ gives $[Mn_2(\mu\text{-}SH)_2(CO)_8]$.[86] An alternative route to these SH complexes involves cleavage of SnPh₃ groups in $[Mn_2(\mu\text{-}SSnPh_3)_2(CO)_8]$ and $[Mn(SSnPh_3)(CO)_{5-n}(PMe_3)_n]$ $(n = 2, 3)$ by HCl (Equation (8)).[87]

$$[Mn(SSnPh_3)(CO)_{5-n}(PMe_3)_n] + HCl \longrightarrow [Mn(SH)(CO)_{5-n}(PMe_3)_n] + Ph_3SnCl \qquad (8)$$
$$n = 2, 3$$

A new synthetic method is described for the preparation of $[Mn_2(\mu\text{-}SCF_3)_2(CO)_8]$, using the reaction of $[Mn_2(CO)_{10}]$ with CF₃S radicals.[88]

Complexes having R₃PCS₂ ligands have been extensively studied. Synthesis of manganese complexes of these ligands were first reported in 1987,[89] with additional papers appearing in the following years[90-1] providing further details. Preparation of $[Mn(Br)(S_2CPCy_3)(CO)_3]$ (16) can be achieved by the reaction between $[Mn(X)(CO)_5]$ and $Cy_3PCS_2$ in refluxing CS₂. In refluxing toluene, however, this reaction takes a different path with $[Mn_2(\mu\text{-}S_2CPCy_3)(CO)_6]$ (17) being isolated instead. The reaction depends on the R groups in this reagent; with Et₃PCS₂ in toluene, $[Mn(S_2CPEt_3)(CO)_2(PEt_3)_2][Br]$ forms; in CS₂, surprisingly, $[Mn(Br)(CO)_{5-n}(PEt_3)_n]$ $(n = 1, 2)$ complexes are the isolated products. Various cationic complexes were obtained from reactions of metal carbonyl halides, AgClO₄, and R₃PCS₂, with the ligand serving either in a monodentate or a bidentate fashion. In addition, NaBH₄ converts $[Mn(S_2CPEt_3)(CO)_2(PEt_3)_2][ClO_4]$ to $[Mn(S_2CH)(CO)_2(PEt_3)_2]$ and $[Mn(S_2CPCy_3)(CO)_4][ClO_4]$ to $[Mn_2(\mu\text{-}S_2CHPCy_3)(CO)_6]$.[91]

(16)

(17)

The dinuclear complexes $[Mn_2(\mu\text{-}S_2CPR_3)(CO)_6]$ (R = Cy, Pr$^i$) are also obtained by the reaction of $[Mn_2(CO)_{10}]$ and R₃PCS₂.[92] Reactions of $[Mn_2(\mu\text{-}S_2CPR_3)(CO)_6]$ species with halogens (giving $[Mn(X)(CO)_3(S_2CPR_3)]$) and with dppm (carbonyl replacement) are also described. In another study, several chemical reaction sequences were reported for these species. For example, reduction with Na/Hg followed by protonation and metathesis gives $[NEt_4][Mn_2(\mu\text{-}S_2CH)(CO)_6]$; in contrast, hydride reduction and alkylation yields $[Mn_2(\mu\text{-}H)\{\mu\text{-}MeSC(PR_3)S\}(CO)_6]$.[93]

Two further papers from the same research group report the preparation of $[Mn(S_2CPR_3)(SnCl_3)(CO)_3]$ (by SnCl₂ addition to the chloromanganese complex)[94] and $[(OC)_3Mn\text{-}(\mu\text{-}Br)(\mu\text{-}S_2CPPr^i_3)Mo(CO)_3]$ (by reaction of $[Mn(Br)(S_2CPPr^i_3)(CO)_3]$ and $[Mo(CO)_3(RCN)_3]$).[95]

Related studies have introduced other RCS₂ ligands by this scenario. The complexes $[Mn(S_2CSnPh_3)(CO)_4]$[96] and $[Mn(X)(S_2CCMe_2PPh_3)(CO)_3]$[97] (X = Cl, Br, I) are products of the reaction of $[Mn(X)(CO)_5]$ with $Li[S_2CSnPh_3]$ and $S_2CCMe_2PPh_3$, respectively. Formation of $[Mn(S_2CSnPh_3)(CO)_3(L)]$ by carbonyl substitution was described separately.[98] A bimetallic complex, $[Mn\{S_2CFe(CO)_2(\eta\text{-}Cp)\}(CO)_4]$, is obtained from a similar reaction with $Na[Fe(CS_2)(CO)_2(\eta\text{-}Cp)]$,[99] while dithiocarbamate complexes, $[Mn(S_2CNHR)(CO)_4]$ (R = Ph, Me) are prepared by the reaction of $[Mn(CONHMe)(NH_2Me)(CO)_4]$ and RNCS.[100]

Studies on $[Mn_2(\mu\text{-}X)_2(CO)_6(P\text{–}P)]$ $(P\text{–}P = (EtO)_2POP(OEt)_2$, dppm) have produced several compounds with sulfur-containing bridging ligands. The complexes $[Mn_2(\mu\text{-}H)(\mu\text{-}X)(CO)_6(P\text{–}P)]$ (X = SH, S-$p$-Tol)[101] are formed by reactions of $[Mn_2(\mu\text{-}H)_2(CO)_6(P\text{–}P)]$ with either sulfur or $p$-TolSH. The SH-containing complex can be deprotonated, and a series of other derivatives (X = SMe, SSnMe₃, SAu(PPh₃))[102] are obtained from the anionic product by reaction with the appropriate organometallic or organic halide.

Complexes with bridging thionyl and sulfuryl groups are known. The reaction between $[Mn(H)(CO)_4(PPh_3)]$ and SOCl₂ gives $[Mn_2(\mu\text{-}SO)(CO)_8(PPh_3)_2]$ and $[Mn_2(\mu\text{-}SO_2)(CO)_8(PPh_3)_2]$.[103] Sulfur dioxide inserts into the metal–metal bond in the bimetallic complex, $[MnRh(\mu\text{-}dppm)_2(CO)_4]$, giving $[MnRh(\mu\text{-}SO_2)(\mu\text{-}dppm)_2(CO)_4]$. Also reported in this study is a reaction of the starting material with H₂S, forming $[MnRh(\mu\text{-}S)(\mu\text{-}dppm)_2(CO)_4]$.[104]

Other sulfur-containing complexes include species with perthiocarbonate (18)[105] and tetrathiosquarate (19) ligands.[106] The reaction between $Na[Mn(CO)_5]$ and RCSCl forms *cis-* and *trans-* $[(OC)_5MnSRC\equiv CRS(Mn(CO)_5]$;[107-8] this reacts with $CyO_2CC\equiv CCO_2Cy$ to form (20).

(18)

(19)

(20)

Minimal attention has been given to manganese complexes containing selenium and tellurium ligands. Depending on stoichiometry, the reaction of $[Mn_2(CO)_{10}]$ and $Se_3^{2-}$ gives either $[Mn_2(Se_2)_2(CO)_6]^{2-}$ or $[Mn_2(Se_4)_2(CO)_6]^{2-}$, isolated and characterized as $PPN^+$ salts.[109] Formation of $[Mn_2(\mu-Te_2)(CO)_6(PEt_3)_4]$ from the reaction of $[Mn_2(CO)_{10}]$ and $Te=PEt_3$ is described,[110] while the reaction of this tellurium reagent and $[Mn(Bz)(CO)_5]$ gives $[Mn(TeBz)(CO)_3(PEt_3)_2]$.[111]

## 3.5 REFERENCES

1. B. Walther, *Coord. Chem. Rev.*, 1984, **60**, 67.
2. C. M. Lukehart, *Adv. Organomet. Chem.*, 1986, **25**, 45.
3. S. P. Schmidt, J. Nitschke and W. C. Trogler, in 'Inorganic Syntheses', ed. H. D. Kaesz, Academic Press, New York, 1989, vol. 26, p. 113.
4. E. Lindner, A. Rau and V. Käss, in 'Inorganic Syntheses', ed. H. D. Kaesz, Academic Press, New York, 1989, vol. 26, p. 162.
5. J. A. Iggo and M. J. Mays, in 'Inorganic Syntheses', ed. H. D. Kaesz, Academic Press, New York, 1989, vol. 26, p. 225.
6. P. M. Treichel, in 'COMC-I', vol. 4, p. 1.
7. M. Appel, K. Schloter, J. Heidrich and W. Beck, *J. Organomet. Chem.*, 1987, **322**, 77.
8. S. P. Mallela and F. Aubke, *Inorg. Chem.*, 1985, **24**, 2969.
9. S. K. Mandal, D. M. Ho and M. Orchin, *Inorg. Chem.*, 1991, **30**, 2244.
10. K. D. Abney, K. M. Long, O. P. Anderson and S. H. Strauss, *Inorg. Chem.*, 1987, **26**, 2638.
11. H. Adams, N. A. Bailey, T. N. Briggs, J. A. McCleverty, H. M. Colquhoun and D. J. Williams, *J. Chem. Soc., Dalton Trans.*, 1986, 813.
12. P. M. Treichel, P. C. Nakagaki and K. J. Haller, *J. Organomet. Chem.*, 1987, **327**, 327.
13. M. P. Gamasa, E. Garcia, J. Gimeno and C. Ballesteros, *J. Organomet. Chem.*, 1986, **307**, 39.
14. W. Dannecker and H. W. Müller, *Z. Naturforsch., Teil B*, 1982, **37**, 318.
15. H. J. Meder and W. Beck, *Z. Naturforsch., Teil B*, 1986, **41**, 1247.
16. P. Jaitner and P. Peringer, *Transition Met. Chem.*, 1984, **9**, 325.
17. E. Lindner and A. Brosamle, *Chem. Ber.*, 1984, **117**, 2730.
18. C. J. Besecker, V. W. Day and W. G. Klemperer, *Organometallics*, 1985, **4**, 564.
19. C. J. Besecker, V. W. Day, W. G. Klemperer and M. R. Thompson, *Inorg. Chem.*, 1985, **24**, 44.
20. W. Klaui, J. Okuda, M. Scotti and H. Valderrama, *J. Organomet. Chem.*, 1985, **280**, C26.
21. F. J. García Alonso, M. G. Sans and V. Riera, *J. Organomet. Chem.*, 1991, **421**, C12.
22. S. B. Copp, S. Subramanian and M. J. Zaworotko, *J. Am. Chem. Soc.*, 1992, **114**, 8719.
23. O. J. Scherer, J. Kerth and W. S. Sheldrick, *Angew. Chem., Int. Ed. Engl.*, 1984, **23**, 156.
24. P. L. Motz, J. J. Alexander and C. F. Campana, *J. Organomet. Chem.*, 1989, **379**, 119.
25. P. L. Motz, J. P. Williams, J. J. Alexander, D. M. Ho, J. S. Ricci and W. T. Miller, *Organometallics*, 1989, **8**, 1523.
26. B. M. Louie, S. J. Rettig, A. Storr and J. Trotter, *Can. J. Chem.*, 1985, **63**, 2261.
27. D. A. Cooper, S. J. Rettig, A. Storr and J. Trotter, *Can. J. Chem.*, 1986, **64**, 1643.
28. G. D. Gracey, S. J. Rettig, A. Storr and J. Trotter, *Can. J. Chem.*, 1987, **65**, 2469.
29. E. C. Onyiriuka and A. Storr, *Can. J. Chem.*, 1987, **65**, 1367.
30. M. Lattman, B. N. Anand, D. R. Garrett and M. A. Whitener, *Inorg. Chim. Acta.*, 1983, **76**, L139.
31. M. Lattman, B. N. Anand, S. S. C. Chu and R. D. Rosenstein, *Phosphorus Sulfur*, 1983, **18**, 303.

32. B. N. Anand, R. Bains, and Km. Usha, *J. Chem. Soc., Dalton Trans.*, 1990, 2315.
33. S. K. Chopra, S. S. C. Chu, P. DeMeester, D. E. Geyer, M. Lattman and S. A. Morse, *J. Organomet. Chem.*, 1985, **294**, 347.
34. M. Lattman, B. N. Anand, S. S. C. Chu and R. D. Rosenstein, *Organometallics*, 1984, **3**, 670.
35. S. S. C. Chu, R. D. Rosenstein, M. Lattman and B. N. Anand, *J. Organomet. Chem.*, 1984, **265**, 45.
36. G. Weiler, G. Huttner, L. Zsolnai and H. Berke, *Z. Naturforsch., Teil B*, 1987, **42**, 203.
37. H. Lang, M. Leise and C. Emmerich, *J. Organomet. Chem.*, 1991, **418**, C9.
38. R. B. King, F.-J. Wu and E. M. Holt, *Inorg. Chem.*, 1988, **27**, 1241.
39. G. Huttner, U. Weber, B. Sigwarth, O. Scheidsteger, H. Lang and L. Zsolnai, *J. Organomet. Chem.*, 1985, **282**, 331.
40. V. Grossbruchhaus and D. Rehder, *Inorg. Chim. Acta*, 1990, **172**, 141.
41. N. A. Compton *et al.*, *J. Chem. Soc., Dalton Trans.*, 1991, 669.
42. J. M. Wallis, G. Müller and H. Schmidbaur, *Inorg. Chem.*, 1987, **26**, 458.
43. J. M. Cassidy and K. H. Whitmire, *Inorg. Chem.* 1991, **30**, 2788.
44. A. Lombard, G. Huttner and L. Zsolnai, *J. Organomet. Chem.*, 1988, **352**, 295.
45. H. Schäfer, J. Zipfel, B. Gutekunst and Y. Lemmert, *Z. Anorg. Allg. Chem.*, 1985, **529**, 157.
46. H. Schäfer, J. Zipfel, B. Migula and D. Binder, *Z. Anorg. Allg. Chem.*, 1983, **501**, 111.
47. C. C. Grimm, P. E. Brotman and R. J. Clark, *Organometallics*, 1990, **9**, 1119.
48. M. P. Brown, J. Buckett, M. M. Harding, R. M. Lynden-Bell, M. J. Mays and K. W. Woulfe, *J. Chem. Soc., Dalton Trans.*, 1991, 3097.
49. E. Lindner, K. Auch, W. Hiller and R. Fawzi, *Z. Naturforsch., Teil B*, 1987, **42**, 454.
50. R. Carreño, V. Riera and M. A. Ruiz, *J. Organomet. Chem.*, 1991, **419**, 163.
51. T. Kawamura *et al.*, *J. Organomet. Chem.*, 1984, **276**, C10.
52. T. Kawamura, S. Enoki, S. Hayashida and T. Yonezawa, *Bull. Chem. Soc. Jpn.*, 1982, **55**, 3417.
53. L. Manojlović-Muir, M. J. Mays, K. W. Muir and K. W. Woulfe, *J. Chem. Soc., Dalton Trans.*, 1992, 1531.
54. J. A. Iggo, M. J. Mays, P. R. Raithby and K. Hendrick, *J. Chem. Soc., Dalton Trans.*, 1983, 205.
55. R. B. King, W.-K. Fu and E. M. Holt, *Inorg. Chem.*, 1986, **25**, 2390.
56. A. M. Arif, R. A. Jones and S. T. Schwab, *J. Organomet. Chem.*, 1986, **307**, 219.
57. J. A. Iggo, M. J. Mays and P. R. Raithby, *J. Chem. Soc., Dalton Trans.*, 1984, 633.
58. J. A. Iggo and M. J. Mays, *J. Chem. Soc., Dalton Trans.*, 1984, 643.
59. J. A. Iggo, M. J. Mays, P. R. Raithby and K. Hendrick, *J. Chem. Soc., Dalton Trans.*, 1983, 205.
60. K. Hendrick, M. McPartlin, J. A. Iggo, A. C. Kemball and M. J. Mays, *J. Chem. Soc., Dalton Trans.*, 1987, 2669.
61. E. Lindner, E. Ossig and M. Darmuth, *J. Organomet. Chem.*, 1989, **379**, 107.
62. E. Lindner, M. Darmuth, R. Fawzi and M. Steimann, *Chem. Ber.*, 1992, **125**, 2713.
63. E. Lindner and V. Käss, *Chem. Ber.*, 1989, **122**, 2269.
64. E. Lindner, V. Käss, W. Hiller and R. Fawzi, *Angew. Chem., Int. Ed. Engl.*, 1989, **28**, 448.
65. E. Lindner, V. Käss and H. A. Mayer, *Chem. Ber.*, 1990, **123**, 783.
66. E. Lindner, K. Auch, W. Hiller and R. Fawzi, *Organometallics*, 1988, **7**, 402.
67. E. Lindner, K. Auch, G. A. Weiss, W. Hiller and R. Fawzi, *Chem. Ber.*, 1986, **119**, 3076.
68. S. Sabo, B. Chaudret and D. Gervais, *J. Organomet. Chem.*, 1983, **258**, C19.
69. M. Müller and H. Vahrenkamp, *Chem. Ber.*, 1983, **116**, 2322.
70. A. J. M. Caffyn, M. J. Mays and P. R. Raithby, *J. Chem. Soc., Dalton Trans.*, 1992, 515.
71. A. D. Horton, M. J. Mays and P. R. Raithby, *J. Chem. Soc., Dalton Trans.*, 1987, 1557.
72. A. J. M. Caffyn, M. J. Mays and P. R. Raithby, *J. Chem. Soc., Dalton Trans.*, 1991, 2349.
73. P. Braunstein, E. de Jésus, A. Tiripicchio and M. Tiripicchio Camellini, *J. Organomet. Chem.*, 1989, **368**, C5.
74. P. Braunstein, E. de Jésus, A. Dedieu, M. Lanfranchi and A. Tiripicchio, *Inorg. Chem.*, 1992, **31**, 399.
75. J. Powell, J. F. Sawyer and M. Shiralian, *Organometallics*, 1989, **8**, 577.
76. C. P. Casey and R. M. Bullock, *Organometallics*, 1984, **3**, 1100.
77. R. P. Rosen, J. B. Hoke, R. R. Whittle, G. L. Geoffroy, J. P. Hutchinson and J. A. Zubieta, *Organometallics*, 1984, **3**, 846.
78. T. Adatia, K. Hendrick, A. D. Norton, M. J. Mays and M. McPartlin, *J. Chem. Soc., Chem. Commun.*, 1986, 1206.
79. A. D. Horton, M. J. Mays and P. R. Raithby, *J. Chem. Soc., Chem. Commun.*, 1985, 247.
80. A. D. Horton, M. J. Mays, T. Adatia, K. Hendrick and M. McPartlin, *J. Chem. Soc., Dalton Trans.*, 1988, 1683.
81. M. J. Doyle, T. J. Duckworth, L. Manojlović-Muir, M. J. Mays, P. R. Raithby and F. J. Robertson, *J. Chem. Soc., Dalton Trans.*, 1992, 2703.
82. G. A. Carriedo, V. Riera, M. L. Rodríguez, P. G. Jones and J. Lautner, *J. Chem. Soc., Dalton Trans.*, 1989, 639.
83. J. W. McDonald, *Inorg. Chem.*, 1985, **24**, 1734.
84. P. M. Treichel and M. H. Tegen, *J. Organomet. Chem.*, 1985, **292**, 385.
85. P. M. Treichel and P. C. Nakagaki, *Organometallics*, 1986, **5**, 711.
86. W. Danzer, W. P. Fehlhammer, A. T. Liu, G. Thiel and W. Beck, *Chem. Ber.*, 1982, **115**, 1682.
87. R. Kury and H. Vahrenkamp, *J. Chem. Res. (S)*, 1982, 30.
88. T. R. Bierschenk and R. J. Lagow, *Inorg. Chem.*, 1983, **22**, 359.
89. D. Miguel, V. Riera, J. A. Miguel, C. Bois, M. Philoche-Levisalles and Y. Jeannin, *J. Chem. Soc., Dalton Trans.*, 1987, 2875.
90. D. Miguel, V. Riera, J. A. Miguel, F. Diego, C. Bois and Y. Jeannin, *J. Chem. Soc., Dalton Trans.*, 1990, 2719.
91. D. Miguel, J. A. Miguel, V. Riera and X. Soláns, *Angew. Chem., Int. Ed. Engl.*, 1989, **28**, 1014.
92. D. Miguel, V. Riera, J. A. Miguel, M. Gómez and X. Soláns, *Organometallics*, 1991, **10**, 1683.
93. B. Alvarez, S. García-Granda, Y. Jeannin, D. Miguel, J. A. Miguel and V. Riera, *Organometallics*, 1991, **10**, 3005.
94. B. Alvarez, D. Miguel, J. A. Pérez-Martínez, V. Riera and S. García-Granda, *J. Organomet. Chem.*, 1992, **427**, C33.
95. D. Miguel, J. A. Pérez-Martínez, V. Riera and S. García-Granda, *J. Organomet. Chem.*, 1991, **420**, C12.
96. T. Hättich and U. Kunze, *Angew. Chem., Int. Ed. Engl.*, 1982, **21**, 364.
97. U. Kunze, R. Merkel and W. Winter, *Angew. Chem., Int. Ed. Engl.*, 1982, **21**, 290.
98. U. Kunze and T. Hättich, *Chem. Ber.*, 1983, **116**, 3071.
99. L. Busetto, A. Palazzi and M. Monari, *J. Chem. Soc., Dalton Trans.*, 1982, 1631.
100. L. Busetto and A. Palazzi, *Inorg. Chim. Acta*, 1982, **64**, L39.

101. F. J. García Alonso, M. García Sanz, V. Riera, M. A. Ruiz, A. Tiripicchio and M. Tiripicchio Camellini, *Angew. Chem., Int. Ed. Engl.*, 1988, **27**, 1167.
102. F. J. García Alonso, M. García Sanz, V. Riera, S. García-Granda and E. Pérez Carreño, *J. Chem. Soc., Dalton Trans.*, 1992, 545.
103. I.-P. Lorenz, J. Messelhäuser, W. Hiller and M. Conrad, *J. Organomet. Chem.*, 1986, **316**, 121.
104. O. Heyke, W. Hiller and I.-P. Lorenz, *Chem. Ber.*, 1991, **124**, 2217.
105. H. Alper, F. Sibtain, F. W. B. Einstein and A. C. Willis, *Organometallics*, 1985, **4**, 604.
106. F. Götzfried, R. Grenz, G. Urban and W. Beck, *Chem. Ber.*, 1985, **118**, 4179.
107. E. Lindner, I. P. Butz, W. Hiller, R. Fawzi and S. Hoehne, *Angew. Chem., Int. Ed. Engl.*, 1983, **22**, 996.
108. E. Lindner, I. P. Butz, S. Hoehne, W. Hiller and R. Fawzi, *J. Organomet. Chem.*, 1983, **259**, 99.
109. S. C. O'Neal, W. T. Pennington and J. W. Kolis, *Inorg. Chem.*, 1990, **29**, 3134.
110. M. L. Steigerwald and C. E. Rice, *J. Am. Chem. Soc.*, 1988, **110**, 4228.
111. K. McGregor, G. B. Deacon, R. S. Dickson, G. D. Fallon, R. S. Rowe and B. O. West, *J. Chem. Soc., Chem. Commun.*, 1990, 1293.

# 4

# Manganese Hydrocarbon Complexes Excluding Cyclopentadienyl

## KEITH F. McDANIEL
*Ohio University, Athens, OH, USA*

## 4.1 INTRODUCTION

The chemistry of manganese hydrocarbon complexes has become increasingly important for organometallic, organic, and inorganic chemists since the 1980s, and the literature contains many papers describing both the preparation and reactions of a variety of complexes. While the majority of manganese $\pi$-complexes investigated incorporate $\eta^5$-cyclopentadienyl ligands, the chemistry of $\eta^2$-alkene, $\eta^2$-alkyne, $\eta^3$-allyl, $\eta^4$-diene, $\eta^5$-dienyl, and $\eta^6$-arene complexes has also been examined. Although the most prominent reaction of many of these manganese $\pi$-complexes involves their interaction with nucleophiles, other notable reactions of these complexes include reactions with electrophiles, photochemical transformations, cycloaddition reactions, and a number of miscellaneous rearrangements.

## 4.2 REACTION WITH NUCLEOPHILES

### 4.2.1 ($\eta^6$-Arene)manganese Complexes

Nucleophilic attack on cationic ($\eta^6$-arene)manganese complex (1), to provide the *exo*-substituted tricarbonyl($\eta^5$-cyclohexadienyl)manganese complex (2) (Equation (1)), was first reported by Jones and Wilkinson in 1964,[1] and has been examined in increasing detail since then. An excellent review[2] of nucleophilic attack on transition metal $\pi$-complexes published in 1984 details many of these results, including examples of the use of hydrides, phosphines, hydroxide, and cyanide. Noteworthy additions to this list include trialkylaluminum reagents,[3] $\alpha$-iminoester, $\alpha$-iminonitrile, $\alpha$-sulfonyl, and $\alpha$-nitro carbanions,[4] $\alpha$-haloester carbanions and nitrile anions,[5] sodium cyclopentadienide,[6] $\alpha$-diazo anion,[7] and Wittig reagents.[8] Use of a Wittig reagent presents the opportunity for further transformation of the resulting ($\eta^5$-cyclohexadienyl)manganese complex (3), as deprotonation of complex (3) followed by reaction with aldehydes generates the expected alkene (4), mainly as the *cis* stereoisomer (Equation (2)).[8] Reaction of cation (1) with the tetracarbonylosmium dianion provides the bis[($\eta^5$-cyclohexadienyl)manganese]tetracarbonylosmium complex (5), a rare example of the reaction of a transition metal nucleophile with a transition metal $\pi$-complex (Equation (3)).[9] Substitution of a carbon monoxide ligand by phosphites has also been reported, and electrocatalytic carbon monoxide substitution by phosphites is chemically irreversible up to a range of 20 V s$^{-1}$.[10]

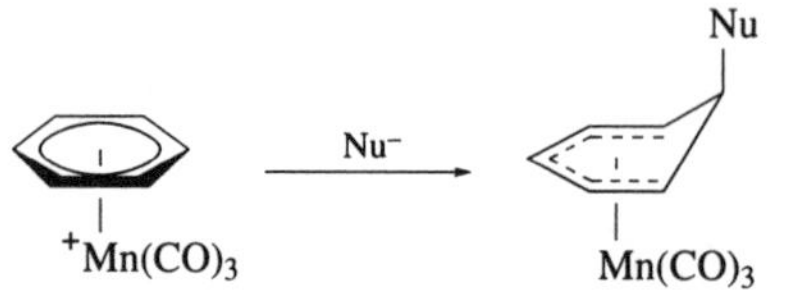

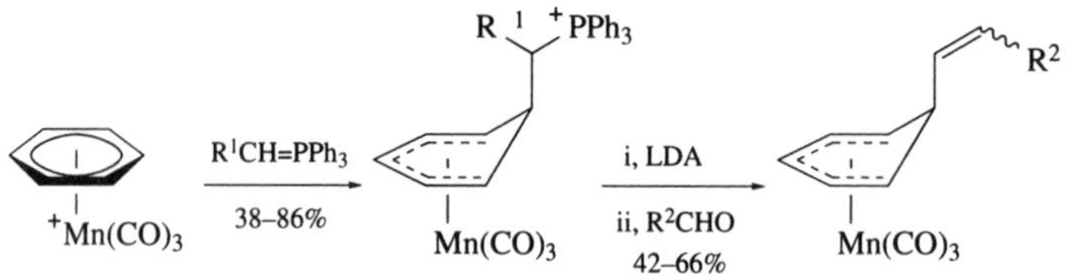

(1)

(2)

$R^1$ = H, Me, Et, OMe, Bz          $R^2$ = Me, Et, CH$_2$=CH, Ph, *p*-MeOC$_6$H$_4$

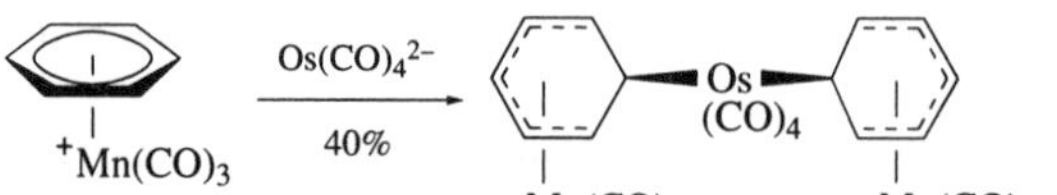

(3)

Although attack of the nucleophile takes place almost exclusively *exo* to the metal, tricarbonyl-($\eta^5$-cyclohexadienyl)manganese complex (8), the result of *endo* hydride attack, has been reported to predominate when bulky hydrides and protic solvents are utilized with the more sterically demanding (hexamethyl-$\eta^6$-benzene)manganese complex (6) (Equation (4)).[11] Reaction of tricarbonyl-($\eta^6$-benzene)manganese cation (1) with PhSLi also gives rise to *endo*-substituted ($\eta^5$-cyclohexadienyl)-manganese complexes, apparently via initial attack at a carbonyl ligand followed by migration to the arene.[12]

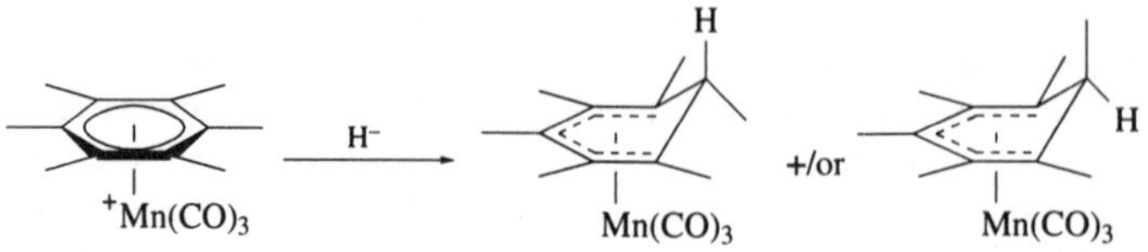

(4)

The regiochemistry of nucleophilic attack on substituted ($\eta^6$-arene)manganese cations has also been investigated.[13-22] Alkoxy substituents have been examined in detail, with attack occurring preferentially *meta* to the electron-rich substituent. For example, reaction of Grignard reagents with ($\eta^6$-arene)manganese complex (9) generates products resulting from attack only at the two *meta* positions (Equation (5)). Surprisingly, reaction at the more hindered *meta* position to produce ($\eta^5$-cyclohexadienyl)manganese complex (10) predominates with methylmagnesium bromide.[14] Other ring substituents are less useful in directing nucleophilic attack regioselectively, although an interesting solvent effect has been reported with the silyl-substituted arene complex (12) (Equation (6)).[17] Whereas attack at the *para* position of complex (12) to produce complexes (15) is favored in tetrahydrofuran, reactions in dichloromethane provide almost exclusively $\eta^5$-cyclohexadienyl complexes (13), from addition at the *ortho* position of complex (12).

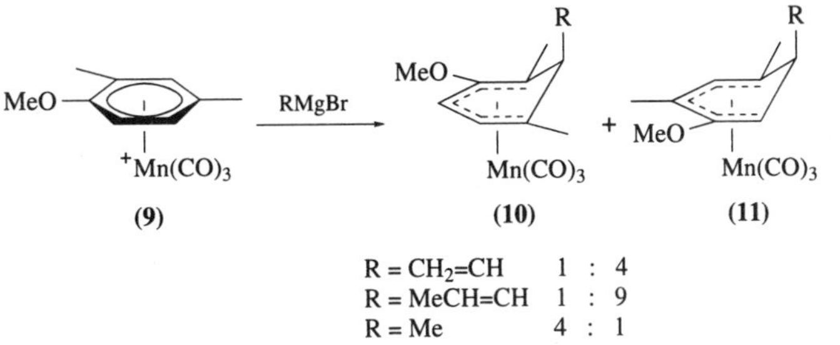

(5)

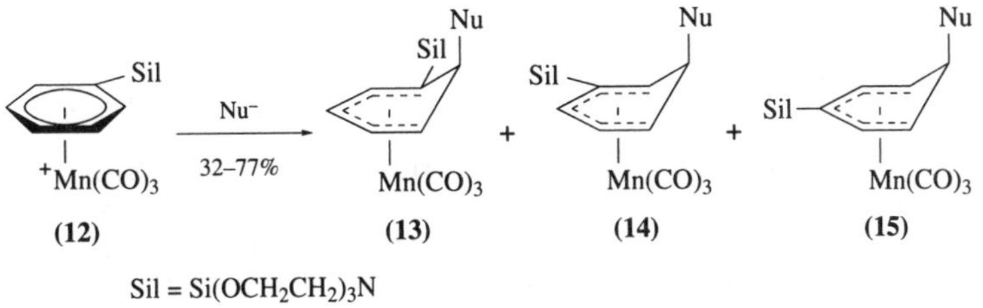

R = CH$_2$=CH 1 : 4
R = MeCH=CH 1 : 9
R = Me 4 : 1

(6)

Sil = Si(OCH$_2$CH$_2$)$_3$N

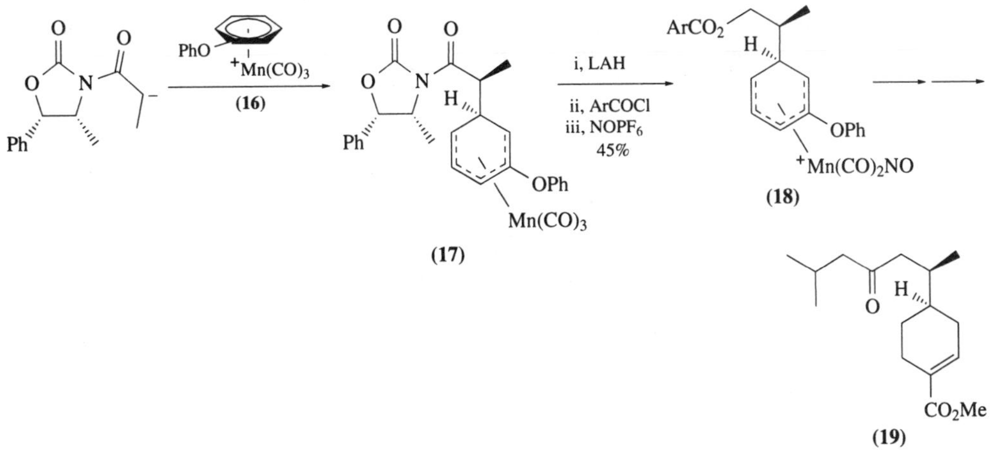

**Scheme 1**

The use of ($\eta^6$-arene)manganese cations in the synthesis of natural products has also been examined, including a formal total synthesis of Juvabione ((19), Scheme 1)[23] and a deoxyristomycinic acid derivative ((24), Scheme 2).[24] In each case, attack of a chiral nucleophile gives a modest degree of asymmetric induction, and further manipulation of the resulting ($\eta^5$-cyclohexadienyl)manganese complexes provides access to functionalized cyclohexadienes or arenes under mild reaction conditions.

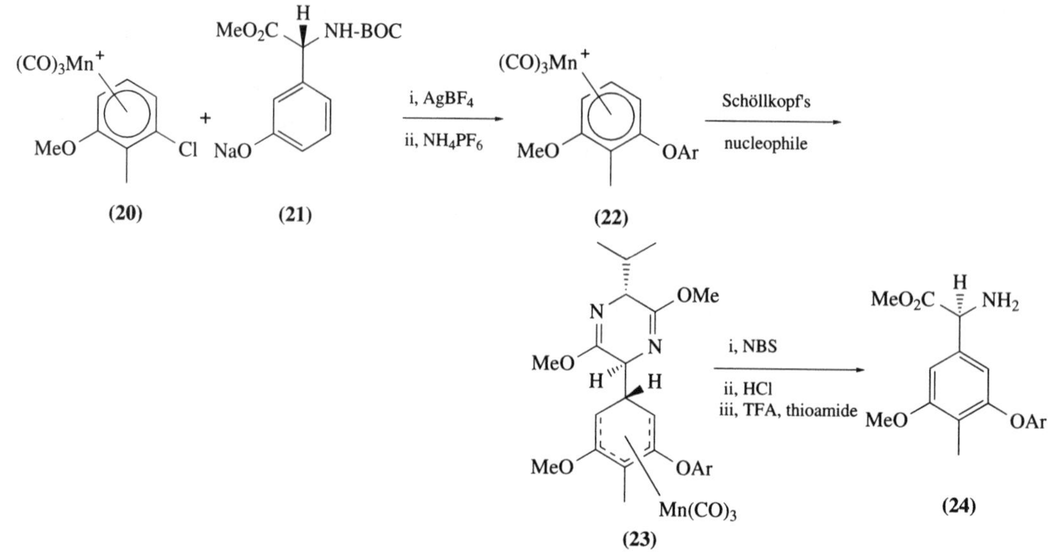

**Scheme 2**

### 4.2.2 Nucleophilic Attack on ($\eta^6$-Cycloheptatrienyl)manganese Complexes

Nucleophilic attack on tricarbonyl($\eta^6$-cycloheptatrienyl)manganese (**25**) has also been examined,[25] and studies indicate that complex (**25**) is approximately 45 000 times more reactive than the related arene complex (**1**).[26] Reformatsky reagents,[27] malonates and acetoacetates,[28] phosphines,[26] sodium pentacarbonylrhenium,[29] and zinc–copper coupling reagents all react with (**25**) to generate (*exo*-substituted-$\eta^5$-cycloheptadienyl)manganese complexes. Substitution of phosphine for one carbon monoxide ligand reduces the potential for complications resulting from electron transfer processes.[26] An example that demonstrates the power of this transformation utilizes nucleophilic attack on ($\eta^6$-cycloheptatrienyl)manganese cation (**25**) followed by oxidatively induced ring closure of the resulting ($\eta^5$-cycloheptadienyl)manganese complexes (**26**) and (**29**) to provide access to bicyclic and tricyclic species (**27**) and (**30**), respectively (Scheme 3).[28]

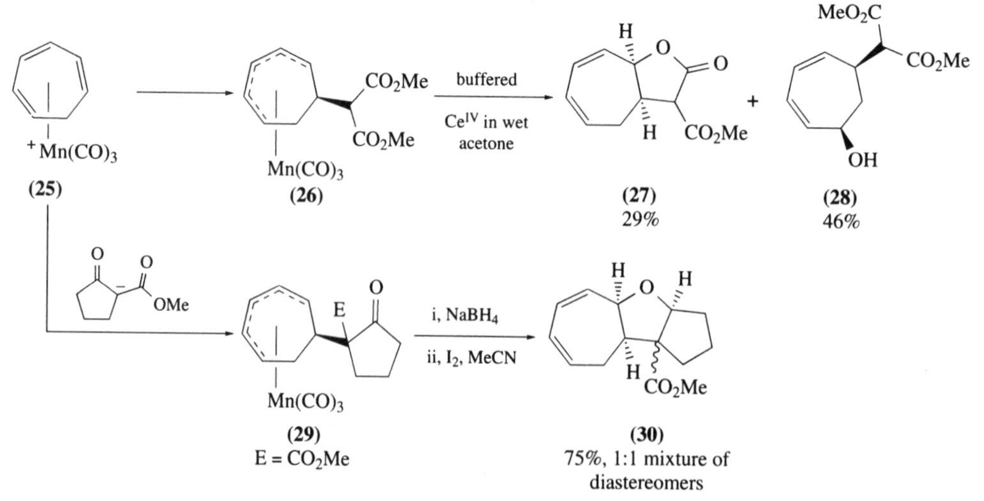

**Scheme 3**

### 4.2.3 Nucleophilic Attack on ($\eta^5$-Dienyl)manganese Complexes

Nucleophilic attack on ($\eta^5$-cyclohexadienyl)- and ($\eta^5$-cycloheptadienyl)manganese complexes has also been examined. The bulk of these investigations have utilized the facile substitution of a carbon monoxide ligand by nitrosyl to generate a cationic ($\eta^5$-dienyl)manganese complex (32) (Equation (7)).[30–5] Although many nucleophiles, including phosphines, malonates, and amines, add to the π-complexed hydrocarbon on the *exo* face, to generate ($\eta^4$-diene)manganese complex (33), nucleophilic attack at a carbon monoxide ligand occurs with alkyl lithium reagents, Grignard reagents, cuprates, and nitro-stabilized carbanions when the complex contains two carbon monoxide ligands. Loss of the nitrosyl ligand by initial electron transfer to regenerate the neutral ($\eta^5$-dienyl)manganese species (31) is also possible. Replacement of a carbon monoxide ligand by a phosphine generally alleviates these complications.[32,33] Removal of the manganese moiety of ($\eta^4$-diene)manganese complexes (33) by reaction with oxidizing agents, such as trimethylamine *N*-oxide,[34] provides access to substituted 1,3-cyclohexadienes.

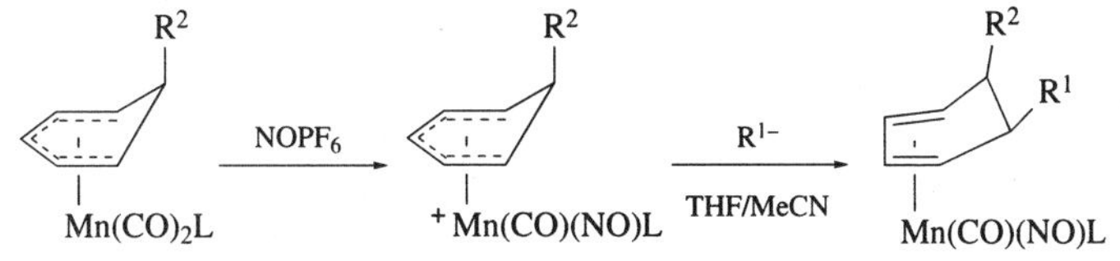

(7)

Detailed examination of the reaction of hydrides and deuterides with cationic ($\eta^5$-cyclohexadienyl)-manganese complexes (32) has demonstrated that *endo* attack to provide complexes (34) predominates in most cases (Equation (8)).[30,35] This attack apparently occurs initially at a carbon monoxide ligand to generate a formyl species, followed by migration of the hydride to the *endo* face of $\eta^5$-dienyl ligand.

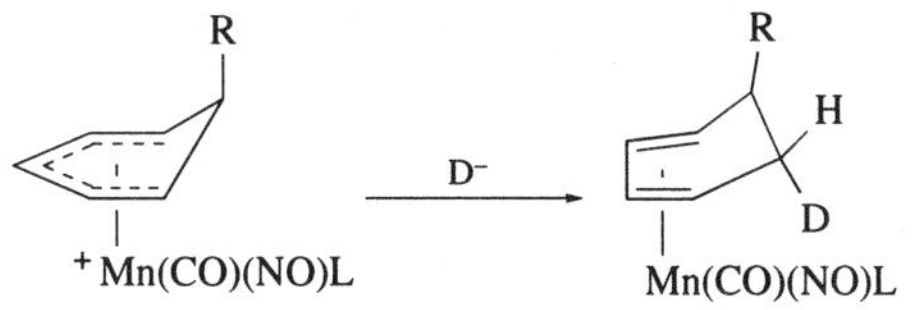

(8)

Nucleophilic attack on chiral ($\eta^5$-cyclohexadienyl)manganese complexes (35) provides the opportunity for asymmetric induction. Reaction of dimethyl sodiomalonate with (35) produces an 80% yield of a 2:1 mixture of diastereomers, where the major isomer arises from attack *cis* to the nitrosyl ligand (Equation (9)). Reaction of sodium borohydride, on the other hand, generates a 2:1 mixture resulting from attack predominantly *cis* to the trimethylphosphine ligand.[36]

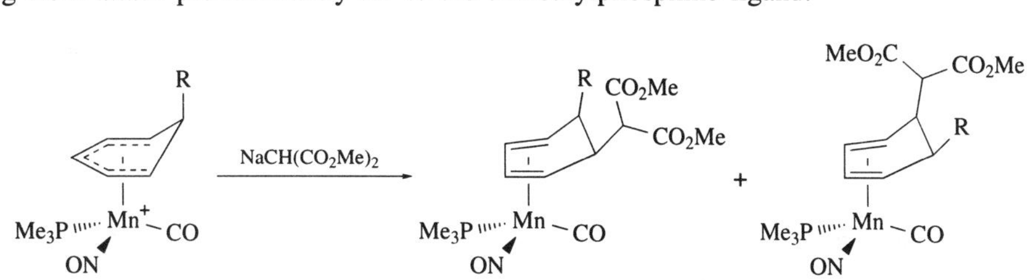

(9)

80% yield, 2:1 mixture of (36):(37)

Reaction of neutral tricarbonyl($\eta^5$-dienyl)manganese complexes with nucleophiles also has been examined, with a variety of results. Reaction of organolithium reagents with ($\eta^5$-cyclohexadienyl)-manganese complexes (2) at 25 °C in diethyl ether produces acyl manganese species (38), via attack at carbon monoxide. Protonation of this anion at −78 °C induces migration of the former nucleophile to the *endo* face of the π-complex, generating a mixture of fluxional ($\eta^3$-cyclohexenyl)manganese complexes (39), (40), and (41), containing agostic hydrogen interactions (Scheme 4).[37] Reaction of phenyllithium with ($\eta^5$-cycloheptadienyl)manganese complex (**42**) also takes place at a carbon monoxide ligand, but attempts to form a manganese carbenoid species, by reaction of anion (**43**) with trimethylsilyl chloride instead produces the ($\eta^6$-arene)manganese complex (**44**) in 84% yield via a novel intramolecular (5 + 1) cycloaddition of the transient phenyl(trimethylsiloxy)carbene group and the C-1 and C-5 dienyl carbons (Scheme 5).[38] Protonation of complex (**44**) generates a 95% yield of cation (**45**), which contains both an $\eta^6$-arene and an $\eta^2$-alkene manganese interaction.[38] Surprisingly, nucleophilic attack on (**45**) takes place at the $\eta^6$-arene in preference to the $\eta^2$-alkene to generate ($\eta^5$-cyclohexadienyl)manganese complex (**46**).

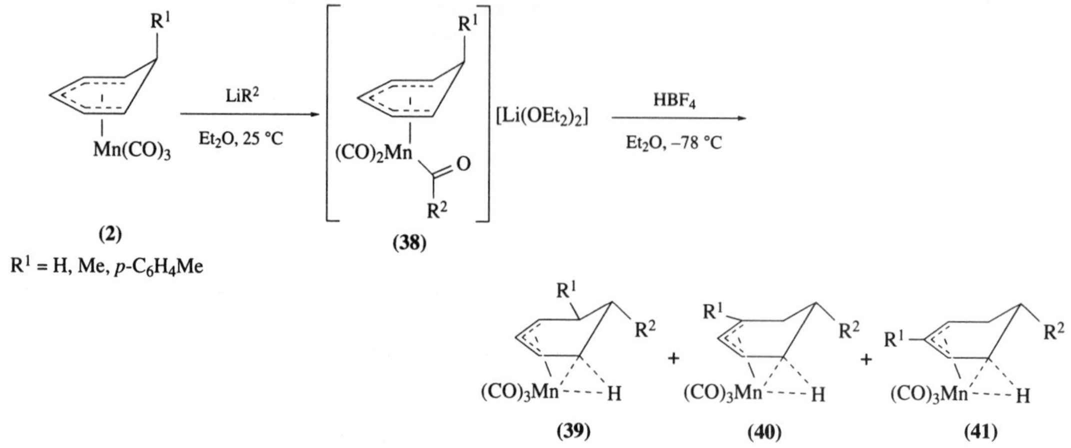

**Scheme 4**

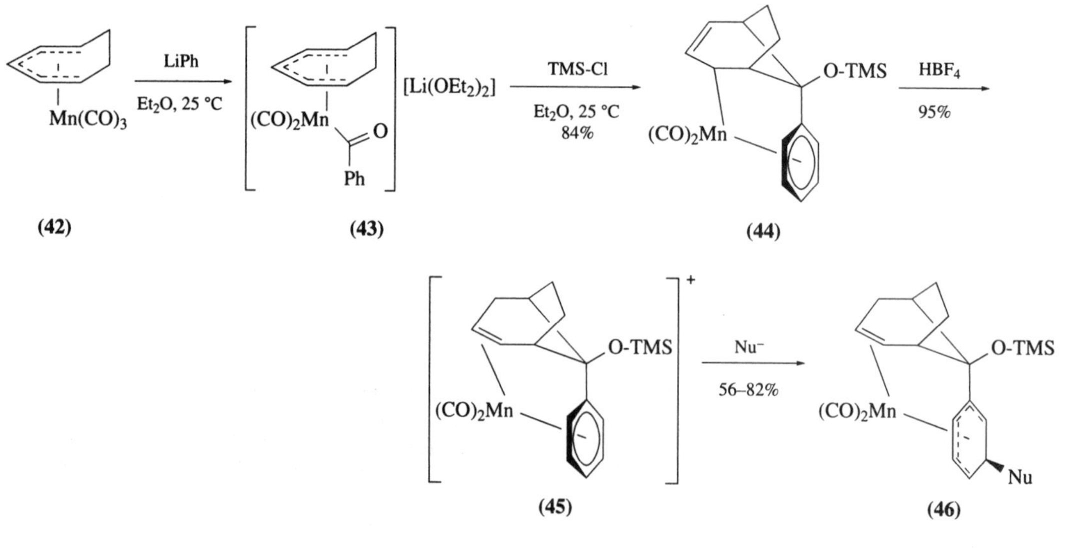

**Scheme 5**

Reactions of tricarbonyl($\eta^5$-cyclohexadienyl)manganese (**2**), tricarbonyl($\eta^5$-cycloheptadienyl)-manganese (**42**), and tricarbonyl($\eta^5$-pentadienyl)manganese (**49**) with slightly more stabilized carbanions such as diphenylmethyllithium and 2-lithio-1,3-dithiane occur at the dienyl ligand, as opposed to Sheridan's results with simple organolithium reagents.[37-9] While nucleophilic attack on cyclic complexes (**2**) and (**42**) occurs exclusively at the terminus of the π-system, producing anionic ($\eta^4$-diene)manganese complexes (**47**) (Equation (10)),[40] reaction of ($\eta^5$-pentadienyl)manganese complex (**49**) with nucleophiles takes place at the internal C-2/C-4 position of the $\eta^5$-pentadienyl system, generating (σ,$\eta^3$-pentenediyl)manganese anion (**50**).[41] Protonation of this anion provides (π-allyl)-manganese complexes (**51**) (Equation (11)).

Reaction of ($\eta^5$-pentadienyl)manganese complex (**49**) with amines and phosphines has also been investigated.[42,43] Use of $R_2NH$ or $R_2PH$ promotes 1,5-addition, followed by isomerization to the most stable (π-allyl)manganese complexes (**53**) and (**55**), respectively (Scheme 6). Reaction of tertiary phosphines with ($\eta^5$-pentadienyl)manganese complex (**49**) induces $\eta^5$ to $\eta^3$ isomerization with

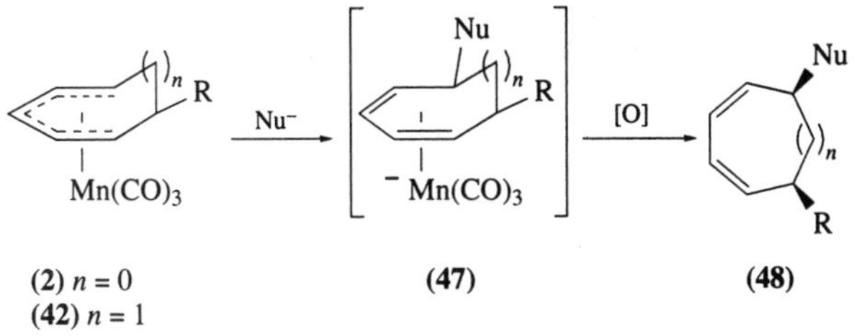

(10)

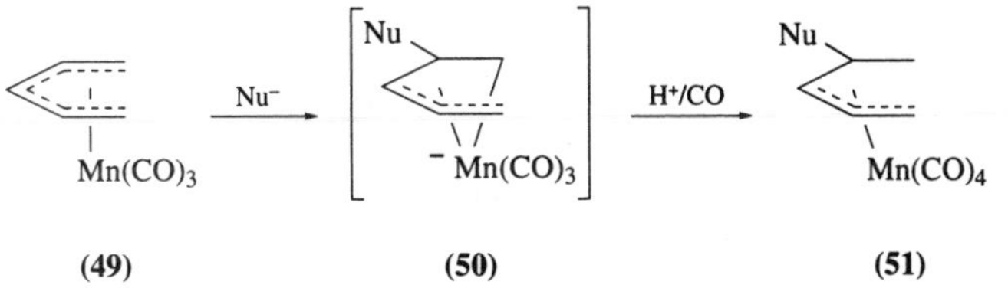

(2) $n = 0$
(42) $n = 1$

(47)

(48)

(11)

(49)

(50)

(51)

accompanying phosphine addition to produce ($\pi$-allyl)manganese complex (56). The phosphine-substituted ($\eta^5$-pentadienyl)manganese complex (57) can be generated from complex (56) by thermally induced dissociation of carbon monoxide followed by $\eta^3$–$\eta^5$ isomerization.

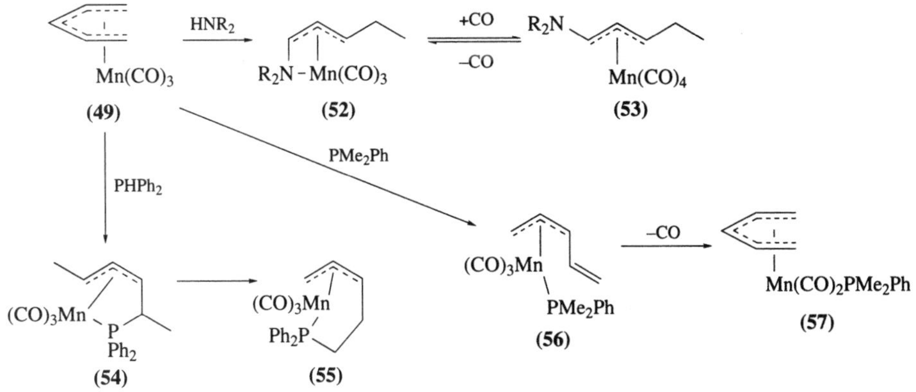

**Scheme 6**

### 4.2.4 Nucleophilic Attack on ($\eta^3$-Allyl)- and ($\eta^2$-Alkene)manganese Complexes

Nucleophilic attack on cationic ($\eta^3$-allyl)- and ($\eta^2$-alkene)manganese complexes has also been reported. Reaction of a variety of nucleophiles with dicarbonyl($\eta^5$-cyclopentadienyl-$\pi$-allyl)manganese cation (58) occurs exclusively at the $\pi$-allyl ligand, generating $\eta^2$-alkene complexes (59) in 20–89% yield (Equation (12)).[44] Reaction of nucleophiles with (61), which contains both an $\eta^6$-arene ligand and an $\eta^2$-alkene ligand, takes place solely at the $\eta^2$-alkene ligand providing alkylmanganese complex (62) (Scheme 7).[45,46] Manganese complex (61) is reported to be approximately 150 times less electrophilic than the analogous [Cp(CO)$_2$Fe–alkene]$^+$ complex.

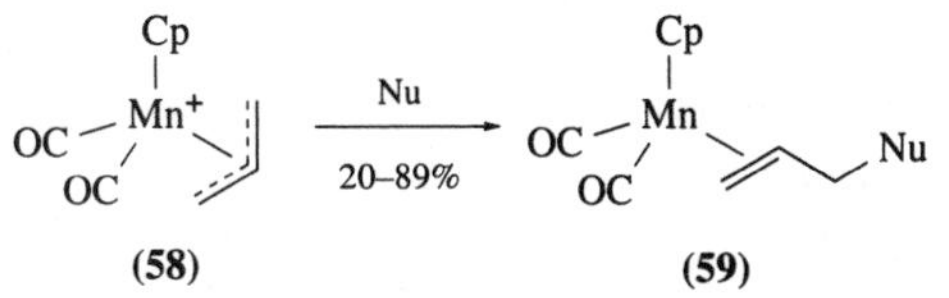

(12)

(58)

(59)

### 4.3 ELECTROPHILIC ATTACK ON MANGANESE $\pi$-COMPLEXES

The chemistry of anionic tricarbonyl($\eta^4$-diene)manganese complexes has been examined in considerable detail by Brookhart and co-workers.[47–53] The most general route to tricarbonyl-($\eta^4$-diene)manganese complexes ((66)–(71)) involves the reaction of ($\pi$-allyl)manganese complex (63)

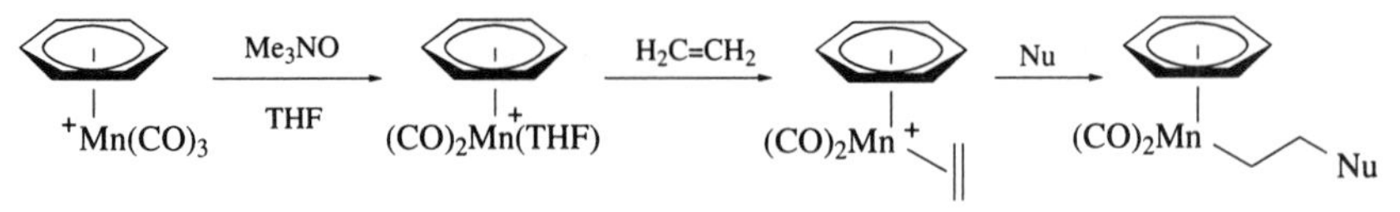

**Scheme 7**

with hydride, followed by substitution of the resulting $\eta^2$-butene ligand with the appropriate 1,3-diene (Equation (13)).[51] The addition of 2 equiv. of hydride to ($\eta^6$-benzene)manganese complex (**1**) also generates ($\eta^4$-cyclohexadiene)manganese anion (**66**).[47] Although these anions are quite sensitive to oxygen, decomposing to generate 1,3-dienes, they are thermally stable up to 120 °C. Protonation of anions (**66**)–(**68**) with water or HBF$_4$ produces ($\pi$-allyl)manganese complexes (**72**), which contain an agostic hydrogen–manganese interaction (Scheme 8). Regeneration of the anionic tricarbonyl-($\eta^4$-diene)manganese complexes takes place on reaction of ($\pi$-allyl)manganese complexes (**72**) with potassium hydride,[47] and a p$K_a$ of 22.2 for the tricarbonyl($\eta^3$-cyclohexenyl)manganese complex (**72**) ($n = 1$) has been reported.[54]

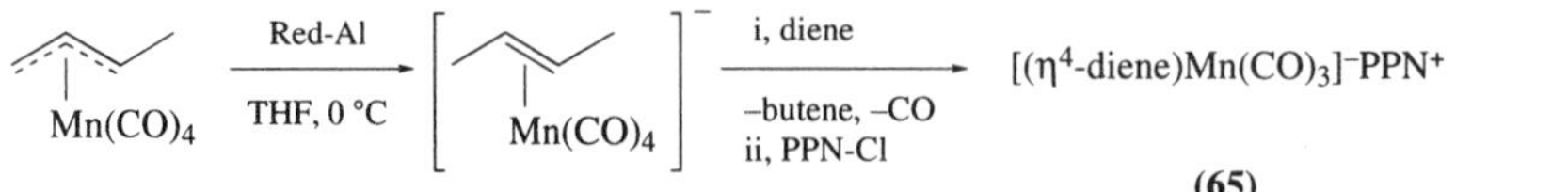

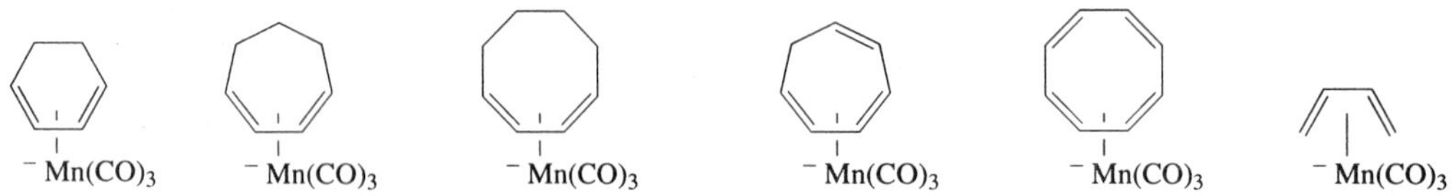

**Scheme 8**

($\pi$-Allyl)manganese complexes (**72**) undergo two separate fluxional processes detectable by $^1$H NMR spectroscopy that differ in the way the agostic C–H–Mn interaction is broken.[47,51] First, a low-energy process ($\Delta G^\ddagger = 33.5$ kJ mol$^{-1}$) proceeding at −10 °C breaks the interaction with manganese to form the unsaturated 16-electron $\pi$-allyl intermediates (**73**), which have a plane of symmetry. Second, a higher energy process ($\Delta G^\ddagger = 66.9$ kJ mol$^{-1}$) proceeding at 10–120 °C transfers hydrogen to manganese to give diene hydride intermediates (**75**). Trapping of the unsaturated ($\pi$-allyl)manganese (**73**) with carbon monoxide produces tetracarbonyl($\pi$-allyl)manganese complex (**74**). Although this complex rapidly loses carbon monoxide to regenerate fluxional complex (**72**) in the cyclohexenyl series ($n = 1$), trapping of carbon monoxide in the cyclooctenyl series ($n = 3$) is irreversible, apparently a result of increased ring strain on flattening of the four-carbon unit in an eight-membered ring.[51]

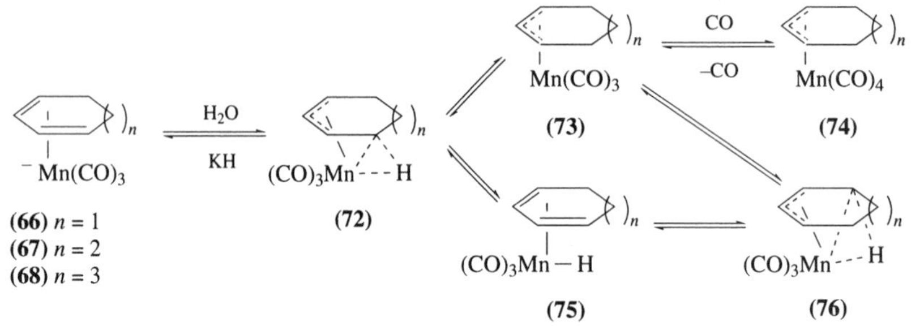

Reaction of ($\eta^4$-diene)manganese anions with electrophiles other than hydrogen also provides ($\pi$-allyl)manganese complexes containing the two electron–three center bonding array. For example, reaction of tricarbonyl($\eta^4$-cyclohexadiene)manganese anion (**66**) with methyltriflate or methyl iodide produces a 22:78 mixture of two *endo*-methyl substituted ($\pi$-allyl)manganese complexes (**78**) and (**79**), by delivery of the methyl group from manganese to the *endo* face of the $\pi$-complex (Equation (14)).[47] Similar results have been reported for analogous acyclic ($\eta^4$-diene)manganese complexes, which react with electrophiles to produce ($\pi$-allyl)manganese complexes (**82**)–(**84**), containing agostic hydrogens

(Scheme 9). With these acyclic complexes, insertion of carbon dioxide is also possible, generating ($\pi$-allyl)manganese complex (**85**).[52]

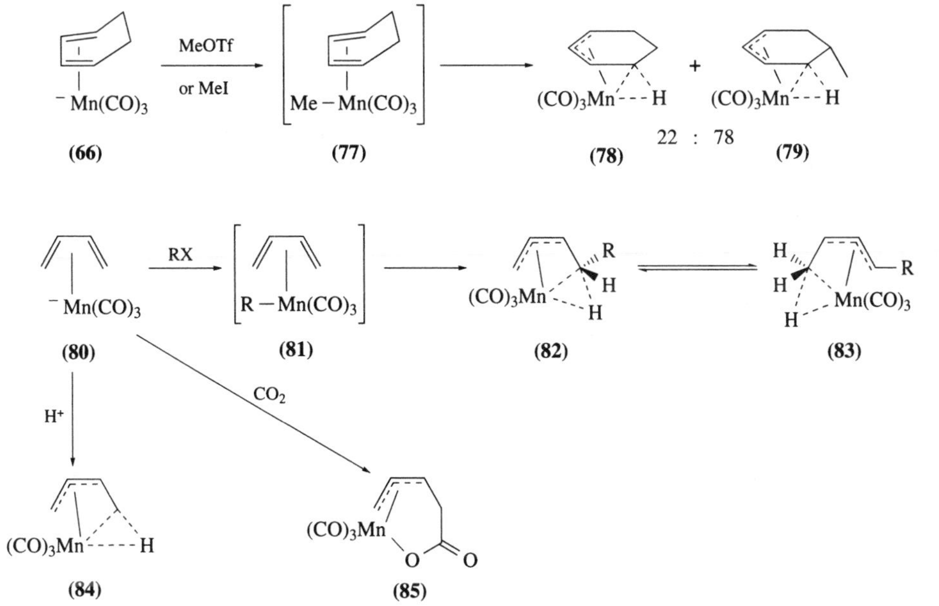

**Scheme 9**

Reaction of electrophiles with anionic ($\eta^4$-diene)manganese complexes which contain at least one additional uncomplexed alkene produces *exo*-substituted ($\eta^5$-dienyl)manganese complexes.[51] This route provides access to novel ($\eta^5$-dienyl)manganese complexes, as is demonstrated by the reaction of manganese complex (**70**) with a series of ($\eta^7$-cycloheptatrienyl)metal cations to produce complexes (**87**) (Equation (15)).[55]

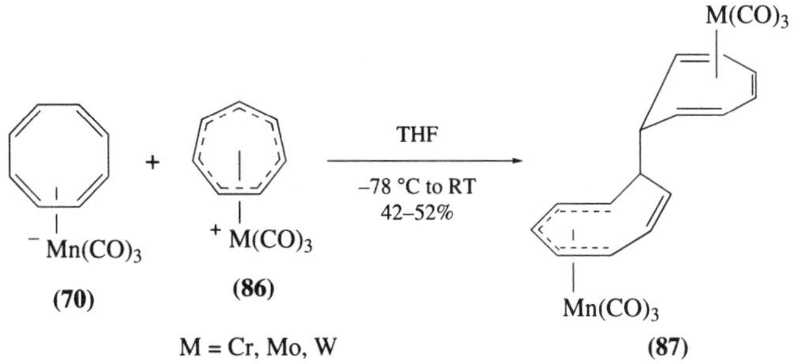

The mechanism of benzene dimerization has also been examined utilizing a two-electron reduction of manganese complex (**1**). Formation of dimer (**88**) apparently is the result of a [2 + 2] thermal cycloaddition of ($\eta^4$-diene)manganese complex (**89**), and not a product of photodimerization.[56] Facile exchange of naphthalene with benzene provides anionic complex (**90**), which undergoes *endo* protonation regiospecifically to provide ($\eta^5$-dienyl)manganese complex (**91**) (Scheme 10).[57]

The reaction of manganese $\pi$-complexes with electrophiles has also been investigated using (*exo*-methylene-$\eta^5$-cyclohexadienyl)manganese complex (**92**) (Scheme 11).[58] Complex (**92**) can be prepared by deprotonation of (hexamethyl-$\eta^6$-benzene)manganese (**6**), and reacts with a wide variety of electrophiles to produce ($\eta^6$-arene)manganese cations (**93**)–(**98**), each containing a functionalized methylene group. These transformations are most likely via a single electron transfer process, although simple nucleophilic displacement may predominate in certain cases. Not surprisingly, substitution of a carbon monoxide ligand by phosphine increases the nucleophilicity of the exocyclic methylene group.[59] Further deprotonation of complex (**92**) produces the (bis-*exo*-methylene-$\eta^4$-diene)manganese anion (**99**), which reacts with 2 mol. of methyl triflate to generate the bisalkylated cation (**100**) (Equation (16)).[60]

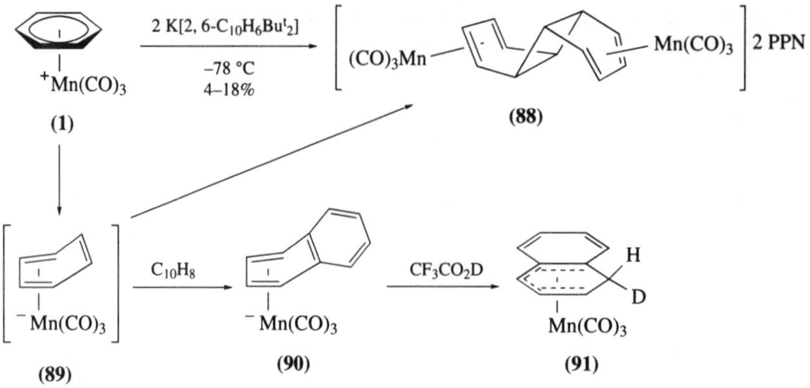

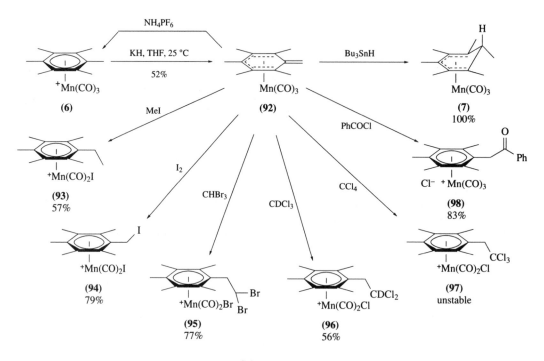

**Scheme 10**

**Scheme 11**

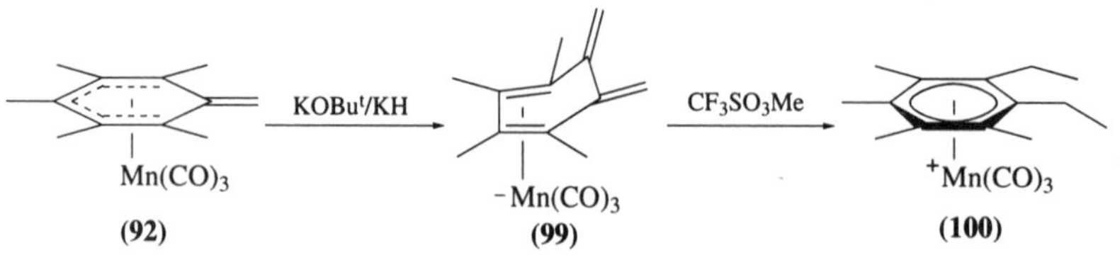

(16)

## 4.4 PHOTOCHEMICAL TRANSFORMATIONS OF MANGANESE π-COMPLEXES

Photochemically induced transformations of manganese π-complexes have been studied in great detail by Kreiter and co-workers.[61–71] The reaction of $Mn_2(CO)_{10}$ or $RMn(CO)_5$ with 1,3-dienes, 1,3,5-trienes, and allenes produces a variety of (π-allyl)manganese complexes, reactions which are initiated by the photochemically induced dissociation of carbon monoxide. A fascinating [5 + 4] cycloaddition reaction between tricarbonyl($\eta^5$-pentadienyl)manganese (**49**) and a series of 1,3-dienes generates ($\eta^2,\eta^3$-cyclononadienyl)manganese complexes (**105**) in 13–68% yield (Scheme 12).[68] Cleavage of the manganese moiety by reaction of (**105**) (R = H) with 2 equiv. of acetic acid under UV irradiation produces 1,5-nonadiene (**106**).[71]

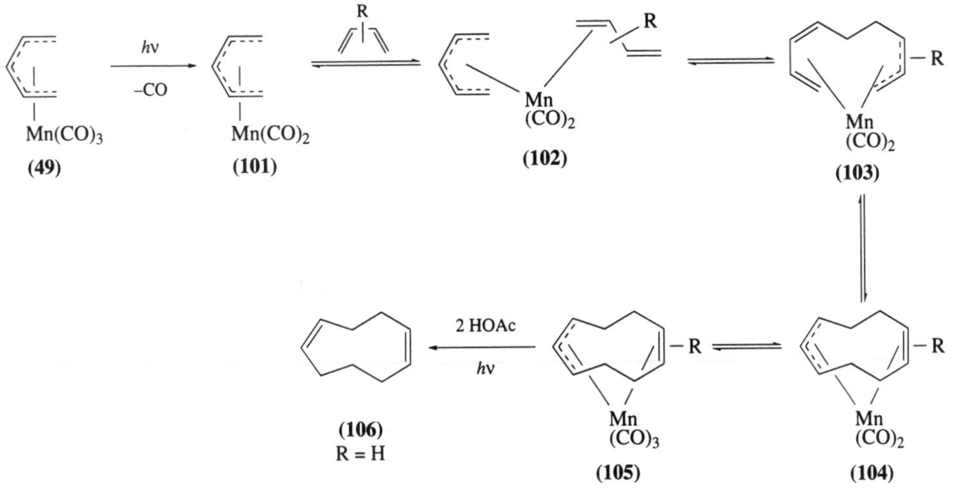

**Scheme 12**

## 4.5 MISCELLANEOUS PREPARATIONS AND REACTIONS OF MANGANESE π-COMPLEXES

### 4.5.1 (η²-Alkene)- and (η²-Alkyne)manganese Complexes

The standard preparative methods for the formation of (η²-alkene)manganese complexes involve substitution of a carbon monoxide ligand under photolytic conditions[72] or substitution of a labile ligand such as tetrahydrofuran[45] or cyclooctene[73] under thermal reaction conditions. A novel approach to (η²-alkene)manganese complexes involves the attack of water on manganese complex (**107**), which induces formation of the amide-substituted (η²-alkene)manganese complex (**109**) (Equation (17)).[74] The formation of (η²-butatriene)manganese complexes (**112**) by reaction of 1-chloro-1-trimethylsilylalkenes with MeCpMn(CO)₃ (Equation (18)) has also been reported.[75]

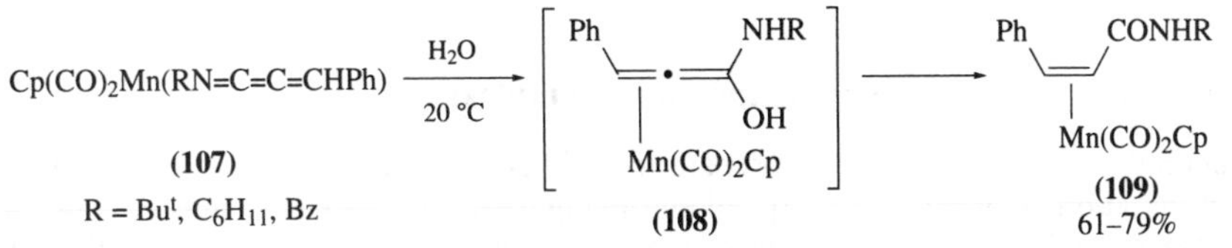

$$(17)$$

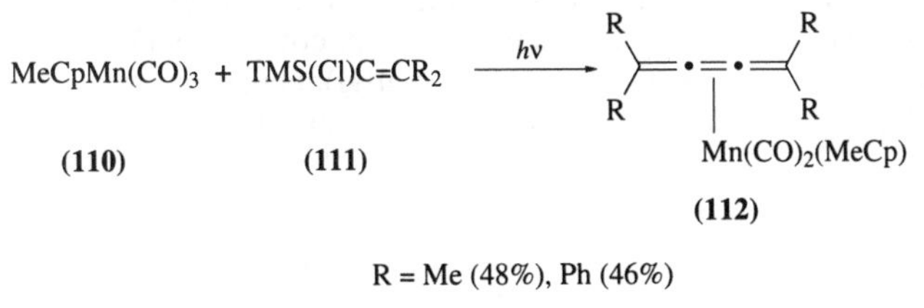

$$(18)$$

R = Me (48%), Ph (46%)

The bis(cyclooctadiene)manganese complex (**114**) was prepared by the reduction of bis(η⁵-cyclopentadienyl)manganese in the presence of 2 equiv. of 1,5-cyclooctadiene (Equation (19)).[76] Similar reactions utilizing the reduction of bis(η⁵-cyclopentadienyl)manganese provide an improved method for the preparation of carbonyl-free organomanganese π-complexes.[76]

The formation of (η²-alkyne)manganese complexes also is most often accomplished by substitution of a labile ligand such as tetrahydrofuran[77,78] or ethyl ether,[79] as is demonstrated by the preparation of η²-alkyne complex (**117**) starting from Cp(CO)₂Mn(THF) (Scheme 13).[77] Reaction of (**117**) with phenyllithium followed by protonation generates the manganese vinylidene complex (**118**), a rearrangement which has also been demonstrated with (1-iodo-η²-alkyne)manganese complexes.[79] The reaction of amines with (η²-acetylene)manganese complex (**119**) provides a useful approach to aminocarbene complexes (**120**) (Equation (20)).[80]

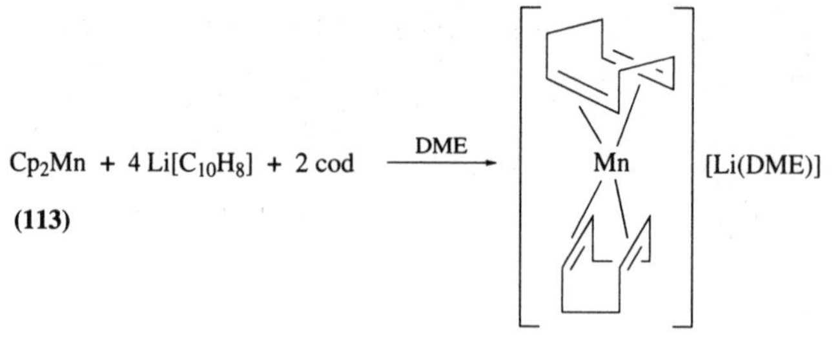

$$Cp_2Mn + 4\,Li[C_{10}H_8] + 2\,cod \xrightarrow{\text{DME}} \left[ \underset{Mn}{\diagdown} \right] [Li(DME)] \qquad (19)$$

**(113)**

**(114)**
30–60%

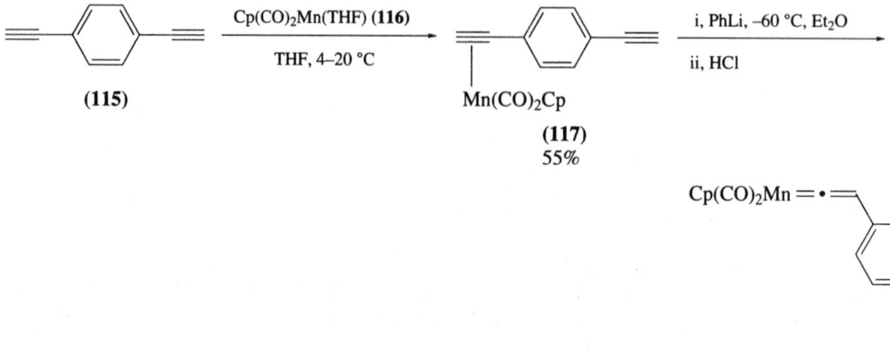

i, PhLi, –60 °C, Et$_2$O

ii, HCl

$$\text{(115)} \xrightarrow[\text{THF, 4–20 °C}]{\text{Cp(CO)}_2\text{Mn(THF) (116)}} \underset{\text{Mn(CO)}_2\text{Cp}}{\text{(117)}}$$

**(117)**
55%

Cp(CO)$_2$Mn =•=

**(118)**

**Scheme 13**

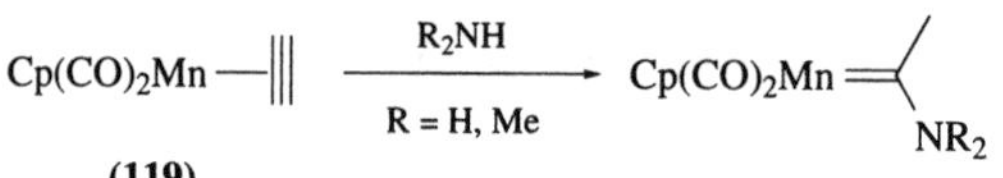

$$Cp(CO)_2Mn—\|\| \xrightarrow[\text{R = H, Me}]{R_2NH} Cp(CO)_2Mn= \underset{NR_2}{\diagup} \qquad (20)$$

**(119)**  **(120)**
70–85%

## 4.5.2 (η³-Allyl)- and (η⁵-Pentadienyl)manganese Complexes

Several methods are available for the preparation of (π-allyl)manganese complexes. Protonation and subsequent elimination of water from (η²-alkene)manganese complexes which contain an α-hydroxy group generates cationic (π-allyl)manganese complexes in moderate overall yield.[81–3] Although both *endo* (**122**) and *exo* (**121**) isomers are often formed, the *exo* isomer typically predominates.[82,84] This transformation can be accomplished in one step for certain complexes (Equation (21)).[83] Protonation of (η²-1,3-diene)manganese complexes also provides (π-allyl)manganese cations, a transformation which is particularly useful for the formation of the cyclic (π-allyl)manganese complexes (**123**) (Equation (22)).[81,82]

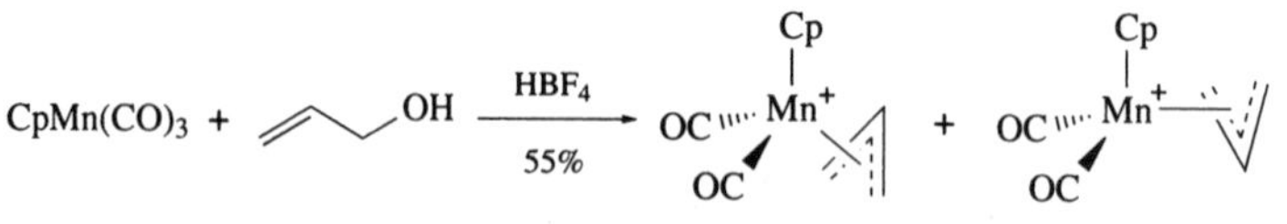

$$CpMn(CO)_3 + \diagdown\diagdown OH \xrightarrow[55\%]{HBF_4} OC^{\text{''''}}\underset{OC}{\overset{Cp}{Mn^+}}\diagdown + OC^{\text{''''}}\underset{OC}{\overset{Cp}{Mn^+}}\diagdown \qquad (21)$$

**(121)**  **(122)**
*exo*  *endo*

$$OC^{\text{''''}}\underset{OC}{\overset{Cp}{Mn}}\diagdown\bigcirc \xrightarrow{HBF_4} OC^{\text{''''}}\underset{OC}{\overset{Cp}{Mn^+}}\bigcirc \qquad (22)$$

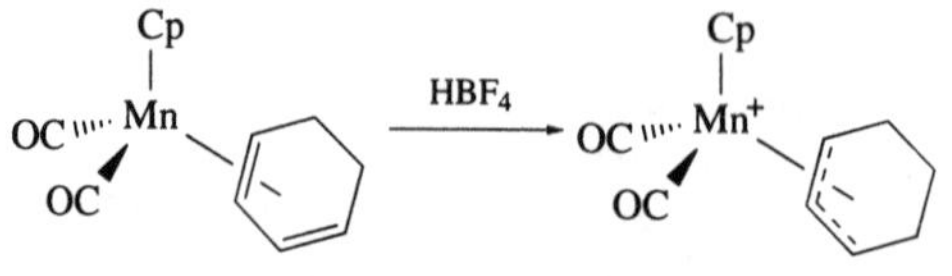

**(123)**

An alternative to these acid catalyzed processes for the preparation of neutral (π-allyl)manganese complexes utilizes the ability of certain (η$^1$-allyl)manganese complexes to undergo facile η$^1$–η$^3$ isomerization. An example of such a preparation involves the formation of (η$^1$-allyl)manganese complex (**124**) by the reaction of sodiopentacarbonylmanganate with the requisite allyl chloride, followed by photolysis to induce dissociation of carbon monoxide and η$^1$–η$^3$ isomerization to form (π-allyl)manganese complex (**125**) (Scheme 14). (η$^5$-Oxapentadienyl)manganese complexes (**126**) are generated by heating (π-allyl)manganese complexes (**125**) in cyclohexane.[85,86]

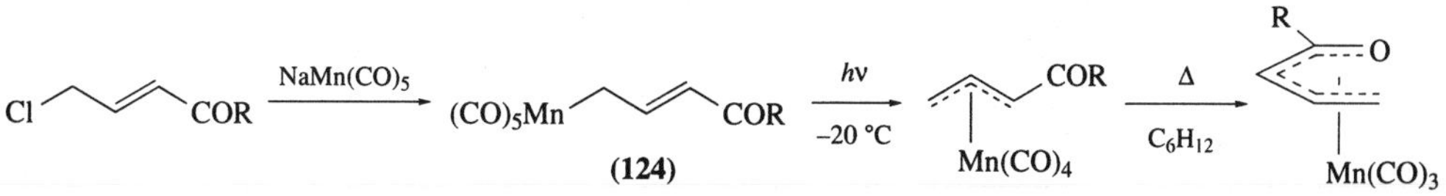

**Scheme 14**

A similar η$^1$–η$^5$ isomerization process has also been used for the preparation of tricarbonyl-(η$^5$-pentadienyl)manganese (**49**).[87] Reaction of 5-bromo-1,3-pentadiene with sodiopentacarbonylmanganate at −78 °C provides (η$^1$-pentadienyl)manganese complex (**127**) (Scheme 15). Irradiation of (**127**) at −20 °C causes η$^1$–η$^3$ isomerization to form (η$^3$-pentadienyl)manganese complex (**128**), whereas thermolysis of (η$^1$-pentadienyl)manganese complex (**127**) in refluxing cyclohexane induces η$^1$–η$^5$ isomerization to generate (η$^5$-pentadienyl)manganese complex (**49**). The preparation of (**49**) utilizing the reaction between 5-bromo-1,3-pentadiene and pentacarbonylmanganese-bromide under phase transfer catalysis conditions has also been reported.[88]

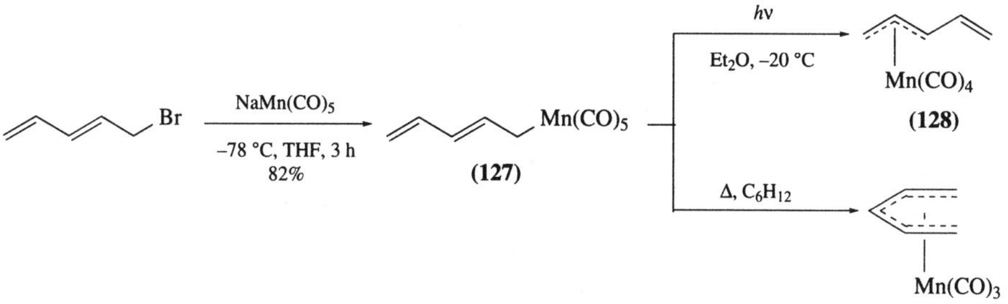

**Scheme 15**

The connection between (η$^3$-allyl)manganese and (η$^5$-pentadienyl)manganese complexes is also demonstrated by the reaction of manganese bromide with pentadienylpotassium (Equation (23)).[89,90] Addition of a chelating bisphosphine fills four coordination sites, allowing for the formation of (η$^3$-pentadienyl)manganese complex (**130**), whereas use of a chelating trisphosphine promotes formation of the (η$^5$-pentadienyl)manganese complex (**131**). Dimerization to form complex (**129**) predominates when a monodentate phosphine is utilized.

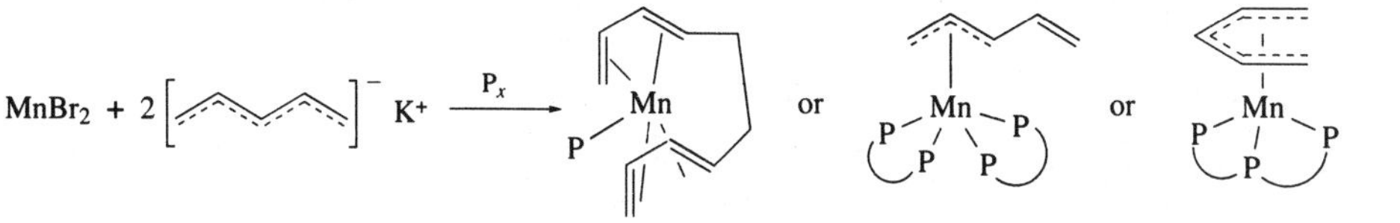

$$MnBr_2 + 2\left[\text{pentadienyl}\right]^- K^+ \xrightarrow{P_x} \quad (23)$$

| (129) | (130) | (131) |
|---|---|---|
| $P_x$ = monodentate phosphine | $P_x$ = chelating bisphosphine | $P_x$ = chelating trisphosphine |

## 4.6 REFERENCES

1. D. Jones and G. Wilkinson, *J. Chem. Soc.*, 1964, 2479.
2. L. A. P. Kane-Maguire, E. D. Honig and D. A. Sweigart, *Chem. Rev.*, 1984, **84**, 525.

3. M. V. Gaudet, A. W. Hanson, P. S. White and M. J. Zaworotko, *Organometallics*, 1989, **8**, 286.
4. F. Rose-Munch and K. Aniss, *Tetrahedron Lett.*, 1990, **31**, 6351.
5. F. Balssa, K. Aniss and F. Rose-Munch, *Tetrahedron Lett.*, 1992, **33**, 1901.
6. T.-M. Chung and Y. K. Chung, *Organometallics*, 1992, **11**, 2822.
7. R. Réau, R. W. Reed, F. Dahan and G. Bertrand, *Organometallics*, 1993, **12**, 1501.
8. S. Lee, Y. K. Chung, T.-S. Yoon and W. Shin, *Organometallics*, 1993, **12**, 2873.
9. B. Niemer, J. Breimair, T. Völkel, B. Wagner, K. Polborn and W. Beck, *Chem. Ber.*, 1991, **124**, 2237.
10. C. C. Neto, C. D. Baer, Y. K. Chung and D. A. Sweigart, *J. Chem. Soc., Chem. Commun.*, 1993, 816.
11. A. M. Morken, D. P. Eyman, M. A. Wolff and S. J. Schauer, *Organometallics*, 1993, **12**, 725.
12. M. Schindehutte, P. H. van Rooyen and S. Lotz, *Organometallics*, 1990, **9**, 293.
13. A. J. Pearson and I. C. Richards, *J. Organomet. Chem.*, 1983, **258**, C41.
14. R. P. Alexander, C. Morley and G. R. Stephenson, *J. Chem. Soc. Perkin Trans. 1*, 1988, 2069.
15. D. A. Brown, W. K. Glass and K. M. Kreddan, *J. Organomet. Chem.*, 1991, **413**, 233.
16. E. Jeong and Y. K. Chung, *J. Organomet. Chem.*, 1992, **434**, 225.
17. Y.-A. Lee, Y. K. Chung, Y. Kim, J. H. Jeong, G. Chung and D. Lee, *Organometallics*, 1991, **10**, 3707.
18. A.-S. Oh, Y. K. Chung and S. Kim, *Organometallics*, 1992, **11**, 1394.
19. S. S. Lee, J.-S. Lee and Y. K. Chung, *Organometallics*, 1993, **12**, 4640.
20. S. R. Stobart and M. J. Zaworotko, *J. Chem. Soc. Chem., Commun.*, 1984, 1700.
21. G. R. Krow, W. H. Miles, P. M. Smiley, W. S. Lester and Y. J. Kim, *J. Org. Chem.*, 1992, **57**, 4040.
22. W. H. Miles, P. M. Smiley and H. R. Brinkman, *J. Chem. Soc., Chem. Commun.*, 1989, 1897.
23. W. H. Miles and H. R. Brinkman, *Tetrahedron Lett.*, 1992, **33**, 589.
24. A. J. Pearson, S.-H. Lee and F. Gonzoules, *J. Chem. Soc. Perkin Trans. 1*, 1990, 2251.
25. F. Haque, J. Miller, P. L. Pauson and J. B. Tripathi, *J. Chem. Soc. C.*, 1971, 743.
26. E. D. Honig, M. Quin-jin, W. T. Robinson, P. G. Williard and D. A. Sweigart, *Organometallics*, 1985, **4**, 871.
27. A. J. Pearson and I. C. Richards, *Tetrahedron Lett.*, 1983, **24**, 2465.
28. A. J. Pearson, P. Bruhn and I. C. Richards, *Tetrahedron Lett.*, 1984, **25**, 387.
29. R. C. Bush, R. A. Jacobson and R. J. Angelici, *J. Organomet. Chem.*, 1987, **323**, C25.
30. Y. K. Chung, E. D. Honig, W. T. Robinson, D. A. Sweigart, N. G. Connelly and S. D. Ittel, *Organometallics*, 1983, **2**, 1479.
31. N. G. Connelly, M. J. Freeman, A. G. Orpen, A. R. Sheehan, J. B. Sheridan and D. A. Sweigart, *J. Chem. Soc., Dalton Trans.*, 1985, 1019.
32. Y. K. Chung, D. A. Sweigart, N. G. Connelly and J. B. Sheridan, *J. Am. Chem. Soc.*, 1985, **107**, 2388.
33. Y. K. Chung and D. A. Sweigart, *J. Organomet. Chem.*, 1986, **308**, 223.
34. T.-H. Hyeon and Y. K. Chung, *J. Organomet. Chem.*, 1989, **372**, C12.
35. R. D. Pike, W. J. Ryan, N. S. Lennhoff, J. Van Epp and D. A. Sweigart, *J. Am. Chem. Soc.*, 1990, **112**, 4798.
36. R. D. Pike, W. J. Ryan, G. B. Carpenter and D. A. Sweigart, *J. Am. Chem. Soc.*, 1989, **111**, 8535.
37. J. B. Sheridan, R. S. Padda, K. Chaffee, C. Wang, Y. Huang and R. Lalancette, *J. Chem. Soc., Dalton Trans.*, 1992, 1539.
38. C. Wang, M. G. Lang, J. B. Sheridan and A. L. Rheingold, *J. Am. Chem. Soc.*, 1990, **112**, 3236.
39. C. Wang, J. B. Sheridan and A. L. Rheingold, *J. Am. Chem. Soc.*, 1993, **115**, 3603.
40. B. C. Roell, Jr., W. S. Vaughan, T. S. Macy and K. F. McDaniel, *Organometallics*, 1993, **12**, 224.
41. B. C. Roell, Jr. and K. F. McDaniel, *J. Am. Chem. Soc.*, 1990, **112**, 9004.
42. M. A. Paz-Sandoval *et al.*, *Organometallics*, 1992, **11**, 2467.
43. N. Z. Villarreal, M. A. Paz-Sandoval, P. Joseph-Nathan and R. O. Esquivel, *Organometallics*, 1991, **10**, 2616.
44. G. R. Knox, P. L. Pauson and J. Rooney, *J. Organomet. Chem.*, 1991, **420**, 379.
45. W. A. Halpin, J. C. Williams, Jr., T. Hanna and D. A. Sweigart, *J. Am. Chem. Soc.*, 1989, **111**, 376.
46. T. Hanna, N. S. Lennhoff, D. A. Sweigart, *J. Organomet. Chem.*, 1989, **377**, 133.
47. M. Brookhart, W. Lamanna and A. R. Pinhas, *Organometallics*, 1983, **2**, 638.
48. M. Brookhart and A. Lukacs, *J. Am. Chem. Soc.*, 1984, **106**, 4161.
49. M. Brookhart and A. Lukacs, *Organometallics*, 1983, **2**, 649.
50. F. Timmers and M. Brookhart, *Organometallics*, 1985, **4**, 1365.
51. M. Brookhart, S. K. Noh, F. J. Timmers and Y. H. Hong, *Organometallics*, 1988, **7**, 2458.
52. M. Brookhart, S. K. Noh and F. J. Timmers, *Organometallics*, 1987, **6**, 1829.
53. P. K. Rush, S. K. Noh and M. Brookhart, *Organometallics*, 1986, **5**, 1745.
54. S. S. Kristjansdottir, A. E. Moody, R. T. Weberg and J. R. Norton, *Organometallics*, 1988, **7**, 1983.
55. M. Wieser, K. Sunkel, C. Robl and W. Beck, *Chem. Ber.*, 1992, **125**, 1369.
56. R. L. Thompson, S. J. Geib and N. J. Cooper, *J. Am. Chem. Soc.*, 1991, **113**, 8961.
57. R. L. Thompson, S. Lee, A. L. Rheingold and N. J. Cooper, *Organometallics*, 1991, **10**, 1657.
58. D. M. LaBrush, D. P. Eyman, N. C. Baenziger and L. M. Mallis, *Organometallics*, 1991, **10**, 1026.
59. J. L. Moler, D. P. Eyman, J. M. Nielson, A. M. Morken, S. J. Schauer and D. B. Snyder, *Organometallics*, 1993, **12**, 3304.
60. J. W. Hull, Jr., K. J. Roesselet and W. L. Gladfelter, *Organometallics*, 1992, **11**, 3630.
61. W. Lipps and C. G. Kreiter, *J. Organomet. Chem.*, 1983, **241**, 185.
62. M. Leyendecker and C. G. Kreiter, *J. Organomet. Chem.*, 1983, **249**, C31.
63. M. Leyendecker, W. S. Sheldrick and C. G. Kreiter, *J. Organomet. Chem.*, 1984, **270**, C37.
64. C. G. Kreiter and M. Leyendecker, *J. Organomet. Chem.*, 1985, **280**, 225.
65. C. G. Kreiter and M. Leyendecker, *J. Organomet. Chem.*, 1985, **292**, C18.
66. C. G. Kreiter, M. Leyendecker and W. S. Sheldrick, *J. Organomet. Chem.*, 1986, **302**, 217.
67. C. G. Kreiter, K. Lehr, M. Leyendecker, W. S. Sheldrick and R. Exner, *Chem. Ber.*, 1991, **124**, 3.
68. C. G. Kreiter and K. Lehr, *J. Organomet. Chem.*, 1991, **406**, 159.
69. C. G. Kreiter, K. Lehr and R. Exner, *J. Organomet. Chem.*, 1991, **411**, 225.
70. M. Leyendecker and C. G. Kreiter, *J. Organomet. Chem.*, 1984, **260**, C67.
71. C. G. Kreiter and K. Lehr, *J. Organomet. Chem.*, 1993, **448**, 107.
72. D. Lentz, J. Kroll and C. Langner, *Chem. Ber.*, 1987, **120**, 303.
73. D. Lentz and R. Marschall, *Chem. Ber.*, 1989, **122**, 1223.
74. V. N. Kalinin, V. V. Derunov, M. A. Lusenkova, P. V. Petrovsky and N. E. Kolobova, *J. Organomet. Chem.*, 1989, **379**, 303.

75. U. Schubert and J. Gronen, *Chem. Ber.*, 1989, **122**, 1237.
76. K. Jonas, C.-C. Haselhoff, R. Goddard and C. Kruger, *Inorg. Chim. Acta*, 1992, **200**, 533.
77. N. E. Kolobova, O. S. Zhvanko, L. L. Ivanov, A. S. Batsanov and Y. T. Struchkov, *J. Organomet. Chem.*, 1986, **302**, 235.
78. S. M. Coughlan and G. K. Yang, *J. Organomet. Chem.*, 1993, **450**, 151.
79. C. Lowe, H.-U. Hund and H. Berke, *J. Organomet. Chem.*, 1989, **371**, 311.
80. H. G. Alt, H. E. Engelhardt, E. Steinlein and R. D. Rogers, *J. Organomet. Chem.*, 1988, **344**, 321.
81. B. Buchmann and A. Salzer, *J. Organomet. Chem.*, 1985, **295**, 63.
82. A. M. Rosan and D. M. Romano, *Organometallics*, 1990, **9**, 1048.
83. V. V. Krivykh, O. V. Gusev and M. I. Rybinskaya, *J. Organomet. Chem.*, 1989, **362**, 351.
84. V. V. Krivykh, O. V. Gusev, P. V. Petrovskii and M. I. Rybinskaya, *J. Organomet. Chem.*, 1989, **366**, 129.
85. M.-H. Cheng, C.-Y. Cheng, S.-L. Wang, S.-M. Peng and R. S. Liu, *Organometallics*, 1990, **9**, 1853.
86. A. P. Masters, J. F. Richardson and T. S. Sorensen, *Can. J. Chem.*, 1990, **68**, 2221.
87. T.-W. Lee and R.-S. Liu, *J. Organomet. Chem.*, 1987, **320**, 211.
88. K. F. McDaniel and A. B. A. A. Abu-Baker, *J. Organomet. Chem.*, 1993, **443**, 107.
89. J. R. Bleeke, G. G. Stanley and J. J. Kotyk, *Organometallics*, 1986, **5**, 1642.
90. J. R. Bleeke, J. J. Kotyk, D. A. Moore and D. J. Rauscher, *J. Am. Chem. Soc.*, 1987, **109**, 417.

# 5

# Cyclopentadienyl Manganese Complexes

## PAUL M. TREICHEL
### University of Wisconsin–Madison, WI, USA

---

---

## 5.1 INTRODUCTION

### 5.1.1 Background

Cyclopentadienyl manganese tricarbonyl was among the earliest organometallic manganese complexes to be prepared and, as workers in this area know, a rich chemistry has been built around this exceedingly stable structural unit. A wide variety of complexes have been created by substitution reactions at the cyclopentadienyl ring, chemistry that parallels behavior seen with ferrocene, and by replacement of one or more carbonyl groups by other ligands. Chemistry involving coordinated ligand groups in these complexes is a subject of intensive investigation, since the cyclopentadienyl metal dicarbonyl moiety modifies the chemical properties of the ligand and provides a stable platform on which to carry out chemical transformations. Finally, the manganese(I) oxidation state in these complexes is the starting point for redox chemistry, including electron transfer and oxidative addition reactions that convert $Mn^I$, $d^6$, species to $Mn^{III}$, $d^4$, complexes.

### 5.1.2 Scope of this Chapter

Coverage in this chapter will focus on new complexes, with emphasis on species with different cyclopentadienyl groups and new ligands replacing carbon monoxide. It specifically excludes nitrosyl complexes (complexes derived from the parent species $[Mn(NO)(CO)_2(Cp)]^+$) and complexes with isonitrile ligands. Derivative chemistry will be largely restricted to ligands having nonmetallic donors (mainly groups 15 and 16), while excluding species with carbon-based ligands (alkyl, acyl, carbene, and carbyne groups) along with their associated chemistry since this is covered elsewhere in this volume.

## 5.2 CYCLOPENTADIENYL MANGANESE TRICARBONYL COMPLEXES

A common synthetic route to cyclopentadienyl manganese tricarbonyl complexes uses the reaction of a cyclopentadienide anion and a manganese carbonyl halide. Early methodology often used an alkali metal salt of the cyclopentadienide anion, usually prepared *in situ*, but recently thallium cyclopentadienide salts have been found to have advantages in these processes. This method has been used to prepare a number of cyclopentadienyl complexes, including $[Mn(CO)_3\{\eta\text{-}C_5(CO_2Me)_5\}]$,[1] $[Mn(CO)_3(\eta\text{-}C_5H_4R)]$ (R = $CF_3$,[2] Ph, $PhCH_2$,[3] CHO, COMe, $CO_2Me$,[4] $OCH_2CH_2CH_2CH=CH_2$,[5] and $CH=C_9H_6$[6]), $[Mn(CO)_3\{\eta\text{-}C_5H_3(Me)(COMe)\}]$ (racemic 1,2- and 1,3-isomers),[7] $[(OC)_3Mn(\mu\text{-}\eta,\eta\text{-}C_5H_4C_5H_4)Mn(CO)_3]$,[1,8,9] and $[(OC)_3Mn(\mu\text{-}\eta,\eta\text{-}C_5H_4COCH_2CH_2COC_5H_4)Mn(CO)_3]$.[10]

The reaction of $[Mn_2(CO)_{10}]$ and cyclopentadiene is also used for synthesis of cyclopentadienyl manganese tricarbonyl species. This method was used for the synthesis of $[Mn(CO)_3\{\eta\text{-}C_5(CH_2Ph)_5\}]$.[11] A detailed study of the reaction between $[Mn_2(CO)_{10}]$ and $C_5Me_5H$ reports that both $[Mn(CO)_3Cp^*]$ and $[Mn(CO)_3(\eta\text{-}C_5Me_4H)]$ are products, formed in a 77:23 ratio.[12]

Chemical reactions at the cyclopentadienyl ring present a further route to new complexes in this category. Friedel–Crafts alkylation using $Bu^tCl$ and $AlCl_3$ forms $[Mn(CO)_3(\eta\text{-}C_5H_{5-n}Bu^t_n)]$ (n = 1, 2).[13] An interesting reaction between $[Mn(CO)_3Cp]$ and $Bu_3SnC\equiv CH$ in the presence of $[PdCl_2(NCMe)_2]$ produces $[Mn(CO)_3(\eta\text{-}C_5H_4C\equiv CH)]$.[14]

Lithiation of the cyclopentadienyl ring gives a versatile reactant for use in further syntheses. Reactions of $[Mn(CO)_3(\eta\text{-}C_5H_4Li)]$ and sulfur, selenium, or tellurium form $[(OC)_3Mn(\mu\text{-}\eta,\eta\text{-}C_5H_4AC_5H_4)Mn(CO)_3]$ (A = $S_2$, $Se_2$, $Te_2$). The sulfide, selenide, and tellurium analogues (one chalcogenide atom linking two rings, A = S, Se, Te) are obtained in reactions of this reagent with $SCl_2$, $SeCl_4$, and $TeCl_4$.[15] Silylcyclopentadienyl derivatives are also prepared by this route; $[Mn(CO)_3(\eta\text{-}C_5H_4SiHMe_2)]$ has been prepared from $[Mn(CO)_3(\eta\text{-}C_5H_4Li)]$ and $Me_2SiHCl$,[16] while $[MeSi\{(\eta\text{-}C_5H_4)Mn(CO)_3\}_3]$ is formed from the lithio species and $MeSiCl_3$.[17]

Lithiation of $[Mn(CO)_3(\eta\text{-}C_5Cl_4Br)]$ using BuLi gives $[Mn(CO)_3(\eta\text{-}C_5Cl_4Li)]$. Chemistry of this reagent has been extensively pursued; reactions with $H^+$, MeI, TMS-Cl, $CO_2$, MeSSMe, and PhSSPh are reported in one paper,[18] while another study produced additional derivatives $[Mn(CO)_3(\eta\text{-}C_5Cl_4X)]$ (X = $Me_3Sn$, $Ph_2P$, CHO, $CONH_2$, CN, NCO, and $NH_2$).[19] Sulfur derivatives (compounds with SR groups, and with S and $S_2$ groups bridging two rings) are reported in two subsequent papers;[20,21] substitution of two or three halogens by SR groups was specifically noted.

Several new complexes containing polycyclic hydrocarbon ligands where manganese is bonded to a five-membered ring have been described. The synthesis of (1) and a second isomer having *cis*-metal tricarbonyl groups is achieved by reaction of the anionic hydrocarbon ($Li_2[C_{12}H_8]$) with $[MnBr(CO)_3(py)_2]$.[22] Mono-, bis-, and tris(manganese)tricarbonyl complexes of anions of truxene,[23] $Na_n[C_{27}H_{18-n}]$ (n = 1–3) are formed in a similar kind of reaction; (2) shows the structure of one of these compounds.

The first example of a haptotropic ($\eta^5 \rightleftharpoons \eta^5$) rearrangement has been reported. Ustynyuk *et al.*[24] have examined the system shown in Equation (1) over the temperature range of $-40\,°C$ to $+20\,°C$. Thermodynamic information ($\Delta H^0 = -0.66$ kJ mol$^{-1}$ and $\Delta S^0 = -2.8$ J mol$^{-1}$ °C$^{-1}$) and activation energy parameters ($\Delta H^{\ddagger} = 54.6 \pm 0.3$ kJ mol$^{-1}$ and $\Delta S^{\ddagger} = -28$ J mol$^{-1}$ °C$^{-1}$) were measured for this process.

(1)

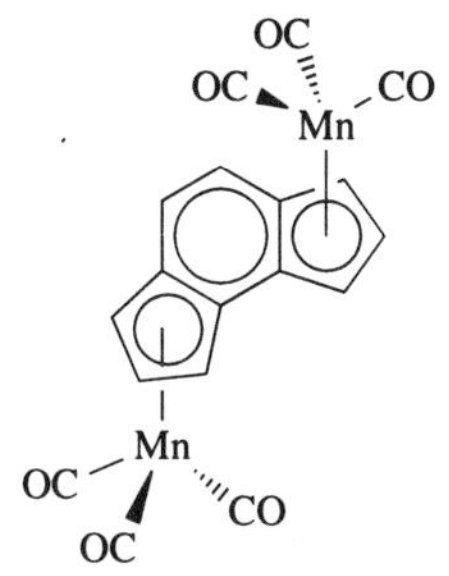

**(1)** 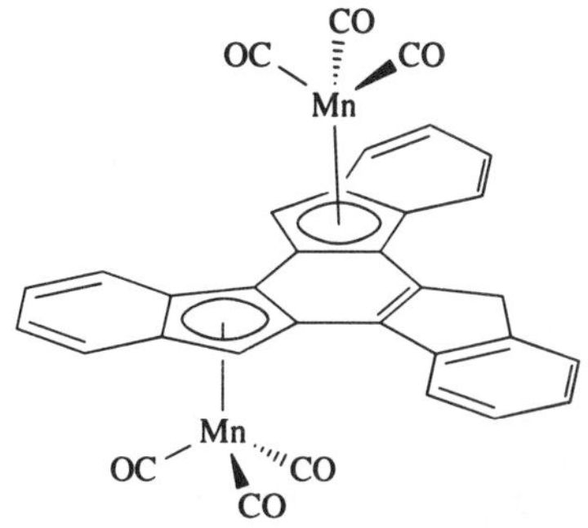          **(2)**

The molecular structure of an $\eta^6$-fluorenylmanganese compound, $[Mn(CO)_3(\eta^6\text{-}C_{13}H_9)]$, has been determined.[25] Protonation of the $\eta^5$-fluorenyl complex, $[Mn(CO)_3(\eta^5\text{-}C_{13}H_9)]$, results in an $\eta^5 \to \eta^6$ rearrangement,[26] while a reaction with $PEt_3$ converts it to a $\eta^1$ species prior to decomposition forming $[Mn(CO)_3(PEt_3)_2]$.[27]

## 5.3 DERIVATIVES OF CYCLOPENTADIENYL MANGANESE TRICARBONYLS, WITH ONE OR MORE LIGANDS SUBSTITUTED FOR CO

A great many complexes in this category have been reported. Often, new complexes are prepared with the intention of further chemical study. This section provides an overview of some of the work since the mid-1980s; as mentioned in Section 5.1, compounds with carbon-based ligands are generally not included.

Two significant review articles are discussed here. The first, an electron transfer processes by Kochi,[28] is of interest since ligand substitution chemistry involving oxidative catalysis is a well-established procedure in this area. Seminal work on this subject was published by Kochi and co-workers in 1983.[29,30] The second article is a review of metal–ligand bond energies by Hoff.[31] This article includes data shown in Table 1 from a study by Yang *et al.* on manganese–ligand bond energies in $[Mn(CO)_2(L)Cp]$ using time-resolved photoacoustic calorimetry.[32] These data relate to the reaction of $[Mn(CO)_3Cp]$ with L in heptane, for which the mechanism in Scheme 1 is proposed.

$$[Mn(CO)_3(\eta\text{-}Cp)] \xrightarrow{\text{i}} [Mn(CO)_2(S)(\eta\text{-}Cp)] \xrightarrow{\text{ii}} [Mn(CO)_2(L)(\eta\text{-}Cp)]$$

i, S = heptane, $h\nu$ = 377 nm $(\Delta H_1)$; ii, +L, –S $(\Delta H_2)$

**Scheme 1**

**Table 1** Metal–ligand bond energies in $[Mn(CO)_2(L)Cp]$.

| L | $\Delta H_1$ (kJ mol$^{-1}$) | $\Delta H_2$ (kJ mol$^{-1}$) |
|---|---|---|
| THF | 193.4 ± 5.0 | −67.4 ± 5.8 |
| Acetone | 200.1 ± 7.1 | −72.8 ± 4.2 |
| cot | 199.2 ± 5.8 | −101.7 ± 9.6 |
| $Bu_2S$ | 189.6 ± 5.8 | −120.1 ± 9.2 |

There are several general references to photochemical substitution reaction processes. Quantum yields and relative rates of reactions of intermediates were measured for photoreactions of CO and $PPh_3$ with $[Mn(PPh_3)_n(CO)_{3-n}(\eta\text{-MeCp})]$.[33] Another study reports the effect of ring strain on the quantum yield of the chelation reaction given in Equation (2).[34] Photolysis of $[Mn(CO)_3Cp]$ at low temperature in a solid matrix produced two species, $[Mn(CO)_nCp]$ ($n$ = 1, 2), characterized by their infrared spectra.[35] Photolysis of a fulvalene complex, $[(OC)_3Mn(\mu\text{-}\eta,\eta\text{-}C_5H_4\text{-}C_5H_4)Mn(CO)_3]$, at 12 K in a matrix results in CO loss, but the *trans*-configuration of this species prevents linkage of the metals via a bridging CO.[36]

$$[Mn(CO)_2\{Me_2P(CH_2)_nPMe_2\}(\eta\text{-}Cp)] \xrightarrow[n = 2, 3, 4]{} [Mn(CO)\{Me_2P(CH_2)_nPMe_2\}(\eta\text{-}Cp)] + CO \qquad (2)$$

References to complexes in the remainder of this section are organized roughly according to structure and reactivity. The most common route of synthesis for such species involves prior formation of $[Mn(CO)_2(THF)(\eta\text{-}CP)]$ by photolysis of the tricarbonyl in THF, and replacement of the labile THF ligand. (In this and subsequent formulas, CP is a generic representation for Cp, Cp*, and MeCp.)

### 5.3.1 Radical Species and Their Precursors

Certain complexes undergo redox reactions to form stable, paramagnetic species. References in this area describe: formation of $[Mn(CO)_2\{NH_2(p\text{-}Tol)\}Cp]$, which can be deprotonated and the anion oxidized to the radical $[Mn(CO)_2\{NH(p\text{-}Tol)\}Cp]$;[37] formation of $[Mn(CO)_2(1,4\text{-pyrazine})Cp]$ and $[\{Mn(CO)_2Cp\}_2(\mu\text{-}1,4\text{-pyrazine})]$ and their reduction to radical anions;[38] formation of $[PPN][Mn(CO)_2\{CO(p\text{-}Tol)\}(\eta\text{-}MeCp)]$ (PPN = $(Ph_3P)_2N$), its oxidation to an uncharged radical species stable at low temperature, and an unusual reaction with NO;[39] formation of $[Mn(CO)_2(H_2NC_6H_4NH_2)(\eta\text{-}MeCp)]$ and oxidative deprotonation of this $p$-phenylenediamine complex to a cation radical.[40]

An extensive study has been published describing dinuclear species with the general formula $[\{Mn(CO)_2(\eta\text{-}CP)\}_2(\mu\text{-}S\text{-}S)]$ (S–S = $S(CH_2)_nS$, $n = 2\text{-}4$; $1,3\text{-}C_6H_4S_2$).[41] The sulfur atoms in the compounds with the bridging ligands $S(CH_2)_2S$ and $S(CH_2)_3S$ are linked by a single bond as shown in (3) and as a result these complexes are diamagnetic. The species with $S(CH_2)_4S$ and $1,3\text{-}C_6H_4S_2$ ligands have two unpaired electrons. Structures of both types of species were confirmed in several single-crystal x-ray diffraction studies. All of these complexes show two reversible one-electron oxidations and one one-electron reduction (reversible for the paramagnetic compounds). A related study, from the same research group, described preparation of the stable mononuclear radical complexes $[Mn(CO)_2(SR)Cp]$ (R = Ph, $C_6H_4NO_2$) and $[Mn(CO)_2(SR)(\eta\text{-}MeCp)]$ (R = $C_6F_5$, $Pr^i$, $Bu^t$) by air oxidation of corresponding RSH complexes.[42] Further oxidation of these complexes using $AgPF_6$ gives dinuclear complexes, $[\{Mn(CO)_2Cp\}_2(\mu\text{-}SR)]^+$.

(3)

### 5.3.2 Monosubstituted Complexes with Metal-containing Ligands

Interest in complexes containing two or more metals, often in different oxidation states or coordination environments, has spawned a rather large number of new compounds. Complexes reported have the general formula $[Mn(CO)_2(L)(\eta\text{-}CP)]$. Metalloligands (L) in this formula include the following examples.

(i) *Ligand bonded through sulfur or selenium*: L = $[FeCp\{\eta\text{-}C_5(SMe)_5\}]$;[43] $[FeCp(\eta\text{-}C_5H_4SMe)]$;[44] $[VCp_2(CS_2)]$;[45] $[Mo_2Cp^*_2(S)_4]$;[46] $[\{FeCp(\eta\text{-}C_5H_4)\}_2Se]$;[47] $[Co(PMe_3)Cp(R_2C=C=S)]$;[48] $[Cr(CO)_5(SMe_2)]$;[49] $[Co_2(\mu\text{-}PMe_2)_2(\mu\text{-}S)Cp^*_2)]$;[50] $[Rh(PMe_3)(\eta^2\text{-}CH_2S)Cp]$.[51]

(ii) *Ligand bonded through nitrogen, phosphorus, or arsenic*: L = PhNO;[52] $[TiCl_2Cp(\eta\text{-}C_5H_4PPh_2)]$;[53] $[CoNi(CO)_3Cp(Bu^tC\equiv P)]$;[54] $[Cr(CO)_3(\eta\text{-}C_6H_5PPh_2)]$;[55] $[Fe(CO)_2(PHNPr^i_2)Cp]$;[56] $[Fe(CO)_2(PR_2)Cp]$ (R = TMS, H);[57,58] $[M(cod)(PBu^t_2)]$ (M = Rh, Ir);[59] $[MH(PPh_2)Cp_2]$ (M = Mo, W);[60] $[Fe_3(CO)_9(\mu\text{-}CH)(\mu\text{-}As)]$;[61] $[Fe_2(\mu\text{-}PR)(CO)_6(L)_2]$;[62] $[Co_3(CO)_9(P)]$, and $[Fe_3(CO)_9(\mu^3\text{-}P)]$ (the latter species also coordinates to two $[Mn(CO)_2Cp]$ groups);[63] and $[Fe_2(CO)_6(P_2)]$, also coordinated to two manganese groups.[64]

(iii) *Ligand bonded via a bridging hydride, vinyl, or cyanide*: L = $[Ta(CO)Cp_2(\mu\text{-}H)]$;[65,66] $[Nb(CO)_3Cp(\mu\text{-}H)]^-$, isolated as a PPN$^+$ salt;[67] $[Os_3(\mu\text{-}H)(CO)_{10}(CH=CHR)]$;[68] $[Ru_3(CO)_{10}(\mu^2\text{-}CN)_2]$ bonded to two $[Mn(CO)_2Cp]$ moieties.[69]

(iv) *Higher nuclearity clusters*: Cyclopentadienyl manganese carbonyl groups can be incorporated into cluster complexes. Since these species bear some similarity to the compounds just listed they are included here. Of specific note are trimetallic compounds of composition $[MnFe_2(CO)_8(\mu\text{-}X)(\eta\text{-}CP)]$ (X = S, Se, Te, PR) (4). Corresponding manganese–ruthenium analogues have also been prepared.

(4)

Analogous phosphido complexes are formed in two ways, from [Mn(CO)$_2$(PRCl$_2$)($\eta$-CP)] and [Fe$_2$(CO)$_9$], or from [Mn(CO)$_2$(PRH$_2$)($\eta$-CP)] and [M$_3$(CO)$_{12}$] (M = Fe, Ru).[70] They will add one or two ligands, reversibly.[62,70] The sulfur complex, [MnFe$_2$($\mu$-S)(CO)$_8$Cp] was obtained, unexpectedly, as the main product when a solution of [Mn(CO)$_3$Cp], [Fe$_3$(CO)$_{12}$], and Bu$^t$SH in toluene was photolyzed.[71] A tellurium analogue is formed from [Te{Mn(CO)$_2$Cp*}$_2$] and [Fe$_2$(CO)$_9$].[72]

### 5.3.3 Complexes with One or More Noncarbonyl Ligands

It is most convenient simply to list new [Mn(L)(CO)$_2$($\eta$-CP)] complexes, with occasional comments. Note that this section does not include the "inidine" complexes, which are covered in Section 5.3.4.

#### 5.3.3.1 Group 15 donors

For example: phosphine substituents bonded to a polyaryloxyphosphazene polymer;[73] PH(X)$_2$ (X = NPr$^i_2$,[74] OMe, Br,[75] the latter species being formed by reactions at the coordinated ligand); P$_2$Me$_4$ (coordinated to two metal groups);[76] *p*-cyanopyridine (three compounds, two with monodentate coordination through py and CN nitrogens, the third with the ligand bridging two metals);[77] dppe, diop (2,3-*o*-isopropylidene-2,3-dihydroxy-1,4-bis(diphenylphosphino)butane) and norphos (monodentate and bidentate complexes); monodentate [Mn(CO)$_2$(diop)Cp] with a cocatalyst, [Rh(Cl)(cod)]$_2$, was used in enantioselective hydrogenation;[78] Ph$_2$PC≡CPPh$_2$ (monodentate and bridging);[79] RPHPHR (R = Bu$^t$, Pr$^i$, Cy, CH$_2$TMS, 3,5-Me$_2$C$_6$H$_3$CH$_2$) bridging two metal groups (formed in two steps, reduction of [Mn(RPCl$_2$)(CO)$_2$Cp] using Bu$^t$Li and protonation);[80] ArN=NAr (Ph, *p*-Tol, C$_6$H$_4$F);[81] vinylpyridine (two species, coordinated via either nitrogen or the alkene group);[82] Bu$^t_2$PN=S=NPBu$^t_2$ and the arsenic analogue, coordinated to two metal moieties;[83] tetracyanoethylene (TCNE) and 7,7,8,8-tetracyanoquino-dimethane (TCNQ)[84] (paramagnetism of the TCNE species,[85] rates of formation,[86] coordination of TCNE to four metals[87]); *p*-C$_6$H$_4$(PR$_2$)$_2$ (coordinated to two metals);[88] 1,2,3-benzothiadiazole and 4-Ph-1,2,3-thiadiazole;[89] (MeAsO)$_4$[90] and *cyclo*-[Bu$^t$AsN=S=N]$_2$,[91] both coordinated to two metals units.

#### 5.3.3.2 Group 16 donors

For example: R$_2$C=C=S (R = Ph,[92] Bu$^t$,[93] the latter bridging two metals); C$_6$Ph$_4$S$_2$, (1,2-diphenylcyclopropenethione dimer);[94] Me$_2$S (bridging two metals);[50] S$_2$, and oxidation products with ligands S$_2$O and S$_2$O$_2$[95] (the complex with an S$_2$O ligand can also be formed in a different reaction[96,97]).

#### 5.3.3.3 Anionic complexes

Complexes with the formula [Mn(X)(CO)$_2$($\eta$-CP)]$^-$, where X is the following anionic ligands: phth$^-$ (phthalimide ion, complex isolated as the NEt$_4^+$ salt;[98] GeH$_3^-$ (K$^+$ salt);[99] Me$^-$ (prepared *in situ* from MeLi and [Mn(N$_2$)(CO)$_2$Cp] below −10 °C, this reacts with [Me$_3$O]$^+$ to give [Mn(Me)$_2$(CO)$_2$Cp]);[100] N$_3^-$ (NEt$_4^+$ salt);[101] SiMePh$_2^-$ (Na$^+$ salt, prepared *in situ*; this anionic complex reacts with metal complexes (*m*L$_n$X = [PhHgCl], [AuCl(PPh$_3$)], and [Fe(I)(CO)$_2$Cp])[102] to give [Mn(*m*L$_n$)(SiMePh$_2$)(CO)$_2$($\eta$-CP)], and it also reacts with Ph$_3$AsCl$_2$ or Ph$_3$SbBr$_2$ to give [Mn(CO)$_2$(APh$_3$)Cp] (A = As, Sb)[103]).

### 5.3.4 Complexes with Multiply Bonded Main Group Elements

Since the mid-1980s, there has been extensive study on complexes with main group elements multiply bonded to cyclopentadienyl manganese dicarbonyl moieties. Complexes having the general formula [{Mn(CO)$_2$(η-CP)}$_2$X] (X = Ge, Sn, Pb, PR, AsR, SbR, BiR, S, SO, Se, Te) (5), are discussed below. Early work on the group 15 species, which often carry the label "inidine" (phosphinidene, arsinidene, etc.), is described in two review articles published in 1986.[104,105]

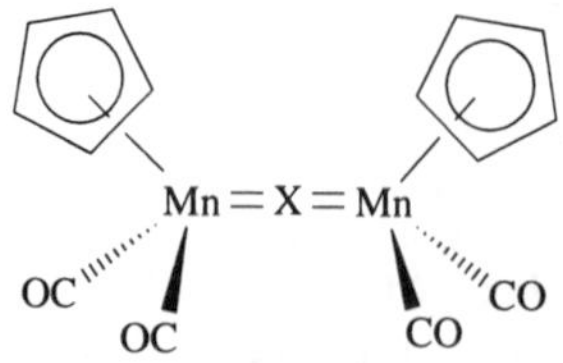

(5)

#### 5.3.4.1 Group 14

Preparative methods for [{Mn(CO)$_2$Cp}$_2$Ge] include the reaction of Na[Mn$_2$(μ-H)(CO)$_4$Cp$_2$] with GeCl$_4$[106] and the acidification of K[Mn(GeH$_3$)(CO)$_2$Cp].[107] The first method is applicable for synthesis of the tin[108] and lead[109] analogues, while [{Mn(CO)$_2$Cp*}$_2$Sn] is also a product of the reaction of SnH$_4$ and [Mn(CO)$_2$(THF)Cp*]. Both the tin and lead complexes are formed in reactions between [Mn(CO)$_2$(THF)Cp] and SnCl$_2$[110] or PbCl$_2$[111] in the presence of a reducing agent (zinc metal); however, a similar reaction with GeCl$_2$ forms [Mn$_2$(μ-GeCl$_2$)(CO)$_4$Cp$_2$].[112] These [{Mn(CO)$_2$Cp}$_2$A] complexes react further with [Mn(CO)$_2$(THF)Cp] forming [{Mn(CO)$_2$(η-MeCp)}$_3$A] (A = Ge, Sn, Pb) (6), and [{Mn(CO)$_2$(η-MeCp)}$_2$Ge] adds bipy to form [{Mn(CO)$_2$(η-MeCp)}$_2$Ge(bipy)].

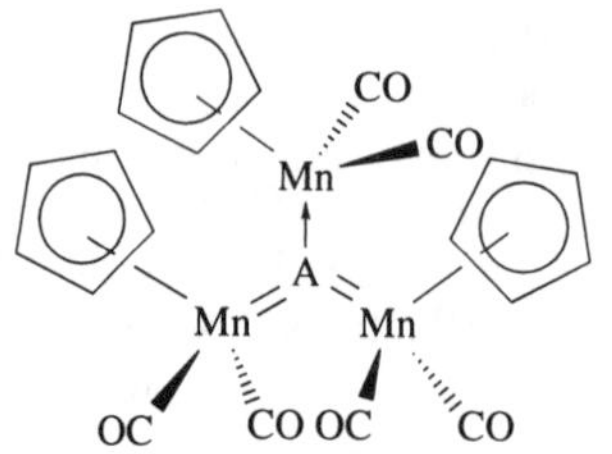

(6)

#### 5.3.4.2 Group 15

Photolysis of [Mn(CO)$_2$(PH$_3$)Cp*] leads to the formation of [{Mn(CO)$_2$Cp*}$_2$PH].[113] Preparation of [{Mn(CO)$_2$Cp*}$_2$P]$^+$ from this complex is accomplished by hydride abstraction using CF$_3$SO$_3$Me.[114]

Arsenic analogues to these species are accessed by two routes. The first involves borohydride reduction of [{Mn(CO)$_2$(η-MeCp)}$_2$AsCl].[115] (This precursor was first reported in 1978, having been prepared by the reaction of AsCl$_3$ and [Mn(CO)$_2$(THF)Cp].) In addition, [{Mn(CO)$_2$Cp*}$_2$AsH] is one of several products in a reaction of [Mn(CO)$_2$(THF)Cp*] with AsH$_3$.[116] Antimony and bismuth derivatives, [{Mn(CO)$_2$Cp}$_2$AX] (A = Sb, Bi) can be prepared from [Mn(CO)$_2$(THF)Cp], zinc, and either SbX$_3$[117] or BiCl$_3$.[118] The reduction of [{Mn(CO)$_2$Cp}$_2$AsCl] with zinc gives a dimeric, diarsinidene species [{Mn(CO)$_2$Cp}$_4$As$_2$].[119] Halide abstraction from [{Mn(CO)$_2$Cp}$_2$AX] (A = As[115,120,121] Bi[118]) occurs, forming [{Mn(CO)$_2$Cp}$_2$A]$^+$ with a variety of Lewis acids, and [{Mn( CO)$_2$Cp*}$_2$A(bipy)] species (A = As, Sb, Bi) are also known.

Additionally, [{Mn(CO)$_2$Cp*}$_2$AsH] loses H$_2$ when heated, forming [{Mn(CO)$_2$Cp*}$_4$(As$_2$)].[115] The antimony complexes [{Mn(CO)$_2$(η-MeCp)}$_2$Sb(SPh)$_n$] (n = 1, 3) have also been prepared.[122]

### 5.3.4.3 Group 16

Formation of $[\{Mn(CO)_2Cp^*\}_2Te]$[123] and $[\{Mn(CO)_2Cp\}_3Te]$[124] is reported in reactions of $H_2Te$ with $[Mn(CO)_2(THF)Cp^*]$ or $[Mn(CO)_2(THF)Cp]$. Methyllithium reacts with $[\{Mn(CO)_2Cp\}_3Te]$, giving $[\{Mn(CO)_2Cp\}_3TeMe]^-$, which is isolable as a $PPN^+$ salt.[125]

Sulfur and selenium complexes with the formula $[\{Mn(CO)_2Cp\}_2AR]^+$ (A = S, Se) are formed by oxidation of $[Mn(AR)(CO)_2Cp]$. With these sulfur and selenium complexes, in contrast to the tellurium species, a delicate energy balance exists between the inidene species and the isomeric thiolate-bridged species, as shown in Equation (3). The choice of the R group and of whether the hydrocarbon bonded to manganese is a Cp, MeCp, or Cp* ring influences which isomer is favored.[126-9]

$$(3)$$

## 5.4 CHEMISTRY OF CYCLOPENTADIENYL MANGANESE–CARBONYL COMPLEXES

This section reviews other areas of the chemistry of these species not covered earlier, including acid–base, redox chemistry, oxidative addition chemistry (a general survey of the species with composition $[Mn(CO_2)(X)(Y)(\eta\text{-}CP)]$), and the formation of dinuclear complexes.

### 5.4.1 Acid–Base Chemistry

The replacement of one or more carbonyls in $[Mn(CO)_3Cp]$ by good donor ligands such as phosphines leads to significant basic character at the metal center. Studies on the basicity of $[Mn(CO)_2(L)Cp]$ (L = CO, phosphines) toward $H^+$, $AlBr_3$, and $GaBr_3$ are reported in two papers.[130,131] The addition of $Hg(O_2CCF_3)_2$ to $[Mn(CO)_2(PPh_3)Cp]$ is viewed as Lewis acid–base reaction chemistry.[132] Protonation of $[Mn(CO)(dppe)Cp]$ is reported to give both *cis*- and *trans*- $[Mn(H)(CO)(dppe)Cp]^+$; there is a kinetic preference for *cis*-protonation while the *trans*-product is preferred on thermodynamic grounds.[133]

### 5.4.2 Redox Chemistry

The highly reduced complex of manganese, $[Mn(CO)_2Cp]^{2-}$ ($v(CO)$ 1870w, 1785w, 1685s, 1600s, and 1550s $cm^{-1}$) is formed by sodium naphthalenide reduction of $[Mn(CO)_2(py)Cp]$. Addition of water to $[Mn(CO)_2Cp]^{2-}$ results in the formation of $[Mn(H)(CO)_2Cp]^-$, while reaction with $Ph_3PbCl$ produces $[Mn(PbPh_3)(CO)_2Cp]^-$ and $[Mn(PbPh_3)_2(CO)_2Cp]$.[134] In contrast, borohydride reduction of $[Mn(CO)_2(THF)Cp]$ leads to $[Mn_2(\mu\text{-}H)(CO)_4Cp_2]^-$, which is isolable as a salt of the $Na(crypt)^+$ cation.[135]

### 5.4.3 $[Mn(X)(Y)(CO)_2Cp]$ Complexes

Room temperature photolysis of $[Mn(CO)_3Cp]$ with $H_2$ in supercritical xenon under pressure gives $[Mn(H)_2(CO)_2Cp]$, a compound stable only under high hydrogen pressure or at low temperature. It reacts with $N_2$, giving $[Mn(CO)_2(N_2)Cp]$.[136]

Photolysis of $[Mn(CO)_2(PR_3)(\eta\text{-}MeCp)]$ in the presence of $R_2SiH_2$ gives the oxidative addition product $[Mn(H)(CO)(SiHR^1{}_2)(PR^2{}_3)(\eta\text{-}MeCp)]$ ($R^1$ = Me, Ph; $R^2$ = Ph, Me, Bu, $C_6H_4X$).[137] These compounds are easily deprotonated giving anionic species $[Mn(SiHR_2)(CO)(PR_3)(\eta\text{-}MeCp)]^-$, which are useful in further synthesis,[102] while reductive elimination of the silane occurs when additional phosphine is provided. Phosphines that are electron donors enhance the oxidative addition reactions and slow reductive elimination. Data from several structural studies indicate that this compound is best viewed as containing a three-centered Si–H–Mn bond.[138,139]

A different study prepared [Mn(H){Si(Me)(Ph)(Np)}(CO)$_2$($\eta$-MeCp)], determining that photo-induced oxidative addition occurs with retention of configuration at silicon. This paper also reported similar photochemical reactions between several germanes and [Mn(CO)$_3$Cp].[140]

Near-UV radiation of [Mn(CO)$_3$($\eta$-C$_5$Cl$_5$)] and Et$_3$SiH in an alkane glass matrix forms [Mn(H)(SiEt$_3$)(CO)$_2$($\eta$-C$_5$Cl$_5$)], via the intermediate [Mn(CO)$_2$($\eta$-C$_5$Cl$_5$)]. Activation parameters were determined for the oxidative addition reaction.[141]

### 5.4.4 Dinuclear Complexes

Photolysis of [Mn(CO)$_3$Cp*] in THF results in the formation of [Mn$_2$($\mu$-CO)$_3$Cp*$_2$] (7), a compound with a short (0.217 nm) metal–metal bond length appropriate to a triple bond.[142,143] It was determined in the study that [Mn(CO)$_2$(THF)Cp*] spontaneously converts to this species. The related species [(Cp*)Mn($\mu$-CO)$_3$Cr($\eta$-C$_6$Me$_6$)] and [(Cp*)Mn($\mu$-CO)$_3$Fe($\eta$-C$_4$Me$_4$)] have also been made.[144] In contrast, photolysis of [Mn(CO)$_3$Cp] in a low-temperature matrix results in the formation of [Mn$_2$($\mu$-CO)(CO)$_4$Cp$_2$].[145]

(7)                                    (8)

M = Co, Rh

It is possible to reduce [Mn$_2$($\mu$-TePh)$_2$(CO)$_4$Cp$_2$] in two one-electron steps. Both anions [Mn$_2$($\mu$-TePh)$_2$(CO)$_4$Cp$_2$]$^{n-}$ ($n$ = 1, 2) have been isolated and structural studies carried out; the metal–metal distance increased on each reduction step, indicating addition of electrons to a metal antibonding ($\sigma^*$) orbital.[146]

Compounds of the formula [CpM($\mu$-CO)($\mu$-CS)Mn(CO)Cp] (M = Co, Rh) (8) are formed in reactions of [M(CS)(PR$_3$)Cp] and [Mn(CO)$_2$(THF)Cp]. Several derivatives of this species were also reported in this work.[147] The complex [Cp(CO)Fe($\mu$-CO)($\mu$-PHNPr$^i_2$)Mn(CO)Cp] forms on irradiation of [(Cp)(CO)$_2$Fe($\mu$-PHNPr$^i_2$)Mn(CO)$_2$Cp].[102]

### 5.5 REFERENCES

1.  C. Arsenault, P. Bougeard, B. G. Sayer, S. Yeroushalmi and M. J. McGlinchney, *J. Organomet. Chem.*, 1984, **265**, 283.
2.  W. C. Spink and M. D. Rausch, *J. Organomet. Chem.*, 1986, **308**, C1.
3.  P. Singh, M. D. Rausch and T. E. Bitterwolf, *J. Organomet. Chem.*, 1988, **352**, 273.
4.  S. S. Jones, M. D. Rausch and T. E. Bitterwolf, *J. Organomet. Chem.*, 1990, **396**, 279.
5.  T. E. Bitterwolf, K. A. Lott and A. J. Rest, *J. Organomet. Chem.*, 1991, **408**, 137.
6.  P. Härter, J. Behm and K. J. Burkert, *J. Organomet. Chem.*, 1992, **438**, 297.
7.  T. E. Bitterwolf, T. L. Hubler and A. L. Rheingold, *J. Organomet. Chem.*, 1992, **431**, 199.
8.  M. D. Rausch, W. C. Spink, B. G. Conway, R. D. Rogers and J. L. Atwood, *J. Organomet. Chem.*, 1990, **383**, 227.
9.  P. Gassman and C. H. Winter, *J. Am. Chem. Soc.*, 1986, **108**, 4228.
10. T. E. Bitterwolf, *J. Organomet. Chem.*, 1990, **386**, 9.
11. M. D. Rausch, W.-M. Tsai, J. W. Chambers, R. D. Rogers and H. G. Alt, *Organometallics*, 1989, **8**, 816.
12. R. C. Hemond, R. P. Hughes and H. B. Locker, *Organometallics*, 1986, **5**, 2391.
13. M. P. Balem, M. le Plouzennec and M. Louër, *Inorg. Chem.*, 1982, **21**, 2573.
14. C. L. Sterzo and J. K. Stille, *Organometallics*, 1990, **9**, 687.
15. M. Herberhold and M. Biersack, *J. Organomet. Chem.*, 1990, **381**, 379.
16. E. Colomer, R. J. P. Corriu and R. Pleixats, *J. Organomet. Chem.*, 1990, **381**, C1.
17. M. E. Wright and V. W. Day, *J. Organomet. Chem.*, 1987, **329**, 43.
18. K. Sünkel and D. Motz, *Chem. Ber.*, 1988, **121**, 799.
19. K. Sünkel and D. Steiner, *J. Organomet. Chem.*, 1989, **368**, 67.
20. K. Sünkel and D. Steiner, *Chem. Ber.*, 1989, **122**, 609.
21. K. Sünkel and A. Blum, *Chem. Ber.*, 1992, **125**, 1605.
22. W. L. Bell *et al.*, *Organometallics*, 1988, **7**, 691.
23. T. L. Tisch, T. J. Lynch and R. Dominguez, *J. Organomet. Chem.*, 1989, **377**, 265.

24. Y. A. Ustynyuk, O. I. Trefonova, Y. F. Oprunenko, V. I. Mstislavskiy, I. P. Gloriozov and N. A. Ustynyuk, *Organometallics*, 1990, **9**, 1707.
25. P. M. Treichel, K. P. Fivizzani and K. J. Haller, *Organometallics*, 1982, **1**, 931.
26. M. G. Yezernitskaya, B. V. Lokshin, V. I. Zdanovich, I. A. Lobonova and N. E. Kolobova, *J. Organomet. Chem.*, 1982, **234**, 329.
27. R. N. Biagioni, I. M. Lorkovic, J. Skelton and J. B. Hartung, *Organometallics*, 1990, **9**, 547.
28. J. K. Kochi, *J. Organomet. Chem.*, 1986, **300**, 139.
29. J. W. Hershberger, R. J. Klingler and J. K. Kochi, *J. Am. Chem. Soc.*, 1983, **105**, 61.
30. J. W. Hershberger and J. K. Kochi, *Polyhedron*, 1983, **2**, 929.
31. C. D. Hoff, *Prog. Inorg. Chem.*, 1992, **40**, 503.
32. J. K. Klassen, M. Selke, A. A. Sorensen and G. K. Yang, *J. Am. Chem. Soc.*, 1990, **112**, 1267.
33. G. Teixeira, T. Avilés, A. R. Dias and F. Pina, *J. Organomet. Chem.*, 1988, **353**, 83.
34. A. A. Sorensen and G. K. Yang, *J. Am. Chem. Soc.*, 1991, **113**, 7061.
35. M. Herberhold, W. Kremnitz, H. Trampisch, R. B. Hitam, A. J. Rest and D. J. Taylor, *J. Chem. Soc., Dalton Trans.*, 1982, 1261.
36. P. E. Bloyce, R. H. Hooker, A. J. Rest, T. E. Bitterwolf, N. J. Fitzpatrick and J. E. Shade, *J. Chem. Soc., Dalton Trans.*, 1990, 833.
37. D. Sellmann and J. Müller, *J. Organomet. Chem.*, 1985, **281**, 249.
38. R. Gross-Lannert and W. Kaim, *Inorg. Chem.*, 1986, **25**, 498.
39. J. B. Sheridan, S.-H. Han and G. L. Geoffroy, *J. Am. Chem. Soc.*, 1987, **109**, 8097.
40. R. Gross-Lannert and W. Kaim, *Inorg. Chem.*, 1987, **26**, 3596.
41. H. Braunwarth *et al.*, *J. Organomet. Chem.*, 1991, **411**, 383.
42. P. Lau *et al.*, *Organometallics*, 1991, **10**, 3861.
43. N. E. Kolobova, L. L. Ivanov, O. S. Zhvanko, V. V. Derunov and I. N. Chechulina, *Bull. Acad. Sci. USSR (Engl. Transl.)*, 1982, **31**, 2328.
44. H. Stolzenberg, W. P. Fehlhammer, M. Monari, V. Zanotti and L. Busetto, *J. Organomet. Chem.*, 1984, **272**, 73.
45. C. Moise, *J. Organomet. Chem.*, 1983, **247**, 27.
46. H. Brunner, H. Kauermann, W. Meier and J. Wachter, *J. Organomet. Chem.*, 1984, **263**, 183.
47. W. A. Herrmann, J. Rohrmann, M. L. Ziegler and T. Zahn, *J. Organomet. Chem.*, 1985, **295**, 175.
48. R. Drews, F. Edelmann and U. Behrens, *J. Organomet. Chem.*, 1986, **315**, 369.
49. G. Bremer, R. Boese, M. Keddo and T. Kruck, *Z. Naturforsch., Teil B*, 1986, **41**, 981.
50. H. Werner, G. Luxenberger, W. Hofmann and M. Nadvornik, *J. Organomet. Chem.*, 1987, **323**, 161.
51. H. Werner, W. Paul, W. Knaup, J. Wolf, G. Müller and J. Riede, *J. Organomet. Chem.*, 1988, **358**, 95.
52. N. Setkina, S. P. Dolgova, D. V. Zagorevskii, V. F. Sizoi and D. N. Kursanov, *Bull. Acad. Sci. USSR (Engl. Transl.)*, 1982, **31**, 1239.
53. N. Suryaprakash, A. C. Kunwar and C. L. Khetrapal, *J. Organomet. Chem.*, 1984, **275**, 53.
54. R. Bartsch, J. F. Nixon and N. Sarjudeen, *J. Organomet. Chem.*, 1985, **294**, 267.
55. V. I. Losilkina, M. N. Estekhina, N. K. Baranetskaya and V. N. Setkina, *J. Organomet. Chem.*, 1986, **299**, 187.
56. R. B. King, W.-K. Fu and E. M. Holt, *Inorg. Chem.*, 1985, **24**, 3094.
57. H. Schäfer and W. Leske, *Z. Anorg. Allg. Chem.*, 1987, **550**, 57.
58. W. A. Herrmann, B. Koumbouris, E. Herdtweck, M. L. Ziegler and P. Weber, *Chem. Ber.*, 1987, **120**, 931.
59. A. M. Arif, D. J. Chandler and R. A. Jones, *Inorg. Chem.*, 1987, **26**, 1780.
60. C. Barre, M. M. Kubicki, J.-C. Leblanc and C. Moise, *Inorg. Chem.*, 1990, **29**, 5244.
61. C. Caballero, B. Nuber and M. L. Ziegler, *J. Organomet. Chem.*, 1990, **386**, 209.
62. J. Schneider and G. Huttner, *Chem. Ber.*, 1983, **116**, 917.
63. H. Lang *et al.*, *J. Organomet. Chem.*, 1986, **304**, 157.
64. H. Lang, L. Zsolnai and G. Huttner, *Angew. Chem., Int. Ed. Engl.*, 1983, **22**, 976.
65. J. C. Leblanc, J. F. Reynoud and C. Moise, *J. Organomet. Chem.*, 1983, **244**, C24.
66. J. F. Reynoud, J. C. Leblanc and C. Moise, *J. Organomet. Chem.*, 1985, **296**, 377.
67. P. Oltmanns and D. Rehder, *J. Organomet. Chem.*, 1988, **345**, 87.
68. A. B. Antonova, S. V. Kovalenko, E. D. Korniyets, A. A. Johansson, Yu. T. Struchkov and A. Y. Yanovsky, *J. Organomet. Chem.*, 1984, **267**, 299.
69. B. Oswald, A. K. Powell, F. Rashwan, J. Heinze and H. Vahrenkamp, *Chem. Ber.*, 1990, **123**, 243.
70. J. Schneider, L. Zsolnai and G. Huttner, *Chem. Ber.*, 1982, **115**, 989.
71. A. Winter, L. Zsolnai and G. Huttner, *J. Organomet. Chem.*, 1984, **269**, C29.
72. W. A. Herrmann, C. Hecht, M. L. Ziegler and T. Zahn, *J. Organomet. Chem.*, 1984, **273**, 323.
73. H. R. Allcock, K. D. Lavin, N. M. Tollefson and T. L. Evans, *Organometallics*, 1983, **2**, 267.
74. R. B. King and W.-K. Fu, *J. Organomet. Chem.*, 1984, **272**, C33.
75. R. B. King and W.-K. Fu, *Inorg. Chem.*, 1986, **25**, 2384.
76. H. Werner, B. Klingert, R. Zolk and P. Thometzek, *J. Organomet. Chem.*, 1984, **266**, 97.
77. R. Gross-Lannert and W. Kaim, *J. Organomet. Chem.*, 1985, **292**, C21.
78. H. Brunner and A. Knott, *Z. Naturforsch., Teil B*, 1985, **40**, 1243.
79. O. Orama, *J. Organomet. Chem.*, 1986, **314**, 273.
80. H. Lang, G. Huttner and I. Jibril, *Z. Naturforsch., Teil B*, 1986, **41**, 473.
81. F. W. B. Einstein, D. Sutton and K. G. Tyler, *Inorg. Chem.*, 1987, **26**, 111.
82. J. M. Kelly and C. Long, *J. Organomet. Chem.*, 1982, **231**, C9.
83. M. Herberhold, W. Buhlmeyer, A. Gieren and T. Hübner, *J. Organomet. Chem.*, 1987, **321**, 37.
84. R. Gross-Lannert and W. Kaim, *Angew. Chem., Int. Ed. Engl.*, 1987, **26**, 251.
85. H. Braunwarth, G. Huttner and L. Zsolnai, *J. Organomet. Chem.*, 1989, **372**, C23.
86. B. Schwederski, W. Kaim, B. Olbrich-Deussner and T. Roth, *J. Organomet. Chem.*, 1990, **440**, 145.
87. R. Gross-Lannert, W. Kaim and B. Olbrich-Deussner, *Inorg. Chem.*, 1990, **29**, 5046.
88. W. Kaim, T. Roth, B. Olbrich-Deussner, R. Gross-Lannert, J. Jordanov and E. K. H. Roth, *J. Am. Chem. Soc.*, 1992, **114**, 5693.

89.  A. J. Myer, B. Carresco-Flores, F. Cervantes-Lee, K. H. Pannell, L. Párkányi and K. RaghuVeer, *J. Organomet. Chem.*, 1991, **405**, 309.
90.  A. J. DiMaio and A. L. Rheingold, *Organometallics*, 1991, **10**, 3764.
91.  M. Herberhold and K. Schamel, *J. Organomet. Chem.*, 1988, **346**, 13.
92.  D. Wormsbächer, F. Edelmann and U. Behrens, *Chem. Ber.*, 1982, **115**, 1332.
93.  D. Wormsbächer, F. Edelmann and U. Behrens, *J. Organomet. Chem.*, 1986, **312**, C53.
94.  F. Edelmann, J. Klimes and E. Weiss, *J. Organomet. Chem.*, 1982, **224**, C31.
95.  M. Herberhold and B. Schmidkonz, *J. Organomet. Chem.*, 1986, **308**, 35.
96.  G. A. Urove, M. E. Welker and B. E. Eaton, *J. Organomet. Chem.*, 1990, **384**, 105.
97.  G. A. Urove and M. E. Welker, *Organometallics*, 1988, **7**, 1013.
98.  E. J. Schier, W. Sacher and W. Beck, *Z. Naturforsch., Teil B*, 1987, **42**, 1424.
99.  D. Melzer and E. Weiss, *Chem. Ber.*, 1984, **117**, 2464.
100. D. Sellmann and P. Klostermann, *Z. Naturforsch., Teil B*, 1983, **38**, 1497.
101. D. Sellmann, W. Weber, G. Liehr and H. P. Beck, *J. Organomet. Chem.*, 1984, **269**, 155.
102. U. Schubert and U. Kunz, *J. Organomet. Chem.*, 1986, **303**, C1.
103. U. Kirchgässner and U. Schubert, *Chem. Ber.*, 1989, **122**, 1481.
104. G. Huttner and K. Evertz, *Acc. Chem. Res.*, 1986, **19**, 406.
105. G. Huttner, *Pure Appl. Chem.*, 1986, **58**, 585.
106. F. Ettel, G. Huttner and W. Imhof, *J. Organomet. Chem.*, 1990, **397**, 299.
107. D. Melzer and E. Weiss, *J. Organomet. Chem.*, 1984, **263**, 67.
108. F. Ettel, G. Huttner, L. Zsolnai and C. Emmerich, *J. Organomet. Chem.*, 1991, **414**, 71.
109. F. Ettel, G. Huttner and L. Zsolnai, *Angew. Chem., Int. Ed. Engl.*, 1989, **28**, 1496.
110. W. A. Herrmann, H.-J. Kneuper and E. Herdtweck, *Chem. Ber.*, 1989, **122**, 437.
111. W. A. Herrmann, H.-J. Kneuper and E. Herdtweck, *Chem. Ber.*, 1989, **122**, 445.
112. W. A. Herrmann, H.-J. Kneuper and E. Herdtweck, *Chem. Ber.*, 1989, **122**, 433.
113. W. A. Herrmann, B. Koumbouris, E. Herdtweck, M. L. Ziegler and P. Weber, *Chem. Ber.*, 1987, **120**, 931.
114. A. Strube, J. Heuser, G. Huttner and H. Lang, *J. Organomet. Chem.*, 1988, **356**, C9.
115. A. Strube, G. Huttner, L. Zsolnai and W. Imhof, *J. Organomet. Chem.*, 1990, **399**, 281.
116. W. A. Herrmann, B. Koumbouris, A. Schäfer, T. Zahn and M. L. Ziegler, *Chem. Ber.*, 1985, **118**, 2472.
117. U. Weber, L. Zsolnai and G. Huttner, *J. Organomet. Chem.*, 1984, **260**, 281.
118. S. J. Davies, N. A. Compton, G. Huttner, L. Zsolnai and S. E. Garner, *Chem. Ber.*, 1991, **124**, 2731.
119. G. Huttner, B. Sigwarth, O. Scheidsteger, H. Lang and L. Zsolnai, *Organometallics*, 1985, **4**, 326.
120. A. Strube, G. Huttner and L. Zsolnai, *J. Organomet. Chem.*, 1990, **399**, 267.
121. A. Strube, G. Huttner and L. Zsolnai, *J. Organomet. Chem.*, 1990, **399**, 255.
122. A. Lombard, G. Huttner and K. Evertz, *J. Organomet. Chem.*, 1988, **350**, 243.
123. W. A. Herrmann, C. Hecht, M. L. Ziegler and B. Balbach, *J. Chem. Soc., Chem. Commun.*, 1984, 686.
124. M. Herberhold, D. Reiner and D. Neugebauer, *Angew. Chem., Int. Ed. Engl.*, 1983, **22**, 59.
125. W. A. Herrmann, J. Rohrmann and C. Hecht, *J. Organomet. Chem.*, 1985, **290**, 53.
126. H. Braunwarth, G. Huttner and L. Zsolnai, *Angew. Chem., Int. Ed. Engl.*, 1988, **27**, 698.
127. H. Braunwarth, F. Ettel and G. Huttner, *J. Organomet. Chem.*, 1988, **355**, 281.
128. P. Lau, G. Huttner and L. Zsolnai, *J. Organomet. Chem.*, 1992, **440**, 41.
129. G. Beuter, S. Drobnik, I.-P. Lorenz and A. Lubik, *Chem. Ber.*, 1992, **125**, 2363.
130. A. G. Ginsburg, D. V. Petrovskii, V. N. Setkina and D. N. Kursanov, *Bull. Acad. Sci. USSR (Engl. Transl.)*, 1985, **34**, 176.
131. A. G. Ginsburg, V. N. Setkina and D. N. Kursanov, *Bull. Acad. Sci. USSR (Engl. Transl.)*, 1985, **34**, 408.
132. L. G. Kuzmina, A. G. Ginsburg, Yu. T. Struchkov and D. N. Kursanov, *J. Organomet. Chem.*, 1983, **253**, 329.
133. B. V. Lokshin and M. G. Yezernitskaya, *J. Organomet. Chem.*, 1983, **256**, 89.
134. V. S. Leong and N. J. Cooper, *Organometallics*, 1988, **7**, 2080.
135. K. Plössl, G. Huttner and L. Zsolnai, *Angew. Chem., Int. Ed. Engl.*, 1989, **28**, 446.
136. S. M. Howdle and M. Poliakoff, *J. Chem. Soc., Chem. Commun.*, 1989, 1099.
137. G. Kraft, C. Kalbas and U. Schubert, *J. Organomet. Chem.*, 1985, **289**, 247.
138. U. Schubert, G. Scholz, J. Muller, K. Ackermann, B. Wörle and R. F. D. Stansfield, *J. Organomet. Chem.*, 1986, **306**, 303.
139. U. Schubert, K. Ackermann and B. Wörle, *J. Am. Chem. Soc.*, 1982, **104**, 7378.
140. F. Carré, E. Colomer, R. J. P. Corriu and A. Vioux, *Organometallics*, 1984, **3**, 1272.
141. K. M. Young and M. S. Wrighton, *Organometallics*, 1989, **8**, 1063.
142. W. A. Herrmann, R. Serrano and J. Weichmann, *J. Organomet. Chem.*, 1983, **246**, C57.
143. I. Bernal, J. D. Korp, W. A. Herrmann and R. Serrano, *Chem. Ber.*, 1984, **117**, 434.
144. W. A. Herrmann, C. E. Barnes, R. Serrano and B. Koumbouris, *J. Organomet. Chem.*, 1983, **256**, C30.
145. B. S. Creaven, A. J. Dixon, J. M. Kelly, C. Long and M. Poliakoff, *Organometallics*, 1987, **6**, 2600.
146. G. Huttner, S. Schuler, L. Zsolnai, M. Gottlieb, H. Braunwarth and M. Minelli, *J. Organomet. Chem.*, 1986, **299**, C4.
147. H. Werner, O. Kolb and T. Thometzek, *J. Organomet. Chem.*, 1988, **347**, 137.

# 6

# Manganese Nitrosyl and Isonitrile Complexes

## THOMAS C. FLOOD
*University of Southern California, Los Angeles, CA, USA*

## 6.1 INTRODUCTION

Since the publication of *COMC-I*,[1] many publications covering a wide range of structural types have appeared on manganese nitrosyl and isonitrile complexes. Both nitrosyls and isonitriles have often been employed as ancillary ligands in organometallic chemistry, including that of manganese. Isonitriles are isoelectronic with carbon monoxide and therefore are often used to investigate the scope of ligand substitution reactions of metal carbonyls. They are different from CO in that they are much less $\pi$-acidic and are capable of providing considerable steric size in the attached R group. When coordinated to a metal, the nitrosyl group exhibits electronic properties similar enough to those of coordinated CO that incorporation of NO can be used as a standard means of mimicking the coordination environment of an adjacent metal carbonyl in the periodic table; for example, $FeCp(CO)_2R$ and $MnCp(CO)(NO)R$ are isostructural and their reactivities are similar although by no means identical.

One usually incorporates CNR and NO into complexes intending to take advantage of their properties as ancillary ligands, often hoping not to involve them directly in the organometallic chemistry. However, in this chapter, emphasis is placed on those instances where chemistry of the ligand itself is central to the report, or where a new type of synthesis for incorporation of the ligand has been reported. For reference, lists of new complexes containing CNR and NO ligands are provided in Tables 1 and 2, respectively.

## 6.2 MANGANESE ISONITRILE COMPLEXES

The most common method for the preparation of isonitrile complexes remains that of simple ligand substitution. Recent developments in this area include the use of catalysts to accelerate such substitutions. For example, reaction of $Mn(Cp)(CO)_3$ with CNR (R = $Bu^t$, Bz, Cy, Me, or 2,6-$Me_2C_6H_3$) in toluene at reflux to form $Mn(Cp)(CO)_2(CNR)$ is greatly accelerated in the presence of PdO,[2] as is substitution at $MnBr(CO)_5$ in THF at room temperature by a variety of CNR.[3]

It has been known for some time that substitution of $Mn(Cp)(CO)_3$ by CNR can be accomplished via the photochemically generated intermediate $Mn(Cp)(CO)_2(sol)$ (sol = THF or cyclooctene). Recent examples of this include treatment of $Mn(Cp^*)(CO)_2(sol)$ with CNR to yield $Mn(Cp^*)(CO)_2(L)$ (L = CNMe or $CNCF_3$).[4] Preparations of $Mn(Cp^*)(CO)(CNCF_3)(L)$ (L = $PPh_3$, $PEt_3$, $PF_3$, CNMe, CNPh, $CNCF_3$, and CS (from $CS_2$ and $PPh_3$)) used similar methodology.[5] Photolysis of $Mn(Cp^*)(CO)_3$ directly in the presence of excess $CNCF_3$ afforded $Mn(Cp^*)(CNCF_3)_3$.[4] From IR evidence for the series $Mn(Cp^*)(CO)_2(L)$ (L = CO, CNMe, CS, and $CNCF_3$), it was concluded that $CNCF_3$ is the strongest $\pi$-acid among these ligands.[4] Synthesis of the unstable isonitrile $CNC_6F_5$ (m.p. 13 °C, decomposes) has been reported, and its reaction with $Mn(Cp^*)(CO)_2(THF)$ at room temperature in THF affords the anticipated $Mn(Cp^*)(CO)_2(CNC_6F_5)$.[6]

Electrophilic attack on coordinated cyanide, most commonly by alkylating agents, is another well-known method for the preparation of isonitrile complexes. An unusual organometallic alkylating agent has been reported recently: treatment of *trans*-$Mn(CN)(CO)(dppm)_2$ with $[Fe(\pi\text{-allyl})[P(OMe)_3](NO)_2][PF_6]$ in THF results in allyl transfer to form the corresponding allylisocyanide complex of manganese $[Mn(CN\text{-allyl})(CO)(dppm)_2][PF_6]$.[7] The 2-methallyl and butenyl iron species also exhibited allyl transfer.

Very few reports have involved reactivity of the isonitrile ligand itself. One extensive study involved reactions of *para*-substituted benzylmanganese carbonyls ((1) Equation (1)) with *p*-tolylisonitrile.[8,9] The initial product was the usual acylmanganese species (2). However, on heating (1) in THF at reflux, two CO ligands per metal were lost and the isonitrile ligands coupled to form dimer (3). With alkyl isonitriles, no coupling was observed; rather, decarbonylation occurred to yield *cis*-$Mn(CH_2PhX)(CNR)(CO)_4$.[8] When two equivalents of CNTolyl were employed in conjunction with PdO catalysis, chemistry arising from double insertion was observed (Equation (2)) with the formation of (4) and (5).[9] With yet another equivalent of CNTolyl or CNCy added to (4), a third insertion (shown for CNCy in Scheme 1) took place giving isomeric products (6) and (7). Highly hindered $CNBu^t$ did displace nitrogen from (4) (product (8), Scheme 1) or CO from (4) (product (9)); however, no subsequent insertion was evident in this case.

(1)
X = H, Cl, OMe
(2)
(3)
(1)

(1)
X = Cl, OMe
(4) major
(5)
(2)

Hexakis(isonitrile)manganese(I) and $Mn^{II}$ cations, $[Mn(CNR)_6]^{+/2+}$ have been the focus of recent physical studies. The reagent $[NO][PF_6]$ causes one-electron oxidation of $[Mn(CNtBu)_6]^+$ with no change in ligation at manganese.[10] The NMR $^{13}C-^{55}Mn$ coupling constant of $[Mn(CNR)_6]^+$ (several R groups) has been found to be ~120 Hz.[11] A study of the NMR spectra of $[Mn(CNR)_6]^{+/2+}$ (R = 9 alkyls or aryls) using $^1H$, $^{13}C$, $^{14}N$, and $^{55}Mn$ revealed, among other things, that in $Mn^{II}$ complexes the ligand atom resonances are sharp and the complex undergoes slow substitution of CNR by MeCN solvent.[12] Line broadening in the $^{55}Mn$ NMR spectrum of $[Mn(CNR)_6]^+$ has been shown to provide a convenient means of measuring rates of self-exchange electron transfer in $Mn^{(+/2+)}$ redox reactions. With this technique, solvent effects on rate constants and $\Delta V^{\ddagger}$ for electron transfer in $[Mn(CNR)_6]^{+/2+}$ (R = Me, Et, $Bu^t$, Cy) were measured,[13,14] as well as solvent, temperature, and electrolyte effects on rate constants, $\Delta S^{\ddagger}$, and $\Delta H^{\ddagger}$ (R = Cy).[15] Rate constants for electron self-exchange spanned a range of more than a

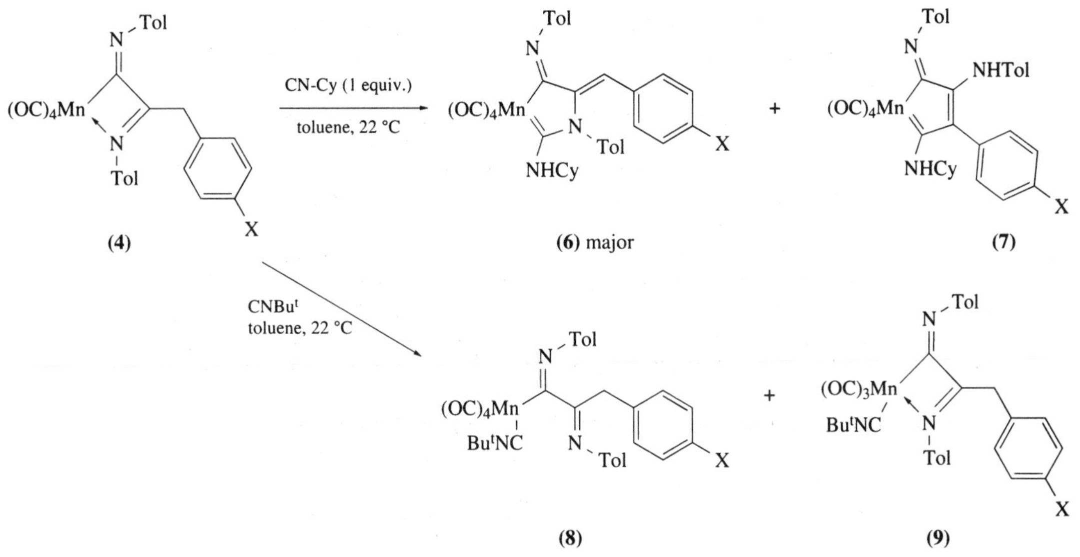

**Scheme 1**

factor 300 as a function of the ligand CNR (R = Me, Et, $Pr^i$, $Bu^t$, Cy, Bz). These rate differences were attributed to manganese–manganese distance effects.[16] The fact that electron self-exchange is faster for aromatic CNAr than for aliphatic CNR was believed to have its origins in arene–arene intermolecular overlap.[17] The electrochemistry of the $Mn^{(+/2+)}$ couple of the complexes $[Mn(CNR)(CO)(dppm)_2][PF_6]$ (R = Me, $Bu^t$) has been investigated.[18]

Table 1 lists CNR-containing manganese complexes reported since the publication of *COMC-I*.

**Table 1** New manganese isonitrile complexes.

| Produced by electrophilic attack on Mn–CN | | |
|---|---|---|
| $Mn(Cp)(CO)_2(C\equiv N-E)$ | E = TMS, $GeMe_3$, $SnMe_3$, $PPh_2$, $AsMe_2$ | 19 |
| $(Cp)(CO)_2Mn-C\equiv N-C(O)-N\equiv C-Mn(Cp)(CO)_2$ | | 19 |
| $Mn(\eta^3-CH_2CRCH_2)(CO)_3(C\equiv N-E)$ | R = H, Me; E = Me, $SnMe_3$ | 20 |
| $[Mn(CNMe)(CO)(dppm)_2][PF_6]$ | | 21 |
| 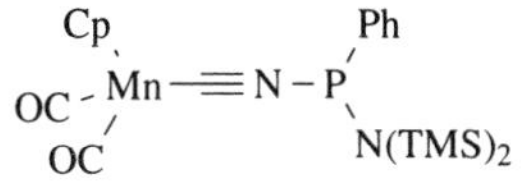 | | 22 |
| 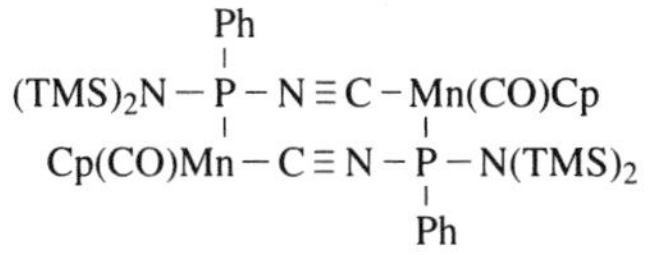 | | 22 |
| 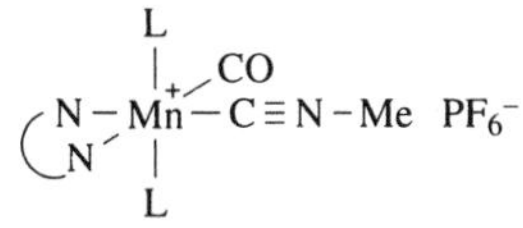 | L = $P(OMe)_3$, $P(OEt)_3$, $P(OPh)_3$; N–N = bipy, phen | 23 |
| Produced by substitution at manganese | | |
| $[Mn(CNBu^t)(CO)(dppm)_2][PF_6]$ | | 24 |
| $[Mn(L-L)_3][PF_6]_2$ | L–L = CN–$CMe_2CH_2CH_2CMe_2$–NC | 25 |
| $[Mn(L-L)(CF_3SO_3)_2(H_2O)_2]$ | L–L = CN–$CMe_2CH_2CH_2CMe_2$–NC | 25 |
| $[Mn(CNBu^t)_4(PPhMe_2)_2][MnBr_4]$ | | 26 |
| $[Mn(CNBu^t)_4(PPhMe_2)_2][PF_6]_2$ | | 26 |
| $[Mn(N-N)(CNPh)_4][ClO_4]$ | N–N = bipy, phen | 27 |

**Table 1** (continued)

[Mn(DiNC)$_3$][PF$_6$]

[Mn(DiNC)$_3$][PF$_6$]$_2$

DiNC =

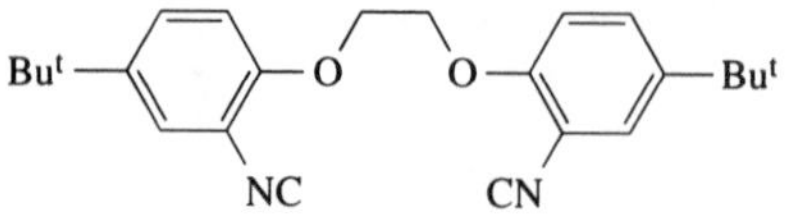

28

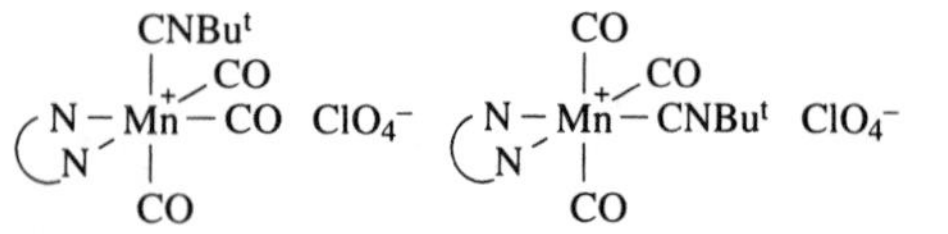

N–N = bipy, phen     27, 29

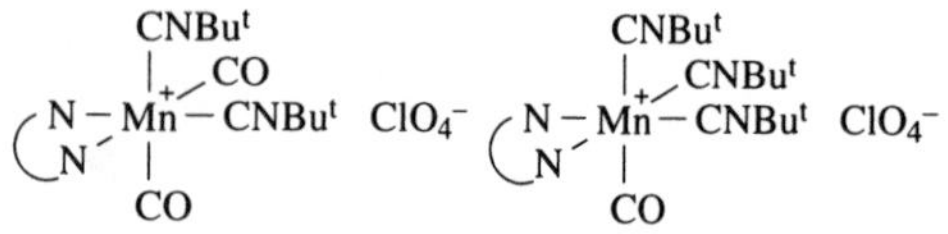

N–N = bipy, phen     27, 29

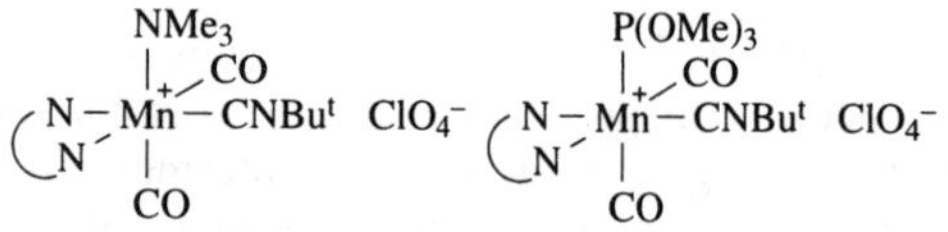

N–N = bipy, TMEDA or Bu$^t$N=CHCH=N–Bu$^t$     30

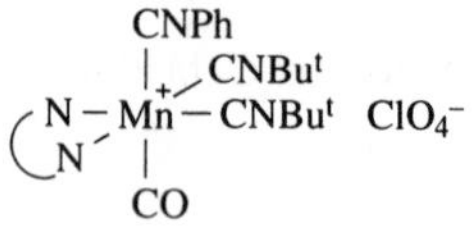

N–N = bipy, TMEDA or Bu$^t$N=CHCH=N–Bu$^t$     30

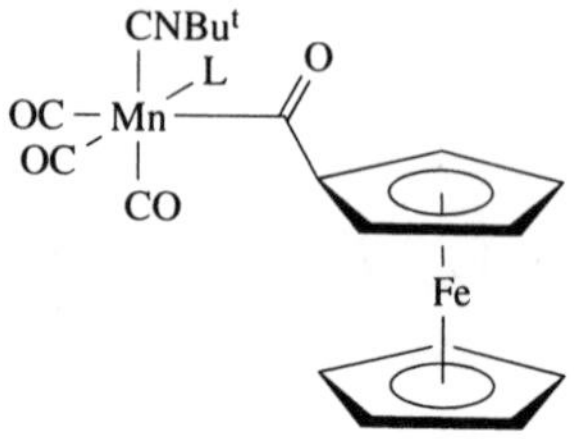

L = CO, CN–Bu$^t$     31

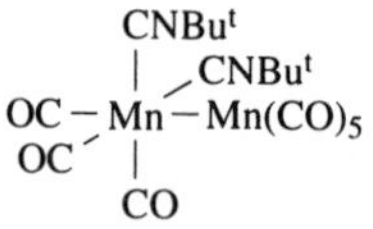

32

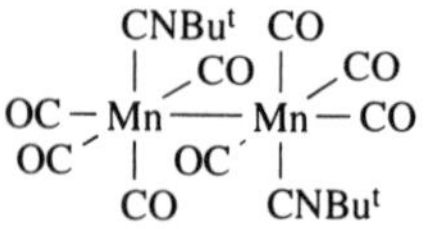

32

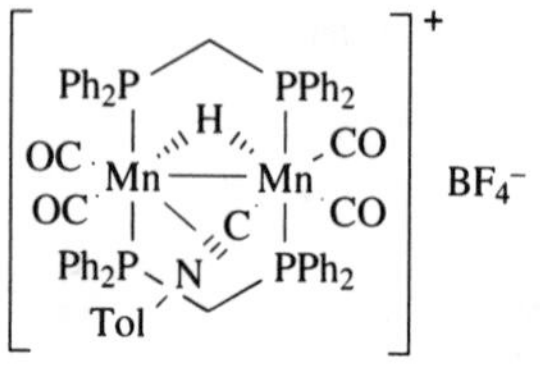

33

## 6.3 MANGANESE NITROSYL COMPLEXES

Treatment of certain manganese carbonyl complexes with $[N(PPh_3)_2][NO_2]$ in polar aprotic solvents generates nitrosyl complexes in high yield. For example, either $[Mn(CO)_6]^+$ or $[Mn(CO)_5(NCMe)]^+$ is converted to $Mn(CO)_4(NO)$. A second equivalent of $[N(PPh_3)_2][NO_2]$ transforms $Mn(CO)_4(NO)$ into $[N(PPh_3)_2][Mn(CO)_2(NO)_2]$.[34] Oxidation of carbyne complex (10) by $[N(PPh_3)_2][NO_2]$ forms the double-CO-insertion product α-ketoacyl (11) (Equation (3)). Complex (11) is also generated by one-electron oxidation of the known anion $[Mn(Cp')(CO)_2(C(O)Tol)]^-$ (where Cp' = $MeC_5H_4$) to form the 17-electron neutral species, followed by its treatment with NO gas.[35] Although a few α-ketoacyl ligands had been known, the NO-induced formation of (11) was the first demonstration of CO insertion into a metal acyl bond.

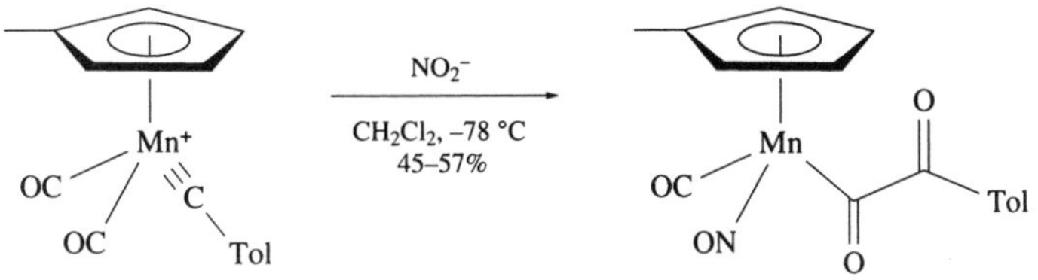

(3)

(10)      (11)

Addition of Cl–NO to $Mn(Cp^*)(CO)_2(HC\equiv CH)$ leads to vinyl species $Mn(Cp^*)(CH=CHCl)(CO)(NO)$ as a mixture of Z and E isomers in 80% yield.[36] NO has been shown to coordinate reversibly to $MnX_2(PR_3)$ either in the solid state or in THF solution when X is Cl or Br and $PR_3$ is $PPr_3$, $PBu_3$, $PPhMe_2$, or $PPhEt_2$. Spectral data indicate that $Mn(NO)X_2(PR_3)$ contains a bent Mn–N–O group and four unpaired electrons.[37] The tricoordinate NO in (12) is protonated to form (13) (Equation (4)). Treatment of (13) with excess acid results in reduction, presumably with oxidation of some of the cluster, to afford amide complex (14).[38] A structure determination of $Mn(CO)(NO)_3$ has been carried out by electron diffraction.[39]

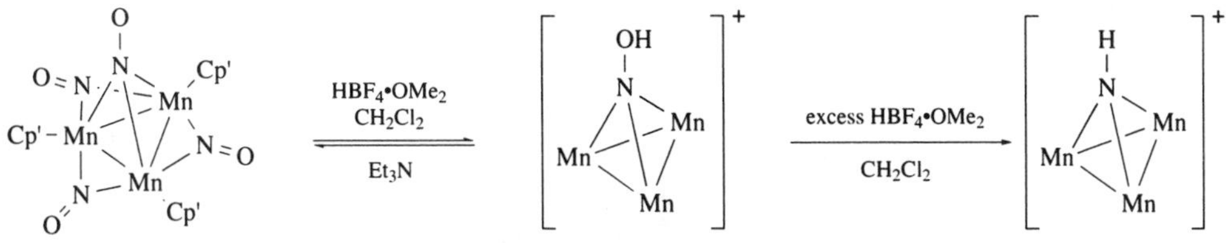

(4)

(12)      (13) 66%      (14) 14%

Table 2 lists NO-containing manganese complexes reported since the publication of *COMC-I*.

**Table 2** New manganese nitrosyl complexes.

| | | |
|---|---|---|
| [Na(crypt(2.2.2))]$_2$[Mn(CO)$_3$(NO)] | | 40 |
| [Mn(CF$_3$)(CO)$_3$(NO)]$^-$ | gas phase | 41 |
| [Mn(Cp)(CO)(NO)(PMe$_3$)][BF$_4$] | | 42 |
| Mn(Cp')(NO)R$_2$ | 17-electron; R = Me, Et, Pr$^n$ | 43 |
| Mn(Cp*)(CO)(NO)[P(TMS)$_2$] | | 44 |
| Mn(Cp')(L)(NO)(COCOTol) | L = PPh$_3$, Bu$^t$NC | 35 |
| Mn(Cp')(CO)(NO)[C(O)CH=PMe$_3$] | | 45 |
| [{Mn(Cp')(NO)$_3$NH][BF$_4$] | Cp analogue Ref. 38 | 46 |
| [(Mn{RCp)(NO)}$_2$(μ-NH$_2$)(μ-CO)][BPh$_4$] | R = Me, H | 46 |

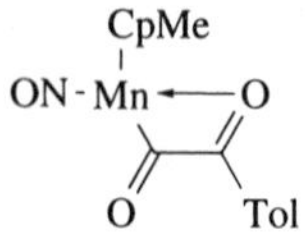

47

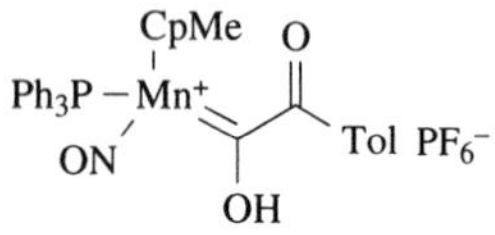

48

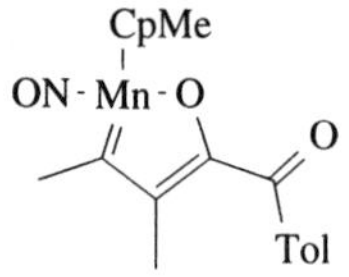

48

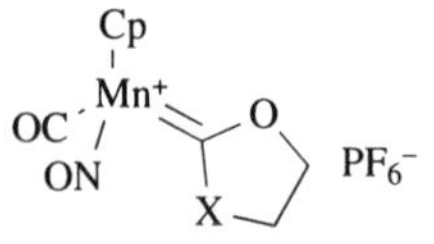

X = O, NH     49

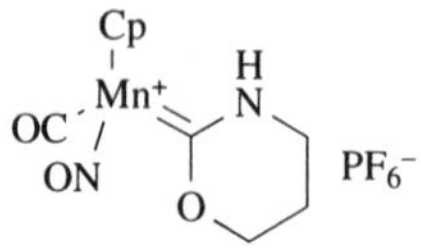

49

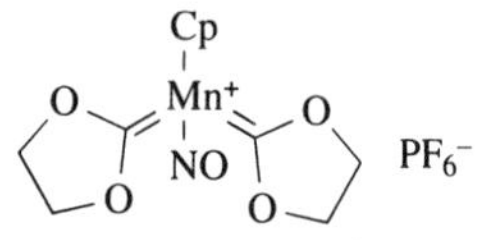

49

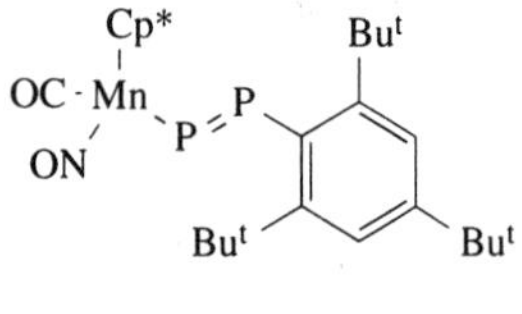

44

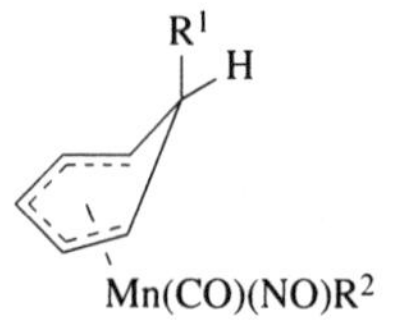

R$^1$ = Ph, R$^2$ = CO$_2$Me    50
R$^1$ = Me, R$^2$ = COPh    51

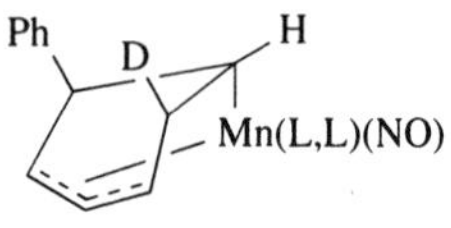

L,L = dppe, dppen     52

**Table 2** (continued)

| | | |
|---|---|---|
| | L = CO, PBu$_3$, P(OMe)$_3$; R = H, Ph, CH(CO$_2$Me)$_2$ | 53 |
| | L = CO, PBu$_3$, P(OMe)$_3$; R = H, Ph, CH(CO$_2$Me)$_2$ | 53 |
| | X = Ph, Me, CH(CO$_2$Me)$_2$ | |

L$^1$ = L$^2$ = CO; R = Ph, Me, H, CN, CH$_2$COBu$^t$; X = H, 2-OMe, 1-Me-4-OMe, 1,2,3,4,5-Me,
1,2,3,4,5,6-Me, 1,2,4,5-Me, 1,2,3,4-Me           50, 52, 54
L$^1$ = L$^2$ = dppe, dppen; R = Ph; X = H           52
L$^1$ = CO, L$^2$ = PMe$_3$, P(OMe)$_3$, PBu$_3$, PPh$_3$, P(C$_2$H$_4$CN)$_3$; R = Me, Ph, H; X = H      50, 52

R$^1$ = Me, Ph; R$^3$ = H, D; L = CO; R$^2$ = H, CH(CO$_2$Me)$_2$, P(O)(OMe)$_2$, PMe$_3$[a], PPh$_3$[a], PBu$_3$[a]      50, 52, 54–57

[a] With PF$_6^-$ counterion.

# 6.4 REFERENCES

1. P. M. Treichel, in 'COMC-I', vol. 4, p. 1.
2. G. W. Harris, J. C. A. Boeyens and N. J. Coville, *J. Organomet. Chem.*, 1983, **255**, 87.
3. N. J. Coville, P. Johnston, A. E. Leins and A. J. Markwell, *J. Organomet. Chem.*, 1989, **378**, 401.
4. D. Lentz, J. Kroll and C. Langner, *Chem. Ber.*, 1987, **120**, 303.
5. D. Lentz and R. Marschall, *Chem. Ber.*, 1989, **122**, 1223.
6. D. Lentz, K. Graske and D. Preugschat, *Chem. Ber.*, 1988, **121**, 1445.
7. N. G. Connelly, A. G. Orpen, G. M. Rosair and G. H. Worth, *J. Chem. Soc., Dalton Trans.*, 1991, 1851.
8. P. L. Motz, J. P. Williams, J. J. Alexander, D. M. Ho, J. S. Ricci and W. T. Miller, Jr., *Organometallics*, 1989, **8**, 1523.
9. P. L. Motz, J. J. Alexander and D. M. Ho, *Organometallics*, 1989, **8**, 2589.
10. K. E. Linder, A. Davison, J. C. Dewan, C. E. Costello and S. Maleknia, *Inorg. Chem.*, 1986, **25**, 2085.
11. R. M. Nielson and S. Wherland, *Inorg. Chem.*, 1984, **23**, 3265.
12. R. M. Nielson and S. Wherland, *Inorg. Chem.*, 1985, **24**, 3458.
13. M. Stebler, R. M. Nielson, W. F. Siems, J. P. Hunt, H. W. Dodgen and S. Wherland, *Inorg. Chem.*, 1988, **27**, 2893.
14. R. M. Nielson, J. P. Hunt, H. W. Dodgen and S. Wherland, *Inorg. Chem.*, 1986, **25**, 1964.
15. R. M. Nielson and S. Wherland, *Inorg. Chem.*, 1984, **23**, 1338.
16. R. M. Nielson and S. Wherland, *J. Am. Chem. Soc.*, 1985, **107**, 1505.
17. R. M. Nielson and S. Wherland, *Inorg. Chem.*, 1986, **25**, 2437.
18. G. A. Carriedo, V. Riera, N. G. Connelly and S. J. Raven, *J. Chem. Soc., Dalton Trans.*, 1987, 1769.
19. H. Behrens, G. Landgraf, P. Merbach, M. Moll and K.-H. Trummer, *J. Organomet. Chem.*, 1983, **253**, 217.
20. M. Moll, H. Behrens, H.-J. Seibold and P. Merbach, *J. Organomet. Chem.*, 1983, **248**, 329.
21. G. A. Carriedo *et al.*, *J. Organomet. Chem.*, 1986, **302**, 47.
22. W. F. McNamara, E. N. Duesler and R. T. Paine, *Organometallics*, 1988, **7**, 384.
23. G. A. Carriedo, C. Carriedo, C. Crespo and P. Gómez, *J. Organomet. Chem.*, 1993, **452**, 91.
24. F. J. Garcia-Alonso, V. Riera and M. J. Misas, *Transition Met. Chem.*, 1985, **10**, 19.
25. M. Dartiguenave, Y. Dartiguenave, A. Guitard, A. Mari and A. L. Beauchamp, *Polyhedron*, 1989, **8**, 317.
26. B. Beagley, C. G. Benson, G. A. Gott, C. A. McAuliffe, R. G. Pritchard and S. P. Tanner, *J. Chem. Soc., Dalton Trans.*, 1988, 2261.
27. F. J. Garcia Alonso, V. Riera, F. Villafañe and M. Vivanco, *J. Organomet. Chem.*, 1984, **276**, 39.
28. D. T. Plummer and R. J. Angelici, *Inorg. Chem.*, 1983, **22**, 4063.

29. F. J. Garcia Alonso, V. Riera, M. L. Valin, D. Moreiras, M. Vivanco and X. Solans, *J. Organomet. Chem.*, 1987, **326**, C71.
30. F. J. Garcia Alonso, V. Riera and M. Vivanco, *J. Organomet. Chem.*, 1990, **398**, 275.
31. M. Herberhold and H. Kniesel, *J. Organomet. Chem.*, 1987, **334**, 347.
32. D. J. Robinson, G. W. Harris, J. C. A. Boeyens and N. J. Coville, *J. Chem. Soc., Chem. Commun.*, 1984, 1307.
33. A. J. Deeming and S. Donovan-Mtunzi, *Organometallics*, 1985, **4**, 693.
34. R. E. Stevens and W. L. Gladfelter, *Inorg. Chem.*, 1983, **22**, 2034.
35. J. B. Sheridan, J. R. Johnson, B. M. Handwerker, G. L. Geoffroy and A. L. Rheingold, *Organometallics*, 1988, **7**, 2404.
36. H. G. Alt and H. E. Engelhardt, *J. Organomet. Chem.*, 1988, **346**, 211.
37. D. S. Barratt and C. A. McAuliffe, *J. Chem. Soc., Dalton Trans.*, 1987, 2497.
38. P. Legzdins, C. R. Nurse and S. J. Rettig, *J. Am. Chem. Soc.*, 1983, **105**, 3727.
39. L. Hedberg, K. Hedberg, S. K. Satija and B. I. Swanson, *Inorg. Chem.*, 1985, **24**, 2766.
40. Y. Chen and J. E. Ellis, *J. Am. Chem. Soc.*, 1983, **105**, 1689.
41. S. K. Shin and J. L. Beauchamp, *J. Am. Chem. Soc.*, 1990, **112**, 2066.
42. W. P. Weiner, F. J. Hollander and R. G. Bergman, *J. Am. Chem. Soc.*, 1984, **106**, 7462.
43. A. Becalska and R. H. Hill, *J. Chem. Soc., Chem. Commun.*, 1989, 1626.
44. L. Weber and G. Meine, *Chem. Ber.*, 1987, **120**, 457.
45. H. Blau, W. Malisch and P. Weickert, *Chem. Ber.*, 1982, **115**, 1488.
46. P. Legzdins, D. T. Martin, C. R. Nurse and B. Wassink, *Organometallics*, 1983, **2**, 1238.
47. D. J. Stufkens, J. B. Sheridan and G. L. Geoffroy, *Inorg. Chem.*, 1990, **29**, 4347.
48. S. L. Bassner, J. B. Sheridan, C. Kelley and G. L. Geoffroy, *Organometallics*, 1989, **8**, 2121.
49. M. M. Singh and R. J. Angelici, *Inorg. Chem.*, 1984, **23**, 2691; 2699.
50. Y. K. Chung, D. A. Sweigart, N. G. Connelly and J. B. Sheridan, *J. Am. Chem. Soc.*, 1985, **107**, 2388.
51. J. B. Sheridan, R. S. Padda, K. Chaffee, C. Wang, Y. Huang and R. Lalancette, *J. Chem. Soc., Dalton Trans.*, 1992, 1539.
52. R. D. Pike, W. J. Ryan, N. S. Lennhoff, J. Van Epp and D. A. Sweigart, *J. Am. Chem. Soc.*, 1990, **112**, 4798.
53. E. D. Honig and D. A. Sweigart, *J. Chem. Soc., Chem. Commun.*, 1986, 691.
54. Y. K. Chung, H. S. Choi, D. A. Sweigart and N. G. Connelly, *J. Am. Chem. Soc.*, 1982, **104**, 4245.
55. R. D. Pike, W. J. Ryan, G. B. Carpenter and D. A. Sweigart, *J. Am. Chem. Soc.*, 1989, **111**, 8535.
56. S. D. Ittel, J. F. Whitney, Y. K. Chung, P. G. Williard and D. A. Sweigart, *Organometallics*, 1988, **7**, 1323.
57. T. H. Hyeon and Y. K. Chung, *J. Organomet. Chem.*, 1989, **372**, C12.

# 7

# High-valent Organomanganese Compounds

GREGORY S. GIROLAMI
*University of Illinois, Urbana, IL, USA*

and

ROBERT J. MORRIS
*Ball State University, Muncie, IN, USA*

## 7.1 INTRODUCTION

Manganese is unusual among the transition elements in that organomanganese(II) species are appropriate to include in a review of the "higher" oxidation states. Although high-valent inorganic complexes of manganese have long been known,[1] until recently the organometallic chemistry of manganese was limited almost entirely to the 0 and +1 oxidation states. At the time of the last review in 1982, a small but growing number of organomanganese(II) compounds were known and organomanganese complexes in oxidation states above +2 were extremely rare. Since 1982, this area has expanded considerably and organomanganese compounds even in the +7 oxidation state now exist.

The slow development of higher-valent organomanganese chemistry, especially in oxidation states above +2, is due in part to the lack of suitable inorganic starting materials; for example, the higher binary halides are either unstable or nonexistent,[2] other compounds such as $K_2MnCl_5$, "$Mn(O_2CMe)_3$," $K_2MnCl_6$, and $MnO_2$ are rather insoluble and unreactive in organic solvents, and permanganate salts and related species are readily reduced by potential organic ligands.[1] The only higher-valent inorganic starting material that has proven generally suitable for the preparation of high-valent organomanganese species is the acetylacetonate complex $Mn(acac)_3$. Higher-valent organomanganese complexes have more commonly been prepared by oxidation of lower-valent species.

In this chapter we have attempted to be both comprehensive and critical; as in *COMC-I*, species where cyanide ligands form the only metal–carbon bonds have been excluded. We refer the reader to a review of the inorganic chemistry of manganese for an excellent summary of the sometimes unbelievable results that have appeared in the manganese literature, and of the pitfalls awaiting those who attempt to draw detailed conclusions about the structures of manganese compounds solely from "sporting methods" such as electronic spectroscopy and magnetic moments.[1]

## 7.2 ORGANOMANGANESE(II) COMPOUNDS

Organomanganese(II) reagents are finding increasing uses in organic synthesis, where they are effective in promoting 1,4 conjugate additions to α,β-unsaturated ketones and esters, the conversion of acid chlorides to ketones, and the highly selective alkylation of aldehydes in the presence of ketones.[3] In these respects, the chemistry resembles that of organocuprates, except that organomanganese(II) compounds are often more chemoselective and more convenient to handle. One drawback, however, is that the yields of the desired product are sometimes lower than those obtainable by using the analogous organocuprate. Cahiez has explored in depth the utility of organomanganese reagents in organic synthesis.[3,4] Intermediates generated from the interaction of organomanganese(II) reagents and organic substrates, were eventually isolated and structurally characterized in 1992.[5]

Generally, organomanganese(II) reagents for use in organic synthesis are prepared by the addition of an alkyl lithium or Grignard reagent to a manganese(II) halide; invariably, the organomanganese products are not isolated but used *in situ*. A variety of stoichiometries have been suggested for the organomanganese species formed in solution (among these are $[MnR_3]^-$, $[MnR_4]^{2-}$, and MnRX), but most of these should be viewed with skepticism, not because they are unreasonable but because no supporting evidence is available. A summary of the well-characterized examples of organomanganese(II) species follows.

### 7.2.1 Manganese(II) Alkyls and Aryls

#### 7.2.1.1 Anionic manganese(II) alkyls and aryls

Early workers suggested that the addition of alkyllithium reagents to manganese halides generated lithium salts of stoichiometry $Li[MnR_3]$ (R = Me, Et, Bu$^n$, etc.).[6,7] These formulations, however, were not established by any experimental means, and it is likely that a mixture of organomanganese species was actually present in the solutions. The decomposition rates of alkylmanganates prepared *in situ* by the addition of alkyllithium reagents to $MnCl_2$ in tetrahydrofuran have also been studied.[8]

In 1976, Wilkinson reported the isolation of the first crystalline samples of alkylmanganate complexes: alkylation of $MnCl_2$ with methyllithium followed by the addition of $N,N,N',N'$-tetramethylethylenediamine (TMEDA) gave the tetramethylmanganate salt $[Li(TMEDA)]_2[MnMe_4]$.[9] A similar trimethylsilylmethyl complex $Li_2Mn(CH_2TMS)_4$ was reported and the EPR spectra of both complexes were described. At about the same time, the (piperidyl)methyl complex $[Li(diox)_2]_2[Mn(CH_2NC_5H_{10})_4]$ was briefly described as a pale pink compound with a magnetic moment of 5.95 $\mu_B$.[10] More recently, this synthetic method has been modified (Equation (1)) to give an entire series of pale-colored 13-electron tetraalkyl manganate(II) complexes, (**1a–f**).[11] A spirocyclic manganacyclopentane complex $Li_2Mn(C_4H_8)_2$ has also been briefly described.[12]

These 13-electron peralkyl species are stable, and even the alkylmanganates that contain β-hydrogen atoms are remarkably resistant to β-elimination. Room temperature magnetic susceptibility measurements in benzene showed that the complexes adopt high-spin $d^5$ configurations with $\mu_{eff} = 5.6$–5.9 $\mu_B$. The high-spin natures of these compounds suggest that they adopt tetrahedral geometries; this conclusion is supported by the near-cubic symmetry observed by EPR spectroscopy.

$$MnCl_2 + 4 LiR + 2 TMEDA \longrightarrow [Li(TMEDA)]_2[MnR_4] + 2 LiCl \qquad (1)$$

(1a) R = Me
(1b) R = Et
(1c) R = $CH_2CH_2Bu^t$
(1d) R = $Bu^n$
(1e) R = $CH_2TMS$
(1f) R = Ph

The EPR data also suggest that interactions of the lithium cations with the $[MnR_4]^{2-}$ anions weaken as the R group becomes larger.[11]

The single crystal x-ray structures of $[Li(TMEDA)]_2[MnMe_4]$ (1a), $[Li(TMEDA)]_2[MnEt_4]$ (1b) and $[Li(TMEDA)]_2[Mn(CH_2CH_2Bu^t)_4]$ (1c) show tetrahedral coordination geometries around the manganese centers, and there are no agostic $Mn \cdots H-C$ interactions. The lithium cations each interact strongly with two α-carbon atoms of the alkyl ligands.[11]

Three triaryl manganates, $[Li(Et_2O)_2]_2[Mn_2Ph_6]$, $[Li(THF)_4]_2[Mn_2Ph_6]$, and $[Li(THF)_4]$-$[Mn(C_6H_2Me_3)_3]$, have also been structurally characterized.[13,14] The hexaphenyldimanganate dianions (2) in the former two salts consist of centrosymmetric dimanganese units with two bridging and four terminal phenyl groups. The manganese centers are tetrahedral. In contrast, the monomeric trimesityl manganate anion (3) adopts a trigonal planar structure of idealized $D_3$ symmetry.[13] Treatment of manganese(II) halides with $o,o'$-dilithiodiphenylether ($Li_2dpa$) in tetrahydrofuran gives a yellow-orange salt $[Li(THF)_3]_2[Mn(dpa)_2]$, which is a high-spin manganese(II) complex as judged from its magnetic moment of 5.84 $\mu_B$.[15]

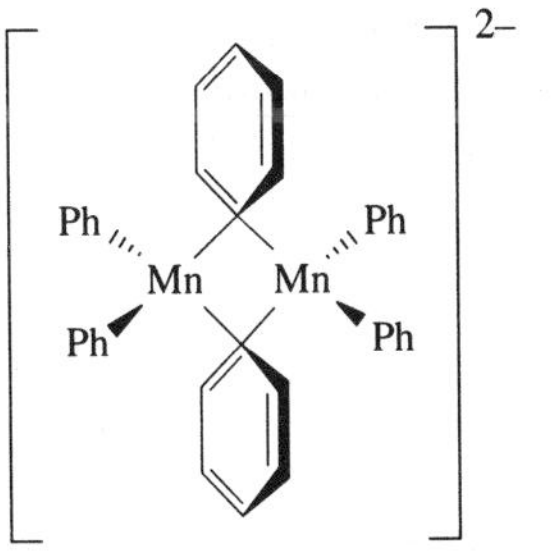

(2)

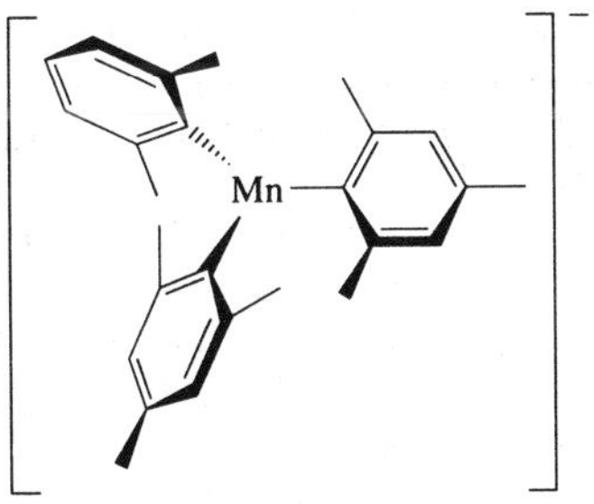

(3)

In addition to the homoleptic organomanganate species above, there are several anionic manganate complexes with mixed ligand sets. For example, treatment of $MnI_2$ with $LiC_6H_4$-2-$CH_2NMe_2$ affords the mixed halo–alkyl manganate salt $[Li(THF)]_2[Mn(C_6H_4$-2-$CH_2NMe_2)_2I_2]$.[16] This pale yellow complex is high-spin ($\mu_{eff} = 5.7$ $\mu_B$). Attempts to prepare the manganese(II) "Grignard reagent" $Mn[C(TMS)_3]Cl$ from $MnCl_2$ and one equivalent $LiC(TMS)_3$ led to the isolation of an unusual anionic cluster of stoichiometry $[Li(THF)_4][Mn_3\{C(TMS)_3\}_3Cl_4(THF)]$ (4), whose core consists of an isosceles triangle of manganese atoms with one terminal alkyl group on each manganese center.[17]

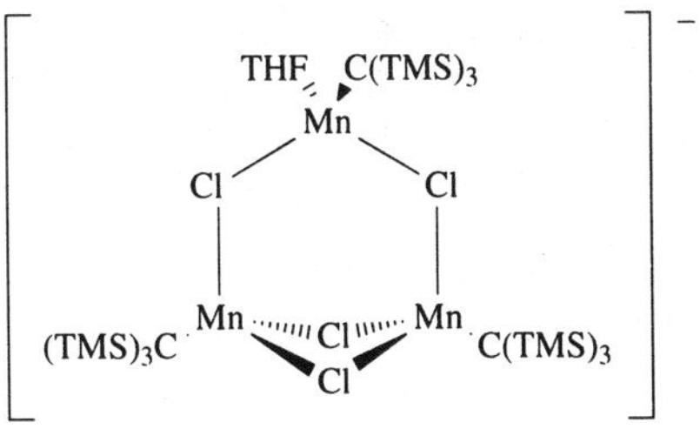

(4)

### 7.2.1.2 Neutral manganese(II) alkyls and aryls

The formation of neutral manganese(II) alkyls by the interaction of manganese(II) halides with alkylating reagents was first reported by Gilman, but the organometallic products were neither isolated nor characterized.[18,19] Later, Kochi compared the rates of decomposition of manganese(II) alkyls prepared *in situ* and obtained evidence that β-elimination followed by alkene dissociation was the principal decomposition route for alkyls that contain β-hydrogen atoms.[8]

Shortly thereafter, the synthesis and structural characterization of several β-stabilized organomanganese species of the type $[MnR_2]_n$ appeared in the literature (Equation (2)).[9] Currently known dialkylmanganese species include monomeric $Mn[C(TMS)_3]_2$[20] and $Mn[CH(TMS)_2]_2$,[21] dimeric $[Mn(CH_2CMe_2Ph)_2]_2$,[9] trimeric $[Mn(C_6H_2Me_3)_2]_3$,[22] tetrameric $[Mn(CH_2Bu^t)_2]_4$,[9] and polymeric $[Mn(CH_2TMS)_2]_n$ (5a–f).[9]

$$MnCl_2 + MgR_2 \longrightarrow 1/x\,[MnR_2]_x + MgCl_2 \qquad (2)$$

(5a) R = C(TMS)$_3$
(5b) R = CH(TMS)$_2$
(5c) R = CH$_2$CMe$_2$Ph
(5d) R = C$_6$H$_2$Me$_3$
(5e) R = CH$_2$Bu$^t$
(5f) R = CH$_2$TMS

The allyl compound $Mn(C_3H_5)_2$ has been claimed,[23] and a monomeric ylide complex $MnL_2$ (6), where L is the 2,2,4,4-tetraphenyl-1H-2$\lambda^5$,4$\lambda^5$-benzodiphosphepinyl group, has also been reported and structurally characterized.[24]

The monomeric complexes have magnetic moments of ca. 5.5 $\mu_B$ that are indicative of high-spin manganese(II) centers;[20,21] in contrast, the oligomeric dialkyls exhibit reduced magnetic moments near 2.4 $\mu_B$ owing to weak antiferromagnetic interactions between adjacent manganese centers.[9,21] Two manganese dialkyls have been characterized in the gas phase by electron diffraction: $Mn(CH_2Bu^t)_2$ and $Mn[CH(TMS)_2]_2$ are monomeric with linear C–Mn–C backbones; in neither complex is there evidence of agostic Mn$\cdots$H–C interactions.[24,25] The He$^I$ and He$^{II}$ photoelectron spectra and variable-temperature magnetic susceptibility have been reported for $Mn[CH(TMS)_2]_2$; a SCF MO calculation on the hypothetical monomer $MnMe_2$ has also been carried out.[21]

(6)

Organic isocyanates insert into the manganese–alkyl bonds of the neopentyl and neophyl complexes $[Mn(CH_2Bu^t)_2]_4$ and $[Mn(CH_2CMe_2Ph)_2]_2$; the products form colorless crystals and are the first organomanganese intermediates to be isolated from the reaction of an organomanganese(II) reagent with an organic substrate.[5] Hydrolysis of the products gives the corresponding organic amide $R^1C(O)NHR^2$. The crystal structures of $Mn_3(CH_2Bu^t)_4$-$[OC(CH_2Bu^t)N(TMS)]_2$ and $Mn_4(CH_2Bu^t)_2(NCO)_2$-$[OC(CH_2Bu^t)N(TMS)]_4$ show that the amide ligands are both chelating and bridging. *t*-Butylisonitrile also inserts into the manganese–alkyl bonds of $[Mn(CH_2Bu^t)_2]_4$ in the presence of $PMe_3$ to afford the iminoacyl $Mn[C(=NBu^t)CH_2Bu^t]_2(PMe_3)$.[5]

Manganese(II) dialkyls react with the weak acids 2,4,6-tri($Bu^t$)phenol and di($Bu^t$)phosphine to yield one equivalent of alkane per metal center and the organomanganese(II) species $Mn_2(CH_2CMe_2Ph)_2$-$(\mu\text{-}OC_6H_2Bu^t_3)_2$ (7) and $Mn_2(CH_2CMe_3)_2(\mu\text{-}PBu^t_2)_2$ (8), respectively (Scheme 1).[26] These complexes are paramagnetic, with magnetic moments of 5.2 and 3.1 $\mu_B$ per metal center, respectively, and have been structurally characterized. These complexes show no agostic interactions despite the formal electron count of 11.

The similar dinuclear monoalkyl manganese(II) compound $Mn_2(CH_2SPh)_2(\mu\text{-}Cl)_2(TMEDA)_2$ (9) has been prepared by treatment of $MnCl_2$ with $[LiCH_2SPh\cdot TMEDA]_2$; the structure of the α-thioalkyl complex consists of two square pyramids joined at a basal edge.[27] The reaction of this complex with benzaldehyde has been described.

Several classes of Lewis base adducts of manganese(II) dialkyls are known.[9,28–37] Addition of a tertiary phosphine to isolated samples of $MnR_2$ species affords dinuclear complexes of stoichiometry $Mn_2R_4(PR_3)_2$;[28–30] these species can also be prepared directly from $MnCl_2$ by addition of a dialkylmagnesium reagent in the presence of the phosphine. Among the species that have been isolated are $Mn_2(CH_2TMS)_4(PR_3)_2$ (where $PR_3 = PMe_3$, $PMe_2Ph$, $PMePh_2$, $PEt_3$, and $P(c\text{-Hx})_3$), $Mn_2(CH_2Bu^t)_4(PMe_3)_2$, $Mn_2(Bz)_4(PMe_3)_2$, and $Mn_2(c\text{-Hx})_4(\mu\text{-dmpe})$. The x-ray crystal structures show that there are two asymmetrically bridging alkyl groups, each of which is involved in an agostic interaction with the manganese centers.[28–30] The Mn–Mn distances in these $Mn_2(\mu\text{-R})_2R_2(PR_3)_2$ molecules range from 0.2667 nm to 0.2828 nm and increase as the alkyl and phosphine groups become larger. A complex with the formula $MnPh_2[P(c\text{-Hx})_3]$ has also been reported that presumably is also

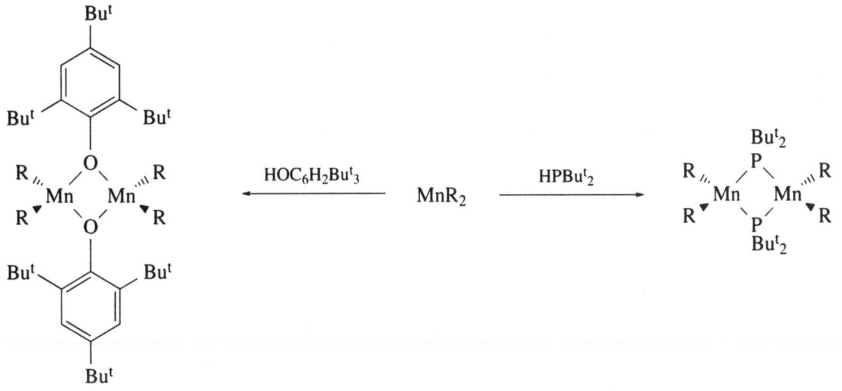

## Scheme 1

(7)　　　　　　　　**Scheme 1**　　　　　　　(8)

dimeric, but no crystallographic data are available.[32] The monoalkyls $MnMeBr(Pr_3)_2$ (where R = Et, Pr[i], and Bu[t]) have been briefly mentioned.[38]

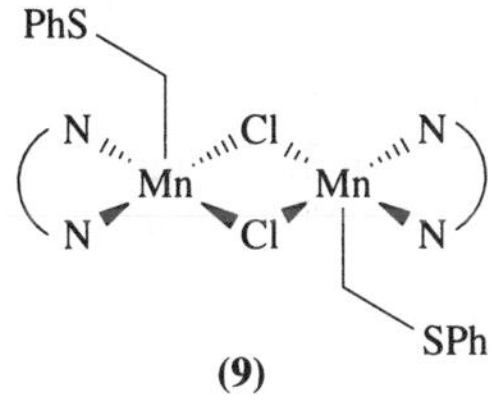

(9)

If excess phosphine is added to the $MnR_2$ dialkyls or to the dinuclear $Mn_2R_4(PR_3)_2$ species, four-coordinate mononuclear $MnR_2(PR_3)_2$ compounds can be prepared.[30,31] When the tertiary phosphine is not potentially chelating, these species are often unstable toward loss of phosphine and only the dinuclear $Mn_2R_4(PR_3)_2$ species can be isolated. In solution, the equilibrium shown in Equation (3) is established.[31] The x-ray crystal structure of the neophyl complex $Mn(CH_2CMe_2Ph)_2(PMe_3)_2$ shows a severely distorted tetrahedral geometry that reflects the relative sizes of the ligands: the C–Mn–C angle is 137.9°, while the P–Mn–P angle is 96.2°.[31] There is no evidence of agostic interactions involving any of the hydrogen atoms of the neophyl ligands.[31]

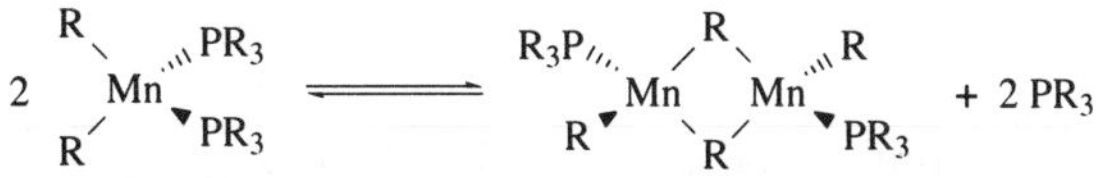

$$2 \quad \underset{R}{\overset{R}{\diagdown}}Mn\underset{PR_3}{\overset{PR_3}{\diagup}} \quad \rightleftharpoons \quad R_3P\cdots\underset{R}{\overset{R}{Mn}}\diagdown\underset{R}{\overset{}{Mn}}\cdots PR_3 \quad + \ 2\ PR_3 \qquad (3)$$

Addition of chelating phosphines to the $MnR_2$ dialkyls yields readily isolable mononuclear Lewis base adducts. With the phosphine 1,2-bis(dimethylphosphino)ethane (dmpe), tetrahedral monomers of stoichiometry $MnR_2(dmpe)$, where R is $CH_2TMS$, $CH_2Bu^t$, or Bz, are readily obtained as crystalline solids.[31] All of the four-coordinate $MnR_2L_2$ complexes exhibit EPR spectra characteristic of high-spin, rhombically-distorted manganese(II) centers.[31] The bis(trimethylsilyl)methyl complexes $Mn[CH(TMS)_2]_2(PMe_3)_2$ and $Mn[CH(TMS)_2]_2(dmpe)$ are also known, and the latter has been structurally characterized.[39]

Perhaps the most remarkable of the mononuclear phosphine adducts of the manganese(II) dialkyls is the *t*-butyl derivative $MnBu^t_2(dmpe)$ (10).[30] This four-coordinate 13-electron species has 18 β-hydrogen atoms and yet is stable thermally up to its melting point of 140 °C. The authors propose that the inability of this complex to undergo β-elimination is a consequence of the lack of an empty *d*-orbital; the high-spin nature of this $d^5$ complex means that every *d*-orbital is half-filled. The kinetic barrier to β-elimination is therefore high, even though by usual criteria this molecule is highly unsaturated both sterically and electronically.

Chelating amines such as TMEDA also react with $MnR_2$ dialkyls to form the four-coordinate mononuclear adducts $MnR_2(TMEDA)$ where R = $CH_2TMS$ and $CH_2Bu^t$.[21] The neopentyl complex $Mn(CH_2Bu^t)_2(TMEDA)$ has also been prepared directly by treating $MnCl_2$ with neopentyllithium in the presence of TMEDA.[11] It exhibits a severely rhombically distorted EPR spectrum and has a smaller zero-field splitting parameter than corresponding phosphine adducts; this result is consistent with the increased ligand field strength of phosphines vs. amines.[11]

Several ether adducts of the manganese dialkyls are also known. The structure of the three-coordinate tetrahydrofuran adduct $Mn[CH(TMS)_2]_2(THF)$ has been determined by x-ray crystallography.[39] The variable-temperature magnetic susceptibility of this latter complex has been studied, and the reactions of $Mn[CH(TMS)_2]_2(THF)$ with $HOC_6H_2MeBu^t_2$ and $Sn[N(TMS)_2]_2$ have been examined. The dioxane adduct of dibenzyl manganese, $Mn(CH_2Ph)_2(diox)_2$, has been prepared and its reactions with $CO_2$ and $CS_2$ described.[34] The pentafluorophenyl complexes $Mn(C_6F_5)_2L_2$ (where L = THF, $\frac{1}{2}MeOCH_2CH_2Me$, or $\frac{1}{2}TMEDA$) are also known.[40]

Related to the above complexes are some organomanganese(II) species with "internal" Lewis bases. For example, treatment of manganese(II) halides with $LiC_6H_4$-2-$CH_2NMe_2$ followed by addition of dioxane affords the 2-[(N,N-dimethylamino)methyl]phenyl complex $Mn(C_6H_4$-2-$CH_2NMe_2)_2$.[16,33] This yellow-green complex is paramagnetic ($\mu_{eff} = 5.48$ $\mu_B$) and evidently monomeric. A 3-(N,N-dimethylamino)propyl complex $Mn(CH_2CH_2CH_2NMe_2)_2$ was prepared by a similar route or by addition of the corresponding lithium alkyl to manganocene.[34,35,41] Related N,N'-dimethylaminonaphthyl and N,N'-dimethylaminoferrocenyl compounds are known.[42,43] The dinuclear complex $[Mn(CH_2C_6H_4$-2-$NMe_2)_2]_2$ was prepared by addition of 2-(N,N'-dimethylamino)benzyllithium to $MnI_2$. This complex has a reduced magnetic moment ($\mu_{eff} = 2.6$ $\mu_B$) and possesses an alkyl-bridged structure with a Mn–Mn distance of 0.281nm.[16] One of the manganese centers is five-coordinate and possesses approximate trigonal-bipyramidal geometry; the other manganese center is tetrahedral.

In addition to the four-coordinate $MnR_2L_2$ adducts, there are several examples of six-coordinate adducts of manganese(II) dialkyls. The first of these to be prepared was *trans*-$MnMe_2(dmpe)_2$ (**11**), which can be synthesized by addition of $MgMe_2$ to the manganese(II) coordination complex $MnBr_2(dmpe)_2$ or directly from $MnCl_2$ and $MgMe_2$ in the presence of dmpe.[36,37] The *ortho*-xylylene complex *cis*-$Mn[(CH_2)_2C_6H_4](dmpe)_2$ (**12**) can be made similarly.[31] Unlike four-coordinate species, which are invariably high-spin, the six-coordinate complexes are low-spin and have magnetic moments near 2 $\mu_B$. Their EPR spectra show well-resolved signals near $g = 2$ with hyperfine coupling to both manganese and phosphorus. Both molecules have been characterized crystallographically.[30,35] The *ortho*-xylylene complex is unusual in that the organic ligand is acting as a pure di-sigma donor; in most other *ortho*-xylylene complexes the ligand is bound as a diene.[31]

(11)                                    (12)

Thermolysis of the four-coordinate benzyl complex $Mn(Bz)_2(dmpe)$ in the presence of excess phosphine (Equation (4)) gives a product which was not isolated as a pure solid but which exhibited an EPR spectrum substantially similar to that of the *ortho*-xylylene complex. On these grounds, it was suggested that one of the benzyl ligands had *ortho*-metallated, one equivalent of toluene had been lost, and a second dmpe ligand had coordinated to give the benzometallacyclobutene complex $\overline{Mn(CH_2C_6H_4)}(dmpe)_2$ (**13**).[31]

(4)

(13)

Finally, attempts to prepare the manganese(II) complex $MnEt_2(dmpe)_2$ from $MnBr_2(dmpe)_2$ and $MgEt_2$ gave instead the manganese(I) hydrido–ethene complex $MnH(CH_2=CH_2)(dmpe)_2$.[30] This result, combined with the inability of the $Bu^t$ complex $MnBu^t_2(dmpe)$ to undergo β-elimination, suggests that $MnEt_2(dmpe)_2$ is formed in the alkylation reaction, but that this intermediate rapidly β-eliminates because (like $MnMe_2(dmpe)_2$) it is presumably low-spin and thus has the requisite empty $d$-orbitals.

### 7.2.2 Manganese(II) Isonitriles

The reaction of $MnX_2$ with *t*-butylisonitrile gives monoadducts $MnX_2(CNBu^t)$, where X = Cl, Br, I, or NCS.[44] The halo complexes are probably polymeric as judged from their magnetic moments in the solid state of ca. 5.1 $\mu_B$, which are slightly lower than the spin-only value for high-spin manganese(II) centers. In contrast, the NCS compound is proposed to be a dimer since its IR spectrum contains bands for both bridging and terminal thiocyanate groups. These complexes all give strong EPR signals near $g = 2$, and a weaker feature with manganese hyperfine coupling near $g = 6$. Unlike some other Lewis base adducts of manganese(II) halides, these complexes do not bind dioxygen reversibly. The isonitrile ligands can, however, be replaced by $PBu^n_3$ in solution.

The reaction of $MnI_2(PPh_3)_2$ with *t*-butylisonitrile in toluene gives species of stoichiometry $MnI_2(CNBu^t)_x$, where $x$ is 1, 1.5, or 2 depending on the amount of isonitrile added.[45] Interestingly, $MnI_2(CNBu^t)$ prepared in this way is proposed to be an isomer of the material prepared directly from $MnI_2$, since different $\nu_{CN}$ stretching frequencies are seen for the isonitrile ligands. Although $MnI_2(CNBu^t)_{1.5}$ was claimed to be a single pure species, it was not considered whether this material could be a 1:1 mixture of the mono- and diadducts. With excess *t*-butylisonitrile, $MnI_2(PPh_3)_2$ reacts in toluene to give two complexes, one of which is the $[Mn(CNBu^t)_6]^+$ salt of the $[MnI_3(CNBu^t)]^-$ anion. This salt has been structurally characterized, and the anion shown to adopt a tetrahedral structure.[41] In pentane, $MnI_2(PPh_3)_2$ reacts with two equivalents of *t*-butylisonitrile to give $MnI_2(PPh_3)(CNBu^t)_2$, which is proposed to be dinuclear as judged from its magnetic moment of 5.2 $\mu_B$ in the solid state.[45]

Similar reactions of $MnBr_2(PMe_2Ph)$ with $CNBu^t$ have afforded the complexes $MnBr_2(CNBu^t)(PMe_2Ph)$ and $[trans\text{-}Mn(CNBu^t)_4(PMe_2Ph)_2][MnBr_4]$ depending on the solvent employed.[46] Analogues of the latter salt with the tertiary phosphines $PMePh_2$, $PEt_2Ph$, or $PPr^n_3$ in place of the $PMe_2Ph$ ligands have been prepared by ligand substitution. Air oxidation of $[trans\text{-}Mn(CNBu^t)_4(PMe_2Ph)_2][MnBr_4]$ affords the salt $[trans\text{-}Mn(CNBu^t)_4(PMe_2Ph)_2][MnBr_3(OPMe_2Ph)]_2$, which has been structurally characterized.[46]

The deep blue manganese(II) isonitrile complex $[Mn(Bu^tDiNC)_3][PF_6]_2$ (**14**), where $Bu^tDiNC$ is the chelating diisonitrile 1,2-bis(2-isocyano-4-*t*-butylphenoxy)ethane, has been prepared by nitric acid oxidation of the corresponding manganese(I) salt and has been characterized by IR, cyclic voltammetry, and UV–visible spectroscopy.[47] The bonding, redox potentials, NMR spectra, EPR spectra, and magnetic susceptibilities of other manganese(II) hexaisonitrile complexes have been studied.[48,49] The self-exchange electron transfer reactions of $[Mn(CNR)_6]^+$ and $[Mn(CNR)_6]^{2+}$ species have been investigated in some detail.[50–4]

(**14**)

### 7.2.3 Mono(cyclopentadienyl)manganese(II) Complexes

A few half-sandwich complexes of manganese(II) free of neutral Lewis bases have been reported. The synthesis of the gasoline additive MMT (methylcyclopentadienyl manganese tricarbonyl) via carbonylation of a mixture of $(\eta\text{-}C_5H_4Me)_2Mn$ and manganese(II) acetate in toluene is thought to proceed via the half-sandwich intermediate "$(\eta\text{-}C_5H_4Me)Mn(O_2CMe)$."[54] Treatment of the manganese(I) carbonyl $(\eta\text{-}C_5H_4Me)Mn(CO)_2(THF)$ with $SiHClBu^i_2$ yields two products, one of which is $[(\eta\text{-}C_5H_4Me)MnCl]_n$. This white pentane-insoluble species gives an EPR signal at $g = 2.001$ and was proposed to be polymeric.[56] The related complex $[(\eta\text{-}C_5Me_4Et)MnCl]_n$ was evidently prepared similarly but not characterized.[56] The unstable organomanganese(II) alkyl CpMnMe has been proposed to be formed during the thermolysis of $CpMnMe_3$.[57]

The first monocyclopentadienyl complexes of manganese(II) to be prepared were Lewis base adducts of the type $[(\eta\text{-}C_5R_5)MnX(PR_3)]_2$ (**15a–h**). These complexes can be synthesized from the reaction of $MnX_2$ (X = Cl, Br, or I) with $(\eta\text{-}C_5R_5)_2Mn$ or $NaC_5R_5$ followed by addition of the appropriate phosphine (Scheme 2).[56,57] These green halide-bridged dimers are antiferromagnetically coupled as is typical for dinuclear high-spin $Mn^{II}$ complexes. Above 230 K, the compounds display magnetic moments of ca. 6.0

$\mu_B$ per $Mn^{II}$ center that are only slightly above the spin-only value for five unpaired electrons. Several of these complexes have also been characterized by x-ray crystallography.

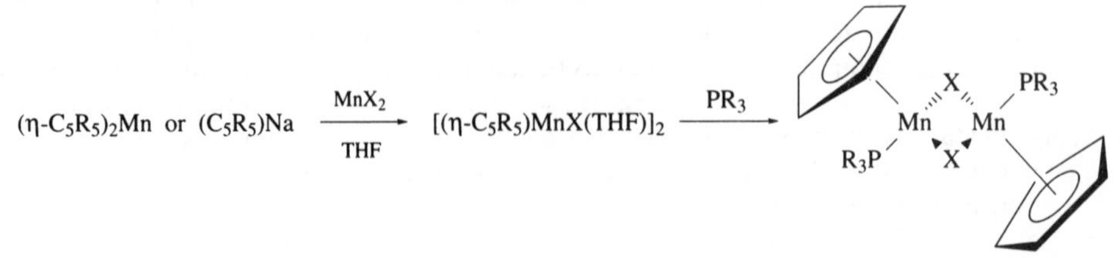

$$(\eta\text{-}C_5R_5)_2Mn \text{ or } (C_5R_5)Na \xrightarrow[\text{THF}]{MnX_2} [(\eta\text{-}C_5R_5)MnX(THF)]_2 \xrightarrow{PR_3}$$

(15a) $C_5R_5 = C_5H_4Me$, X = Cl, $PR_3 = PEt_3$
(15b) $C_5R_5 = C_5H_4Me$, X = Br, $PR_3 = PEt_3$
(15c) $C_5R_5 = C_5H_4Me$, X = I, $PR_3 = PEt_3$
(15d) $C_5R_5 = C_5H_4TMS$, X = I, $PR_3 = PEt_3$
(15e) $C_5R_5 = 1,2\text{-}C_5H_3Me_2$, X = I, $PR_3 = PEt_3$
(15f) $C_5R_5 = C_5H_4Me$, X = I, $PR_3 = AsEt_3$
(15g) $C_5R_5 = Cp$, X = I, $PR_3 = PEt_3$
(15h) $C_5R_5 = Cp$, X = I, $PR_3 = PMe_3$

**Scheme 2**

The mononuclear complex $CpMnI(PMe_3)_2$ has been obtained from the reaction of $[CpMnI(THF)]_2$ with excess trimethylphosphine, and the related complexes $(\eta\text{-}C_5H_4Me)MnCl(PMe_3)_2$ and $(\eta\text{-}C_5H_4Me)MnI(dmpe)$ can be prepared similarly (16a–c) (Equation (5)). The mononuclear $N,N,N',N'$-tetramethylethylenediamine complex $CpMnCl(TMEDA)$ (17) has been prepared from the reaction of $MnCl_2$ with NaCp in the presence of TMEDA (Equation (6)).[60] This species exhibits a magnetic moment of 5.84 $\mu_B$ and has been structurally characterized.

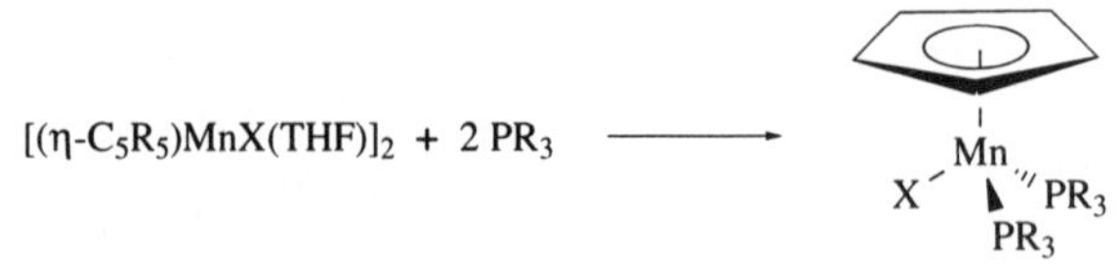

$$[(\eta\text{-}C_5R_5)MnX(THF)]_2 + 2 PR_3 \longrightarrow \qquad\qquad (5)$$

(16a) $C_5R_5 = Cp$, X = I, $PR_3 = PMe_3$
(16b) $C_5R_5 = C_5H_4Me$, X = Cl, $PR_3 = PMe_3$
(16c) $C_5R_5 = C_5H_4Me$, X = I, $PR_3 = 1/2$ dmpe

$$MnCl_2 + TMEDA + NaCp \longrightarrow \qquad\qquad + NaCl \qquad\qquad (6)$$

(17)

Treatment of $(\eta\text{-}Cp^*)Mn(CO)_2(THF)$ with sulfur radicals ·SR gives the low-spin manganese(II) half-sandwich complexes $(\eta\text{-}Cp^*)Mn(CO)_2(SR)$, where R is *t*-butyl or adamantyl.[61,62] The synthesis of related thiolate complexes $(\eta\text{-}C_5R_5)Mn(CO)_2(SR)$, where R is Ph, 4-$C_6H_4NO_2$, 4-$C_6H_4Me$, 4-$C_6H_4OMe$, 2,4,6-$C_6H_2Me_3$, $C_6F_5$, Et, $Pr^i$, and $Bu^t$, have been reported.[63,64] The cyclic voltammograms and the IR, EPR, and UV–visible spectra of these complexes have been described. Some selenolate and tellurolate analogues of these manganese(II) species have also been described.[62,65]

## 7.2.4 Bis(cyclopentadienyl)manganese(II) Complexes

### 7.2.4.1 Manganocenes

Since the original synthesis of manganocene, $Cp_2Mn$, from the reaction of NaCp with $MnCl_2$,[66] a number of ring-substituted analogues including (18a–e) have been reported (Equation (7)); these complexes are all liquids at or near room temperature and are typically purified by distillation.[67–70] More highly substituted manganocenes such as $(\eta\text{-}C_5Me_4H)_2Mn$ (18f), $(\eta\text{-}Cp^*)_2Mn$ (18g), and $(\eta\text{-}C_5Me_4Et)_2Mn$ (18h) have also been prepared.[69]

$$MnCl_2 + 2\,C_5R_5^- \longrightarrow (\eta\text{-}C_5R_5)_2Mn + 2\,Cl^- \tag{7}$$

**(18a)** $C_5R_5 = C_5H_4Et$

**(18b)** $C_5R_5 = C_5H_4Pr^i$

**(18c)** $C_5R_5 = C_5H_4TMS$

**(18d)** $C_5R_5 = C_5H_4Bu^t$

**(18e)** $C_5R_5 = 1,2\text{-}C_5H_3Me_2$

**(18f)** $C_5R_5 = C_5Me_4H$

**(18g)** $C_5R_5 = Cp^*$

**(18h)** $C_5R_5 = C_5Me_4Et$

The spin state of these substituted manganocenes at room temperature depends on the number (and kind) of substituents on the cyclopentadienyl rings. Thus, $(\eta\text{-}C_5Me_4H)_2Mn$, $(\eta\text{-}Cp^*)_2Mn$, and $(\eta\text{-}C_5Me_4Et)_2Mn$ are all low-spin at room temperature, while less substituted manganocenes such as **(18a–e)** exhibit a temperature-dependent equilibrium between the high-spin and low-spin states. Separate resonances for molecules with these spin states can be observed by NMR spectroscopy. At 127 °C in toluene, the equilibrium amounts of the low-spin form are 13% and 30% for $(\eta\text{-}C_5H_4Et)_2Mn$ and $(\eta\text{-}C_5H_3Me_2)_2Mn$, respectively. The low-spin state is favored by donor substituents on the ring.[69]

The silyl-substituted manganocene **(18c)** has been structurally characterized and the two $C_5H_4TMS$ rings are both $\eta^5$ with the trimethylsilyl groups in staggered positions.[69] Electron diffraction studies of $Cp_2Mn$, which is high-spin in the gas phase,[71] and $(\eta\text{-}C_5H_4Me)_2Mn$, which exists as a mixture of high- and low-spin molecules,[72] have been reported. The x-ray crystal structure of low-spin $(\eta\text{-}Cp^*)_2Mn$ has been determined at three different temperatures in order to provide additional insights into the molecular distortions seen in earlier studies.[73–6] The $Cp^*$ rings in this complex are staggered and slightly slipped.

Manganocene and ring-substituted manganocenes are strongly paramagnetic but they have been investigated by $^1H$ and $^2H$ NMR spectroscopy, the latter with both natural abundance and deuterium-enriched samples.[69,77–9] The advantage of examining the $^2H$ NMR spectrum is that the signals are approximately 30 times narrower than the corresponding $^1H$ NMR signals. Proton NMR spectroscopy has been used to determine a spin-crossover enthalpy of $\Delta H^\circ = 21 \pm 5$ kJ mol$^{-1}$ and entropy of $\Delta S^\circ = 100 \pm 20$ J mol$^{-1}$ K$^{-1}$ for the slow spin exchange involving the $^6A_{1g}$ and $^2E_{2g}$ electronic states of manganocene.[70,80] For $(\eta\text{-}C_5H_4Me)_2Mn$, the corresponding values are approximately $\Delta H^\circ = 20$ kJ mol$^{-1}$ and $\Delta S^\circ = 20$ J mol$^{-1}$ K$^{-1}$. Carbon-13 NMR spectra have also been recorded for $Cp_2Mn$ and other manganocenes.[67,70]

The vapor-phase electronic spectra of $Cp_2Mn$ and $(\eta\text{-}Cp^*)_2Mn$ have been studied in the near-ultraviolet region and a Rydberg series corresponding to the $3d(a_{1g}) \rightarrow Rnp$ ($n = 4$–6) transitions is observed. For $(\eta\text{-}Cp^*)_2Mn$, an ionization energy of 5.4 eV has been determined as a convergence limit of the series.[81,82] Photoelectron spectra of manganocene and its substituted analogues have also been measured.[83,84] Ion cyclotron resonance studies of the $Cp_2Mn^{0/+}$ redox couple and the reactions of manganocene with first-row metallocenes have been performed and self-exchange and cross reaction rates have been measured in the gas phase.[85–7] The electrochemical rate constants and activation parameters for the $(\eta\text{-}Cp^*)_2Mn^{0/+}$ redox couple have been measured and an $E_f$ of $-540$ m V vs. SCE has been observed in acetonitrile.[88]

Attempts to prepare "open" manganocenes with 2,4-pentadienyl ligands have been uniformly unsuccessful to date. For example, treatment of $MnCl_2$ with 2 equiv. of the sterically-hindered pentadienyl derivative 3-Me-1,5-$C_5H_4(TMS)_2$ gives the manganate salt $K[Mn\{3\text{-Me-1,4-}C_5H_4(TMS)_2\}_3]$ **(19)**, which is a three-coordinate high-spin manganese(II) complex with three unidentate pentadienyl ligands.[89] Attempts to prepare an open manganocene with the less sterically-hindered 3-methyl-2,4-pentadienyl anion led instead to the isolation of the trinuclear complex $Mn_3(3\text{-Me}C_5H_6)_4$ **(20)**.[90] An x-ray crystallographic study shows that this cluster has a nearly linear Mn–Mn–Mn backbone (Mn–Mn = 0.2316 nm), and magnetic susceptibility measurements show that there are five unpaired electrons per cluster. Theoretical calculations of the electronic structure of this complex have been carried out.[90]

The reaction of $MnCl_2$ with 1 equiv. each of (2,4-dimethylpentadienyl)potassium and cyclopentadienylsodium yields the unusual mixed-valent complex $Cp_2Mn_2(2,4\text{-Me}_2C_5H_5)$ **(21)**.[89] This compound may be thought of as an associated salt of the $[CpMn]^+$ cation and the 18-electron $[CpMn(2,4\text{-Me}_2C_5H_5)]^-$ anion. The structurally related manganaborane cluster $[Mn_3\{2,3\text{-}C_2B_4H_4\text{-2,3-}TMS_2\}_4]^{3-}$ has also been prepared and crystallographically characterized.[91]

Attempts to prepare open manganocenes stabilized by Lewis bases such as tertiary phosphines have also been unsuccessful; these reactions result instead in coupling of the pentadienyl ligands and the isolation of manganese(0) products with decatetraene ligands.[92,93]

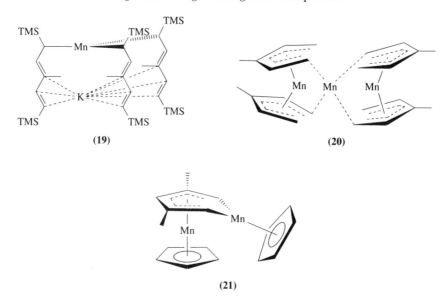

**(19)**                  **(20)**

**(21)**

### 7.2.4.2 Lewis base adducts of manganocenes

Since manganocenes are 17-electron molecules, one might expect that they would be relatively unreactive toward two-electron donors. A number of Lewis base adducts of stoichiometry $Cp_2MnL$ and $Cp_2MnL_2$ are known that have electron counts of 19 and 21, respectively. Surprisingly, these high valence electron counts do not always result in a change in the hapticity of the cyclopentadienyl rings; in most cases both rings are pentahapto but occasionally "tilted" so that the five Mn–C distances are not all equal. Although tilted cyclopentadienyl rings have been observed in other organotransition metal complexes with formal electron counts above 18 (and the tilting attributed to the effect of electrons in metal–ligand antibonding orbitals), for the Lewis base adducts of manganocene it is quite clear that steric effects rather than electronic effects are responsible for the tilting.[94,95] The relative unimportance of electronic effects is directly related to the high-spin nature of the $Cp_2MnL_x$ adducts; the bonding in these species is largely ionic and the overlap between the metal and ligand orbitals is small compared with that in other cyclopentadienyl complexes.

### (i) Carbon ligands

Addition of *t*-butylisonitrile to $Cp_2Mn$ initially gives an adduct of unknown stoichiometry which was characterized by its EPR spectrum; subsequent reduction to the manganese(I) product $CpMn(CNBu^t)_3$ is observed.[94] The reaction of manganocene with $V(CO)_6$ affords the manganese(I) tricarbonyl $(\eta\text{-}Cp)Mn(CO)_3$.[96] IR studies of the reactions of CO with $Cp_2Mn$ adsorbed onto $AlPO_4$ and $SiO_2$[96] and in frozen solution[98] have been reported. Interaction of $CH_4$ with $(\eta\text{-}Cp^*)_2Mn$ in a matrix has also been claimed.[99]

### (ii) Nitrogen ligands

Addition of $NHEt_2$ to $Cp_2Mn$ (Equation (8)) yields the colorless, sublimable secondary amine complex $Cp_2Mn(NHEt_2)$ **(22)**.[100] Although refinement of the crystal structure was unsatisfactory, the Cp groups were clearly tilted. When the chelating Lewis base TMEDA is added to a toluene suspension of manganocene, the adduct $Cp_2Mn(TMEDA)$ **(23)** is isolated in 95% yield as colorless crystals (Equation (9)).[60] The structural analysis reveals that one of the Cp ligands is still $\eta^5$, while the other ligand has slipped to $\eta^1$. The $^1H$ NMR spectrum and magnetic susceptibility of **(23)** were reported.

Heating an intimately ground mixture of $MnBr_2$ and solid $Li_2[C_5H_4(CH_2)_3C_5H_4]$ leads to a distillable red-orange oil which is presumably the [3]-manganocenophane $[\eta^5,\eta^5\text{-}C_5H_4(CH_2)_3C_5H_4]Mn$. Treatment of this oil with 3,5-dichloropyridine (Equation (10)) gives the 1:1 adduct $[\eta^5,\eta^5\text{-}C_5H_4(CH_2)_3C_5H_4]Mn(3,5\text{-}Cl_2C_5H_3N)$ **(24)**.[101] The x-ray crystallographic study of **(24)** shows that both rings are bound in a symmetrical $\eta^5$ fashion to the manganese center. The formation of a bis(3,5-dichloropyridine) adduct of the manganocenophane was also mentioned.[101]

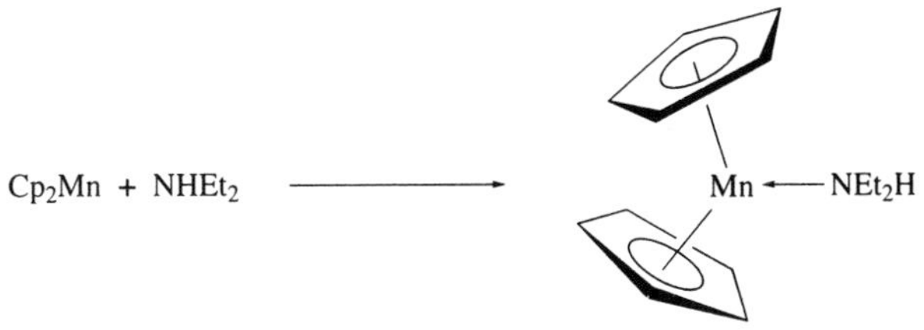

$$Cp_2Mn + NHEt_2 \longrightarrow \quad \text{(22)}$$

(8)

$$Cp_2Mn + \quad Me_2N \overset{\frown}{\quad} NMe_2 \longrightarrow \quad \text{(23)}$$

(9)

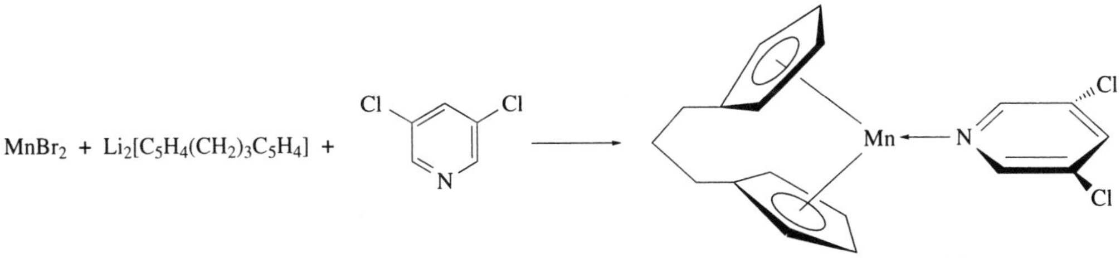

$$MnBr_2 + Li_2[C_5H_4(CH_2)_3C_5H_4] + \quad \text{(24)}$$

(10)

Claims of $N_2$ adducts of $Cp_2Mn$ and $(\eta$-$Cp^*)_2Mn$ have been made based on matrix isolation studies.[102,103]

### (iii) Phosphorus ligands

The reactions of tertiary phosphines with $Cp_2Mn$ have been explored in some detail. The interaction of $Cp_2Mn$ with tertiary phosphines in toluene (Equation (11)) yields the monoadducts $Cp_2Mn(PR_3)$, where $PR_3 = PMe_3$, $PMe_2Ph$, $PMePh_2$, or $PEt_3$ (25a–d).[94] These compounds (and those described in the next paragraph contain high-spin $d^5$ $Mn^{II}$ centers as shown by their magnetic moments and EPR spectra. For the $PMe_3$ and $PMe_2Ph$ adducts, x-ray crystallography shows one symmetric Cp ring and one "tilted" Cp ring. The tilting of the rings in these 19-electron complexes was attributed to steric repulsions between the Cp rings and the phosphine ligand.[94]

The interaction of $Cp_2Mn$ or $(\eta$-$C_5H_4Me)_2Mn$ with the chelating tertiary phosphine dmpe gives 21-electron adducts of stoichiometry $(\eta$-$C_5H_4R)_2Mn(dmpe)$ (R = H, (26a); R = Me, (26b)) according to Equation (12).[94] The structure of (26a) is pseudotetrahedral; both cyclopentadienyl rings are tilted.

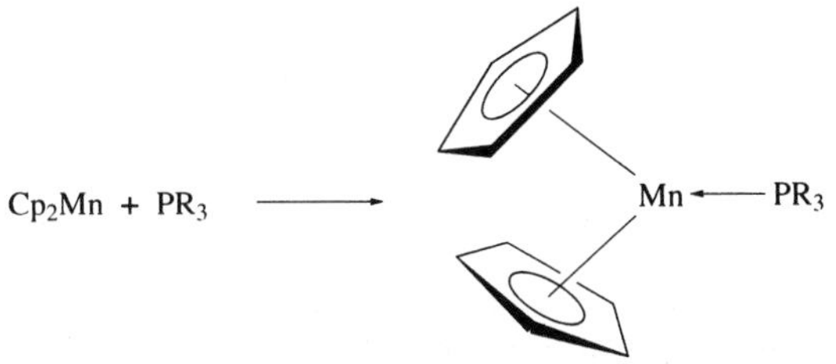

$$Cp_2Mn + PR_3 \longrightarrow \quad \text{(11)}$$

(11)

**(25a)** $PR_3 = PMe_3$
**(25b)** $PR_3 = PMe_2Ph$
**(25c)** $PR_3 = PMePh_2$
**(25d)** $PR_3 = PEt_3$

Finally, manganocene reacts with $P(OMe)_3$ to give the manganese(I) product $CpMn[P(OMe)_3]_3$.[104]

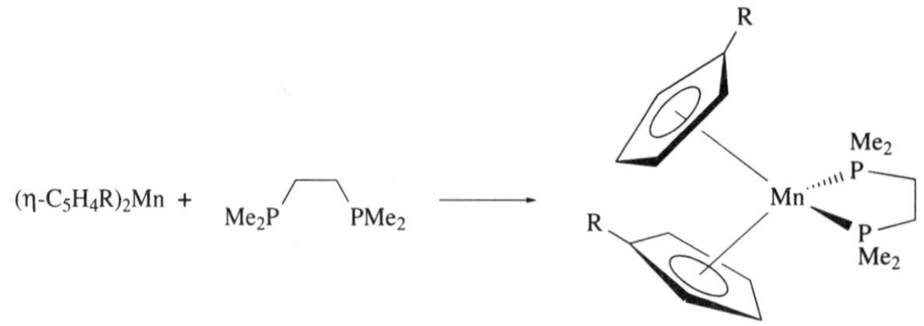

$$(\eta\text{-}C_5H_4R)_2Mn \;+\; Me_2P\text{\textbackslash}\text{\textbackslash}PMe_2 \longrightarrow \tag{12}$$

(26a) R = H
(26b) R = Me

*(iv) Oxygen ligands*

Cp$_2$Mn(THF) is the only reported ether adduct of Cp$_2$Mn.[95] The single crystal x-ray structure reveals that the Cp rings are only slightly tilted since the THF ligand is not very sterically demanding. This complex is high-spin.

Chemiluminescence has been observed from the interaction of Cp$_2$Mn with O$_2$ at room temperature,[105] and infrared studies of the vapor-phase reaction of Cp$_2$Mn with O$_2$ at low temperatures suggested that an unstable O$_2$ adduct of manganocene had formed ($v_{OO}$ = ca. 1255 cm$^{-1}$).[106]

*(v) Hydrogen ligands*

The formation of adducts of Cp$_2$Mn and ($\eta$-Cp*)$_2$Mn with H$_2$, HD, and D$_2$ at 12–16 K has been asserted without experimental detail.[107]

### 7.2.4.3 Other reactions of manganocenes

The interaction of a toluene suspension of Cp$_2$Mn with CO$_2$ and NHEt$_2$ yields the manganese(II) carbamate [Mn(O$_2$CNEt$_2$)$_2$]$_6$ in nearly quantitative yield.[100] Manganocene has been shown to be a useful precursor for the synthesis of other organomanganese compounds.[41,108,109] The reaction of Cp$_2$Mn with the dinuclear chromium complex ($\eta$-C$_5$H$_4$Me)$_2$Cr$_2$($\mu$-SBu$^t$)$_2$S in boiling toluene gives the metallospirane cluster [($\eta$-C$_5$H$_4$Me)$_2$Cr$_2$($\mu$-SBu$^t$)($\mu_3$-S)$_2$]$_2$Mn (27) as dark green crystals in 31% yield.[110–12] Treatment of Cp$_2$Mn with Hg[Ge(C$_6$F$_5$)$_3$]$_2$ yielded a bright green product formulated as the salt [Mn(THF)$_2$][HgCp$_2${Ge(C$_6$F$_5$)$_3$}$_2$].[113] The oxidation of manganocenes to the corresponding cations will be discussed in Section 7.3.3.

## 7.3 ORGANOMANGANESE(III) COMPOUNDS

### 7.3.1 Manganese(III) Alkyls and Aryls

The first arylmanganese(III) compound was prepared in 1987, and shortly thereafter the first alkylmanganese(III) species were described. Often, attempts to synthesize Mn$^{III}$ alkyls and aryls result instead in reduction to Mn$^{II}$ or disproportionation to Mn$^{II}$ and Mn$^{IV}$ products. For example, treatment of Mn(acac)$_3$ with excess ethyllithium yields the Mn$^{II}$ species [Li(TMEDA)]$_2$[MnEt$_4$],[114] while treatment with LiCH$_2$TMS results in disproportionation to the thermally unstable Mn$^{IV}$ alkyl Mn(CH$_2$TMS)$_4$ and a precipitate identified as the polymeric Mn$^{II}$ alkyl [Mn(CH$_2$TMS)$_2$]$_n$.[115] The tendency of Mn$^{III}$ species to reduce or disproportionate is not universal, however, and several classes of manganese(III) alkyls and aryls are known.

### 7.3.1.1 Anionic manganese(III) alkyls and aryls

The first binary alkylmanganate(III) compound was prepared by the reaction of $Mn(acac)_3$ with excess methyllithium in diethyl ether; subsequent addition of TMEDA gave the pentamethylmanganate(III) complex $[Li(TMEDA)]_2[MnMe_5]$ (**28**).[114,116] Two alternative preparations of this manganese(III) alkyl were also described: air oxidation of the divalent tetramethylmanganate salt $[Li(TMEDA)]_2[MnMe_4]$ in the presence of excess methyllithium, and comproportionation from $Mn^{II}$ and $Mn^{IV}$ species. The pentamethylmanganate(III) species is EPR silent and exhibits a solution magnetic moment of 4.9 $\mu_B$ corresponding to a high-spin configuration; although it has not been characterized crystallographically, it probably adopts a square-pyramidal geometry.[114]

Whereas compound (**28**) can also be obtained in a comproportionation reaction of $Mn^{II}$ and $Mn^{IV}$ species in diethyl ether (Equation (13)), if the same comproportionation reaction is carried out in toluene, a different $Mn^{III}$ product, the tetramethylmanganate(III) salt $[Li(TMEDA)_2][MnMe_4]$ (**29**) is isolated (Equation (14)).[114] This observation suggests that there is a solvent-dependent equilibrium between the two trivalent permethylmanganate species in solution.

$$1/2\ [MnMe_4]^{2-} + 1/2\ [MnMe_6]^{2-} \xrightarrow{\text{ether}} \begin{bmatrix} Me\ \ \ \ Me \\ |\ \ \ \ \ \ \text{,,}Me \\ Me\text{-}Mn\text{-}Me \\ Me \end{bmatrix}^{2-} \tag{13}$$

(**28**)

$$1/2\ [MnMe_4]^{2-} + 1/2\ [MnMe_6]^{2-} \xrightarrow[-Me^-]{\text{toluene}} \begin{bmatrix} Me \\ \text{,,}Me \\ Me\text{-}Mn\text{-}Me \\ Me \end{bmatrix}^{-} \tag{14}$$

(**29**)

The 12-electron tetramethylmanganate(III) complex (**29**) is EPR silent and high-spin (four unpaired electrons). Crystals of (**29**) are composed of charge-separated $[Li(TMEDA)_2]^+$ cations and distorted square-planar $[MnMe_4]^-$ ions.[114] Interestingly, the $[MnMe_4]^-$ anion shows no tendency to dimerize to $[Mn_2Me_8]^{2-}$, which, if it existed, would possess a metal–metal quadruple bond like that in $[Re_2Me_8]^{2-}$. *Ab initio* calculations have suggested that $[Mn_2Me_8]^{2-}$ should be stable and should possess a diamagnetic $\sigma^2\pi^4\delta^2$ ground state.[117] Several alternative explanations have been advanced to account for the monomeric nature of $[MnMe_4]^-$.[114]

The reaction of $Mn(acac)_3$ with only 5 equiv. of methyllithium (Equation (15)) yielded an EPR-silent, high-spin $Mn^{III}$ complex with the unusual stoichiometry $\{Li_2[MnOMe_3]\cdot2Li_2 - (OCMe=CHCMe_2O)\cdot TMEDA\}_2$ (**30**).[118] The 4-methyl-2-penten-2,4-diolate dianions arise from C-alkylation of 2,4-pentanedionate anions at a carbonyl carbon. An x-ray crystal structure analysis reveals the presence of $[MnOMe_3]^{2-}$ units: the manganese(III) center in this anion, like that in $[MnMe_4]^-$, adopts a tetrahedrally distorted square-planar geometry in which the "*trans*" C–Mn–C and C–Mn–O angles are 149.0(2)° and 152.5(2)°, respectively.[118]

$$Mn(acac)_3 \xrightarrow[\text{LiOH?}]{\text{MeLi}} \begin{bmatrix} Me \\ \text{,,}Me \\ Me\text{-}Mn\text{-}O \\ Me \end{bmatrix}^{2-} \tag{15}$$

(**30**)

Although alkylation of $Mn(acac)_3$ has not yielded $[MnR_5]^{2-}$ or $[MnR_4]^-$ species with alkyl ligands other than methyl (see Section 7.3.1), one other binary alkylmanganate(III) complex has been prepared by a different route.[108,109] Treatment of $CpMn(\eta^6$-biphenyl) with excess potassium under an ethene atmosphere yields a yellow oil of stoichiometry $[K(THF)_x]7thinsp;[Mn(C_4H_8)_2]$ (**31a**). The same complex can also be prepared from $Cp_2Mn$, potassium naphthalenide, and ethene. In these reactions, the ethene molecules have coupled to form two manganacyclopentane rings. Addition of triglyme to (**31a**) yields the crystalline complex $[K(triglyme)][Mn(C_4H_8)_2]$ (**31b**), while addition of pyridine yields the red crystalline complex $[K(py)_2][Mn(C_4H_8)_2(py)]$ (**32**) (Scheme 3). Addition of methanol to the manganacyclopentane complexes results in the liberation of butane and small amounts of butene.[109]

The crystal structure of the pyridine complex (**32**) reveals the presence of a manganese(III) center that adopts a distorted square-pyramidal structure; the pyridine ligand occupies the axial position and the two butanediyl groups form the square base.[108]

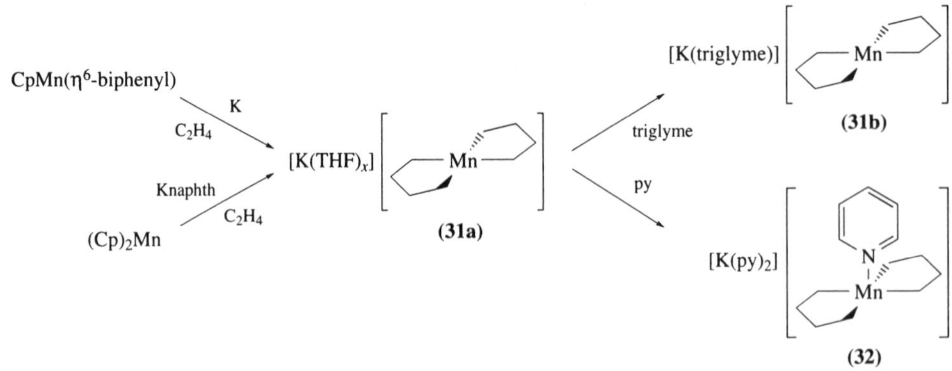

**Scheme 3**

Treatment of manganese(II) halides with $o,o'$-dilithiodiphenylether (Li$_2$dpa) in tetrahydrofuran yields a yellow-brown product of stoichiometry [Li(THF)$_2$][Mn(dpa)$_2$].[15] The experimental conditions that led to the isolation of this complex (instead of the manganese(II) species [Li(THF)$_3$]$_2$[Mn(dpa)$_2$] described in Section 7.2.1) were not described. If the species isolated is an Mn$^{III}$ aryl complex, it probably arises via adventitious air oxidation (the authors suggest a rather improbable disproportionation reaction). The only property of this product given is its magnetic moment of 2.95 $\mu_B$. Although structures might be drawn that are consistent with this observation, such speculations are unwarranted without further data.

Addition of sodium cyclopentadienide to Mn(acac)$_3$ in tetrahydrofuran followed by removal of the solvent has been reported to yield the diamagnetic manganese(III) salt Na[MnCp$_4$] as a purple microcrystalline solid.[119] The IR, NMR, and electronic spectra of this product were described, and it was reported to be partly dissociated to charge-separated ions in tetrahydrofuran. Later attempts to reproduce this work found instead that addition of NaCp to Mn(acac)$_3$ resulted in the immediate decolorization of the solution and the formation of a colorless precipitate, and these later workers concluded that NaCp reduces Mn$^{III}$ to Mn$^{II}$.[120] Red-violet precipitates could be obtained only if O$_2$ was added to the reaction mixtures; however, the color generated was ascribed to organic oxidation products and the purple solid, which contained only traces of manganese, had a stoichiometry near NaCp·2THF.[120]

### 7.3.1.2 Neutral manganese(III) alkyls and aryls

Treatment of manganese(II) halides with 0.5 equiv. of the mesityl (2,4,6-trimethylphenyl) reagent Mg(C$_6$H$_2$Me$_3$)$_2$(THF)$_2$ and excess PMe$_3$ gives Mn$^{II}$ intermediates that turn bright red on exposure to dry dioxygen.[120,121] The Mn$^{III}$ aryls Mn(C$_6$H$_2$Me$_3$)X$_2$(PMe$_3$)$_2$ (**33a–c**), where X is Cl, Br, or I, were isolated from the red solutions (Equation (16)). Attempts to prepare analogues of these complexes with other alkyl or aryl groups or with other trialkylphosphines were unsuccessful. Although the Mn$^{II}$ intermediates present in these reactions before addition of O$_2$ could not be isolated, it was suggested on the basis of EPR studies that they had stoichiometries of either Mn(C$_6$H$_2$Me$_3$)X(PMe$_3$)$_2$ or [Mn(C$_6$H$_2$Me$_3$)X$_2$(PMe$_3$)]$^-$.[120]

$$MnX_2 \xrightarrow[\text{ii, O}_2]{\text{i, 0.5 Mg(Mes)}_2\text{(THF)}_2, \text{PMe}_3} \qquad (16)$$

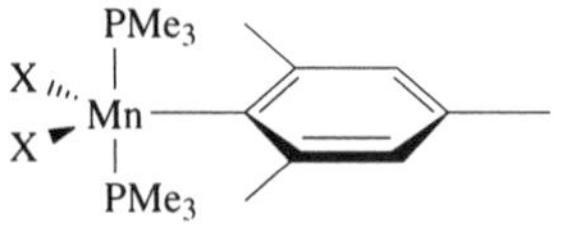

(**33a**) X = Cl
(**33b**) X = Br
(**33c**) X = I

These 14-electron compounds (**33a–c**) are high-spin in solution ($\mu_{eff} = 4.8\ \mu_B$) and are EPR silent. The $^1$H NMR spectra of these species show shifted and broadened peaks as expected from the paramagnetism. The molecular geometry of Mn(C$_6$H$_2$Me$_3$)Br$_2$(PMe$_3$)$_2$ is best described as a distorted trigonal bipyramid with the phosphine ligands occupying the axial sites.[120]

Treatment of the Mn$^{II}$ ($N,N$-dimethylaminomethyl)phenyl complex Mn(C$_6$H$_4$CH$_2$NMe$_2$)$_2$ with 1 equiv. of methyllithium followed by oxidation with AgBF$_4$ yields yellow crystals of the Mn$^{III}$ mixed alkyl–aryl complex MnMe(C$_6$H$_4$CH$_2$NMe$_2$)$_2$ (**34**).[41] The crystal structure of this high-spin complex shows a trigonal-bipyramidal structure in which the two aryl groups occupy the axial positions while the two amine groups and the methyl ligand occupy the equatorial sites. An *N,N*-dimethylaminopropyl complex MnMe(CH$_2$CH$_2$CH$_2$NMe$_2$)$_2$ (**35**) can be prepared by a similar route.[41]

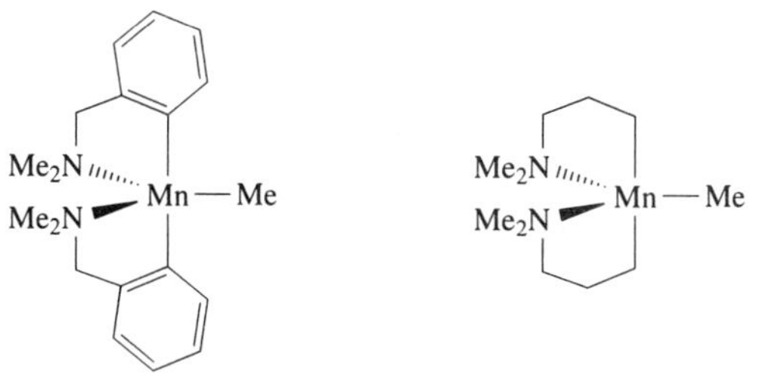

(34)          (35)

Treatment of the manganese(II) *N,N'*-ethylene bis(salicylideneaminato) complex Mn(salen) with organic hydrazines followed by addition of dioxygen and water is claimed to yield five-coordinate organomanganese(III) species of stoichiometry Mn(salen)R, where R is Me or Ph.[121] Similar complexes where the Schiff base is *N,N'-o*-phenylene bis(salicylideneaminato) or certain dimethylglyoxime derivatives are also claimed.[122] The 4.98 $\mu_B$ effective magnetic moments of these compounds are consistent with the presence of high-spin $d^4$ manganese centers. None of these species has been crystallographically characterized, and the formulation of these products as high-spin Mn$^{III}$ alkyls and aryls is extremely doubtful since such species should be instantly hydrolyzed or oxidized under the workup conditions.

Finally, photolysis of the 1,2-bis(dimethylphosphino)ethane complex MnH$_3$(dmpe)$_2$ in the presence of C$_6$D$_6$ results in hydrogen–deuterium exchange with the hydride ligands; the authors suggested that a manganese(III) aryl MnHD(C$_6$D$_5$)(dmpe)$_2$ was formed as an intermediate in this exchange process.[123]

### 7.3.1.3 *Cationic manganese(III) alkyls and aryls*

Alkylation of Mn(acac)$_3$ with methyllithium in the presence of dmpe yields only Mn$^{II}$ and Mn$^{IV}$ products (see Section 7.4).[115] In contrast (Scheme 4), alkylation of Mn(acac)$_3$ with trimethylaluminum in the presence of dmpe gives a manganese(III) cation identified as [MnMe$_2$(dmpe)$_2$][AlMe$_4$] (**36a**); treatment of this cation with NaBPh$_4$ gives the related Mn$^{III}$ salt [MnMe$_2$(dmpe)$_2$][BPh$_4$] (**36b**).[124,125] This red crystalline species is water stable and is a 1:1 electrolyte in tetrahydrofuran. It exhibits an effective magnetic moment of 3.01 $\mu_B$ that is consistent with a low-spin $d^4$ manganese center in an octahedral environment. The x-ray crystal structure of this complex confirms the *trans*-octahedral geometry.[125]

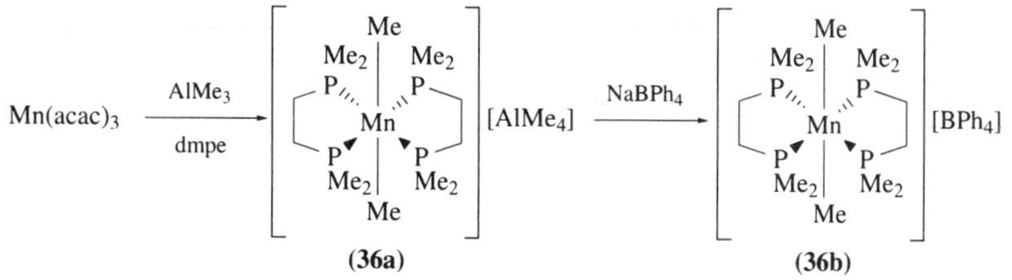

(36a)          (36b)

**Scheme 4**

Thermolysis of [MnMe$_2$(dmpe)$_2$][BPh$_4$] resulted in the liberation of methane and smaller amounts of ethene and ethane.[124,125] All of these organic products are thought to arise from the manganese-bound methyl groups via $\alpha$-hydrogen elimination. The reaction of [MnMe$_2$(dmpe)$_2$][BPh$_4$] with carbon monoxide in benzonitrile results in the loss of both methyl groups (as acetone) and reduction to the manganese(I) carbonyl complex [Mn(CO)(NCPh)(dmpe)$_2$][BPh$_4$].[125]

### 7.3.2 Cyclopentadienyl Manganese(III) Carbonyls

There are numerous studies of molecules of the general form [CpMn(CO)$_2$]$_n$(EX$_m$), especially where E is a heavier main group element.[126,127] Although some of these molecules can be considered to be manganese(III) species, in our view it is more appropriate to regard all of them as manganese(I) derivatives, where the $EX_m$ fragment serves as a carbene analogue. We have excluded all such complexes from the following discussion.

Several classes of cyclopentadienyl manganese(III) complexes that contain carbonyl groups are known. Compounds of the type $[CpMn(CO)_2(\eta^3\text{-enyl})]^+$ have been obtained by one of two routes (Equations (17) and (18)): protonation of $CpMn(CO)_2(\eta^2\text{-cyclohexadiene})$ with $HBF_4$ gives the $\eta^3$-cyclohexenyl complex (37), and protonation of the appropriate $CpMn(CO)_2(\eta^2\text{-allyl-alcohol})$ compound with $HBF_4$ or $HPF_6$ gives the $\eta^3$-propenyl, $\eta^3$-butenyl, and $\eta^3$-2-methylpropenyl complexes (38a–c).[128,129] Attempts to prepare $[CpMn(CO)_2(\eta^3\text{-cyclooctenyl})]^+$ by hydride abstraction from $CpMn(CO)_2(\eta^2\text{-cyclooctene})$ using $[CPh_3][BF_4]$ were unsuccessful.[129]

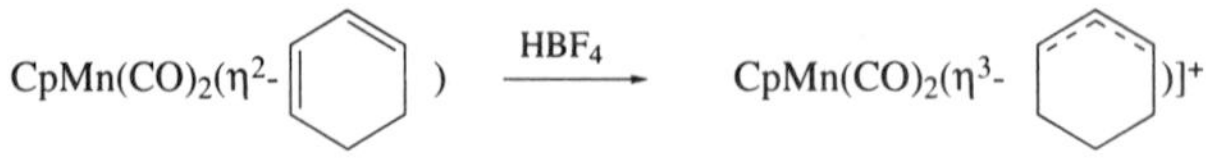

$$CpMn(CO)_2(\eta^2\text{-}\hexagon) \xrightarrow{\ HBF_4\ } CpMn(CO)_2(\eta^3\text{-}\hexagon)]^+ \qquad (17)$$

$$(37)$$

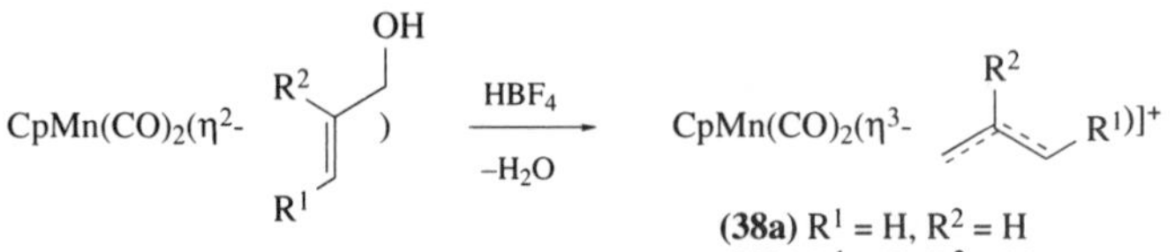

$$CpMn(CO)_2(\eta^2\text{-}) \xrightarrow[-H_2O]{\ HBF_4\ } CpMn(CO)_2(\eta^3\text{-}\ R^1)]^+ \qquad (18)$$

(38a) $R^1 = H$, $R^2 = H$
(38b) $R^1 = H$, $R^2 = Me$
(38c) $R^1 = Me$, $R^2 = H$

These diamagnetic seven-coordinate allylmanganese cations were characterized by IR and $^{13}C$ NMR spectroscopy. They are soluble without decomposition in polar solvents, are only moderately air sensitive, and react only slowly with water to regenerate the allyl alcohol complexes from which they were made. Addition of lithium iodide to the cationic $\eta^3$-propenyl and $\eta^3$-butenyl complexes affords the neutral $CpMn(CO)I(\eta^3\text{-allyl})$ compounds (39a) and (39b), while addition of tri(isopropyl)phosphine or triphenylphosphine to the cationic $\eta^3$-propenyl complex gives coordinated allylphosphonium complexes of $Mn^I$ of stoichiometry $[CpMn(CO)_2(\eta^2\text{-}CH_2=CHCH_2PR_3)]^+$ (Scheme 5). If excess $PPr^i_3$ is added, the free allylphosphonium salt $[C_3H_5PPr^i_3][BF_4]$ can be isolated.[129]

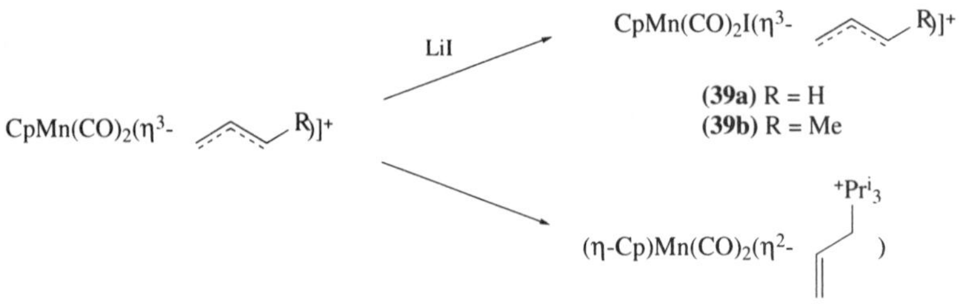

$$CpMn(CO)_2(\eta^3\text{-}\ R)]^+ \xrightarrow{\ LiI\ } CpMn(CO)_2I(\eta^3\text{-}\ R)]^+$$

(39a) $R = H$
(39b) $R = Me$

$$\xrightarrow{\ ^+Pr^i_3\ } (\eta\text{-Cp})Mn(CO)_2(\eta^2\text{-}\ )$$

**Scheme 5**

Other seven-coordinate manganese(III) carbonyl compounds are known. Protonation of $CpMn(CO)_{3-n}(PR_3)_n$ complexes yields hydrido manganese(III) cations of stoichiometry $[CpMnH(CO)_{3-n}(PR_3)_n]^+$ in solution (Equations (19) and (20)); complexes with unidentate phosphines $PR_3$ (where $R = Ph$, Tol, Cy, $Pr^i$, or OPh) and with bidentate phosphines $Ph_2P(CH_2)_nPPh_2$ (where $n = 1$, 2, or 3) were studied, and some analogous complexes with $\eta\text{-}C_5Et_4H$ and $\eta\text{-}C_5Et_5$ rings and with other Lewis bases such as $AsPh_3$ and $SbPh_3$ were also examined.[130-2] These hydrido cations probably adopt four-legged piano stool geometries. In most cases, the favored isomer has the hydride ligand *trans* to a CO group; an exception to this rule is found for the bis(diphenylphosphino)methane complexes, where this geometry is not possible since the phosphine cannot span *trans* positions in the square base. In more recent work, it was found that the complexes $[(\eta\text{-}C_5H_4R)MnH(CO)_2(PMe_2Ph)]^+$, where R is H or Me, actually exist as mixtures of isomers in solution: the *cis* and *trans* isomers are present in a 3:1 ratio, respectively, and interconvert with a barrier of $\Delta G^{\ddagger} = $ ca. 60 kJ mol$^{-1}$.[132] The hydridomanganese(III) cations have been studied by IR, $^1H$, $^{13}C\{^1H\}$ and $^{31}P\{^1H\}$ NMR spectroscopy, and high-field resonances for the metal-bound hydride ligands are seen between $\delta$ −3.9 and $\delta$ −7.0 in the $^1H$ NMR spectra.[130-2] Interestingly, the two-bond couplings between the hydride and phosphine ligands are ca. 30 Hz if these ligands are mutually *trans*, but between 70 Hz and 90 Hz if they are mutually *cis* in the square base of the piano stool. In all of these compounds, the hydridomanganese(III) cations exist in equilibrium with their neutral manganese(I) precursors; the equilibrium is shifted in favor of the cations for compounds containing alkyl-substituted cyclopentadienyl rings and for compounds with two phosphine ligands.

$$\qquad (19)$$

$$\qquad (20)$$

One of the largest classes of cyclopentadienyl manganese(III) carbonyls is obtained by photolysis of $(\eta\text{-}C_5R_5)Mn(CO)_3$ in the presence of $HSiR_3$ and related group 14 species.[134] Photolysis of $(\eta\text{-}C_5H_4Me)Mn(CO)_3$ in the presence of $HSiCl_3$ has long been known to give the "oxidative addition" product $(\eta\text{-}C_5H_4Me)MnH(SiCl_3)(CO)_2$.[135-9] Similar complexes can be obtained by photolysis in the presence of other silanes such as $HSiPh_3$, $H_2SiPh_2$, and $H_2SiEt_2$.[136,140-2] Photolysis of $(\eta\text{-}C_5H_4Me)Mn(CO)_2(PR_3)$ in the presence of $H_2SiPh_2$ or $H_2SiEt_2$ gives the analogous $(\eta\text{-}C_5H_4Me)MnH(SiHR_2)(CO)(PR_3)$ products, where $PR_3$ is $PMe_3$, $PBu^n_3$, $PPh_3$, $P(p\text{-}ClC_6H_4)_3$, $P(p\text{-}MeC_6H_4)_3$, or $PMe_2Ph$.[143] Complexes with chiral silicon[144] and germanium[145] centers have also been prepared, and a tin analogue of stoichiometry $(\eta\text{-}C_5H_4Me)Mn(CO)_2H(SnPh_3)$ has been described.[146]

In all of these complexes, the hydride and silyl groups are mutually *cis*, and neutron and x-ray crystallographic determinations show that there are significant $Si\cdots H$ interactions between the silyl and hydride ligands.[135,147-50] The bonding between the manganese, silicon, and hydrogen atoms was much discussed in the early literature, and it was generally concluded that these compounds should be regarded as "silane adducts" of $Mn^I$.[137,151-3] Proton, carbon-13, and silicon-29 NMR studies of these silane adducts have also provided evidence of a direct interaction between the hydrogen and the silicon atoms; in some there is a slow intramolecular interconversion between the nonclassical "silane adduct" and the classical hydrido–silyl manganese(III) complex in solution.[149,150] Similar conclusions were drawn from studies of silane dissociation and related chemical behavior.[154-8]

Extended Hückel and Fenske–Hall MO calculations of the electronic structures of a series of $CpMnH(SiR_3)L_2$ complexes suggest that they adopt nonclassical structures best described as "silane adducts" or "sigma complexes." Thus, the manganese atom is best regarded as a $d^6$ $Mn^I$ center with a three-center bond involving the silicon and hydrogen atoms.[159,160]

Valence photoelectron spectra for $CpMnH(SiCl_3)(CO)_2$, $(\eta\text{-}C_5H_4Me)MnH(SiCl_3)(CO)(PMe_3)$, and $(\eta\text{-}C_5H_4Me)MnH(SiHPh_2)L_2$ have been measured to provide further information about the nature of the interaction between the metal and silane, and these studies support a different conclusion.[160-3] The $He^I$ and $He^{II}$ data suggest that for the $SiCl_3$ complexes, there is a formal $d^4$ electron count on the manganese center, which indicates that oxidative addition of the silane molecule has occurred to give manganese(III) products. In contrast, the $SiHPh_2$ complex is best described as a $d^6$ manganese(I) center with a silane ligand.

The trichlorosilane complex $(\eta\text{-}C_5H_4Me)MnH(SiCl_3)(CO)_2$ can be deprotonated by the addition of $NEt_3$, and the resulting manganese(I) anion, $[(\eta\text{-}C_5H_4Me)Mn(SiCl_3)(CO)_2]^-$, reacts with $SnCl_4$, $SnCl_3Ph$, and $SnCl_2Ph_2$ to afford manganese(III) species of stoichiometry $(\eta\text{-}C_5H_4Me)Mn(SiCl_3)(SnCl_{3-n}Ph_n)(CO)_2$ (Scheme 6).[164] In the reaction with excess $SnCl_4$, a second $Mn^{III}$ product, $Cl_2Sn\{Mn(SiCl_3)(CO)_2(\eta\text{-}C_5H_4Me)\}_2$ (**40**), is also obtained. Related complexes of stoichiometry $(\eta\text{-}C_5H_4Me)Mn(SiR_3)(X)(CO)_2$, where X is TMS, $GePh_3$, $SnMe_3$, $PbMe_3$, $AuPPh_3$, HgPh, HgBr, CdCl, ZnCl, or $Fe(CO)_2Cp$ have been isolated by this same route.[165,166] In addition, analogous germyl complexes have been obtained by treating the anionic germyl complex $[(\eta\text{-}C_5H_4Me)Mn(GeCl_3)(CO)_2]^-$ with the appropriate electrophiles.[166] Interestingly, treatment of the anionic stannyl complex $[(\eta\text{-}C_5H_4Me)Mn(SnPh_3)(CO)_2]^-$ with HgBrPh or AuCl(PPh_3) results in loss of the stannyl group and isolation of $(\eta\text{-}C_5H_4Me)Mn(HgPh)_2(CO)_2$ and $(\eta\text{-}C_5H_4Me)MnH(AuPPh_3)(CO)_2$, respectively.[165] An unusual trinuclear complex $[(\eta\text{-}C_5H_4Me)Mn(CO)_2(\mu\text{-}H)]_2Pt(picoline)_2$ has also been prepared by treatment of $[(\eta\text{-}C_5H_4Me)Mn(SiMePh)_2(CO)_2]^-$ with $PtCl_2(picoline)_2$.[165] Treatment of the cadmium and mercury derivatives $(\eta\text{-}C_5H_4Me)Mn(SiMePh_2)(MX)(CO)_2$ with a second equivalent of the anionic silyl species yields the polynuclear complexes $[(\eta\text{-}C_5H_4Me)Mn(SiMePh_2)(CO)_2]_2M$, where M is cadmium or mercury.[166]

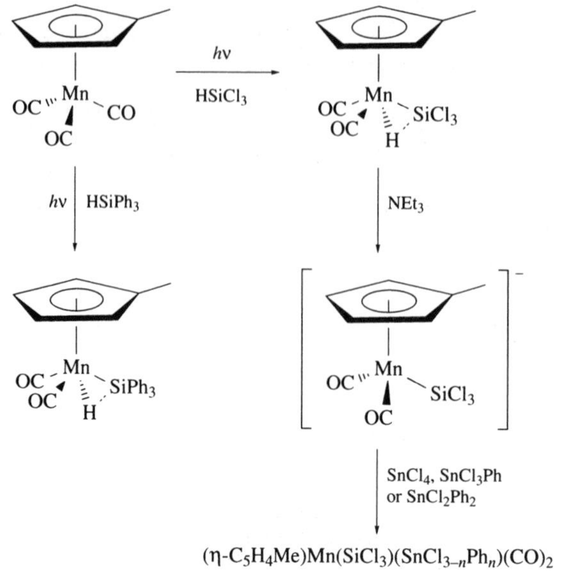

$(\eta\text{-}C_5H_4Me)Mn(SiCl_3)(SnCl_{3-n}Ph_n)(CO)_2$

**Scheme 6**

One related complex, $[CpMn(SnCl_3)(CO)_2(PPh_3)][SnCl_5]$ (**41**), has been obtained by treatment of $CpMn(CO)_2(PPh_3)$ with excess $SnCl_4$, and its structure has been established crystallographically.[167] Other reactions of $(\eta\text{-}C_5R_5)Mn(CO)_{3-n}(PR_3)_n$ complexes with Lewis acids such as $SnCl_4$, $SbCl_3$, $HgCl_2$, and $GeCl_4$ have also been studied by IR spectroscopy.[168] The crystal structure of $(\eta\text{-}C_5H_4Me)\text{-}Mn(GeCl_3)_2(CO)_2$ shows that this species adopts a four-legged piano stool structure with mutually *trans* germyl groups.[169]

### 7.3.3 Other Organomanganese(III) Species

Air oxidation of the *N,N'*-ethylenebis(salicylideneaminato) complex $Mn(salen)(H_2O)_2$ in ethanol/methanol in the presence of perchlorate and hydroxide is reported to yield a cationic manganese(III) carbonyl of stoichiometry $[Mn_2(salen)_2(\mu\text{-}CO)_2][ClO_4]_2$.[170] The IR spectrum ($\nu_{CO} = 1710$ cm$^{-1}$), fast atom bombardment mass spectrum, and magnetic moment (4.48 $\mu_B$ per manganese center) are consistent with this formulation. Treatment of the complex with pyridine/$I_2$ gives a gaseous product, which was identified as CO by GC–MS after differentiating from background $N_2$. High-spin manganese(III) centers should be very poor $\pi$-donors, however, and the binding of carbon monoxide to such a center is extraordinarily surprising. Accordingly, the identity of the complex must be viewed with some suspicion, especially in view of the conditions under which it was prepared.

Whereas the oxidation of manganocene is an irreversible process, the oxidation of decamethylmanganocene readily yields stable salts. The decamethylmanganocenium salt $[(\eta\text{-}Cp^*)_2Mn][PF_6]$[171] was first prepared in the late 1970s, and more recently salts with cyanoalkene and bis(dithiolene)metal anions have been studied owing to their magnetic properties.[172] The salt $[\eta\text{-}Cp^*)_2Mn][TCNQ]$ (TCNQ = 7,7,8,8-tetracyanoquinodimethane) is a rare example of a molecular ferromagnet,[173,174] whereas $[(\eta\text{-}Cp^*)_2Mn][Ni(S_2C_2(CF_3)_2)_2]$ is a metamagnet below 3 K.[175] The reaction of $(\eta\text{-}Cp^*)_2Mn$ with 2,5-dimethyl-*N,N'*dicyanoquinonediimine, $Me_2DCNQI$, gives a charge transfer salt $[(\eta\text{-}Cp^*)_2Mn][Me_2DCNQI]$ that obeys the Curie–Weiss law with $\theta = 15$ K and $\mu_{eff} = 4.16$ $\mu_B$. The donors and acceptors are ferromagnetically coupled and probably form one-dimensional ···ADADA··· stacks.[176]

One other class of organomanganese(III) species, the unstable isonitrile complexes of stoichiometry $[Mn(CNR)_6][X]_3$,[177] was discussed in *COMC-I*.

Finally, manganese(III) species such as $[MnH(\eta^3\text{-}C_5H_7)(dmpe)_2]^+$ and $[MnH(\eta^5\text{-}C_5H_7)\text{-}\{(Me_2PCH_2)_3CMe\}]^+$ have been proposed to be intermediates in the reactions of certain manganese(I) pentadienyl complexes.[93]

## 7.4 ORGANOMANGANESE(IV) COMPOUNDS

Organomanganese compounds in oxidation state +4 are considerably rarer than those in the oxidation state +3. The first example was prepared in 1972, when Bower and Tennent reported the synthesis of the $Mn^{IV}$ compound Mn(bicyclo[2.2.1]hept-1-yl)$_4$ (**42a**) via treatment of $MnBr_2$ with bicyclo[2.2.1]hept-1-yllithium.[178] Although it was not stated how a $Mn^{IV}$ product could have been obtained from $MnBr_2$, it is likely that the initial reaction gave a $Mn^{II}$ alkyl which was subsequently oxidized by adventitious $O_2$. In 1976, Wilkinson reported that deliberate oxidations of binary $Mn^{II}$ alkyls, [MnR$_2$]$_x$, with $O_2$ led to thermally unstable green products that were identified as the analogous $Mn^{IV}$ alkyls Mn(CH$_2$TMS)$_4$, Mn(CH$_2$Bu$^t$)$_4$, and Mn(CH$_2$CPhMe$_2$)$_4$ (**42b–d**) on the basis of their EPR spectra.[9] These $Mn^{IV}$ alkyls can also be prepared in the absence of $O_2$ by disproportionation of a $Mn^{III}$ intermediate obtained by alkylating Mn(acac)$_3$.[115] Apart from certain cyclopentadienyl and carbonyl complexes of $Mn^{III}$, the compounds mentioned above were the only isolable high-valent organomanganese species reported in *COMC-I*.

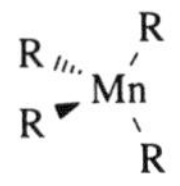

(**42a**) R =

(**42b**) R = CH$_2$TMS
(**42c**) R = CH$_2$Bu$^t$
(**42d**) R = CH$_2$CMe$_2$Ph

In work since 1982 it has been found that alkylation of Mn(acac)$_3$ with methyllithium in the presence of dmpe yields a mixture of the $Mn^{II}$ alkyl MnMe$_2$(dmpe)$_2$ and the octahedral $Mn^{IV}$ alkyl MnMe$_4$(dmpe).[115] The disproportionation reaction is essentially quantitative, and the two products can easily be separated by fractional crystallization. The PMe$_3$ analogue MnMe$_4$(PMe$_3$)$_2$ (**43**) can be synthesized similarly, although in this case the $Mn^{II}$ product is unstable and isolation of the $Mn^{IV}$ alkyl is even simpler. The complexes MnMe$_4$(dmpe) and MnMe$_4$(PMe$_3$)$_2$ are orange-yellow and paramagnetic; they give EPR spectra characteristic of rhombically-distorted $S = \frac{3}{2}$ ions.[115] The $^1$H NMR spectra of these species in C$_6$D$_6$ at 25 °C are shifted and broadened as expected: the dmpe compound exhibits resonances at $\delta$ −23.2 (FWHM = 550 Hz) and at $\delta$ −45.7 (FWHM = 1150 Hz) for the PMe and PCH$_2$ protons, respectively, while the PMe$_3$ compound shows a single resonance at $\delta$ −18.1 (FWHM = 950 Hz) for the phosphine protons.[179] In neither case were the manganese-bound methyl resonances located. The x-ray crystal structure of MnMe$_4$(dmpe) was the first of an organomanganese(IV) compound: the Mn–C distances of 0.212(1) nm for the mutually *trans* methyl groups and 0.207(1) nm for the methyl groups *trans* to phosphorus reflect the differing *trans* influences of the ligands.[115] A preliminary x-ray crystal structure of MnMe$_4$(PMe$_3$)$_2$ confirmed the *cis* geometry.[115]

These $Mn^{IV}$ alkyls can serve as starting materials for the preparation of other organomanganese(IV) compounds. Alkylation of MnMe$_4$(PMe$_3$)$_2$ with 2 equiv. of methyllithium followed by the addition of 2 equiv. of TMEDA gives an orange solution from which the hexamethylmanganate(IV) complex, [Li(TMEDA)]$_2$[MnMe$_6$] (**44**), may be isolated (Scheme 7).[114,116] This salt can also be synthesized from the manganese(III) starting materials Mn(acac)$_3$ or [Li(TMEDA)]$_2$[MnMe$_5$] by $O_2$ oxidation in the presence of excess methyllithium. An extremely pyrophoric salt, [Li(Et$_2$O)$_x$]$_2$[MnMe$_6$], can also be isolated if TMEDA is not added after the alkylation step.[116]

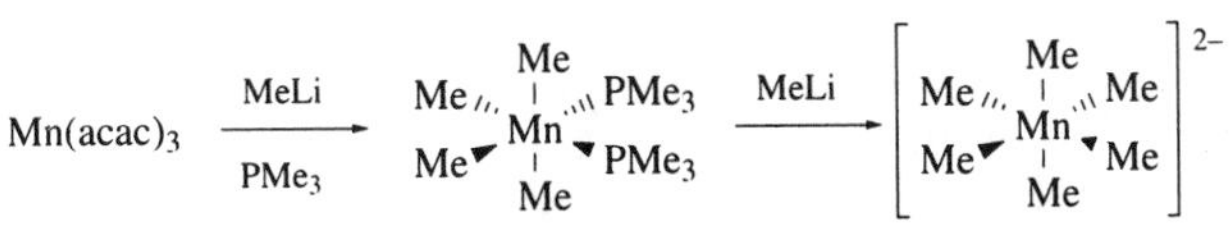

**Scheme 7**

The hexamethylmanganate complex has a magnetic moment of 3.9 $\mu_B$ and its EPR spectrum shows a strong sextet at 340 mT ($A_{Mn}$ = 0.0060 cm$^{-1}$) and a relatively weak half-field transition at 145 mT that are consistent with the presence of a nearly cubic $S = \frac{3}{2}$ ion. The [MnMe$_6$]$^{2-}$ anion is a nearly-ideal octahedron.[114,116]

The reaction of manganocene with dimethylcadmium gives an intermediate of stoichiometry $\{Cp_2Mn\}_2\{CdMe_2\}_3$ as judged from low-temperature (80 K) IR studies; this intermediate undergoes electron and alkyl group transfer at higher temperatures to give the manganese(IV) alkyl $CpMnMe_3$. Reactions of $Cp_2Mn$ with a variety of dialkylcadmium reagents (Equation (21)) have been studied, and $CpMnR_3$ species where R = Me, Et, $Pr^n$, $Bu^n$, and $Bu^i$ (45a–e) have been isolated.[57,180-5] The thermal stabilities of the $CpMnR_3$ complexes decrease in the order listed; the phenyl and triethylgermyl analogues could not be prepared.[182,183] The methyl derivative can be obtained as amber crystals while the other compounds are orange liquids.[183] All are volatile and monomeric in benzene.[41,182] The $CpMnR_3$ compounds react with iodine to give the corresponding iodoalkane, with $HgCl_2$ to give $HgRCl$, and with DCl to give $MnCl_2$, $C_5H_5D$, RD, and some undeuterated alkane and alkene.[183] The ethyl complex $CpMnR_3$ has been investigated as a potential precursor for the chemical vapor deposition of manganese.[41]

$$(Cp)_2Mn \; + \; 3/2 \, CdR_2 \; \longrightarrow \; \underset{\underset{R}{\overset{R-Mn\cdots R}{|}}}{\text{Cp}} \; + \; \text{"}[CdCp]_n\text{"} \; + \; 1/2 \, Cd \qquad (21)$$

(45a) R = Me
(45b) R = Et
(45c) R = $Pr^n$
(45d) R = $Bu^n$
(45e) R = $Bu^i$

Thermolysis of $CpMnMe_3$ yields 1.5 equiv. of methane and 0.5 equiv. of ethane.[184] The mechanism of this reaction has been explored by investigating the thermolysis of the specifically labeled compound $(\eta\text{-}C_5D_5)Mn(Me)_3$. Heating $(\eta\text{-}C_5D_5)Mn(Me)_3$ gives $CH_4$, MeD, and $CH_2D_2$ in a 2:1:1 ratio, so that approximately half of the methane is formed by extracting of hydrogen atoms from the cyclopentadienyl ligand. Thermolysis of the other $CpMnR_3$ complexes gives ~1.5 equiv. of alkene and ~1.0 equiv. of alkene, and the alkanes formed by thermolysis of $(\eta\text{-}C_5D_5)MnEt_3$ and $(\eta\text{-}C_5D_5)Mn(Pr^n)_3$ are mostly undeuterated; this suggests that these complexes decompose via a β-elimination pathway.[184] The manganese-containing solid formed in these thermolysis experiments has also been investigated.[184]

The $Mn^{IV}$ alkyl $Mn(\text{bicyclo}[2.2.1]\text{hept-1-yl})_4$ does not react with cyclopentadiene, but in refluxing tetrahydrofuran it decomposes to yield bicyclo[2.2.1]heptane and a brown-red precipitate which has been identified as the $Mn^{IV}$ vinyloxide complex, $Mn(OCH=CH_2)_4\cdot\frac{1}{2}THF$ (46).[186] The authors propose that tetrahydrofuran molecules are deprotonated by the alkyl groups on manganese, and that subsequent elimination of ethene affords the vinyloxide ligands (as has been seen in certain other systems).

## 7.5 ORGANOMANGANESE(V) AND (VII) COMPOUNDS

Two recent developments have opened up the organometallic chemistry of manganese in oxidation states above +4. When samples of $CpMn(CO)_3$ are condensed with $O_2$ and argon onto a CsI window at 20 K and the frozen matrix irradiated with broad-band UV–visible light, the carbonyl bands disappear from the IR spectrum and new IR bands appear that have been attributed to the organomanganese(V) compound $CpMnO_2$.[187] This complex exhibits IR absorptions at 938 and 893 cm$^{-1}$ that can be assigned to antisymmetric and symmetric O–Mn–O stretching vibrations of the $MnO_2$ unit; the O–Mn–O angle was estimated to be $110 \pm 5°$ from the relative intensities of these bands. The authors were careful to point out, however, that they have only circumstantial evidence that the cyclopentadienyl ring remains attached to the manganese center after photolysis, and it could not be determined whether the ring was bound as an $\eta^1$ or an $\eta^5$ ligand.

In very recent work, the organomanganese(VII) pentafluorophenyl compound $Mn(C_6F_5)(NBu^t)_3$ (47) has been isolated as a brown-green oil by treatment of the manganese(VII) species $MnCl(NBu^t)_3$ with $AgC_6F_5$ (Equation (22)).[188] The pentafluorophenyl complex decomposes slowly at room temperature and is diamagnetic as judged from its $^1H$ NMR spectrum. The existence of these species attests to the strong σ- and π-donor properties of the *t*-butylimido group, which is able to stabilize a variety of transition metals in unusually high oxidation states.

$$\text{MnCl(NBu}^t)_3 + \text{AgC}_6\text{F}_5 \longrightarrow \quad (47) \quad + \text{AgCl} \tag{22}$$

(47)

## 7.6 REFERENCES

1. B. Chiswell, E. D. McKenzie and L. F. Lindoy, in 'Comprehensive Coordination Chemistry', eds. G. Wilkinson, R. D. Gillard and J. A. McCleverty, Pergamon, Oxford, 1987, chap. 41.
2. R. Colton and J. H. Canterford, in 'Halides of the First Row Transition Metals', Wiley, London, 1969, p. 212.
3. G. Cahiez and J. F. Normant, in 'Modern Synthetic Methods', ed. R. Scheffold, Wiley, New York, 1983, vol. 3, p. 173; G. Cahiez, *L'Actualité Chimique*, 1984, 24.
4. G. Cahiez and B. Laboue, *Tetrahedron Lett.*, 1992, **33**, 4439; G. Cahiez, P.-Y. Chavant and E. Metais, *ibid.*, 1992, **33**, 5245.
5. S. U. Koschmieder, G. Wilkinson, B. Hussain-Bates and M. B. Hursthouse, *J. Chem. Soc., Dalton Trans.*, 1992, 19.
6. C. Beerman and K. Clauss, *Angew. Chem.*, 1959, **71**, 627.
7. R. Riemschneider, H.-G. Kassahn and W. Schneider, *Z. Naturforsch., Teil B*, 1960, **15**, 547.
8. M. Tamura and J. Kochi, *J. Organomet. Chem.*, 1971, **29**, 111.
9. R. A. Andersen, E. Carmona-Guzman, J. F. Gibson and G. Wilkinson, *J. Chem. Soc., Dalton Trans.*, 1976, 2204.
10. D. Steinborn, *Z. Chem.*, 1976, **16**, 328.
11. R. J. Morris and G. S. Girolami, *Organometallics*, 1989, **8**, 1478.
12. V. Griehl, E. Anton, H. O. Froelich and U. Boessneck, *East Ger. Pat.* DD278802 (1990) (*Chem. Abstr.*, 1991, **114**, 82 137j).
13. R. A. Bartlett, M. M. Olmstead, P. P. Power and S. C. Shoner, *Organometallics*, 1988, **7**, 1801.
14. W. Seidel and I. Burger, *Z. Chem.*, 1977, **17**, 31.
15. H. Drevs, *ibid.*, 1975, **15**, 451.
16. L. E. Manzer and L. J. Guggenberger, *J. Organomet. Chem.*, 1977, **139**, C34.
17. N. H. Buttrus *et al.*, *J. Chem. Soc., Dalton Trans.*, 1988, 381.
18. H. Gilman and J. C. Bailie, *J. Org. Chem.*, 1937, **2**, 84.
19. H. Gilman and R. H. Kirby, *J. Am. Chem. Soc.*, 1941, **63**, 2046.
20. N. H. Buttrus, C. Eaborn, P. B. Hitchcock, J. D. Smith and A. C. Sullivan, *J. Chem. Soc., Chem. Commun.*, 1985, 1380.
21. R. A. Andersen *et al.*, *Acta Chem. Scand. Ser. A*, 1988, **42**, 554.
22. S. Gambarotta, C. Floriani, A. Chiesi-Villa and C. Guastini, *J. Chem. Soc., Chem. Commun.*, 1983, 1128.
23. T. Kauffmann *et al.*, *Chem. Ber.*, 1993, **126**, 2093.
24. H. Schmidbaur *et al.*, *Organometallics*, 1982, **1**, 1266.
25. R. A. Andersen, A. Haaland, K. Rypdal and H. V. Volden, *J. Chem. Soc., Chem. Commun.*, 1985, 1807.
26. R. A. Jones, S. U. Koschmieder and C. M. Nunn, *Inorg. Chem.*, 1988, **27**, 4524.
27. E. M. Meyer and C. Floriani, *Angew. Chem.*, 1986, **98**, 376; *Angew. Chem., Int. Ed. Engl.*, 1986, **25**, 356.
28. J. I. Davies, C. G. Howard, A. C. Skapski and G. Wilkinson, *J. Chem. Soc., Chem. Commun.*, 1982, 1077.
29. C. G. Howard, G. Wilkinson, M. Thornton-Pett and M. B. Hursthouse, *J. Chem. Soc., Dalton Trans.*, 1983, 2025.
30. G. S. Girolami *et al.*, *J. Chem. Soc., Dalton Trans.*, 1985, 921.
31. C. G. Howard, G. S. Girolami, G. Wilkinson, M. Thornton-Pett and M. B. Hursthouse, *J. Chem. Soc., Dalton Trans.*, 1983, 2631.
32. K. Maruyama, T. Ito and A. Yamamoto, *Bull. Chem. Soc. Jpn.*, 1979, **52**, 849.
33. G. Burkart, Ph.D. Thesis, Ruhr-Universität Bochum, 1985.
34. K. Jacob and K.-H. Thiele, *Z. Anorg. Allg. Chem.*, 1979, **455**, 3.
35. V. A. Shakoor, K. Jacob and K.-H. Thiele, *ibid.*, 1983, **498**, 115.
36. G. S. Girolami, G. Wilkinson, M. Thornton-Pett and M. B. Hursthouse, *J. Am. Chem. Soc.*, 1983, **105**, 6752.
37. G. S. Girolami, G. Wilkinson, A. M. R. Galas, M. Thornton-Pett and M. B. Hursthouse, *J. Chem. Soc., Dalton Trans.*, 1985, 1339.
38. M. T. Reetz, H. Haning and S. Stanchev, *Tetrahedron Lett.*, 1992, **33**, 6963.
39. P. B. Hitchcock, M. F. Lappert, W.-P. Leung and N. H. Buttrus, *J. Organomet. Chem.*, 1990, **394**, 57.
40. G. B. Deacon, R. S. Dickson, J. Latten and B. O. West, *Polyhedron*, 1993, **12**, 497.
41. J. L. Latten, R. S. Dickson, G. B. Deacon, B. O. West and E. R. T. Tiekink, *ibid.*, 1992, **435**, 101.
42. H. Drevs, *J. Organomet. Chem.*, 1992, **433**, C1.
43. K. Jacob, W. Kretschmer, K. H. Thiele, I. Pavlik, A. Lycka and J. Holocek, *Z. Anorg. Allg. Chem.*, 1991, **606**, 133.
44. C. G. Benson, C. A. McAuliffe, A. G. Makie and S. P. Tanner, *Inorg. Chim. Acta*, 1987, **128**, 191.
45. S. M. Godfrey, G. Q. Li, C. A. McAuliffe, P. T. Ndifon and R. G. Pritchard, *ibid.*, 1992, **198–200**, 23.
46. B. Beagley, C. G. Benson, G. A. Gott, C. A. McAuliffe, R. G. Pritchard and S. P. Tanner, *J. Chem. Soc, Dalton Trans.*, 1988, 2261.
47. D. T. Plummer and R. J. Angelici, *Inorg. Chem.*, 1983, **22**, 4063.
48. R. M. Nielson and S. Wherland, *ibid.*, 1985, **24**, 1803.
49. R. M. Nielson and S. Wherland, *ibid.*, 1985, **24**, 3458.
50. R. M. Nielson and S. Wherland, *ibid.*, 1984, **23**, 1338.
51. R. M. Nielson and S. Wherland, *J. Am. Chem. Soc.*, 1985, **107**, 1505.
52. R. M. Nielson, J. P. Hunt, H. W. Dodgen and S. Wherland, *Inorg. Chem.*, 1986, **25**, 1964.

53. R. M. Nielson and S. Wherland, *ibid.*, 1986, **25**, 2437.
54. M. Stebler, R. M. Nielson, W. F. Siems, J. P. Hunt, H. W. Dodgen and S. Wherland, *ibid.*, 1988, **27**, 2893.
55. F. J. Wu, B. C. Berris and D. R. Bell, *US Pat.* 4 946 975 (1990) (*Chem. Abstr.*, 1991, **114**, 82 136h).
56. H. Handwerker, H. Beruda, M. Kleine and C. Zybill, *Organometallics*, 1992, **11**, 3542.
57. V. P. Maryin, *J. Organomet. Chem.*, 1992, **441**, 241.
58. F. H. Köhler, N. Hebendanz, U. Thewalt, B. Kanellakopulos and R. Klenze, *Angew. Chem.*, 1984, **96**, 697; *Angew. Chem., Int. Ed. Engl.*, 1984, **23**, 721.
59. F. H. Köhler, N. Hebendanz, G. Müller and U. Thewalt, *Organometallics*, 1987, **6**, 115.
60. J. Heck, W. Massa and P. Weinig, *Angew. Chem.*, 1984, **96**, 699; *Angew. Chem., Int. Ed. Engl.*, 1984, **23**, 722.
61. A. Winter, G. Huttner, L. Zsolnai, P. Kroneck and M. Gottlieb, *Angew. Chem.*, 1984, **96**, 986; *Angew. Chem., Int. Ed. Engl.*, 1984, **23**, 975.
62. A. Winter, G. Huttner, M. Gottlieb and I. Jibril, *J. Organomet. Chem.* 1985, **286**, 317.
63. P. Lau *et al.*, *Organometallics*, 1991, **10**, 3861.
64. H. Braunwarth *et al.*, *J. Organomet. Chem.*, 1991, **411**, 383.
65. P. Lau, G. Huttner and L. Zsolnai, *ibid.*, 1992, **440**, 41.
66. G. Wilkinson, F. A. Cotton and J. M. Birmingham, *J. Inorg. Nucl. Chem.*, 1956, **2**, 95.
67. F. H. Köhler and N. Hebendanz, *Chem. Ber.*, 1983, **116**, 1261.
68. J. H. Ammeter *et al.*, *Helv. Chim. Acta*, 1981, **64**, 1063.
69. N. Hebendanz, F. H. Köhler, G. Müller and J. Riede, *J. Am. Chem. Soc.*, 1986, **108**, 3281.
70. D. Cozak, F. Gauvin and J. Demers, *Can. J. Chem.*, 1986, **64**, 71.
71. A. Haaland, *Inorg. Nucl. Chem. Lett.*, 1979, **15**, 267.
72. A. Almenningen, A. Haaland and S. Samdal, *J. Organomet. Chem.*, 1978, **149**, 219.
73. N. Augart, R. Boese and G. Schmid, *Z. Anorg. Allg. Chem.*, 1991, **595**, 27.
74. D. P. Freyberg, J. L. Robbins, K. N. Raymond and J. C. Smart, *J. Am. Chem. Soc.*, 1979, **101**, 892.
75. E. König, V. P. Desai, B. Kanellakopulos and R. Klenze, *Chem. Phys.*, 1980, **54**, 109.
76. L. Fernholt, A. Haaland, R. Seip, J. L. Robbins and J. C. Smart, *J. Organomet. Chem.*, 1980, **194**, 351.
77. J. Blümel, P. Hofmann and F. H. Köhler, *Magn. Res. Chem.*, 1993, **31**, 2.
78. N. Hebendanz, F. H. Köhler, F. Scherbaum and B. Schlesinger, *ibid.*, 1989, **27**, 798.
79. F. H. Köhler and B. Schlesinger, *Inorg. Chem.*, 1992, **31**, 2853.
80. D. Cozak and F. Gauvin, *Organometallics*, 1987, **6**, 1912.
81. S. Y. Ketkov, G. A. Domrachev and V. P. Mar'in, *Metalloorg. Khim.*, 1990, **3**, 394; *Organomet. Chem. USSR*, 1990, **3**, 192.
82. S. Y. Ketkov and G. A. Domrachev, *Inorg. Chim. Acta*, 1990, **178**, 233.
83. C. Cauletti *et al.*, *J. Electron Spectrosc. Relat. Phenom.*, 1980, **19**, 327.
84. J. C. Green, *Struct. Bonding (Berlin)*, 1981, **43**, 37.
85. M. F. Ryan, J. R. Eyler and D. E. Richardson, *J. Am. Chem. Soc.*, 1992, **114**, 8611.
86. D. E. Richardson, C. S. Christ, P. Sharpe and J. R. Eyler, *ibid.*, 1987, **109**, 3894.
87. M. S. Tunuli, *Int. J. Mass Spectrom. Ion Proc.*, 1986, **72**, 249.
88. T. Gennett, D. F. Milner and M. J. Weaver, *J. Phys. Chem.*, 1985, **89**, 2787.
89. M. S. Kralik, L. Stahl, A. M. Arif, C. E. Strouse and R. D. Ernst, *Organometallics*, 1992, **11**, 3617.
90. M. C. Böhm, R. D. Ernst, R. Gleiter and D. R. Wilson, *Inorg. Chem.*, 1983, **22**, 3815.
91. A. R. Oki, H. Zhang, N. S. Hosmane, H. Ro and W. E. Hatfield, *J. Am. Chem. Soc.*, 1991, **113**, 8531.
92. J. R. Bleeke and J. J. Kotyk, *Organometallics*, 1983, **2**, 1263.
93. J. R. Bleeke and J. J. Kotyk, *ibid.*, 1985, **4**, 194.
94. C. G. Howard, G. S. Girolami, G. Wilkinson, M. Thornton-Pett and M. B. Hursthouse, *J. Am. Chem. Soc.*, 1984, **106**, 2033.
95. F. Bottomley, P. N. Keizer and P. S. White, *J. Am. Chem. Soc.*, 1988, **110**, 137.
96. F. Calderazzo, G. Pampaloni and P. F. Zanazzi, *Chem. Ber.*, 1986, **119**, 2796.
97. T. Lindblad and B. Rebenstorf, *J. Chem. Soc., Faraday Trans.*, 1991, **87**, 2473.
98. I. I. Grinval'd, B. V. Lokshin, N. K. Rudnevskii and V. P. Mar'in, *Izv. Akad. Nauk SSSR, Ser. Khim.*, 1988, 2298; *Bull. Acad. Sci. USSR, Div. Chem. Sci.*, 1988, **37**, 2068.
99. I. I. Grinval'd, B. V. Lokshin, N. K. Rudnevskii, V. P. Mar'in, A. A. Alad'in and G. A. Razuvaev, *Dokl. Akad. Nauk SSSR*, 1988, **298**, 366; *Proc. Acad. Sci. USSR*, 1988, **298**, 21.
100. A. Belforte, F. Calderazzo and P. F. Zanazzi, *J. Chem. Soc., Dalton Trans.*, 1988, 2921.
101. J. T. Weed, M. F. Rettig and R. M. Wing, *J. Am. Chem. Soc.*, 1983, **105**, 6510.
102. I. I. Grinval'd, B. V. Lokshin, V. P. Mar'in, L. I. Vyshinskaya and G. A. Razuvaev, *Dokl. Akad. Nauk SSSR*, 1987, **296**, 111; *Proc. Acad. Sci. USSR*, 1987, **296**, 397.
103. I. I. Grinval'd, B. V. Lokshin, O. A. Usacheva, I. A. Abronin, L. M. Golubinskaya and V. I. Bregadze, *Metalloorg. Khim.*, 1990, **3**, 166; *Organomet. Chem. USSR*, 1990, **3**, 90.
104. H. Werner and B. Juthani, *J. Organomet. Chem.*, 1977, **129**, C39.
105. R. G. Bulgakov, S. P. Kuleshov, V. N. Yakovlev, G. Y. Maistrenko, G. A. Tolstikov and V. P. Kazakov, *Izv. Akad. Nauk SSSR, Ser. Khim.*, 1986, 2216; *Bull. Acad. Sci. USSR., Div. Chem. Sci.*, 1986, **35**, 2022.
106. I. I. Grinval'd, B. V. Lokshin, N. K. Rudnevskii and V. M. Fomin, *Izv. Akad. Nauk SSSR, Ser. Khim.*, 1988, 58; *Bull. Acad. Sci. USSR., Div. Chem. Sci.* 1988, **37**, 49.
107. I. I. Grinval'd, B. V. Lokshin, N. K. Rudnevskii and V. P. Mar'in, *Dokl. Akad. Nauk SSSR*, 1988, **298**, 1142; *Proc. Acad. Sci. USSR*, 1988, **298**, 52.
108. K. Jonas, G. Burkart, C. Häselhoff, P. Betz and C. Krüger, *Angew. Chem.*, 1990, **102**, 291; *Angew. Chem., Int. Ed. Engl.*, 1990, **29**, 322.
109. K. Jonas, C.-C. Häselhoff, R. Goddard and C. Krüger, *Inorg. Chim. Acta*, 1992, **198–200**, 533.
110. A. A. Pasynskii, I. L. Eremenko, G. S. Gasanov, Y. T. Struchkov and V. E. Shklover, *J. Organomet. Chem.*, 1984, **276**, 349.
111. I. L. Eremenko, A. A. Pasynskii, G. S. Gasanov, B. Orazsakhatov, Y. T. Struchkov and V. E. Shklover, *ibid.* 1984, **275**, 183.
112. A. A. Pasynskii *et al.*, *Koord. Khim.*, 1984, **10**, 634; *Sov. J. Coord. Chem.*, 1984, **10**, 347.
113. L. V. Pankratov *et al.*, *Izv. Akad. Nauk SSSR, Ser. Khim.*, 1986, 2548; *Bull. Acad. Sci. USSR, Div. Chem. Sci.*, 1986, **35**, 2334.
114. R. J. Morris and G. S. Girolami, *Organometallics*, 1991, **10**, 792.

115. C. G. Howard, G. S. Girolami, G. Wilkinson, M. Thornton-Pett and M. B. Hursthouse, *J. Chem. Soc., Chem. Commun.*, 1983, 1163.
116. R. J. Morris and G. S. Girolami, *J. Am. Chem. Soc.*, 1988, **110**, 6245.
117. M. Benard, *J. Am. Chem. Soc.*, 1978, **100**, 2354.
118. R. J. Morris and G. S. Girolami, *Polyhedron*, 1988, **7**, 2001.
119. G. L. Tembe, K. Jacob, W. Brüser, D. Schulz and K.-H. Thiele, *Z. Anorg. Allg. Chem.*, 1987, **554**, 132.
120. R. J. Morris and G. S. Girolami, *Organometallics*, 1991, **10**, 799.
121. R. J. Morris and G. S. Girolami, *ibid.*, 1987, **6**, 1815.
122. K. Dey and R. L. De, *J. Inorg. Nucl. Chem.*, 1977, **39**, 153.
123. C. Perthuisot, M. Fan and W. D. Jones, *Organometallics*, 1992, **11**, 3622.
124. S. Komiya and M. Kaneda, *J. Organomet. Chem.*, 1988, **340**, C8
125. S. Komiya, T. Kimura and M. Kaneda, *Organometallics*, 1991, **10**, 1311.
126. G. Huttner and K. Evertz, *Acc. Chem. Res.*, 1986, **19**, 406.
127. G. Huttner, *Pure Appl. Chem.*, 1986, **58**, 585.
128. A. M. Rosan, *J. Chem. Soc., Chem. Commun.*, 1981, 311.
129. B. Buchmann and A. Salzer, *J. Organomet. Chem.*, 1985, **295**, 63.
130. B. V. Lokshin, A. G. Ginzburg, V. N. Setkina, D. N. Kursanov and I. B. Nemirovskaya, *ibid.*, 1972, **37**, 347.
131. A. G. Ginzburg, L. A. Fedorov, P. V. Petrovskii, E. I. Fedin, V. N. Setkina and D. N. Kursanov, *ibid.*, 1974, **73**, 77.
132. A. G. Ginzburg, P. O. Okulevich, V. N. Setkina, G. A. Panosyan and D. N. Kursanov, *ibid.*, 1974, **81**, 201.
133. T. C. Flood, E. Rosenberg and A. Sarhangi, *J. Am. Chem. Soc.*, 1977, **99**, 4334.
134. U. Schubert, *Adv. Organomet. Chem.*, 1990, **30**, 151.
135. W. L. Hutcheon, Ph. D. Thesis, University of Alberta, Edmonton, 1971; *Chem. Eng. News*, 1970, **48**(24), 75.
136. W. Jetz and W. A. G. Graham, *Inorg. Chem.*, 1971, **10**, 4.
137. A. J. Hart-Davis and W. A. G. Graham, *J. Am. Chem. Soc.*, 1971, **93**, 4388.
138. D. F. Dong, J. K. Hoyano and W. A. G. Graham, *Can. J. Chem.*, 1981, **59**, 1455.
139. W. A. G. Graham, *J. Organomet. Chem.*, 1986, **300**, 81.
140. E. Colomer, R. Corriu and A. Vioux, *J. Chem. Res. (S)*, 1977, 168; *J. Chem. Res. (M)*, 1977, 1939.
141. F. Carré, E. Colomer, R. J. P. Corriu and A. Vioux, *Organometallics*, 1984, **3**, 1272.
142. U. Schubert, G. Kraft and C. Kalbas, *Transition Met. Chem. (Weinheim)*, 1984, **9**, 161.
143. G. Kraft, C. Kalbas and U. Schubert, *J. Organomet. Chem.*, 1985, **289**, 247.
144. E. Colomer, R. J. P. Corriu and A. Vioux, *Inorg. Chem.*, 1979, **18**, 695.
145. E. Colomer, R. J. P. Corriu and A. Vioux, *Angew. Chem.*, 1981, **93**, 488; *Angew. Chem., Int. Ed. Engl.*, 1981, **20**, 476.
146. U. Schubert, E. Kunz, B. Harkers, J. Willnecker and J. Meyer, *J. Am. Chem. Soc.*, 1989, **111**, 2572.
147. U. Schubert, K. Ackermann and B. Wörle, *ibid.*, 1982, **104**, 7378.
148. U. Schubert, K. Ackermann, G. Kraft and B. Wörle, *Z. Naturforsch., Teil B, Anorg. Chem. Org. Chem.*, 1983, **38**, 1488.
149. U. Schubert, G. Kraft and E. Walther, *Z. Anorg. Allg. Chem.*, 1984, **519**, 96.
150. U. Schubert, G. Scholz, J. Müller, K. Ackermann, B. Wörle and R. F. D. Stansfield, *J. Organomet. Chem.*, 1986, **306**, 303.
151. R. A. Smith and M. J. Bennett, *Acta Crystallogr., Sect. B*, 1977, **33**, 1113.
152. R. A. Smith and M. J. Bennett, *ibid.*, 1977, **33**, 1118.
153. M. A. Andrews, S. W. Kirtley and H. Kaesz, *Adv. Chem. Ser.*, 1978, **167**, 215.
154. E. Colomer, R. J. P. Corriu, C. Marzin and A. Vioux, *Inorg. Chem.*, 1982, **21**, 368.
155. E. Colomer and R. J. P. Corriu, *Topics Curr. Chem.*, 1981, **96**, 79.
156. M. Cowie and M. J. Bennett, *Inorg. Chem.*, 1977, **16**, 2325.
157. D. M. Hester, J. Sun, A. W. Harper and G. K. Yang, *J. Am. Chem. Soc.*, 1992, **114**, 5234.
158. R. H. Hill and M. S. Wrighton, *Organometallics*, 1987, **6**, 632.
159. H. Rabaâ, J.-Y. Saillard and U. Schubert, *J. Organomet. Chem.*, 1987, **330**, 397.
160. D. L. Lichtenberger and A. Rai-Chaudhuri, *J. Am. Chem. Soc.*, 1989, **111**, 3583.
161. D. L. Lichtenberger and A. Rai-Chaudhuri, *Inorg. Chem.*, 1990, **29**, 975.
162. D. L. Lichtenberger and A. Rai-Chaudhuri, *J. Am. Chem. Soc.*, 1990, **112**, 2492.
163. D. L. Lichtenberger and A. Rai-Chaudhuri, *Organometallics*, 1990, **9**, 1686.
164. W. Jetz and W. A. G. Graham, *Inorg. Chem.*, 1971, **10**, 1647.
165. E. Kunz, M. Knorr, J. Willnecker and U. Schubert, *New J. Chem.*, 1988, **12**, 467.
166. E. Kunz and U. Schubert, *Chem. Ber.*, 1989, **122**, 231.
167. A. G. Ginzburg, N. G. Bokyi, A. I. Yanovsky, Y. T. Struchkov, V. N. Setkina and D. N. Kursanov, *J. Organomet. Chem.*, 1977, **136**, 45.
168. A. G. Ginzburg, B. V. Lokshin, V. N. Setkina and D. N. Kursanov, *J. Organomet. Chem.*, 1973, **55**, 357.
169. W. Gäde and E. Weiss, *Chem. Ber.*, 1981, **114**, 2399.
170. F. M. Ashmawy, C. A. McAuliffe, K. L. Minten, R. V. Parish and T. James, *J. Chem. Soc., Chem. Commun.*, 1983, 436.
171. J. L. Robbins, N. Edelstein, B. Spencer and J. C. Smart, *J. Am. Chem. Soc.*, 1982, **104**, 1882.
172. J. S. Miller and A. J. Epstein, *Angew. Chem.*, 1994, **106**, 399; *Angew. Chem., Int. Ed. Engl.*, 1994, **33**, 385.
173. W. E. Broderick, J. A. Thompson, E. P. Day and B. M. Hoffman, *Science*, 1990, **249**, 401.
174. G. T. Yee *et al.*, *Adv. Mater.*, 1991, **3**, 309.
175. W. E. Broderick, J. A. Thompson and B. M. Hoffman, *Inorg. Chem.*, 1991, **30**, 2958.
176. J. S. Miller, C. Vazquez, R. S. McLean, W. M. Reiff, A. Aumüller and S. Hünig, *Adv. Mater.*, 1993, **5**, 448.
177. P. M. Treichel and H. J. Mueh, *Inorg. Chem.*, 1977, **16**, 1167.
178. B. K. Bower and H. G. Tennent, *J. Am. Chem. Soc.*, 1972, **94**, 2512.
179. R. J. Morris and G. S. Girolami, Ph.D. Thesis, University of Illinois, Urbana-Champaign, 1990.
180. V. P. Mar'in and L. I. Vyshinskaya, *Metalloorg. Khim.*, 1992, **5**, 331; *Organomet. Chem. USSR*, 1992, **5**, 155.
181. G. A. Razuvaev, S. G. Yudenich, O. N. Druzhkov and V. A. Dodonov, *Zh. Obshch. Khim.*, 1986, **56**, 1666; *J. Gen. Chem. USSR*, 1986, **56**, 1477.
182. G. A. Razuvaev, V. P. Mar'in, L. I. Vyshinskaya, I. I. Grinval'd and N. N. Spiridonova, *Zh. Obshch. Khim.*, 1987, **57**, 1773; *J. Gen. Chem. USSR*, 1987, **57**, 1583.

183. G. A. Razuvaev, V. P. Mar'in, N. N. Spiridonova, L. I. Vyshinskaya and I. I. Grinval'd, *Dokl. Akad. Nauk SSSR*, 1986, **289**, 378; *Proc. Acad. Sci. USSR*, 1986, **289**, 276.
184. G. A. Razuvaev, L. I. Vyshinskaya and V. P. Mar'in, *Dokl. Akad. Nauk SSSR*, 1986, **289**, 1388; *Proc. Acad. Sci. USSR*, 1986, **289**, 339.
185. G. A. Razuvaev, V. P. Mar'in and L. I. Vyshinskaya, Abstract Proceedings of the XIIIth International Conference on Organometallic Chemistry, Torino, 1988, p. 420.
186. K. Jacob and K.-H. Thiele, *Z. Anorg. Allg. Chem.*, 1988, **564**, 81.
187. M. J. Almond, R. W. Atkins and R. H. Orrin, *J. Chem. Soc., Dalton Trans.*, 1994, 311.
188. A. A. Danopoulos, G. Wilkinson, T. K. N. Sweet and M. B. Hursthouse, *J. Chem. Soc., Dalton Trans.*, 1994, 1037.

# 8
# Technetium

ALFRED P. SATTELBERGER
*Los Alamos National Laboratory, NM, USA*
and
JEFFREY C. BRYAN
*Oak Ridge National Laboratory, TN, USA*

## 8.1 INTRODUCTION

The chemistry of technetium is decidedly underdeveloped relative to the chemistry of the other transition elements. All isotopes of element 43 are unstable with respect to β decay or electron capture and possess half-lives that are short on the geologic timescale, thereby precluding the existence of primordial technetium. Three isotopes have half-lives greater than $10^5$ y, but the only isotope available in macroscopic quantities is $^{99}$Tc ($t_{1/2} = 2.12 \times 10^5$ y). The latter isotope is recovered during the processing of spent nuclear fuel. It is also found in the earth's crust due to spontaneous fission of $^{238}$U as well as by slow neutron-induced fission of $^{235}$U. $^{99}$Tc is a weak β emitter ($E_{max} = 0.292$ MeV) and decays to stable $^{99}$Ru; the β emission is relatively pure, but may be accompanied by secondary x rays (bremsstrahlung), particularly when dealing with multimilligram quantities of the isotope. The nuclear spin of $^{99}$Tc is +9/2 and the nuclear magnetic moment is +5.66 nuclear magnetons. The handling of $^{99}$Tc on a small scale (<100 mg) does not present a significant health hazard provided some elementary precautions are taken. These include working in well-ventilated fume hoods or inert-atmosphere glove boxes and wearing safety glasses. The walls of ordinary laboratory glassware are sufficient to stop most of the β emission. If practical, it is a good idea to carry out reactions with technetium compounds in closed systems. Special precautions are required when dealing with volatile technetium compounds such as [Tc(CO)$_3$(η-Cp)] or finely divided solids, since inhalation or ingestion are the primary radiological hazards in working with this element. We also note that it is incredibly easy to contaminate an average-sized laboratory if established radiation-handling procedures are not followed. Researchers should consult local safety regulations dealing with ionizing radiation before considering experimental work with $^{99}$Tc.

Much of the technetium research reported in the literature is coupled to applications. The most important of these is nuclear medicine. Compounds containing the metastable isotope $^{99m}$Tc, which decays to $^{99}$Tc with a 6 h half-life, are used millions of times each year to γ image various organs of the human body. Such imaging is considered far safer to the patient than exploratory surgery or other invasive techniques. Not all of the body's organs are accessible by $^{99m}$Tc agents currently available, and many existing agents could be substantially improved if their degree of localization in the target organ (biodistribution) could be increased. As a consequence, there is considerable industrial interest in developing, testing, and marketing new agents. There is also considerable interest in the environmental chemistry of technetium as well as the behavior and distribution of technetium in nuclear waste-processing schemes. Due to the high fission yield (> 6%), appreciable quantities of $^{99}$Tc are isolated from uranium fission product mixtures. Nuclear reactors with a power of 100 MW produce over 2 g of $^{99}$Tc per day. The first gram of technetium was isolated in the 1950s.

Due to the aforementioned safety considerations as well as the expense associated with setting up and maintaining radiation laboratories (including waste disposal!), research into fundamental technetium chemistry has generally been a luxury few could afford to indulge in. Organotechnetium chemistry has been particularly slow to develop. On average only four organotechnetium references per year appeared in the literature from 1980 through to 1987. Since then the average has increased to roughly 13 per year. This is due, at least in part, to the discovery by the Davison group at the Massachusetts Institute of Technology that hexakis(isocyanide)technetium(I) cations are good heart-imaging agents.[1] One derivative, marketed by E. I. duPont under the trade name Cardiolite, is now commonly used by radiopharmacies. Since the discovery of Cardiolite, other organotechnetium compounds have been patented as imaging agents.

Our review was facilitated by a number of review articles published primarily in the early 1980s.[2–24] Many of these focused on nuclear medicine applications[2–6] or specific aspects of technetium coordination chemistry,[7–21] although a few provide a general overview of technetium chemistry.[22–4] All of these reviews appeared within a few years of each other, a fact that led one author to write "Despite the fact that the chemistry of the element [Tc] is still poorly developed ..., there is no shortage of review articles."[19] Three volumes of symposia proceedings on technetium in chemistry and nuclear medicine have also been published.[25–7] Since the work contained in these volumes has generally been presented elsewhere in more accessible journals, we have chosen not to reference specific papers from these proceedings. Our review of organotechnetium chemistry primarily covers the years 1980 through to 1993.

## 8.2 TECHNETIUM CARBONYLS AND THEIR HALIDE DERIVATIVES

### 8.2.1 Binary and Mixed-metal Carbonyls

Ditechnetium decacarbonyl remains a key starting material for low-valent technetium chemistry. Several new syntheses of [Tc$_2$(CO)$_{10}$] (**1**) have been reported[28–32] attempting to improve on the rather severe reaction conditions (350 atm, 220 °C) and low yields (10–50%) of the original preparation.[33] Reaction conditions and yields for the new preparations of (**1**) are summarized in Table 1.

(**1**)

**Table 1** Reaction conditions for the preparation of [Tc$_2$(CO)$_{10}$] (1).

| Reagents | Solvent | CO pressure (atm) | Temperature (°C) | Time (h) | Yield (%) | Ref. |
|---|---|---|---|---|---|---|
| NaTcO$_4$ | MeOH | 190 | 230 | 48 | 84 | 28 |
| NH$_4$TcO$_4$, Na/Hg | THF | 100 | 120 | 72 | 70 | 29 |
| NH$_4$TcO$_4$ | None | 420 | 210 | 48 | 80–90 | 30 |
| NH$_4$TcO$_4$, H$_2$ | Toluene | 140 | 200 | 4 | 96 | 31 |
| TcO$_2$ on Cu | None | 300 | 200 | NA[a] | NA[a] | 32 |
| Tc$_2$O$_7$ | None | 350 | 220 | 20 | 10 | 33 |
| TcO$_2$ | None | 35 | 250 | 24 | 50 | 33 |

[a] NA, not available.

Enthalpic data for (1) were estimated based on values obtained for other dimetal decacarbonyl compounds of the manganese triad. The Tc–CO bond dissociation energy is estimated to be 133 kJ mol$^{-1}$.[34] More recently, several physical measurements were made on [Tc$_2$(CO)$_{10}$] (1). Thermal gravimetric analysis of (1) at 220–260 °C indicates loss of CO, leaving a residue formulated as "Tc(TcC)."[35] The gas-phase IR spectrum of (1) has been reported. The vibrational frequencies obtained were very similar to those observed in solution, suggesting that the gas-phase and solution structures are closely related. Good intensities were obtained with temperatures as low as 50–55 °C, demonstrating reasonable volatility.[36] Volatilities of (1) and related compounds were quantified by examining the temperature dependence of the vapor pressure. The vapor pressure at room temperature for (1) is estimated to be of the order of 0.1 Pa.[37]

[Tc$_2$(CO)$_{10}$] (1) was examined as a potential alkene epoxidation catalyst, using O$_2$[38] or Bu$^t$O$_2$H[32] as the oxygen source. Low yields of epoxide (13% and 32%, respectively) were blamed on catalytic, radical decomposition of the epoxide[38] or metal-catalyzed decomposition of the peroxide.[32]

Reduction of (1) with sodium amalgam gives the pentacarbonyl anion [Tc(CO)$_5$]$^-$, which reacts with rhenium or manganese pentacarbonyl bromides ([MBr(CO)$_5$]; M = Mn, Re) to give heteronuclear decacarbonyl compounds: [TcM(CO)$_{10}$]. These compounds were characterized by IR and mass spectra.[39] In a related study the IR spectra were calculated for [MCo(CO)$_9$] (M = Mn, Re, Tc). The technetium-containing compound exhibits the highest C–O stretching force constant and has the greatest degree of M–Co polarization.[40]

### 8.2.2 Halo and Hydrido Technetium Carbonyls

The only mention of a hydrido technetium carbonyl compound is a theoretical study of the protonation of [Tc(CO)$_5$]$^-$ (along with other $d^8$ pentacarbonyl and $d^{10}$ tetracarbonyl complexes) to form [TcH(CO)$_5$]. The total protonation energy for [Tc(CO)$_5$]$^-$ is given as 1260 kJ mol$^{-1}$, a value intermediate between those calculated for [Mn(CO)$_5$]$^-$ and [Re(CO)$_5$]$^-$.[41]

Halocarbonyls constitute an important route into organotechnetium chemistry. A new synthesis of the known [TcBr(CO)$_5$] was reported: [Tc$_2$(CO)$_{10}$] (1) reacts with Br$_2$ in CS$_2$, providing the bromocarbonyl in a 55% yield.[39] A few physical measurements have been reported for technetium halocarbonyl compounds. The chemical shift of the technetium x-ray $K$ lines was investigated for a wide variety of compounds. Two organometallic compounds, [TcCl(CO)$_5$] and [TcBr(CO)$_5$], were included in this study, exhibiting shifts of $K\alpha_1$ of $-46 \pm 12$ meV and $-50 \pm 12$ meV relative to technetium metal.[42] Volatility of [TcI(CO)$_5$], [TcCl(CO)$_3$]$_4$, and related compounds was quantified by examining the temperature dependence of the vapor pressure. The vapor pressures at room temperature for [TcI(CO)$_5$] and [TcCl(CO)$_3$]$_4$ are estimated to be approximately 0.15 Pa.[37]

The pentacarbonyl halides (2) are decarbonylated in a stepwise manner by heating in noncoordinating solvents or by vacuum sublimation. Decarbonylation initially results in formation of the dimer [TcX(CO)$_4$]$_2$ (3) (X = Cl, Br, I), and ultimately the tetramer [TcX(CO)$_3$]$_4$ (4) (Scheme 1). Ease of decarbonylation decreases in the order Cl > Br > I. This correlates with an increase in the CO stretching frequency observed in the IR spectra of (2). The higher $\nu_{CO}$, the more labile the carbonyl ligand becomes. Comparisons to other [MX(CO)$_5$] (M = Mn, Re) indicate that the technetium compounds are the most susceptible to decarbonylation.[43,44] Fragmentation patterns in the mass spectra of (2) (X = Cl, Br, I) underscore their facile decarbonylation.[45] *Ab initio* SCF calculations probing the mechanism of photolysis of [TcCl(CO)$_5$] to *cis*-[TcCl(CO)$_4$L'] (L' = incident nucleophile) suggests decarbonylation of the CO *trans* to the chloride as the first step. Subsequent pseudorotation and reaction with L' yields the observed *cis* product.[46] The structure of the tetramer (**4**) was determined.[47] The structure of the dimer (**3**) is based on spectroscopic and analytical data, and by analogy to known rhenium structures.[43,44]

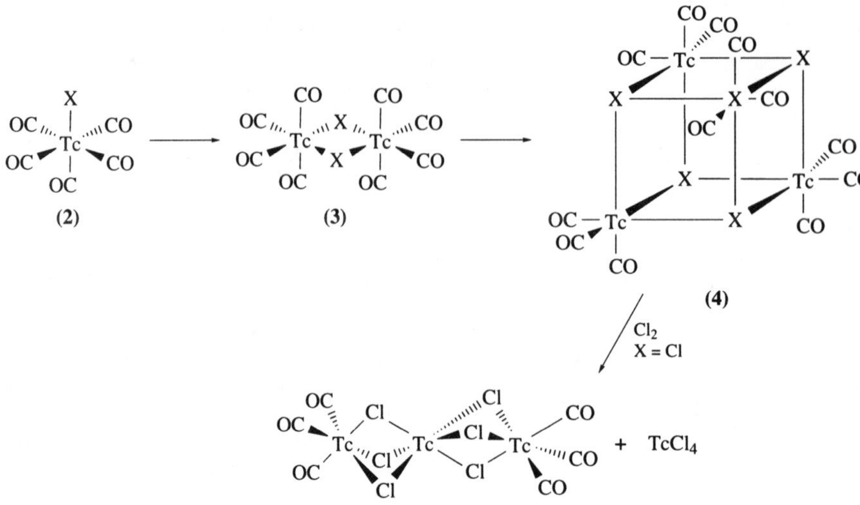

**Scheme 1**

The reactivities of [TcX(CO)$_3$]$_4$ (**4**) toward X$_2$ (X = Cl, Br, I) were studied by subliming the technetium tetramer in a stream of halogen. The reaction of [TcCl(CO)$_3$]$_4$ (**4**) (X = Cl) with Cl$_2$ forms TcCl$_4$ and the mixed-valence trimer [(CO)$_3$Tc($\mu$-Cl)$_3$Tc($\mu$-Cl)$_3$Tc(CO)$_3$] (**5**) as depicted in Scheme 1. Reaction of [TcBr(CO)$_3$]$_4$ with Br$_2$ yields only [TcBr$_3$]$_n$, and the reaction of [TcI(CO)$_3$]$_4$ with I$_2$ gives uncharacterized products.[48]

## 8.3 OTHER TECHNETIUM CARBONYL DERIVATIVES

### 8.3.1 Oxygen and Sulfur

When Na[TcO$_4$] is carbonylated in methanol under conditions that are too mild (135 atm CO, 150 °C, 2 h) to produce [Tc$_2$(CO)$_{10}$] (**1**), a novel trinuclear technetium(I) cluster, Na[Tc$_3$(OMe)$_4$(CO)$_9$] (**6**), is formed, as shown in Equation (1).[28] Subsequent carbonylation of (**6**), under more severe conditions (190 atm CO, 230 °C, 2 d), gives (**1**). This cluster (**6**) is an attractive starting material for compounds containing the "Tc(CO)$_3$$^+$" fragment (see Section 8.5) as it is easily broken up by a proton source.[28,49] The closely related compound [Tc(OH)(CO)$_3$]$_4$ was prepared by heating K[TcO$_4$] in an excess of formic acid and was assigned a cubane structure (**4**) (X = OH) based on elemental analysis, IR spectroscopy, MS, powder diffraction, and analogy to related rhenium compounds.[45,50]

$$\text{NaTcO}_4 \ + \ \text{CO} \ \xrightarrow[\text{[Cu]}]{\text{MeOH}} \quad (6) \tag{1}$$

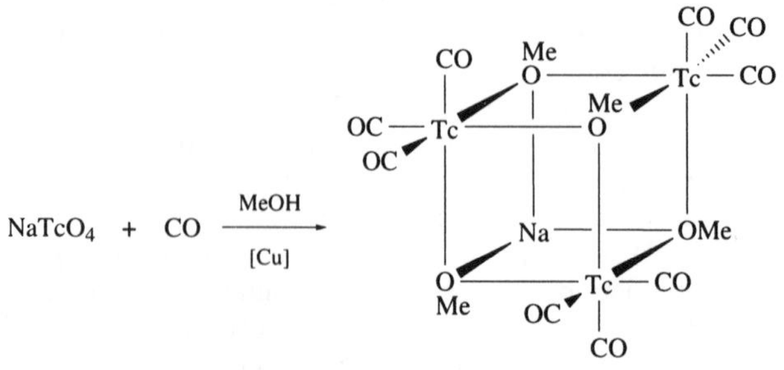

A series of technetium(I) carbonyl β-diketonates have been prepared and their behavior under vacuum sublimation examined. Three types of compounds were prepared: [Tc(O⌢O)(CO)$_4$], [Tc(O⌢O)(CO)$_3$]$_2$, and *fac*-[Tc(O⌢O)(CO)$_3$L] (O⌢O = acac, tfacac, hfacac, Bu$^t$C(O)CHC(O)CF$_3$; L = H$_2$O, EtOH, HNEt$_2$). The tetracarbonyl compounds are prepared by grinding [TcCl(CO)$_5$] with

K(O⌢O) under $CCl_4$ and are very unstable with respect to decarbonylation. This is attributed to the inherent instability of *trans* CO ligands *cis* to a σ donor. The tricarbonyl solvate compounds *fac*-[Tc(O⌢O)(CO)$_3$L] are prepared by grinding the reagents in the presence of the donor ligand L. The dimers are prepared by heating the corresponding ethanol adduct [Tc(O⌢O)(CO)$_3$(EtOH)] under vacuum, and are believed to be bridged by the β-diketonate ligands. The amine adducts [Tc(O⌢O)(CO)$_3$(NHEt$_2$)] were found to exhibit the greatest thermal stability.[36,37,45,51]

In a related study, carbonyl carboxylate complexes of technetium(I) were prepared and their thermal stability examined. [Tc(η$^1$-O$_2$CCF$_3$)(CO)$_5$] was prepared by grinding [TcCl(CO)$_5$] with AgO$_2$CCF$_3$ under CHCl$_3$. Additionally, *fac*-[Tc(η$^1$-O$_2$CCF$_3$)(CO)$_3$(en)] and *fac*-[Tc(O$_2$CMe)(CO)$_3$L] (L = en, H$_2$O, EtOH) were prepared by grinding the reagents in the presence of the donor ligand. These compounds generally decomposed under vacuum sublimation.[52]

14-electron technetium(III) complexes containing bulky arylthiolate and carbonyl ligands were reported recently by the Davison group. The acetonitrile ligands in [Tc(SAr)$_3$(NCMe)$_2$] (Ar = 2,3,5,6-C$_6$HMe$_4$, 2,4,6-C$_6$H$_2$Pr$^i_3$) do not undergo self-exchange, but are readily substituted by π acid ligands forming [Tc(SAr)$_3$L$^1_2$] (7) (L$^1$ = CO, CNPr$^i$) (Scheme 2). The bis(carbonyl) complex (7) (L$^1$ = CO) can lose one one CO ligand to donor solvents, yielding [Tc(SAr)$_3$(CO)L$^2$] (8) (L$^2$ = py, NCMe) (Scheme 2). The x-ray structures of the monocarbonyl compounds (8) were determined. The analogous rhenium compounds were also prepared, and comparison of the IR data indicate that π backbonding is stronger for rhenium, implying greater overlap of the 5d orbitals (rhenium).[53]

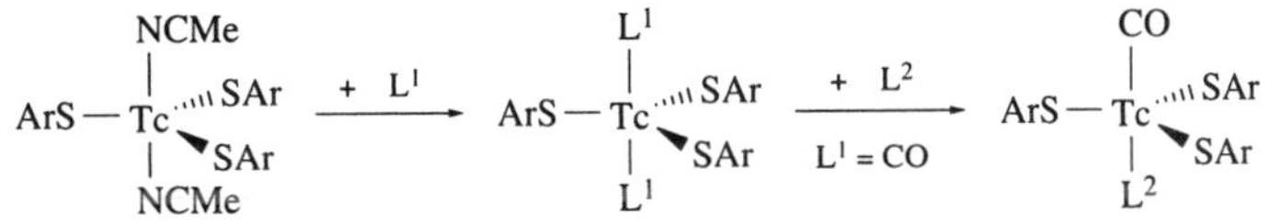

**Scheme 2**

A similar comparison of IR and x-ray structural data for [M(S$_2$CNEt$_2$)$_3$(CO)] (9) (M = Tc, Re) also suggests that technetium is a poorer π donor than rhenium. Seven-coordinate (9) is prepared by the reaction of [TcO$_4$]$^-$ with [S$_2$CNEt$_2$]$^-$ and formamidinesulfinic acid ((NH$_2$)(NH)CSO$_2$H). The CO ligand is thought to originate from the acid, which also functions as the reductant (Equation (2)).[54] [$^{99m}$Tc(S$_2$CNEt$_2$)$_3$(CO)] (9) is an efficient imaging agent for the liver and bile.[55]

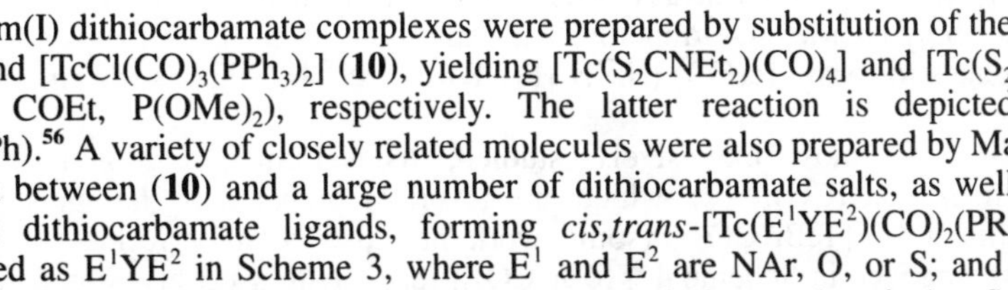

(2)

Other technetium(I) dithiocarbamate complexes were prepared by substitution of the halide ligand of [TcBr(CO)$_5$] (2) and [TcCl(CO)$_3$(PPh$_3$)$_2$] (10), yielding [Tc(S$_2$CNEt$_2$)(CO)$_4$] and [Tc(S$_2$Y)(CO)$_2$(PPh$_3$)$_2$] (11) (Y = CNEt$_2$, COEt, P(OMe)$_2$), respectively. The latter reaction is depicted in Scheme 3 (E$^1$ = E$^2$ = S, R = Ph).[56] A variety of closely related molecules were also prepared by Magon *et al.*. They report the reaction between (10) and a large number of dithiocarbamate salts, as well as other anions isoelectronic with dithiocarbamate ligands, forming *cis,trans*-[Tc(E$^1$YE$^2$)(CO)$_2$(PR$_3$)$_2$] (11). These ligands are depicted as E$^1$YE$^2$ in Scheme 3, where E$^1$ and E$^2$ are NAr, O, or S; and Y is CR, COEt, CNEt$_2$, or N.[57–60] Compound (10) also reacts with the lithium salt of the Schiff base *N-o*-hydroxybenzylidene-2-thiazolylimine to form *cis,trans*-[Tc(Schiff)(CO)$_2$(PR$_3$)$_2$] (12) (Scheme 3).[61]

### 8.3.2 Nitrogen and Phosphorus

Many of the complexes depicted in Scheme 3 and described in the final paragraph of the preceding section, (10)–(12), should also be mentioned here since they contain nitrogen and phosphorus donor ligands.[57–61]

The photolytic and thermal substitution reactions of $[Tc_2(CO)_{10}]$ (1) with $PF_3$ were studied. All together 24 different isomers of the general formula $[Tc_2(CO)_{10-x}(PF_3)_x]$, where $x = 0$–8, were observed by GC–MS and GC–IR spectroscopy. Substitution at axial sites is preferred under thermolytic conditions, and substitution at both axial and equatorial sites was observed under photolysis.[30]

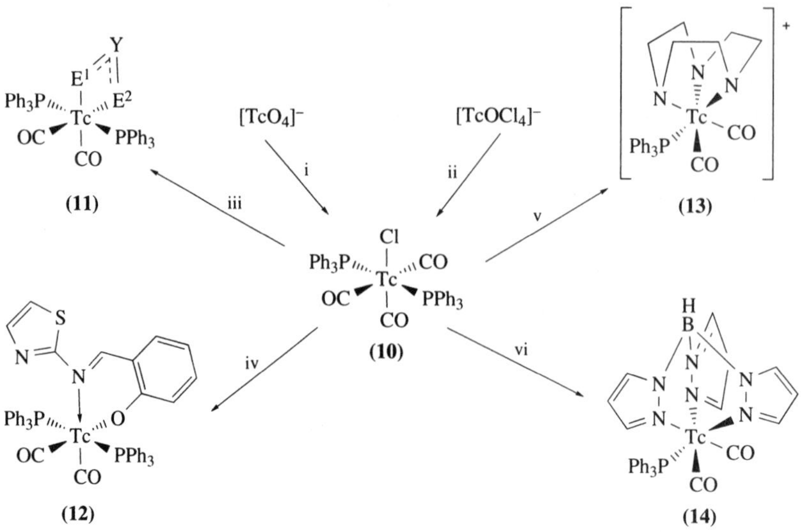

i, HCl, $PPh_3$, CO; ii, $PPh_3$, CO; iii, $[E^1YE^2]^-$; iv, lithium salt of the Schiff base *N-o*-hydroxybenzylidene-2-thiazolylimine; v, 1,4,7-triazacyclononane; vi, $K[HBpz_3]$

**Scheme 3**

Several synthetically useful precursors for technetium(I) carbonyl compounds containing nitrogen and/or phosphorus donor ligands have been reported.[49,62–4] The most practical precursors are those described by Alberto *et al.*, which are prepared from common starting materials such as $[TcO_4]^-$ or $[TcOCl_4]^-$ under mild conditions and do not involve the intermediate formation of volatile compounds.[49,62] Pertechnetate or $[TcOCl_4]^-$ is reduced by triphenylphosphine and carbon monoxide, yielding *mer*-$[TcCl(CO)_3(PPh_3)_2]$ (10) (Scheme 3). The x-ray crystal structure of (10) shows a shorter Tc–C bond length and a longer C≡O bond length for the carbonyl *trans* to chloride relative to the mutually *trans* carbonyl ligands. The tricarbonyl (10) reacts with tripodal nitrogen donors such as 1,4,7-triazacyclononane and hydrotris(pyrazolyl)borate to give $[Tc(CO)_2(N_3C_6H_{15})(PPh_3)]^+$ (13) and $[Tc(HBpz_3)(CO)_2(PPh_3)]$ (14), respectively (Scheme 3).[49,62]

Another precursor to technetium(I) carbonyl compounds is $[Tc(CO)_5(NCMe)]^+$, prepared from $[Tc_2(CO)_{10}]$ (1) and $NO^+$ in acetonitrile solution. The pentacarbonyl cation reacts with a variety of neutral nitrogen donor ligands to form adducts of technetium(I) tricarbonyl cations with the general formula $[Tc(CO)_3L]^+$ (L = 1,4,7-triazacyclononane, 1,4,7-trimethyl-1,4,7-triazacyclononane, 1,4,7-trithiacyclononane, bipy/NCMe, 4,4'-dimethyl-2,2'-bipyridine/NCMe). The bipyridyl complexes react with phosphite ligands to form $[Tc(CO)_2(N^\frown N)[P(OR)_3]_2]^+$ ($N^\frown N$ = bipy, 4,4'-dimethyl-2,2'-bipyridine; R = Me, Pr$^i$).[63]

The traditional, and most extensively studied, technetium(I) carbonyl precursors are the pentacarbonyl halides $[TcX(CO)_5]$ (2), which are generally prepared by halogenation of $[Tc_2(CO)_{10}]$ (1).[29,39] Recent studies of $[TcBr(CO)_5]$ (some work has also been performed for $[TcCl(CO)_5]$ and $[TcI(CO)_5]$) are summarized in Scheme 4. $[TcBr(CO)_5]$ reacts with a variety of σ donor ligands to form *fac*-$[TcBr(CO)_3L_2]$ complexes, where $L_2$ represents two monodentate ligands or a bidentate ligand.[65–7] The mass spectrum,[45] gas phase IR spectrum,[36] and x-ray structure[47] of $[TcBr(CO)_3(en)]$ are reported separately. $[TcBr(CO)_5]$ also reacts with silver ions in acetonitrile to form *fac*-$[Tc(CO)_3(NCMe)_3]^+$, which reacts with phosphine ligands in different ways, as illustrated in Scheme 4, dependent on the steric requirements of the phosphine and the solvent.[64,68] The pyrazolylborate tricarbonyl complexes $[Tc(HBpz_3)(CO)_3]$ and $[Tc\{HB(3,5-Me_2C_3HN_2)_3\}(CO)_3]$ are also prepared by substitution onto $[TcBr(CO)_5]$. Good correlation of the charge transfer bands for these compounds and their manganese and rhenium analogues, with the summations of the dipole moments involving the pyrazolylborate ligands was observed.[69] $[Tc\{HB(3,5-Me_2C_3HN_2)_3\}(CO)_3]$ loses a carbonyl ligand upon photolysis under $N_2$, forming the dimer $[Tc\{HB(3,5-Me_2C_3HN_2)_3\}(CO)_2]_2(\mu-N_2)$ (Scheme 4).[70]

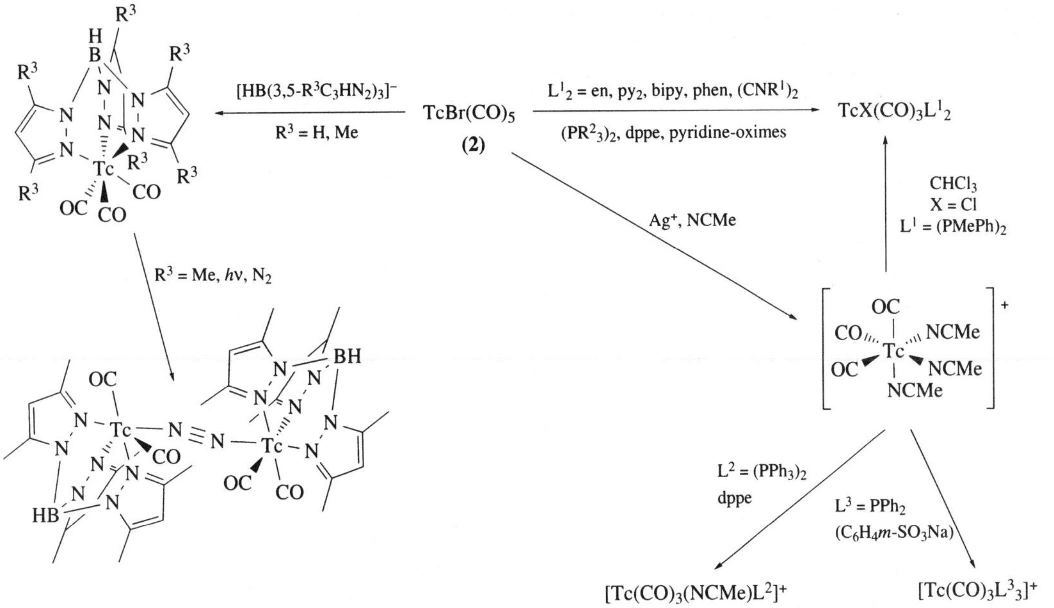

**Scheme 4**

Electrochemical oxidation of compounds related to compound (**10**), that is, [TcCl(CO)$_3$(PMe$_2$Ph)$_2$] and [TcCl(CO)$_2$(PMe$_2$Ph)$_3$], was studied in acetonitrile solvent. Both complexes exhibit overall two-electron oxidation processes. A technetium(III) dication, [TcCl(NCMe)$_2$(PMe$_2$Ph)$_3$]$^{2+}$, was isolated from bulk electrolysis of an acetonitrile solution of [TcCl(CO)$_2$(PMe$_2$Ph)$_3$].[71] A related technetium(III) carbonyl phosphine compound was formed by substitution of the acetonitrile ligand in [TcCl$_3$(NCMe)(PPh$_3$)$_2$] with CO, yielding [TcCl$_3$(CO)(PPh$_3$)$_2$]. IR ($v_{CO}$ = 2054 cm$^{-1}$) and x-ray crystal (Tc–C = 0.1985(9) nm, see Table 2) data suggest there is little back bonding to the carbonyl ligand.[72]

The hydride analogue of (**10**), *mer,trans*-[TcH(CO)$_3$(PPh$_3$)$_2$], was prepared from [TcH$_3$(PPh$_3$)$_4$] and CO.[73] Another technetium(I) hydrido carbonyl, [TcH(CO)(dppe)$_2$] (**15**) (E = O), is prepared by substitution of the N$_2$ in [TcH(dppe)$_2$(N$_2$)] (Scheme 5). Remarkably, this substitution can also be accomplished by refluxing the dinitrogen complex in MeOH. The hydride ligand in (**15**) was replaced by MeCN on reaction with NH$_4$PF$_6$ in refluxing acetonitrile, giving [Tc(CO)(dppe)$_2$(NCMe)]$^+$ (**16**) (Scheme 5).[74]

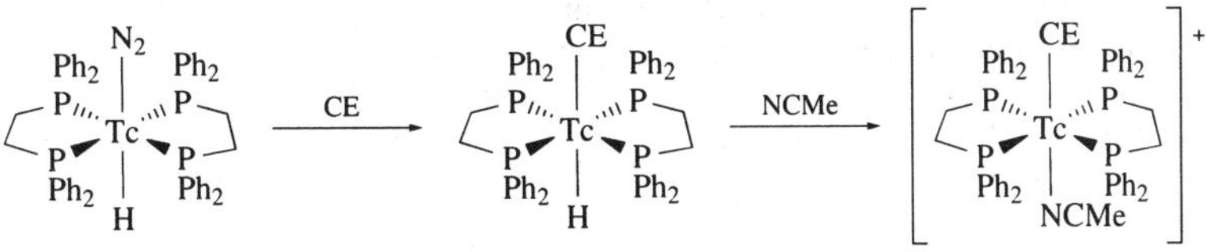

**Scheme 5**

A theoretical investigation into the relative stability of classical (hydride) and nonclassical ($\eta^2$-H$_2$) isomers of metal polyhydride complexes suggests that as strong $\pi$ acid ligands (e.g., CO) are added to the classical polyhydride complexes [TcH$_n$(PR$_3$)$_{8-n}$] and [TcH$_n$(PR$_3$)$_{7-n}$] (n = 2–7), the hydride ligands become increasingly nonclassical. For example, [TcH$_2$(H$_2$)(CO)(PR$_3$)$_3$]$^+$ is predicted to have two classical hydride ligands and one dihydrogen ligand.[75]

An interesting technetium nitrene complex, "Tc(=NEt)(tfacac)(CO)$_2$$^+$," is observed to be relatively stable in the MS fragmentation pattern for [Tc(tfacac)(CO)$_2$(NHEt$_2$)].[45]

## 8.4 TECHNETIUM ISOCYANIDES AND THEIR DERIVATIVES

### 8.4.1 Binary Technetium Isocyanides

As mentioned in Section 8.1, the recent surge of work in organotechnetium chemistry is due, at least in part, to the discovery that hexakis(isocyanide)technetium(I) compounds are good heart-imaging agents.[1] Much of the work published on these systems appears in nuclear medicine journals, and is not covered in detail by this review.[2,76-8]

The first mention of a hexakis(isocyanide)technetium(I) compound was in a review article.[19] Two key synthetic routes to $[Tc(CNR)_6]^+$ (17) have since been published for a wide variety of R groups, as illustrated in Scheme 6: (i) dithionate reduction of pertechnetate in the presence of isocyanide, and (ii) reductive substitution of the hexakis(thiourea)technetium(III) cation by isocyanides.[79-82] A minor, and relatively unsuccessful, variation on the first route involves bulk electrolysis of $[TcO_4]^-$ in the presence of acetate or formate. The resulting "technetium–carboxylate" complex is reacted with isocyanide to form (17).[83]

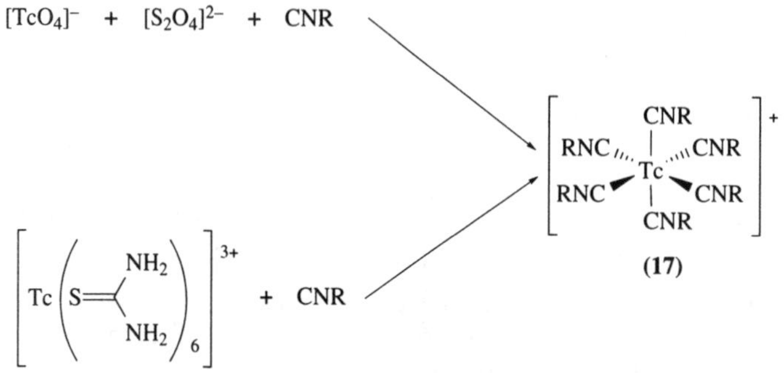

$$[TcO_4]^- \ + \ [S_2O_4]^{2-} \ + \ CNR$$

$$\begin{bmatrix} Tc \begin{pmatrix} S={\scriptstyle <}\begin{matrix} NH_2 \\ NH_2 \end{matrix} \end{pmatrix}_6 \end{bmatrix}^{3+} \ + \ CNR$$

$$\begin{bmatrix} & CNR & \\ RNC & | & CNR \\ & Tc & \\ RNC & | & CNR \\ & CNR & \end{bmatrix}^+$$

(17)

**Scheme 6**

The hexakis(isocyanide)technetium(I) cations (17) are inert toward substitution by other isocyanide ligands. Attempts to prepare compounds containing different isocyanide ligands by reacting the hexakis(thiourea)technetium(III) cation with mixtures of isocyanides resulted in statistical combinations of isocyanide ligands being bound to technetium.[84]

### 8.4.2 Technetium Isocyanide Derivatives

Attempts to substitute neutral phosphine or nitrogen donor ligands into $[Tc(CNR)_6]^+$ (17) generally result in poor product yields. The best routes involve reduction of $[TcO_4]^-$ in the presence of the appropriate ligand mixture. Even then, yields range from 10% to 25%, and separation from other products (most often (17)) proved difficult. Despite these difficulties, complexes of the type $[Tc(CNR)_4L_2]^+$ (18) (L = PPh$_3$, or 1/2 bidentate aromatic amine such as bipy, phen, or their derivatives) and $[Tc(CNBu^t)_5(PPh_3)]^+$ were isolated (Scheme 7).[85,86] The x-ray crystal structure of $[Tc(bipy)(CNR)_4]^+$ reveals a bent isocyanide ligand (<C–N–C = 148°) with a shortened Tc–C bond length of 0.190 (2) nm (Table 3). Additionally, spectroscopic data suggests that this compound may have undergone "internal oxidation," resulting in the formation of a technetium(III) carbene (see also Section 8.7.2).[85]

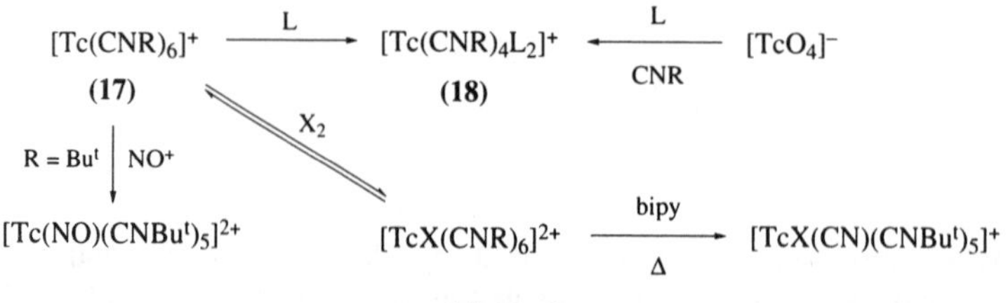

$$[Tc(CNR)_6]^+ \xrightarrow{\ L\ } [Tc(CNR)_4L_2]^+ \xleftarrow[CNR]{\ L\ } [TcO_4]^-$$

(17)          (18)

$$R = Bu^t \Big\downarrow NO^+$$

$$[Tc(NO)(CNBu^t)_5]^{2+} \qquad [TcX(CNR)_6]^{2+} \xrightarrow[\Delta]{bipy} [TcX(CN)(CNBu^t)_5]^+$$

$X_2$

**Scheme 7**

Reaction of $[Tc(CNBu^t)_6]^+$ (17) (R = Bu$^t$) with NO$^+$ results in the formation of $[Tc(NO)(CNBu^t)_5]^{2+}$ in high yield (Scheme 7). Reaction of CNBu$^t$ with $[TcBr_4(NO)]^-$ gives *mer*-$[TcBr_2(NO)(CNBu^t)_3]$ in 30% yield.[87] Oxidation of $[Tc(CNR)_6]^+$ (17) (R = Bu$^t$, Me) with halogens provides seven-coordinate

$[TcX(CNR)_6]^{2+}$ in high yield (Scheme 7). Attempts to couple two of the isocyanide ligands in $[TcX(CNR)_6]^{2+}$ results in reduction to $[Tc(CNR)_6]^+$ (**17**). The *t*-butylisocyanide technetium(III) cations are readily dealkylated on heating in acetonitrile in the presence of 2,2'-bipyridine, forming the cyanide complex $[Tc(CN)X(CNBu^t)_5]^+$ (Scheme 7). Dealkylation was also observed in the field desorption mass spectra.[88]

Substitution of isocyanides into other complexes has generally been a more fruitful approach to the preparation of technetium isocyanide derivatives. For example, the mixed-ligand complexes $[TcBr(CNR)_2(CO)_3]$ are readily prepared from $[TcBr(CO)_5]$ and CNR (Scheme 4, $L^1 = (CNR)_2$).[67] Also the dinitrogen ligand in $[TcH(dppe)_2(N_2)]$ can be displaced by isocyanide, forming $[TcH(CNR)(dppe)_2]$ (**15**) (R = $Bu^t$, Cy) (Scheme 5). The hydride ligand in (**15**) was replaced by MeCN on reaction with $NH_4PF_6$ in refluxing acetonitrile, giving $[Tc(CNR)(dppe)_2(NCMe)]^+$ (**16**) (Scheme 5).[74,89] A related complex, $[Tc(CNBu^t)_2(depe)_2]^+$ (depe = 1,2-diethylphosphinoethane), was evaluated as a potential heart-imaging agent, but did not demonstrate sufficient myocardial uptake in humans.[90] Reaction of $[TcX_4]$ (X = Cl, Br) in neat $CNBu^t$ gives *cis*-$[TcX_4(CNBu^t)_2]$ (Equation (3)).[91]

$$TcCl_4 \ + \ CNBu^t \ \longrightarrow \ \begin{array}{c} Cl \\ Cl_{\prime\prime\prime\prime}\!\!\!\diagdown\!\!\!\mid\!\!\!\diagup^{\prime\prime\prime}CNBu^t \\ Tc \\ Cl \diagup\mid\diagdown CNBu^t \\ Cl \end{array} \qquad (3)$$

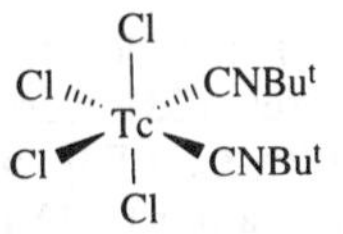

Compounds containing isocyanide and bulky arylthiolate ligands can be prepared by substitution of the acetonitrile ligand in $[Tc(SAr)_3(NCMe)_2]$, giving $[Tc(SAr)_3(CNPr^i)_2]$ (**7**) (Ar = 2,3,5,6-$C_6HMe_4$, 2,4,6-$C_6H_2Pr^i_3$; L = $CNPr^i$) (Scheme 2).[53,92] Closely related 14-electron technetium(III) isocyanide compounds containing the tetradentate tris(*o*-mercaptophenyl)phosphinate (PS$_3$) ligand $[Tc(PS_3)(CNR)]$ (**19**) (R = Me, $Pr^i$) were also prepared by substitution of the acetonitrile ligand in $[Tc(PS_3)(NCMe)]$ (Scheme 8). In the presence of a large excess of isocyanide, (**19**) can bind a sixth ligand, forming $[Tc(PS_3)(CNR)_2]$ (**20**). The bis(isocyanide) complex (**20**) is unstable towards loss of one isocyanide, suggesting that ligand substitution in complexes (**7**) and (**19**) and their acetonitrile precursors proceeds by an associative mechanism.[93,94]

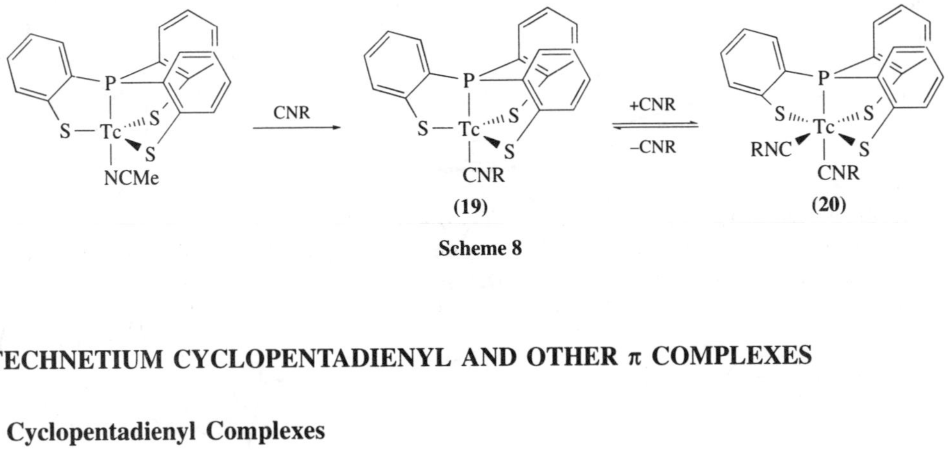

**Scheme 8**

# 8.5 TECHNETIUM CYCLOPENTADIENYL AND OTHER π COMPLEXES

## 8.5.1 Cyclopentadienyl Complexes

The study of cyclopentadienyl technetium complexes has received considerable attention since 1989. The volatility of the unsubstituted cyclopentadienyl compound $[Tc(CO)_3(\eta\text{-}C_5R_5)]$ (**23**) (R = H) at room temperature was determined to be about 1 Pa.[37] Researchers investigating $[Tc(CO)_3(\eta\text{-}Cp)]$ and related compounds should exercise due caution!

As indicated in Scheme 9, technetium tetrachloride reacts with KCp to form the sandwich complex $[TcCl(\eta\text{-}Cp)_2]$ (**21**) and with an excess of KCp to form the tris(cyclopentadienyl) complex $[Tc(\eta\text{-}Cp)_2(\eta^1\text{-}Cp)]$ (**22**). These two compounds are related in the sense that KCp converts the chloride (**21**) to the tris(Cp) (**22**) and HCl/$NH_4Cl$ converts (**22**) to (**21**). The dimer $[Tc(\eta\text{-}Cp)_2]_2$ is produced by reduction of (**21**) with potassium naphthalide, or by thermolysis of (**22**), which produces the hydride $[TcH(\eta\text{-}Cp)_2]$ as a side product (Scheme 9).[95] The x-ray structures of compounds (**21**) and (**22**) were determined. The bond distances and angles for the $\eta^1$-Cp ligand in the x-ray structure of (**22**) are unusual. The charge distribution in (**22**) and the effective charge of the σ-bonded carbon are discussed in comparison to analogous vanadium and rhenium compounds.[95,96]

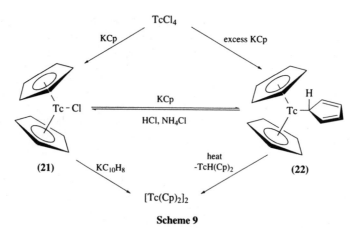

**Scheme 9**

Several new preparations of a wide variety of substituted (i.e., Cp\*, $C_5Me_4Et$) cyclopentadienyl tricarbonyltechnetium complexes (**23**) were reported almost simultaneously (Scheme 10).[28,29,49,97,98] Derivatives of (**23**), [Tc(CO)$_3$($\eta$-C$_5$H$_4$R)], were examined as potential brain-imaging agents. These compounds were prepared by heating [TcO$_4$]$^-$ with [Fe($\eta$-C$_5$H$_4$R)$_2$] and [MnBr(CO)$_5$] in a sealed glass vessel to 150 °C.[98,99] Reactivity studies have focused primarily on the pentamethylcyclopentadienyl derivative [Tc(CO)$_3$(Cp\*)] (**23**) (R = Me). As shown in Scheme 10, photolysis of (**23**) in cyclohexane forms a mixture of dimers, [(Cp\*)Tc($\mu$-CO)$_3$Tc(Cp\*)] (**24**) and [(Cp\*)(CO)$_2$Tc($\mu$-CO)Tc(CO)$_2$(Cp\*)] (**25**). An x-ray crystal structure of (**24**) reveals a Tc≡Tc triple bond of 0.2413(3) nm. The structure of (**25**) is inferred from spectroscopic data and analogy to known rhenium compounds.[100] Compound (**23**) is not oxidized by NO$^+$, but reacts to form [Tc(NO)(CO)$_2$(Cp\*)]$^+$. However, (**23**) is oxidized by Br$_2$ to form [TcBr$_2$(CO)$_2$(Cp\*)] (Scheme 10).[29]

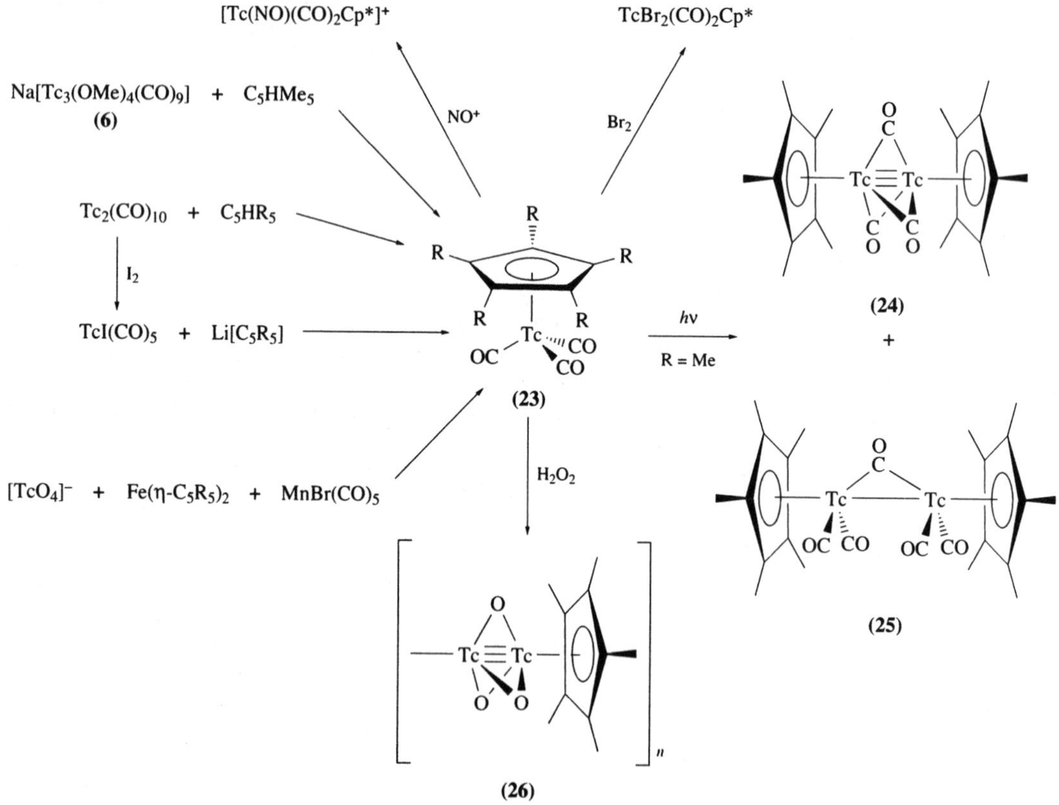

**Scheme 10**

Oxidation of (**23**) (R = Me) in benzene with aqueous hydrogen peroxide is purported to give the mixed-valence polymer [Tc($\mu$-O)$_3$Tc(Cp\*)]$_n$ (**26**), instead of the technetium(VII) monomer [TcO$_3$(Cp\*)], which is expected based on analogy to rhenium chemistry.[101] The characterization (including an x-ray structure determination) of (26) as a polymer has caused some controversy, because of the "supershort" Tc–Tc bond length of 0.1867(4) nm (compare with 0.241 nm for (24) and because the spectroscopic data (IR, NMR) reported for (26) are more in line with those expected for the trioxo monomer. Herrmann *et al.* suggested that the apparent polymeric structure is a result of disorder, since crystals of monomeric [ReO$_3$(Cp*)] have similar cell constants and was "unsatisfactorily solved."[102] Attempts to repeat the preparation of (26) following the synthesis outlined by Kanellakopulos *et al.*[101] failed to give (26) or the monomer [TcO$_3$(Cp*)]. The replacement of H$_2$O$_2$ with other oxidants such as [Mn$_2$O$_7$] and dimethyldioxirane, which are more efficient reagents in oxidizing [Re(CO)$_3$(Cp*)] to [ReO$_3$(Cp*)], generally failed to oxidize (23).[49] A theoretical paper has been published describing the electronic structure of the putative polymer [Tc($\mu$-O)$_3$Tc(Cp*)]$_n$ (26). This paper suggests that the Tc–Tc bond is best described as $\sigma^2(\pi\delta)^4\delta^*$, with a bond order of 2.5. The very short Tc–Tc distance of 0.1867(4) nm in (26) was suggested to be a result of the metal–metal multiple bond superimposed on an already short Tc–Tc contact enforced by the bridging oxo ligands.[103]

[TcO$_3$(Cp*)] remains an elusive synthetic target. An *ab initio* SCF study predicts that it and other "[TcO$_3$R]" compounds should be stable (see Section 8.6), although M–Cp $\pi$ bonding is predicted to be relatively weak ($D_{Cp-Tc} = 68$ kJ mol$^{-1}$).[104] Weak M–Cp $\pi$ bonding may explain the $\eta^1$ disposition of the cyclopentadienyl ligand in the isoelectronic imido compound [Tc(NAr)$_3$($\eta^1$-Cp)] (27) (Ar = 2,6-diisopropylphenyl) (Scheme 11). Compound (27) was prepared from [TcI(NAr)$_3$] and KCp, and reacts reversibly with an additional equivalent of KCp to form what is believed to be [Tc(NAr)$_3$(Cp)$_2$]$^-$ (28) (Scheme 11). The hapticity of the Cp ligands in (28) has not been determined.[105]

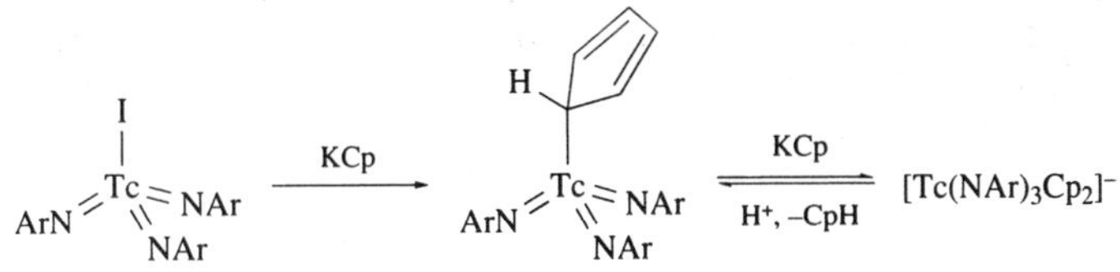

**Scheme 11**

### 8.5.2 Arene and Other $\pi$ Complexes

A new method for the preparation of bis(arene)technetium(I) cations was reported. Reduction of Na[TcO$_4$] with Zn/HCl or Al/AlCl$_3$/ultrasound in the presence of a wide variety of arenes in cyclohexane gives [Tc($\eta$-Ar)$_2$]$^+$ cations (29) in high yield. If Al/AlCl$_3$ was used as the reductant in the *absence* of ultrasound, partially substituted arenes underwent transalkylation, leading to product mixtures. The bis(arene) cations (29) were also evaluated as potential heart-imaging agents.[106] The reaction of the cubane cluster Na[Tc$_3$(OMe)$_4$(CO)$_9$] (6) with HCl and benzene yields the benzenetricarbonyl complex [Tc(CO)$_3$($\eta$-C$_6$H$_6$)]Cl (30).[28]

Photolysis of [Tc$_2$(CO)$_{10}$] (1) in the presence of butadiene at low temperatures gives the bridging *trans*-butadiene complex [Tc(CO)$_4$]$_2$($\mu$-C$_4$H$_6$) (31) in low yield.[107]

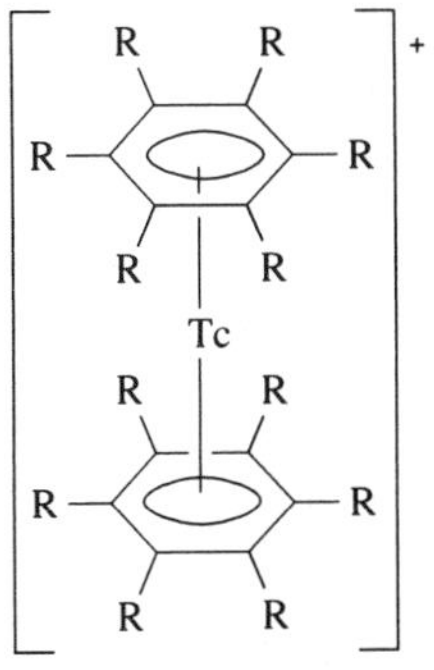

(29) 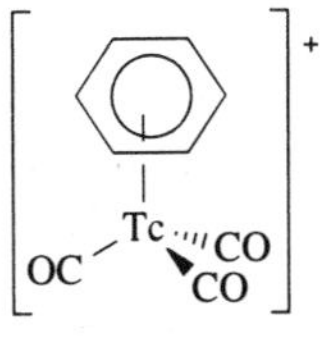　　　　　(30) 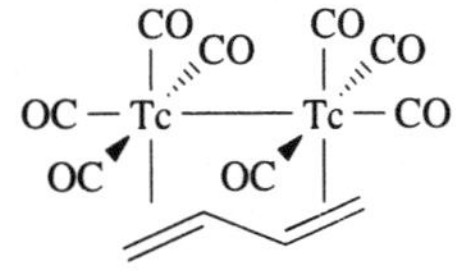　　　　　(31)

## 8.6 DERIVATIVES CONTAINING SINGLE- OR MULTIPLE-BONDED $\eta^1$ CARBON GROUPS

The first examples of technetium alkyl complexes were reported in 1990 by Herrmann and co-workers. Technetium heptoxide reacts with tetramethyltin at low temperatures to form [TcMe(O)$_3$] (32) (Scheme 12). Exceptional care must be exercised in preparing and handling (32) as it is extremely volatile (it sublimes at room temperature and atmospheric pressure). If the reaction is carried out at or above room temperature, mixtures of the stannylester [TcO$_3$(OSnMe$_3$)] and the technetium(VI) dimer [Me$_2$(O)Tc($\mu$-O)$_2$Tc(O)Me$_2$] (33) are formed (Scheme 12).[102,108]

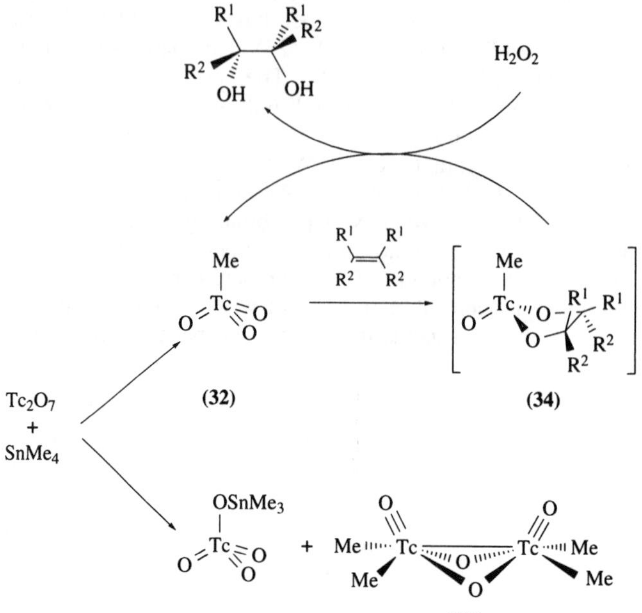

**Scheme 12**

Catalytic oxidation of alkenes to *cis*-diols by (32) and H$_2$O$_2$ was investigated. In this respect, (32) behaves more like high-valent osmium–oxo complexes than its rhenium analogue [ReMe(O)$_3$], which is an epoxidation catalyst toward alkenes. The catalytic cycle presumably runs through a glycolate complex similar to (34) (Scheme 12).[49]

Tris(arylimido) alkyl complexes are much easier to prepare, and are more stable and less volatile than the isoelectronic trioxo compound (32). Tris(arylimido)trimethylsiloxytechnetium(VII) reacts with Grignard reagents to form [Tc(NAr)$_3$R] (35) (Ar = 2,6-diisopropylphenyl, 2,6-dimethylphenyl; R = Me, Et, $\eta^1$-allyl, CH$_2$TMS) (Equation (4)).[109]

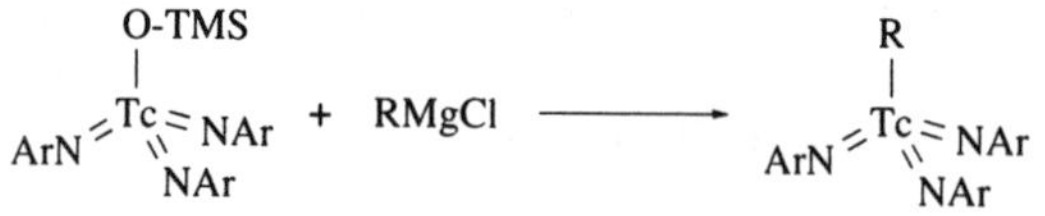

(4)

An imido analogue to (33), [Me$_2$(ArN)Tc($\mu$-NAr)$_2$Tc(NAr)Me$_2$] (36), was prepared by exhaustive methylation of the homoleptic imido complex [Tc$_2$(NAr)$_4$($\mu$-NAr)$_2$] (Scheme 13). An intermediate dimethyl complex, [(ArN)$_2$Tc($\mu$-NAr)$_2$Tc(NAr)Me$_2$] (37), can be isolated from the reaction of two equivalents of the methyl Grignard with [Tc$_2$(NAr)$_4$($\mu$-NAr)$_2$] (Ar = 2,6-dimethylphenyl) (Scheme 13). These reactions involve the unprecedented substitution of an imido ligand with two methyl ligands.[110]

Compounds containing technetium–carbon multiple bonds are virtually unknown. As mentioned in Section 8.4.2, one of the isocyanide ligands in [Tc(bipy)(CNR)$_4$]$^+$ (18) exhibits characteristics consistent with a carbene. When protonated, an unstable aminocarbyne complex is formed.[85] More recently, vinylidene complexes, [TcCl(=C=CHR)(dppe)$_2$] (38) (R = Me, Bu$^t$, Ph), have been prepared from the reaction of terminal acetylenes with the 16-electron technetium(I) complex [TcCl(dppe)$_2$] (Scheme 14). Protonation of the $\beta$ carbon of the vinylidene yields the carbyne complexes [Tc($\equiv$CCH$_2$R)Cldppe$_2$]$^+$ (39) R = Bu$^t$, Ph) (Scheme 14).[111]

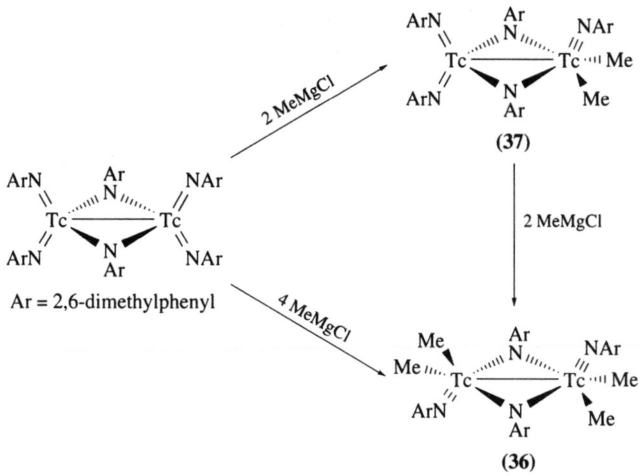

**Scheme 13**

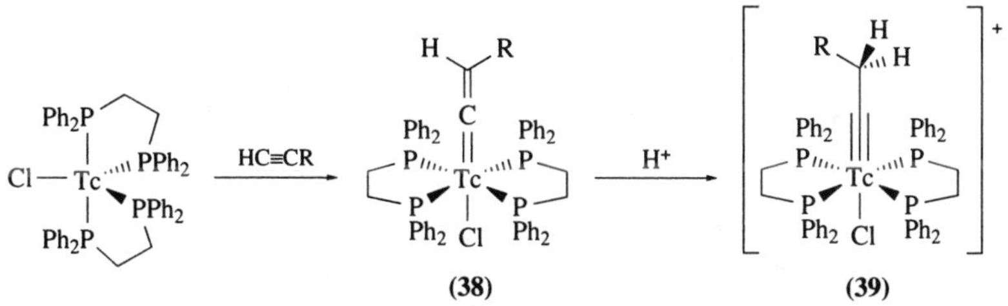

**Scheme 14**

## 8.7 STRUCTURAL DATA AND $^{99}$Tc NMR STUDIES

### 8.7.1 Structural Data

Most of the structures described in this chapter were published after two very good reviews of technetium x-ray crystal structures.[16,17] A statistical summary of technetium–carbon bond lengths for all structurally characterized organotechnetium complexes is presented in Table 2. High, low, and average technetium bond lengths are given for technetium–carbonyl, technetium–isocyanide, technetium–alkyl, technetium–cyclopentadienyl (ring centroid), and technetium–butadiene (*trans*, π) complexes. At the bottom of each column is a count of the number of bond lengths used in determining the values above them. The reader should note that the low value for Tc–CNR is unusually short. If this value is ignored, the shortest Tc–CNR bond length is 0.196 nm and the average is 0.205 nm.

**Table 2** Statistical summary of technetium–carbon bond lengths (nm) for organotechnetium complexes.

| | *Tc–CO* | *Tc–CNR* | *Tc–Calkyl* | *Tc–Cp* | *Tc–C₄H₆* |
|---|---|---|---|---|---|
| High value | 0.202 | 0.214 | 0.216 | 0.196 | 0.242 |
| Low value | 0.182 | 0.189 | 0.209 | 0.178 | 0.236 |
| Average | 0.190 | 0.204 | 0.214 | 0.191 | 0.239 |
| Count | 70 | 12 | 12 | 6 | 4 |

Technetium–alkyl, technetium–carbene, and technetium–carbyne bond lengths are presented individually in Table 3. They represent a significant class of organotechnetium structures that have been determined very recently.

The Tc–C$_{alkyl}$ single bond lengths listed in Table 3 agree reasonably well with the range of 0.200–0.213 nm predicted by *ab initio* SCF calculations before any of these compounds were prepared.[113] It is also clear from Table 3 that the technetium–carbon multiple bonds listed are significantly shorter than the technetium–carbon single bonds.

**Table 3** Technetium–carbon bond lengths (nm) for alkyl, carbene, and carbyne complexes.

| Line formula | Compound | Tc–C | Tc=C | Tc≡C | Ref. |
|---|---|---|---|---|---|
| [Tc(NAr)₃(η¹-Cp)] | (**27**) | 0.215 6(3) | | | 105 |
| [TcMe₂(O)]₂(μ-O)₂ | (**33**) | 0.213 3(2), 0.212 9(3) 0.212 8(3), 0.208 6(3) | | | 102 |
| [Tc(NAr)₃Me] | (**35**) | 0.213 6(17) | | | 112 |
| [Tc(NAr)Me₂]₂(μ-NAr)₂ | (**36**) | 0.213 4(2), 0.214 9(2) 0.215 3(2), 0.215 9(2) | | | 110 |
| [(ArN)₂Tc(μ-NAr)₂Tc(NAr)Me₂] | (**37**) | 0.211 9(17), 0.214 4(17) | | | 110 |
| [Tc(bipy)(CNR)₄]⁺ | (**18**) | | 0.190 (2) | | 85 |
| [TcCl(=C=CHR)(dppe)₂] | (**38**) | | 0.186 1(9) | | 111 |
| [Tc(≡CCH₂R)Cl(dppe)₂]⁺ | (**39**) | | | 0.172 4(7) | 111 |

## 8.7.2 ⁹⁹Tc NMR

⁹⁹Tc ($I = 9/2$) has a high sensitivity, which allows for rapid NMR data acquisition. It also has a significant electric quadrapole moment, which limits determination of ⁹⁹Tc NMR to molecules of reasonably high symmetry. This limitation can also be an advantage for qualitatively assessing the degree of symmetry about technetium in a molecule of unknown structure. ⁹⁹Tc also exhibits a very broad chemical shift range of over 9000 ppm. With the exception of technetium(VII), the chemical shift decreases as the oxidation state decreases. Technetium(VII) is thought to deviate from this trend due to its lack of $d$ electrons. Chemical shift also varies significantly with subtle changes in the electronic structure.[74,85,114–16]

The relationship between oxidation state and chemical shift has allowed some researchers to question the formal oxidation state of technetium in their compounds. Compounds (**6**), (**13**), (**14**), and (**18**) would formally be considered to contain technetium(I), but the ⁹⁹Tc resonance for these complexes are observed in the range −1200 to −620 ppm, which is well within the range (−1330 to −80 ppm) normally observed for technetium(III) and well outside the technetium(I) range (−3520 to −1460 ppm). This "internal oxidation" of technetium is substantiated in the x-ray crystal structures of these complexes, which all exhibit relatively short technetium–carbon bonds to π acid ligands, suggesting "Tc=CE" (E = O, NBuᵗ) is the most accurate representation of the metal–ligand bond.[62,85] Extensive additional data on (**18**) also supports "internal oxidation" to technetium(III) (see Section 8.4.2).[85]

The broad chemical shift range of ⁹⁹Tc was exploited to analyze mixtures of the mixed-ligand isocyanide complexes [Tc(CNR¹)ₙ(CNR²)₆₋ₙ]⁺ (R¹, R² = Buᵗ, Cy, CH₂CO₂Et). The differences between R¹ and R² did not need to be great in order to observe significant shifts in the ⁹⁹Tc resonance.[84]

Several other papers mentioned in this review also report ⁹⁹Tc NMR spectra for compounds reported therein.[28,56,66,67]

## 8.8 REFERENCES

1. A. Davison *et al.*, *J. Nucl. Med.*, 1984, **25**, 1350.
2. M. J. Clarke and L. Podbielski, *Coord. Chem. Rev.*, 1987, **78**, 253.
3. E. Deutsch and K. Libson, *Commun. Inorg. Chem.*, 1984, **3**, 83.
4. S. C. Srivastava and P. Richards, in 'Radiotracers for Medical Applications', ed. G. V. S. Rayudu, CRC Press, Boca Raton, FL, 1983, vol. 1, p. 107.
5. S. C. Srivastava and P. Richards, *J. Indian Chem. Soc.*, 1982, **59**, 1262.
6. E. Deutsch and B. L. Barnett, in 'Inorganic Chemistry in Biology and Medicine', ed. A. E. Martell, American Chemical Society, Washington, DC, 1980, p. 103.
7. J. B. Raynor, T. J. Kemp and A. M. Thyer, *Inorg. Chim. Acta*, 1992, **193**, 191.
8. R. Kirmse and U. Abram, *Isotopenpraxis*, 1990, **26**, 151.
9. U. Mazzi, *Polyhedron*, 1989, **8**, 1683.
10. N. Turp and J. E. Turp, *Coord. Chem. Rev.*, 1987, **80**, 157.
11. J. E. Turp, *Coord. Chem. Rev.*, 1986, **71**, 389.
12. V. I. Spitsyn, A. F. Kuzina, A. A. Oblova and S. V. Kryuchkov, *Usp. Khim.*, 1985, **54**, 637.
13. J. E. Turp, *Coord. Chem. Rev.*, 1983, **52**, 241.
14. J. E. Turp, *Coord. Chem. Rev.*, 1982, **45**, 281.
15. C. D. Russell, *Int. J. Appl. Radiat. Isot.*, 1982, **33**, 883.
16. M. Melník and J. E. Van Lier, *Coord. Chem. Rev.*, 1987, **77**, 275.
17. G. Bandoli, U. Mazzi, E. Roncari and E. Deutsch, *Coord. Chem. Rev.*, 1982, **44**, 191.
18. A. Davison and A. G. Jones, *Int. J. Appl. Radiat. Isot.*, 1982, **33**, 875.
19. A. G. Jones and A. Davison, *Int. J. Appl. Radiat. Isot.*, 1982, **33**, 867.
20. V. I. Spitzin, *Z. Chem.*, 1981, **21**, 131.

21. K. Schwochau and U. Pleger, *Radiochim. Acta*, 1993, **63**, 103.
22. E. Deutsch, K. Libson, S. Jurisson and L. F. Lindoy, *Prog. Inorg. Chem.*, 1983, **30**, 75.
23. K. Schwochau, *Radiochim. Acta*, 1983, **32**, 139.
24. M. J. Clarke and P. H. Fackler, *Struct. Bonding*, 1982, **50**, 57.
25. E. Deutsch, M. Nicolini and H. N. Wagner, Jr., (eds.), 'Technetium in Chemistry and Nuclear Medicine', Cortina International, Verona, 1983.
26. M. Nicolini, G. Bandoli and U. Mazzi, (eds.), 'Technetium in Chemistry and Nuclear Medicine 2', Cortina International, Verona, 1985.
27. M. Nicolini, G. Bandoli and U. Mazzi, (eds.), 'Technetium and Rhenium in Chemistry and Nuclear Medicine', Cortina International, Verona, 1990.
28. W. A. Herrmann, R. Alberto, J. C. Bryan and A. P. Sattelberger, *Chem. Ber.*, 1991, **124**, 1107.
29. H. H. Knight Castro, A. Meetsma, J. H. Teuben, W. Vaalburg, K. Panek and G. Ensing, *J. Organomet. Chem.*, 1991, **410**, 63.
30. C. C. Grimm and R. J. Clark, *Organometallics*, 1990, **9**, 1123.
31. F. Calderazzo, U. Mazzi, G. Pampaloni, R. Poli, F. Tisato and P. F. Zanazzi, *Gazz. Chem. Ital.*, 1989, **119**, 241.
32. S. Rummel and D. Schnurpfeil, *Oxid. Commun.*, 1984, **6**, 319.
33. J. C. Hileman, in 'Preparative Inorganic Reactions', ed. W. J. Jolly, Wiley, New York, 1964, vol. 1, p. 77.
34. A. K. Baev, *Zh. Fiz. Khim.*, 1980, **54**, 2169.
35. S. V. Kryuchkov, K. E. German and A. E. Simonov, *Koord. Khim.*, 1991, **17**, 480.
36. A. E. Miroslavov, I. V. Borisova, G. V. Sidorenko and D. N. Suglobov, *Radiokhimiya*, 1991, **33**(6), 20.
37. I. V. Borisova, A. E. Miroslavov, G. V. Sidorenko and D. N. Suglobov, *Radiokhimiya*, 1991, **33**(6), 9.
38. S. Rummel and D. Schnurpfeil, *J. Prakt. Chem.*, 1987, **329**, 10.
39. G. D. Michels and H. J. Svec, *Inorg. Chem.*, 1981, **20**, 3445.
40. G. Sbrignadello, *Inorg. Chim. Acta*, 1981, **48**, 237.
41. T. Ziegler, *Organometallics*, 1985, **4**, 675.
42. V. I. Spitsyn *et al.*, *Dokl. Akad. Nauk SSSR*, 1989, **307**, 644.
43. A. E. Miroslavov, G. V. Sidorenko, I. V. Borisova, E. K. Legin, A. A. Lychev and D. N. Suglobov, *Radiokhimiya*, 1990, **32**(6), 14.
44. A. E. Miroslavov *et al.*, *Radiokhimiya*, 1990, **32**(4), 6.
45. V. M. Adamov, B. N. Belyaev, I. V. Borisova, A. E. Miroslavov, G. V. Sidorenko and D. N. Suglobov, *Radiokhimiya*, 1991, **33**(4), 38.
46. C. Daniel and A. Veillard, *Nouv. J. Chim.*, 1986, **10**, 83.
47. M. S. Grigoriev and S. V. Kryutchkov, *Radiochim. Acta*, 1993, **63**, 187.
48. A. E. Miroslavov, I. V. Borisova, G. V. Sidorenko and D. N. Suglobov, *Radiokhimiya*, 1991, **33**(6), 14.
49. R. Alberto, W. A. Herrmann, J. C. Bryan, P. A. Schubiger, F. Baumgärtner and D. Mihalios, *Radiochim. Acta*, 1993, **63**, 153.
50. A. E. Miroslavov, G. V. Sidorenko, I. V. Borisova, E. A. Legin, A. A. Lychev and D. N. Suglobov, *Radiokhimiya*, 1989, **31**(6), 33.
51. I. V. Borisova, A. E. Miroslavov, G. V. Sidorenko, D. N. Suglobov and L. L. Shcherbakova, *Radiokhimiya*, 1991, **33**(4), 27.
52. I. V. Borisova, A. E. Miroslavov, G. V. Sidorenko and D. N. Suglobov, *Radiokhimiya*, 1991, **33**(6), 1.
53. N. de Vries, J. C. Dewan, A. G. Jones and A. Davison, *Inorg. Chem.*, 1988, **27**, 1574.
54. J. Baldas, J. Bonnyman, P. M. Pojer, G. A. Williams and M. F. Mackay, *J. Chem. Soc., Dalton Trans.*, 1982, 451.
55. J. Baldas and J. Bonnyman, *Nucl. Med. Biol.*, 1992, **19**, 741.
56. B. Lorenz, M. Findeisen and K. Schmidt, *Isotopenpraxis*, 1991, **27**, 266.
57. R. Rossi, A. Marchi, L. Magon, U. Casellato and R. Graziani, *J. Chem. Res., Synop.*, 1990, 78.
58. R. Rossi *et al.*, *J. Chem. Soc., Dalton Trans.*, 1990, 477.
59. A. Marchi, R. Rossi, A. Duatti, G. L. Zucchini and L. Magon, *Trans. Met. Chem.*, 1986, **11**, 164.
60. L. Magon *et al.*, *Inorg. Chem.*, 1985, **24**, 4744.
61. L. Magon, A. Marchi, L. Magon, A. Duatti, U. Casellato and R. Graziani, *Inorg. Chim. Acta*, 1989, **160**, 23.
62. R. Alberto, W. A. Herrmann, P. Kiprof and F. Baumgärtner, *Inorg. Chem.*, 1992, **31**, 895.
63. H. H. Knight Castro, C. E. Hissink, J. H. Teuben, W. Vaalburg and K. Panek, *Recl. Trav. Chim. Pays-Bas*, 1992, **111**, 105.
64. L. Kaden, B. Lorenz, S. Rummel, K. Schmidt and M. Wahren, *Inorg. Chim. Acta*, 1988, **142**, 1.
65. I. V. Borisova, A. E. Miroslavov, G. V. Sidorenko and D. N. Suglobov, *Radiokhimiya*, 1991, **33**(3), 1.
66. L. Kaden, M. Findeisen, B. Lorenz, K. Schmidt and M. Wahren, *Isotopenpraxis*, 1991, **27**, 265.
67. B. Lorenz, M. Findeisen, B. Olk and K. Schmidt, *Z. Anorg. Allg. Chem.*, 1988, **566**, 160.
68. L. Kaden, B. Lorenz, S. Rummel, K. Schmidt and M. Wahren, *German Pat.* 2 600 55 (1987) (*Chem. Abst.*, 1988, **111**, 2, 16–730).
69. B. Vanellakopulos *et al.*, *J. Organomet. Chem.*, 1993, **448**, 119.
70. B. Vanellakopulos *et al.*, *J. Organomet. Chem.*, 1993, **455**, 137.
71. U. Mazzi, E. Roncari, R. Seeber and G. A. Mazzocchin, *Inorg. Chim. Acta*, 1980, **41**, 95.
72. R. M. Pearlstein, W. M. Davis, A. G. Jones and A. Davison, *Inorg. Chem.*, 1989, **28**, 3332.
73. J. A. Cook and A. Davison, Presented at the 206th National Meeting of the American Chemical Society, Chicago, IL, August 1993, paper INOR 193.
74. L. Kaden, M. Findeisen, B. Lorenz, K. Schmidt and M. Wahren, *Inorg. Chim. Acta*, 1992, **193**, 213.
75. Z. Lin and M. B. Hall, *J. Am. Chem. Soc.*, 1992, **114**, 6102.
76. A. G. Jones *et al.*, *Int. J. Nucl. Med. Biol.*, 1984, **11**, 225.
77. J. C. Maublant, P. Gachon and N. Moins, *J. Nucl. Med.*, 1988, **29**, 48.
78. J. F. Kronauge, A. Davison, A. M. Roseberry, C. E. Costello, S. Maleknia and A. G. Jones, *Inorg. Chem.*, 1991, **30**, 4265.
79. M. J. Abrams, A. Davison, A. G. Jones, C. E. Costello and H. Pang, *Inorg. Chem.*, 1983, **22**, 2798.
80. M. J. Abrams, A. Davison, R. Faggiani, A. G. Jones and C. J. L. Lock, *Inorg. Chem.*, 1984, **23**, 3284.
81. U. Abram and G. Knop, *Z. Chem.*, 1988, **28**, 106.
82. F. E. Hahn, M. Tamm, A. Dittler-Klingemann and R. Neumeier, *Chem. Ber.*, 1991, **124**, 1683.

83. C. M. Kennedy and T. C. Pinkerton, *Appl. Radiat. Isot.*, 1988, **39**, 1179.
84. U. Abram, R. Beyer, R. Münze, M. Findeisen and B. Lorenz, *Inorg. Chim. Acta*, 1989, **160**, 139.
85. L. A. O'Connell, J. Dewan, A. G. Jones and A. Davison, *Inorg. Chem.*, 1990, **29**, 3539.
86. L. A. O'Connell and A. Davison, *Inorg. Chim. Acta*, 1990, **176**, 7.
87. K. E. Linder, A. Davison, J. C. Dewan, C. E. Costello and S. Maleknia, *Inorg. Chem.*, 1986, **25**, 2085.
88. J. P. Farr, M. J. Abrams, C. E. Costello, A. Davison, S. J. Lippard and A. G. Jones, *Organometallics*, 1985, **4**, 139.
89. U. Abram *et al.*, *Inorg. Chim. Acta*, 1988, **148**, 141.
90. G. D. Zanelli, N. Cook, A. Lahiri, D. Ellison, P. Webbon and G. Woolley, *J. Nucl. Med.*, 1988, **29**, 62.
91. U. Abram, R. Wollert and W. Hiller, *Radiochim. Acta*, 1993, **63**, 145.
92. A. Davison, N. de Vries, J. Dewan and A. Jones, *Inorg. Chim. Acta*, 1986, **120**, L15.
93. N. de Vries, A. Davison and A. G. Jones, *Inorg. Chim. Acta*, 1989, **165**, 9.
94. N. de Vries, J. Cook, A. G. Jones and A. Davison, *Inorg. Chem.*, 1991, **30**, 2662.
95. C. Apostolidis, B. Kanellakopulos, R. Maier, J. Rebizant and M. L. Ziegler, *J. Organomet. Chem.*, 1990, **396**, 315.
96. C. Apostolidis, B. Kanellakopulos, R. Maier, J. Rebizant and M. L. Ziegler, *J. Organomet. Chem.*, 1991, **411**, 171.
97. K. Raptis, E. Dornberger, B. Kanellakopulos, B. Nuber and M. L. Ziegler, *J. Organomet. Chem.*, 1991, **408**, 61.
98. M. Wenzel, *J. Labelled Comp. Radiopharm.*, 1992, **31**, 641.
99. M. Wenzel and M. Saidi, *J. Labelled Comp. Radiopharm.*, 1993, **33**, 77.
100. K. Raptis, B. Kanellakopulos, B. Nuber and M. L. Ziegler, *J. Organomet. Chem.*, 1991, **405**, 323.
101. B. Kanellakopulos, B. Nuber, K. Raptis and M. L. Ziegler, *Angew. Chem., Int. Ed. Engl.*, 1989, **28**, 1055.
102. W. A. Herrmann, R. Alberto, P. Kiprof and F. Baumgärtner, *Angew. Chem., Int. Ed. Engl.*, 1990, **29**, 189.
103. A. W. E. Chan, R. Hoffmann and S. Alvarez, *Inorg. Chem.*, 1991, **30**, 1086.
104. T. Szyperski and P. Schwerdtfeger, *Angew. Chem., Int. Ed. Engl.*, 1989, **28**, 1228.
105. A. K. Burrell and J. C. Bryan, *Organometallics*, 1992, **11**, 3501.
106. D. W. Wester, J. R. Coveney, D. L. Nosco, M. S. Robbins and R. T. Dean, *J. Med. Chem.*, 1991, **34**, 3284.
107. B. Kanellakopulos, B. Nuber, K. Raptis and M. L. Ziegler, *Z. Naturforsch., Teil B, Chem. Sci.*, 1991, **46b**, 55.
108. B. Kanellakopulos, K. Raptis, B. Nuber and M. L. Ziegler, *Z. Naturforsch., Teil B, Chem. Sci.*, 1991, **46b**, 15.
109. J. C. Bryan, A. K. Burrell, M. M. Miller, W. H. Smith, C. J. Burns and A. P. Sattelberger, *Polyhedron*, 1993, **12**, 1769.
110. A. K. Burrell and J. C. Bryan, *Organometallics*, 1993, **12**, 2426.
111. A. K. Burrell, J. C. Bryan and G. J. Kubas, *Organometallics*, 1994, **13**, 1067.
112. J. C. Bryan and A. P. Sattelberger, paper presented at the 207th National Meeting of the American Chemical Society, San Diego, CA, March 1994, Paper INOR 503.
113. C. W. Bauschlicher Jr., S. R. Langhoff, H. Partridge and L. A. Barnes, *J. Chem. Phys.*, 1989, **91**, 2399.
114. L. A. O'Connell, R. M. Pearlstein, A. Davison, J. R. Thornback, J. F. Kronauge and A. G. Jones, *Inorg. Chim. Acta*, 1989, **161**, 39.
115. M. Findeisen, L. Kaden, B. Lorenz and M. Wahren, *Inorg. Chim. Acta*, 1988, **142**, 3.
116. M. Findeisen, L. Kaden, B. Lorenz, S. Rummel and M. Wahren, *Inorg. Chim. Acta*, 1987, **128**, L15.

# 9

# Low-valent Organorhenium Compounds

## JOSEPH M. O'CONNOR
*University of California, San Diego, CA, USA*

## 9.1  INTRODUCTION

Research in the field of organorhenium chemistry has evolved dramatically since the mid-1980s. The high rhenium-to-element bond strengths and wide range of available oxidation states continue to produce unusual and fascinating structural motifs. Whereas synthetic and structural studies remain as the cornerstone for new advances, it is research into mechanism and reactivity that increasingly attracts organometallic chemists to rhenium. The combination of high bond strengths and the pronounced reluctance of rhenium to form coordinatively unsaturated complexes permits the observation of intermediates which may have only a fleeting existence in the related reactions of other metals. In addition, these characteristic properties of high bond strength and maximum coordination number provide for unusual, often seemingly unique, reaction pathways for rhenium complexes. Thus, contemporary organorhenium chemistry is significantly advancing our understanding of fundamental, as well as exotic, organometallic reactions.[1]

Due to the incredible growth of the field and the space restrictions for this review, only two-thirds of the published work which falls within the scope of this chapter has been included. The low-valent organometallic chemistry of rhenium described in this chapter therefore focuses largely on studies which address the chemistry of rhenium–carbon and rhenium–hydrogen bonds.

## 9.2  RHENIUM CARBONYLS

### 9.2.1  Dirhenium Decacarbonyl

Dirhenium decacarbonyl continues to serve as a key starting material for organometallic complexes of rhenium, although $[NMe_4][Re_2(\mu\text{-}OMe)_3(CO)_6]$ is a promising new starting material, accessible in 80% yield from $[ReO_4][NMe_4]$.[2] The classic preparative route toward $[Re_2(CO)_{10}]$ gives yields of 80–85% from $[Re_2O_7]$ on a small scale, but requires high CO pressures (36 MPa), temperatures of 250 °C, and the use of an autoclave lined with a Cu–Ag alloy.[3] Low temperature (150 °C)[4] and relatively low pressure (10 MPa CO)[5] routes toward $[Re_2(CO)_{10}]$ have been reported, but the yields are 80% and 69% respectively. Heinekey and co-workers have developed a convenient procedure in which thermolysis of $[ReO_4][Na]$ at 230 °C for 48 h in the presence of copper metal, under 22 MPa of CO, provides $[Re_2(CO)_{10}]$ in 90% yield.[6]

The reaction between $[Re(CO)_5]^-$ and $[Re(CO)_6]^+$ to give $[Re_2(CO)_{10}]$, although of no synthetic value, is very interesting from a mechanistic perspective. In the related case of $[Re(CO)_5]^-$ and $[Mn(CO)_6]^+$, the reaction proceeds by a rapid two-electron/one-carbonyl transfer process followed by a slower one-electron transfer and dimerization (Scheme 1).[7]

### *9.2.1.1  Rhenium–rhenium bond homolysis and 17-electron complexes*

As expected, photolysis of $[^{185}Re_2(CO)_{10}]$ and $[^{187}Re_2(CO)_{10}]$ mixtures leads to rapid and complete crossover of the rhenium isotopes as a result of homolytic rhenium–rhenium bond cleavage.[3] However,

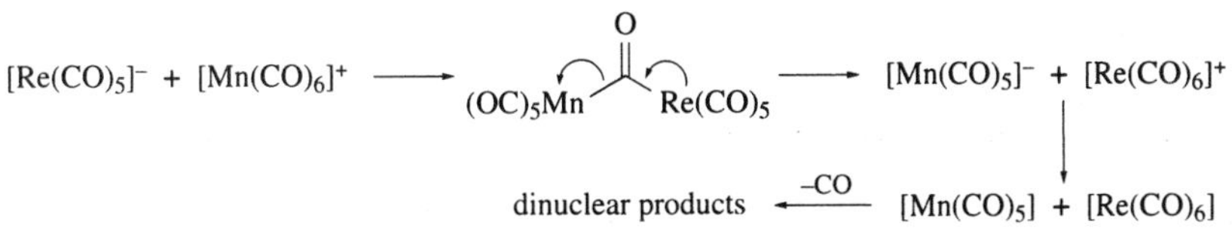

$$[Re(CO)_5]^- + [Mn(CO)_6]^+ \longrightarrow (OC)_5Mn \overset{\overset{O}{\|}}{\diagup} Re(CO)_5 \longrightarrow [Mn(CO)_5]^- + [Re(CO)_6]^+$$

$$\text{dinuclear products} \xleftarrow{-CO} [Mn(CO)_5] + [Re(CO)_6]$$

**Scheme 1**

there is now spectroscopic evidence that in solution,[8] and in the gas phase,[9] CO loss from [Re$_2$(CO)$_{10}$] may compete with rhenium–rhenium bond scission as the primary photochemical event.

Interest in 17-electron carbonyl complexes stems from the importance of radical intermediates in the reactions of both mononuclear and dinuclear rhenium complexes. At 22 °C in ethanol [Re(CO)$_5$] abstracts a chlorine atom from CCl$_4$ 65 times faster than does [Mn(CO)$_5$] ($k = 3.9 \times 10^7$ M$^{-1}$ s$^{-1}$ and $6.1 \times 10^5$ M$^{-1}$ s$^{-1}$ respectively),[10] and in hexane (22 °C) recombination of [M(CO)$_5$] occurs with $k = 9.5 \times 10^8$ M$^{-1}$ s$^{-1}$ and $3.7 \times 10^9$ M$^{-1}$ s$^{-1}$ for manganese and rhenium, respectively.[11]

The substitution reactions of [Re(CO)$_5$] proceed via a second-order associative pathway, and not the dissociative pathway previously proposed.[11] Substituted analogues, [Re(CO)$_4$L] (L = CO, P(OR)$_3$, PR$_3$, AsEt$_3$), have been generated by laser flash photolysis of the corresponding dimers, and subsequent reaction with halogen atom sources (CH$_2$Br$_2$, CHCl$_3$, CCl$_4$) gives *cis*-[Re(CO)$_4$LX] (X = Cl, Br).[12] The rate constants for reaction of [Re(CO)$_4$L] (L = PMe$_3$, P(OPr$^i$)$_3$) and different organic halogen atom donors do not correlate with C–X bond strengths and the relationship of rate data to half-wave reduction potentials for the organic halides suggests that electron transfer plays a role.[13]

Rhenium radicals are stabilized by ligands with large cone angles such as PCy$_3$. The deep blue, persistent radical, [Re(CO)$_3$(PCy$_3$)$_2$] (**1**) ($\nu_{CO}$ 1849 cm$^{-1}$) is generated upon irradiation of [Re$_2$(CO)$_8$(PCy$_3$)$_2$] and PCy$_3$ in benzene.[14] The EPR spectrum at 77 K indicates the presence of a major and a minor species, each with coupling to rhenium. The major species was assigned a square-pyramidal structure with *trans* basal phosphines and the minor species was suggested to contain *cis* basal phosphines in a distorted square-pyramidal structure.

Heinekey and co-workers have used homolysis of weak rhenium–carbon σ-bonds to thermally generate rhenium(0) radicals. The η$^3$-benzyl complex (**2**), generated from Na[Re(CO)$_5$] and Ph$_3$CPF$_6$, serves as a useful precursor to rhenium radicals (Equation (1)). Upon heating (**2**) is converted to the η$^5$-cyclohexadienyl complex (**3**), which reacts with added phosphine to give the persistent rhenium radical [Re(CO)$_3$L$_2$] (L = PCy$_3$, PPr$^i$$_3$) (Equation (1)).[15,16] Complex (**1**) was structurally characterized and found to be a distorted square pyramid.[17]

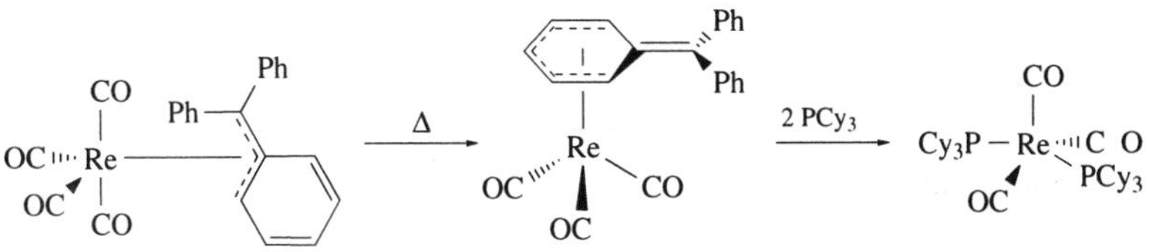

(1)

One-electron oxidation of [Re(CO)$_5$]$^-$ and substituted analogues provide an alternative route toward 17-electron complexes including [Re(CO)$_5$], [Re(CO)$_4$(PPh$_3$)], [Re(CO)$_3$(PPh$_3$)$_2$], and [Re(CO)$_3$\{P(OC$_6$H$_4$Me-2)$_3$\}$_2$].[18] A variety of spin traps such as α,β-diimine α,β-diketone, nitrosodurene, and thiocarbonyls react with rhenium-centered radicals.[19-23]

### 9.2.1.2 Substitution reactions

Photolysis of wet THF solutions containing [Re$_2$(CO)$_{10}$] at 313 nm cleanly generates the labile aquo species *eq*-[Re$_2$(CO)$_9$(OH$_2$)], which decomposes under continued irradiation to give [ReH(CO)$_5$] and [Re$_4$(OH)$_4$(CO)$_{12}$].[24] The aquo ligand is readily displaced by CO, MeCN, and PPh$_3$. Mechanistic studies led to the proposal that the aquo complex is formed by initial homolysis of the rhenium–rhenium bond, thermal substitution of CO by H$_2$O, and radical recombination. When bridging phosphorus ligands (e.g., dmpm) are incorporated into the starting complex stable dinuclear substitution products result.[25] Table 1 provides selected spectroscopic data for representative dirhenium carbonyl derivatives.

**Table 1** Spectroscopic data for selected dinuclear carbonyl complexes.

| Compound | IR (cm$^{-1}$) | $^1H$ NMR hydride region (ppm) | Ref. |
|---|---|---|---|
| [Re$_2$(CO)$_8$(dppm)] | 2073(m), 2020(m), 1980(s), 1956(w), 1940(w), 1915(m)[a] | | 25 |
| [Re$_2$(CO)$_8$(dmpm)] | 2067(m), 2012(s), 1980(sh), 1971(vs), 1947(m), 1934(sh), 1914(s)[a] | | 25 |
| [Re$_2$(CO)$_8$(dppe)] | 2070(m), 2017(m), 1983(s), 1944(m), 1943(m), 1915(s)[a] | | 25 |
| [Re$_2$(CO)$_8$(dmpe)] | 2066(w), 2010(m), 1973(s), 1938(m), 1911(s)[a] | | 25 |
| [Re$_2$(CO)$_6$(dppm)(μ-H)(μ-OH)] | 2037(s), 2012(s), 1940(m), 1917(sh), 1908(s), 3566(w,br)[a] | 0.02 (t, 1H), −9.18 (t, 1H)[e] | 25 |
| [Re$_2$(CO)$_6$(dppm)(μ-OH)$_2$] | 2029(s), 2011(m), 1924(m), 1908(sh), 1890(s), 3586(wbr)[a] | −0.35(t, 2H)[e] | 25 |
| [Re$_2$(CO)$_6$(dppm)(μ-H)(μ-OMe)] | 2038(s), 2013(s), 1943(m), 1923(sh), 1912(w)[a] | −9.11(t, 1H)[e] | 25 |
| [Re$_2$(CO)$_6$(dppm)(μ-OMe)$_2$] | 2027(s), 2010(m), 1924(m), 1893(w), 1885(sh)[a] | | 25 |
| [Re$_2$(CO)$_6$(dmpm)(μ-H)(μ-OH)] | 2035(m), 2010(m), 1936(m), 1918(m), 1902(s), 3560(wbr)[a] | −0.66 (t, 1H)[f], −9.98 (t, 1H) | 25 |
| [Re$_2$(CO)$_6$(dmpm)(μ-OH)$_2$] | 2027(m), 2010(m), 1917(m), 1903(sh), 1883(s), 3575(wbr)[a] | −0.95 (t, 1H)[f] | 25 |
| [Re$_2$(CO)$_6$(dppm)(μ-H)(μ-Cl)] | 2051(s), 2026(m), 1960(m), 1940(m), 1922(s)[a] | −10.87 (t, 1H)[f] | 25 |
| [Re$_2$(CO)$_6$(dppm)(μ-Cl)$_2$] | 2050(s), 2037(m), 1960(m), 1945(m), 1918(s)[a] | | 25 |
| [Re$_2$(CO)$_6$(dmpm)(μ-H)(μ-Cl)] | 2049(m), 2023(m), 1957(m), 1938(m), 1917(s)[a] | −11.49(t, 1H)[f] | 25 |
| [Re$_2$(CO)$_6$(dmpm)(μ-Cl)$_2$] | 2051(s), 2036(m), 1955(m), 1940(m), 1914(s)[a] | | 25 |
| [Re$_2$(CO)$_9$(PPh$_3$)] | 2113(w,sp), 2040(w,sp), 1998(vs), 1964(sh), 1940(m)[d] | | 27 |
| [Re$_2$(CO)$_9$(PMe$_3$)] | 2108(w,sp), 2040(m,sp), 1990(vs), 1956(sh), 1927(m)[d] | | 27 |
| [Re$_2$(CO)$_8$(PPh$_3$)$_2$] | 2000(sh) 1958(vs)[d] | | 27 |
| [Re$_2$(CO)$_8$(PMe$_3$)$_2$] | 2000(sh) 1943(vs)[d] | | 27 |
| [Re$_2$(CO)$_9$(MeCN)] | 2104(w), 2046(vst), 1988(sh), 1960(m), 1928(m)[b] | | 28 |
| [Re$_2$(CO)$_9$(Pr$^n$CN)] | 2104(w), 2046(m), 1994(vst), 1987(sh), 1960(m), 1928(m)[b] | | 28 |
| [Re$_2$(CO)$_9$(PhCN)] | 2103(w), 2047(m), 1993(vst), 1987(sh), 1960(m), 1931(m)[b] | | 28 |
| [Re$_2$(CO)$_9${P(O-$o$-Tol)$_3$}] | 2112(w), 2046(m), 2001(vst), 1960(m), 1942(sh)[b] | | 28 |
| [(2,6)-Re$_2$(CO)$_8$(MeCN)]$_2$ | 2072(w), 2016(st), 1969(vst), 1933(m), 1906(st)[b] | | 28 |
| [(2,3)-Re$_2$(CO)$_8$(MeCN)]$_2$ | 2077(w), 2002(st), 1962(vst), 1906(sh), 1892(m)[b] | | 28 |
| [Re$_2$(CO)$_9$(CNMe)] | 2197(m), 2105(m), 2076(w), 2053(m), 2019(m), 1998(vs)[c] | | 29 |
| [Re$_2$(CO)$_8$(CNMe)$_2$] | 2190(m), 2072(w), 2032(m), 1980(vs), 1947(sh), 1922(m)[d] | | 29 |
| [Re$_2$(CO)$_7$(CNMe)$_3$] | 2180(m), 2040(w), 2032(sh), 1980(vs), 1959(vs), 1919(sh), 1989(m)[d] | | 29 |
| [Re$_2$(CO)$_6$(CNMe)$_4$] | 2168(m), 1971(vs), 1928(m), 1880(m)[d] | | 29 |
| [Re$_2$(CO)$_8$(μ-dppm)$_2$] | 1959(m), 1927(vs), 1871(s)[a] | | 30 |
| [Re$_2$(CO)$_8$(μ-dppm)(μ-dmpm)] | 1952(m), 1920(vs), 1868(s)[a] | | 30 |
| [Re$_2$(CO)$_8$(μ-dmpm)$_2$] | 1943(m), 1912(vs), 1861(s)[a] | | 30 |
| [Re$_2$(CO)$_4$(μ-H)(μ-dppm)$_2$(μ-OH)] | 1923(s), 1852(s)[b] | −0.97 (quintet, 1H)[e] −8.34 (quintet, 1H)[e] | 30 |
| [Re$_2$(CO)$_4$(μ-H)(μ-dppm)$_2$(μ-OMe)] | 1923(s), 1851(s)[b] | −8.93 (quintet, 1H)[e] | 30 |
| [Re$_2$(CO)$_4$(μ-H)(μ-dmpm)(μ-dmpm)(μ-OH)] | 1913(s), 1842(s)[b] | −9.06 (quintet, 1H)[e] | 30 |

[a] toluene. [b] CH$_2$Cl$_2$. [c] hexane. [d] CHCl$_3$. [e] CD$_2$Cl$_2$. [f] CDCl$_3$.

Thermal CO substitution reactions of $[Re_2(CO)_{10}]$ have been conclusively demonstrated not to involve rhenium–rhenium bond scission as the primary activation step.[3,26] Experimentally, no crossover is observed in the reactions of $[^{185}Re_2(CO)_{10}]$ and $[^{187}Re_2(CO)_{10}]$ with $^{13}CO$ at 150 °C after 14 half-lives in *n*-octane.[3] Similarly, conversion of $[^{185}Re_2(CO)_{10}]/[^{187}Re_2(CO)_{10}]$ and $PPh_3$ to $[Re_2(CO)_9(PPh_3)]$ proceeds under a CO atmosphere with essentially no crossover. Suppression of crossover by CO or phosphine indicates that homolytic rhenium–rhenium bond cleavage occurs subsequent to formation of $[Re_2(CO)_9]$. The available evidence supports a reversible CO dissociation mechanism, although a heterolytic rhenium–rhenium bond cleavage to give 16- and 18-electron metal centers bridged by two CO ligands represents an intriguing alternative.[31]

In addition to thermal and photochemical substitution reactions, a variety of catalysts have been found to accelerate carbonyl substitution in $[Re_2(CO)_{10}]$. In benzene, reaction of $[Re_2(CO)_{10}]$ with $CNBu^t$ to give $[Re_2(CO)_9(CNBu^t)]$ is slow under photochemical conditions and fails to proceed in the dark at 55 °C. However, in the presence of PdO or Pd–CuCO$_3$ (5%) reaction is rapid. Similarly, Pd–C(40%), Pd–Al$_2$O$_3$ (5%), $[Fe(CO)_2(Cp)]_2$, and $[Mn_2(CO)_{10}]$ all accelerate substitution.[32,33] The use of $[Pt(PPh_3)_4]$ as a catalyst gives a 72% yield of $[Re_2(CO)_9(PPh_3)]$ after 2 d in refluxing benzene,[34] whereas Pd–C (10%) gives a 1:1 ratio of $[Re_2(CO)_9(PPh_3)]$ and $[Re_2(CO)_8(PPh_3)_2]$ in 40% combined yield (xylene, 140 °C, 9 h).[27] The PdO- and $Me_3NO$-mediated substitution reactions of $[Re_2(CO)_9L]$ (L = $CNBu^t$, P(O-*o*-Tol)$_3$, PBz$_3$, RCN (R = Me, Et, Pr, Ph)) with $NCBu^t$ and RCN have been studied with an emphasis on isomer distribution.[28] A series of isocyanide and phosphine derivatives of $[Re_2(CO)_{10}]$ have been structurally characterized.[27,29]

Oxidation of $[Re_2(CO)_{10}]$ by a variety of neutral and cationic oxidants in acetonitrile generates the mononuclear cation $[Re(CO)_5(NCMe)]^+$. The failure of the strong anionic oxidants $[Fe(CN)_6]^{3-}$ and $[IrCl_6]^{2-}$ to oxidize $[Re_2(CO)_{10}]$ may be due to an electrostatic barrier to outer-sphere electron transfer in aprotic solvents.[35]

### 9.2.1.3 Reactions with alkenes

Photolysis of $[Re_2(CO)_{10}]$ in the presence of alkenes leads to the μ-vinyl hydride complexes $[Re_2$-$(μ-H)(μ-η^1,η^2$-CH=CHR)(CO)$_8]$, (e.g., (4), R = alkyl), which are isolated in excellent (terminal alkenes) to moderate (internal alkenes) yields.[36,37] Evidence for initial rhenium–rhenium bond homolysis has been obtained from studies on the related reactions of $[Re_2(CO)_9L]$, $[1,2-Re_2(CO)_8L_2]$, and $[Re_2(CO)_8(μ-LL)]$ with alkenes (L = PMe$_3$, PPh$_3$; LL = dmpm, dppe, dmpe).[36]

The μ-vinyl ligands in (4) are fluxional at 25 °C via a process in which the σ- and π-bonds of the vinyl ligand are interchanged between the rhenium atoms. The substituted vinyl ligands also undergo thermal and photochemical *cis–trans* alkene isomerization.[39]

A fascinating catalytic conversion of ethene to butene and hexene has been uncovered by Nubel and Brown.[39] Although the efficiency of the process is inadequate for practical applications, the mechanism provides a rare example of a catalytic process requiring two adjacent metal sites. At 25 °C the rate is 2–3 turnovers per day and catalyst death appears to be related to formation of $[Re(CO)_3(OH)]_4$.[39]

The photochemical reactions of $[Re_2(CO)_{10}]$ with cyclic alkenes and dienes are more complicated than for acyclic alkenes. For example, 1,3-butadiene gives rise to six complexes including $[Re_2(μ-η^2,η^2$-CH$_2$=CHCH=CH$_2$)(CO)$_8]$, which has been shown to arise from rearrangement of the vinyl hydride $[Re_2(μ-H)(μ-η^1,η^2$-CH=CHCH=CH$_2$)(CO)$_8]$ at 25 °C.[40,41] Reaction of $[Re_2(μ-H)(μ-η^1,η^2$-CH=CHEt)-(CO)$_8]$ with 3,3-dimethylcyclopropene leads to C–C bond cleavage and formation of the μ-carbene $[Re_2(μ-η^1,η^3$-CHCH=CMe$_2$)(CO)$_8]$.[42]

Photolysis of $[Re_2(CO)_{10}]$ in the presence of $C_{60}$ results in 90% conversion to $[C_{60}\{Re(CO)_5\}_2]$, which was not isolated, but appears to be a 1,4-disubstituted $C_{60}$ adduct.[43]

### 9.2.1.4 Reactions with alkynes

The products of reaction of $[Re_2(CO)_{10}]$ with alkyne depend dramatically on the nature of the alkyne substrate. Thermolysis of $[Re_2(CO)_{10}]$ at 190 °C in the presence of excess PhC≡CPh gives rise to three new complexes containing coupled alkyne ligands, one of which reacts with isocyanide at 25 °C to give crystallographically characterized $[Re_2(CNR)_2(CO)_4(PhCCPh)_3]$ (R = $CH_2SO_2C_6H_4Me$) (5).[44] Photolysis of $[Re_3H_3(CO)_{12}]$ and PhC≡CPh also gives rise to a number of products including the first structurally characterized rhenacyclopentadiene and $η^2$-alkyne rhenium complexes, (6) and (7). The rhenium-($η^2$-alkyne) bond distances in (6) and (7) imply four-electron donor $η^2$-alkyne ligands.[45]

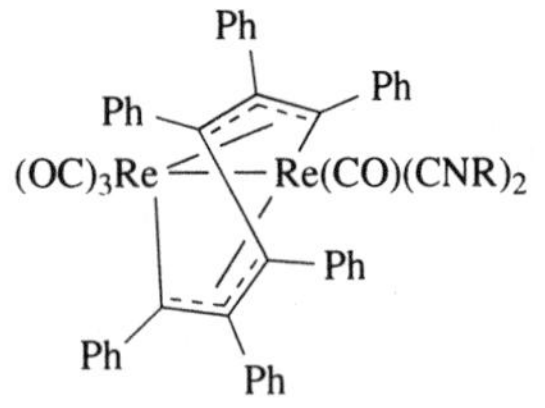

(5)

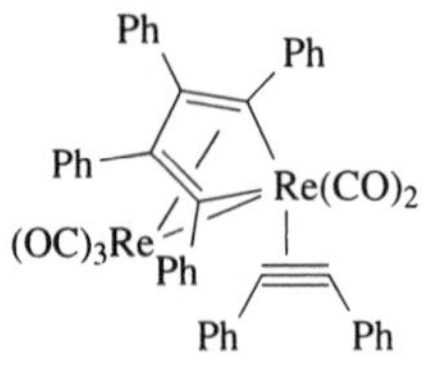

(6)

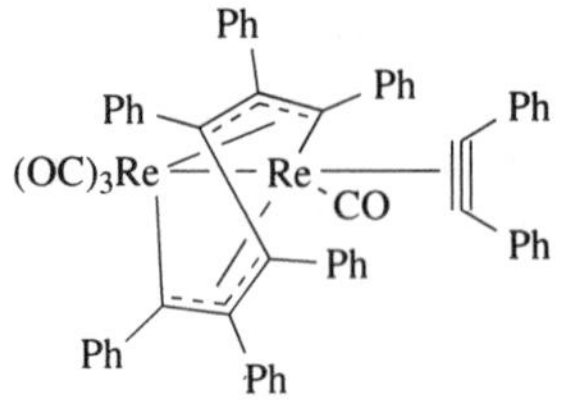

(7)

Photolysis of $[Re_2(CO)_{10}]$ and ethyne in hexane solution gives rise to five complexes with the major product, $[Re_2(\mu\text{-}H)(\mu\text{-}\eta^1,\eta^2\text{-}C{\equiv}CH)(CO)_8]$, formed in 40% yield.[46,47] In a similar fashion, $[Re_2(CO)_8(NCMe)_2]$ reacts thermally with phenylethyne to give $[Re_2(\mu\text{-}H)(\mu\text{-}\eta^1,\eta^2\text{-}C{\equiv}CPh)(CO)_7(NCMe)]$.[48,49] When the rhenium carbonyl reagent contains bridging diphosphorus ligands (e.g., dppm, dmpm, dppe, dmpe), rhenium–rhenium bond homolysis is minimized and CO dissociation becomes the major pathway to alkyne-derived products. Photolysis of $[Re_2(CO)_8(\mu\text{-}dppm)]$ and PhC≡CH in toluene at 25 °C (12 h) gives rise to (8) (17%), (9) (46%) and (10) (19%) (Equation (2)).[30,50]

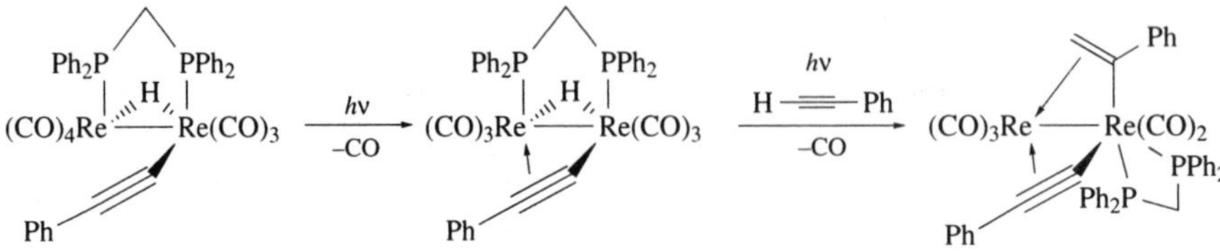

(2)

(8)                              (9)                              (10)

Whereas the $\mu$-hydrido, $\mu$-vinyl complexes $[Re_2(\mu\text{-}H)(\mu\text{-}CH{=}CHR)(CO)_8]$ react at 25 °C with phenylethyne to give $[Re_2(\mu\text{-}H)(\mu\text{-}\eta^1,\eta^2\text{-}C{\equiv}CPh)(CO)_9]$,[45] electron-rich ynamines give an $\eta^3$-ynamine complex (11).[51] Further reaction of (11) with ynamine at 68 °C gives dinuclear metallacyclopentadiene products (12), resulting from head-to-head, head-to-tail, and tail-to-tail coupling of two ynamines (Scheme 2). Reduction of (11) with $H_2$ (10 atm, 70 °C, 70 min) gives $[Re_2(\mu\text{-}H)\{\mu\text{-}Me\text{-}(H)C{=}CNMe_2\}(CO)_8]$, which cleanly isomerizes to the *trans*-alkene complex $[Re_2\{trans\text{-}\mu\text{-}Me\text{-}(H)C{=}C(H)NMe_2\}(CO)_8]$ (13) at 60 °C.[52] The net reaction is *trans*-addition of $H_2$ across the alkyne ligand. The ynamine ligand in (11) undergoes a number of thermally induced transformations which proceed via loss of CO and oxidative addition to a carbon–hydrogen bond of the ynamine ligand.[53]

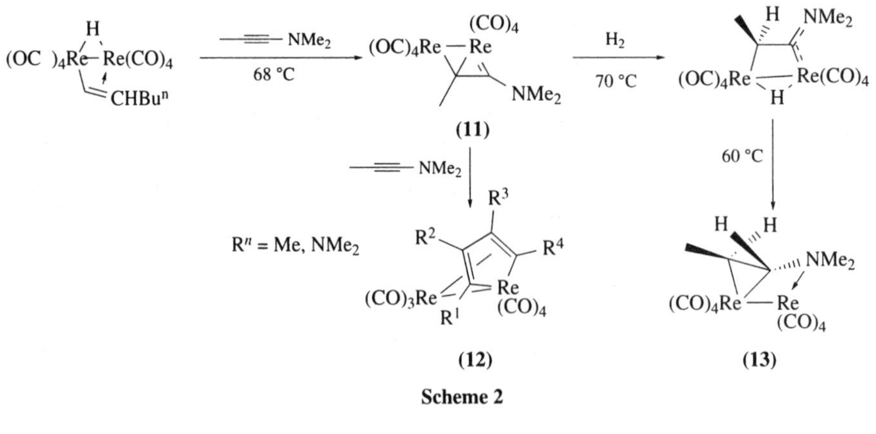

Scheme 2

Reaction of $[Re_2(CO)_9(NCMe)]$ with electrophilic alkynes results in alkyne insertion into the rhenium–rhenium bond to give dimetallated alkene complexes (14) (Equation (3)).[54] The reaction is proposed to occur by initial alkyne substitution for the nitrile ligand, migration of the $\alpha$-carbon to the adjacent rhenium, cleavage of the metal–metal bond and coordination of the ester oxygen. Compound (14) reacts with CO (8.28 MPa) at 25 °C to give $[Re_2\{(E)\text{-}\mu\text{-}\eta^1,\eta^1\text{-}HC{=}C(CO_2Me)\}(CO)_{10}]$, whereas irradiation of (14) at 25 °C leads to an alkenyl hydrogen shift and formation of the bridging vinylidene complex (15). The rhenium–rhenium bond in (15) is reversibly cleaved by addition of CO.[54]

The rhenium(II) A-Frame-like complex $[Re_2(\mu\text{-}Cl)(Cl)_3(CO)(dppm)_2]$ reacts with both terminal and internal alkynes, in the presence of $TIPF_6$, to give the first $\eta^2$-alkyne complexes from an electron-rich metal–metal triply bonded complex.[55] The related dirhenium complexes $[Re_2(\mu\text{-}Cl)Cl_3(\mu\text{-}CO)(L)\text{-}$

$$(3)$$

**(14)**          **(15)**

($\mu$-dppm)$_2$] (**16**) (L = CO, XylNC) react with terminal alkynes, RC≡CH (R = H, Pr$^n$, Bu$^n$, Ph, or *p*-Tol) in the presence of TlPF$_6$ to give metallafuran products (**17**) (Equation (4)).[56] This unprecedented cycloaddition reaction involves regioselective addition of the alkyne across a ReCO unit.

$$(4)$$

**(16)**                    **(17)**

### 9.2.2 Clusters

For rhenium carbonyl cluster chemistry up to and including 1988, readers are directed to an excellent review by Henly, which appeared in 1989.[57] Table 2 lists the clusters described in the Henly review and Table 3 provides spectroscopic data for selected clusters.

**Table 2** Rhenium carbonyl clusters described in a review by Henly.[57]

| | | |
|---|---|---|
| [Re$_3$(H)$_3$(CO)$_{12}$] | [Re$_3$(H)$_4$(CO)$_9$L]$^-$ (L = MeCN, PPh$_3$, py) | [Re$_4$(H)$_4$(CO)$_{12}$] |
| [Re$_3$(H)$_3$(CO)$_{10}$(py)$_2$] | [Re$_3$(CO)$_9$($\mu_3$-PPh$_2$)$_3$] | [Re$_4$(H)$_4$(CO)$_{15}$]$^{2-}$ |
| [Re$_3$(H)$_3$(CO)$_{10}$(NCPh)$_2$] | [Re$_3$(H)$_2$(CO)$_6$($\mu$-PPh$_2$)$_3$] | [Re$_4$(H)$_4$(CO)$_{13}$]$^{2-}$ |
| [Re$_3$(H)$_3$(CO)$_{11}$(PPh$_3$)] | [Re$_3$(CO)$_{13}$($\mu$-$\eta^{1,2}$-C$_5$(H)$_7$)] | [Re$_4$(H)$_6$(CO)$_{12}$]$^{2-}$ |
| [Re$_3$(H)$_3$(CO)$_{10}$(PPh$_3$)$_2$] | [Re$_3$(H)$_3$(CO)$_{10}${$\mu$-$\eta^2$-OC(Ph)NH}]$^-$ | [Re$_4$(H)$_5$(CO)$_{12}$]$^-$ |
| [Re$_3$(H)$_3$(CO)$_9$(L)$_3$] (L = PPh$_3$, P(OPh)$_3$) | [Re$_3$(H)$_3$($\mu_3$-OR)(CO)$_9$]$^-$ | [Re$_4$(H)$_5$(CO)$_{14}$]$^-$ |
| [Re$_3$(H)$_3$(CO)$_8${(EtO)$_2$POP(OEt)$_2$}$_2$] | [Re$_3$(H)$_3$($\mu_3$-SBu$^t$)(CO)$_9$]$^-$ | [Re$_4$(CO)$_{16}$]$^{2-}$ |
| [Re$_3$(H)$_3$(CO)$_8$(Ph$_2$PCH$_2$PPh$_2$)$_2$] | [Re$_3$(H)$_3$($\mu$-X)(CO)$_{10}$]$^-$ (X = Cl, Br, I, OPh, OPr$^i$) | [Re$_4$Cl$_2$(CO)$_{15}$(MePPMePMe)] |
| [Re$_3$(H)$_3$(CO)$_{10}${(EtO)$_2$POP(OEt)$_2$}] | [Re$_3$(H)$_2$($\mu$-I)$_2$(CO)$_{10}$]$^-$ | [Re$_4$(H)$_4$(CO)$_{15}$I]$^-$ |
| [Re$_3$(H)$_2$(CO)$_{12}$]$^-$ | [Re$_3$(H)$_3$($\mu_3$-OR)(CO)$_9$]$^-$ (R = Pr$^i$, Ph) | [Re$_4$(C)(CO)$_{15}$I]$^-$ |
| [Re$_3$(H)(CO)$_{12}$]$^{2-}$ | [Re$_3$(H)$_3$($\mu$-O$_2$CR)(CO)$_{10}$]$^-$ (R = H, Me, CF$_3$) | [Re$_5$(H)(C)(CO)$_{16}$]$^{2-}$ |
| [Re$_3$(H)$_2$(CO)$_{10}$(PPh$_3$)$_2$]$^-$ | [Re$_3$(H)$_3$(CO)$_{10}$(NCMe)$_2$] | [Re$_5$(CO)$_{14}$($\mu_4$-PMe)($\mu$-PMe$_2$){$\mu_3$-P(Re(CO)$_5$)}] |
| | | |
| [Re$_3$(CO)$_{13}$(OCSiPh$_3$)] | [Re$_3$(H)$_3$($\mu_3$-$\eta^2$-CH$_2$O(CO)$_9$]$^{2-}$ | [Re$_6$(H)$_2$(C)(CO)$_{18}$]$^{2-}$ |
| [Re$_3$(H)(CO)$_{14}$] | [{Re$_3$(H)(CO)$_{10}$}$_2${$\mu_4$-$\eta^2$-NO)]$^-$ | [Re$_6$(H)(C)(CO)$_{18}$]$^{3-}$ |
| [Re$_3$(H)$_3$(CO)$_{10}$]$^{2-}$ | [Re$_3$(H)$_2$(CO)$_{10}$($\eta^5$-C$_7$H$_9$)] | [Re$_6$(CO)$_{18}$($\mu_4$-PMe)$_3$] |
| [Re$_3$(H)$_3$(CO)$_9$($\mu_3$-O)]$^{2-}$ | [Re$_3$(H)$_4$(CO)$_9$(NMe$_3$)]$^-$ | [Re$_7$C(CO)$_{21}$]$^{3-}$ |
| [Re$_3$(H)$_3$(CO)$_9$($\mu_3$-OH)]$^-$ | [Re$_3$(H)$_4$(CO)$_9$(ONMe$_3$)]$^-$ | [Re$_7$(H)(C)(CO)$_{21}$]$^{2-}$ |
| [Re$_3$(H)$_4$(CO)$_{13}$]$^{2-}$ | [Re$_3$(H)$_3$(CO)$_9$($\mu_3$-O···H···NMe$_3$)]$^-$ | [Re$_7$C(CO)$_{22}$]$^-$ |
| [Re$_3$(H)$_4$(CO)$_{10}$]$^-$ | [Re$_3$(H)$_3$(CO)$_9$($\mu_3$-AuPPh$_3$)]$^-$ | [Re$_8$C(CO)$_{24}$]$^{2-}$ |

#### 9.2.2.1 Tri- and tetranuclear clusters

The mono(acetonitrile) and bis(acetonitrile) clusters, [Re$_3$(H)$_3$(CO)$_{11}$(NCMe)][57] and [Re$_3$(H)$_3$(CO)$_{10}$(NCMe)$_2$][58] are useful precursors to a range of phosphine-substituted trinuclear complexes. Deprotonation of [Re$_3$(H)$_3$(CO)$_{12}$] gives the monoanion [Re$_3$(H)$_2$(CO)$_{12}$]$^-$, which undergoes hydrogenation (THF, 100 °C, 100 atm H$_2$, 24 h) to the unsaturated tetrahydride [Re$_3$(H)$_4$(CO)$_{10}$]$^-$ (**18**).[57] This 46 valence electron complex may be viewed as containing a doubly protonated rhenium–rhenium double bond, or a four-center four-electron Re$_2$H$_2$ bond. The reactivity of (**18**) has been extensively developed by Ciani, D'Alfonso, and co-workers.[59–62] One of the hydrogen ligands bridging the Re=Re double bond is hydridic and constitutes the primary site of reactivity with acids, whereas the hydride ligands bridging Re–Re single bonds are acidic.

**Table 3**  IR spectroscopic data for selected rhenium clusters.

| Compound | IR (cm⁻¹) | ¹H NMR hydride region (ppm) | Ref. |
|---|---|---|---|
| $[Re_3(\mu\text{-}H)_3(CO)_9\{\mu_3\text{-}\eta^2\text{-}C(H)(Me)O\}][NEt_4]_2$ | 2003(m), 1977(s), 1886(s), 1872(s)ᶜ | −9.40ᵉ, −12.64, −13.04 | 60 |
| $[Re_3(\mu\text{-}H)_3(\mu\text{-}\eta^2\text{-}CHNC_6H_{11})(CO)_{10}][NEt_4]$ | 2091(w), 2015(m), 1994(vs), 1938(m), 1909(s), 1893(ms)ᶜ | −12.95 (1H)ᵈ, −13.36 (1H), −15.44 (1H) | 61 |
| $[Re_3(\mu\text{-}H)_3(CO)_9\{\mu\text{-}\eta^2\text{-}C(H)NC_6H_{11}\}][NEt_4]$ | 2042(w), 2006(s), 1995(s), 1938(sh), 1917(s), 1890(sh)ᶜ | −14.47 (3H)ᵇ | 61 |
| $[Re_3(\mu\text{-}H)_4(CO)_9(CNC_6H_{11})]$ | 2187(w)(CN), 2034(m), 2009(s), 1998(vs), 1938(ms), 1910(s)ᶜ | −8.36 (1H)ᵈ, −8.48 (1H), −12.86 (2H)ᵇ | 62 |
| $[Re_3(CO)_6(\mu\text{-}PPh_2)_3(\mu_3\text{-}H)_2]$ | 1974(s), 1938(s)ᵃ | — | 64 |
| $[Re_3(\mu\text{-}H)_3(CO)_9(\eta^3\text{-}\mu_3\text{-}HC(PPh_2)_3]$ | 1928(s), 1953(m), 1961(m), 2018(m), 2044(s)ᵇ | −18.2 (q, 2H)ᵇ | 65 |
| $[Re_7C(CO)_{21}][NEt_4]_3$ | 1977(sh), 1968(vs), 1940(mw), 1920(vw), 1890(m), 1855(w)ᵇ | | 66 |
| $[Re_8C(CO)_{24}][NEt_4]_2$ | 2010(sh), 2004(s), 1950(w), 1930(mw), 1890(w)ᶠ | | 67 |
| $[Re_7C(CO)_{21}][PPN]_3$ | 1993(vs), 1923(mw), 1880(sh)ᵇ | | 68 |
| $[Re_6C(CO)_{19}][PPN]_2$, $[Re_6C(CO)_{19}][NEt_4]_2$ | 2051(vw), 1991(vs), 1976(s), 1905(w), 1895(w, sh), 1815(vw)ᵍ | | 69 |
| $[Re_7C(CO)_{22}][PPN]$ | 2080(vw), 2030(vs), 2017(s), 2005(m,sh), 1996(mw,sh), 1953(w), 1913(w), 1843(vw)ᵍ | | 69 |
| $[Re_7C(CO)_{21}\{P(OPh)_3\}][PPN]$ | 2067(w), 2011(vs), 1984(w), 1942(w)ᵍ | | 70 |
| $[(H)Re_7C(CO)_{21}][PPN]_2$ | 1998(vs), 1982(m,sh), 1952(vw), 1924(w), 1887(w)ᶠ | −19.27, −19.87ᵈ | 71 |
| $[(H)Re_7C(CO)_{21}][NEt_4]_2$ | | −19.30, −19.92ᵈ | 72 |
| $[Re_7(H)_2C(CO)_{21}][NEt_4]$ | 2085(vw), 2025(vs), 2000(sh), 1950(m)ᵍ | −17.05, −18.75, −20.27, −21.60ᵈ | 73 |

ᵃ KBr.  ᵇ Not specified.  ᶜ THF.  ᵈ CD₂Cl₂.  ᵉ THF-d₈.  ᶠ acetone.  ᵍ CH₂Cl₂.

Treatment of (**18**) with $NOBF_4$ generates a 25% yield of a novel nitric oxide complex in which a nitrosyl ligand serves as an $\eta^4$-ligand which bridges two triangular clusters (Equation (5)).[63]

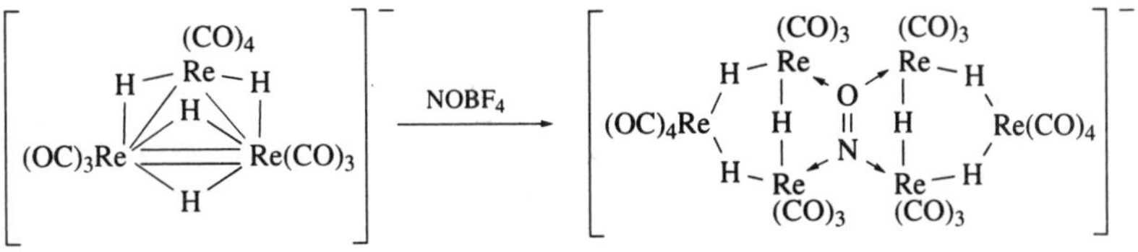

$$(5)$$

(**18**)

The unusual trinuclear carbonyl cluster $[Re_3(\mu\text{-}H)_2(CO)_6(\mu\text{-}PPh_2)_3]$, with triply bridging hydrides and 44 cluster valence electrons, has been prepared in 25% yield from thermolysis of $[Re_2(\mu\text{-}H)(CO)_8(\mu\text{-}PPh_2)]$ in xylene.[64] Crystallographic analysis indicates that the four $\pi$-electrons in the trinuclear core are delocalized, resulting in rhenium–rhenium bonds with some double bond character (average Re–Re 0.273 0(1) nm).[74]

Donor molecules such as CO, $PPh_3$, MeCN, and $H_2O$ readily add to the unsaturated tetranuclear cluster $[Re_4(H)_5(CO)_{12}]^-$ to give the adducts $[Re_4(H)_5(CO)_{14}]^-$, $[Re_4(H)_5(CO)_{12}(PPh_3)]^-$, and $[Re_4(H)_5(CO)_{12}(MeCN)_2]^-$.[75] The neutral unsaturated tetranuclear cluster $[Re_4(H)_4(CO)_{12}]$ reacts with $HC(PPh_2)_3$ in refluxing dichloromethane to give a 42% yield of the trinuclear complex (**19**) as well as mono-, di-, and tetranuclear products.[65]

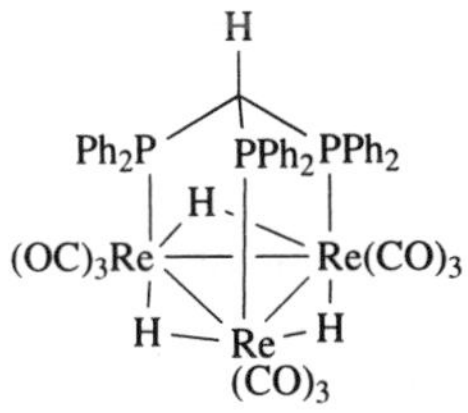

(**19**)

### 9.2.2.2 High nuclearity clusters

In contrast to the later transition metals, high nuclearity clusters of rhenium ($Re_n$, $n > 4$) were unknown until 1982 when Ciani, D'Alfonso, and co-workers reported the hydrolysis of $[Re(H)_2(CO)_4][NEt_4]$ in *n*-tetradecane at 250 °C to give the carbido clusters $[Re_8C(CO)_{24}]^{2-}$ (**20**) (30%) and $[Re_7C(CO)_{21}]^{3-}$ (**21**) (50%).[66,67] For (**20**) and (**21**) the carbide ligand is observed in the $^{13}C$ NMR spectrum (THF-$d_8$) at 431.3 ppm and 423.6 ppm, respectively. The interstitial carbon atom stabilizes high nuclearity clusters of rhenium by increasing the number of valence electrons without an increase in steric congestion. In an octahedral cluster with 86 valence electrons, the ligands and charges must account for 44 electrons, whereas in the case of osmium and platinum, only 38 and 26 electrons, respectively, are required from ligands and charge.[66]

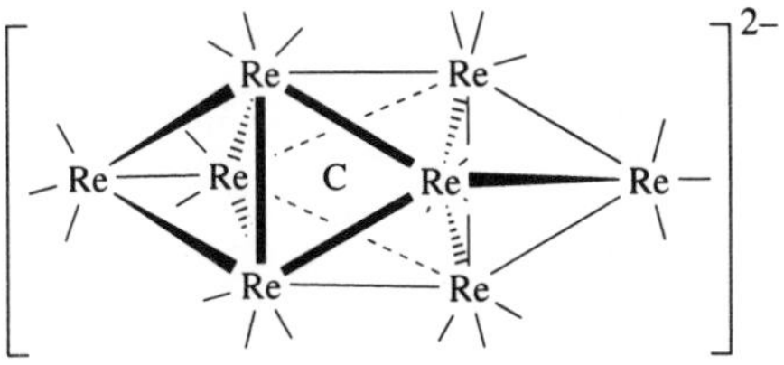

(**20**)

One-electron oxidation of (**21**) with $C_7H_7BF_4$ gives the air sensitive, paramagnetic 97 valence electron complex $[Re_7C(CO)_{21}]^{2-}$, which reacts with CO to give high yields of $[Re_7C(CO)_{22}]^-$ (**22**), a rare example of a rhenium cluster containing a bridging CO ligand (Equation (6)).[68] Oxidation of (21) with excess [FeCp$_2$][PF$_6$] under CO gives an 89% yield of (22), whereas oxidation with 2 equiv. [FeCp$_2$][PF$_6$] in the presence of PPh$_3$ and CO gives a 90% yield of [Re$_6$C(CO)$_{19}$]$^{2-}$, in addition to [Re(CO)$_4$(PPh$_3$)$_2$]$^+$.[71]

$$(6)$$

In the presence of P(OPh)$_3$ and 2 equiv. [FeCp$_2$][PF$_6$], (21) is oxidized to [Re$_7$C(CO)$_{21}${P(OPh)$_3$}]$^-$ (23) in 77% yield (Equation (6)). The $^{13}$C NMR spectrum of (23) suggests a structure with a relatively positive capping moiety [Re(CO)$_3${P(OPh)$_3$}] (+1) and a relatively negative [Re$_6$C(CO)$_{18}$] (−2) core. As such, the complex may represent a polar intermediate in cluster-decapping reactions.[70]

Reaction of the trianion (21) with acids gives a 44% yield of the monohydride [Re$_7$(H)(C)(CO)$_{21}$]$^{2-}$, which exists as a mixture of two isomers in solution.[71] On the basis of x-ray data, one isomer appears to have the hydride ligand bridging an edge of the basal triangle, whereas the less symmetric isomer has the hydride at an interlayer edge.[72] Protonation of the dianion mixture with CF$_3$SO$_3$(H) gives the dihydride [Re$_7$(H)$_2$(C)(CO)$_{21}$]$^-$ (35% yield), which loses the capping [Re(CO)$_3$] fragment in MeCN solvent to give [Re$_6$H$_2$(C)(CO)$_{18}$]$^{2-}$ as a mixture of three isomers.[73]

The noncarbonyl-containing trirhenium cluster [Re$_3$(μ-Cl)$_9$(THF)$_3$] reacts with TMS-CH$_2$MgCl, PhMgBr, and MeLi to give clusters of the type [Re$_3$(R)$_6$(μ-Cl)$_3$] and their reactivity toward H$_2$ and alkenes has been investigated.[76]

## 9.3 HALIDE AND NITRIDE COMPLEXES

The carbonyl halides of rhenium continue to be best generated from [Re$_2$(CO)$_{10}$]. The thermal reaction of halogens with [Re$_2$(CO)$_{10}$] to give [Re(CO)$_5$X] and [Re(CO)$_5$(MeCN)]$^+$ (X = I, Br, Cl) has been proposed to occur via a rate-limiting end-on attack of X$_2$ at the rhenium–rhenium bond to give a halogenium intermediate, (24), which dissociates X$^-$ to give a halogenium ion (25) (Equation (7)).[77] The formation of halogenium intermediates similar to (24) is also consistent with results on the cleavage reactions of [Re$_2$(CO)$_{10-n}$(L)$_n$], where $n = 1$, 2 and L = group 15 donor ligand. In the case of [Re$_2$(CO)$_9$(PR$_3$)], the halogenium intermediate is unsymmetrical and capable of either a concerted decomposition pathway to give cis-[ReX(CO)$_4$(PR$_3$)] and [ReX(CO)$_5$] or a dissociative pathway to give both cis- and trans-[ReX(CO)$_4$(PR$_3$)] in addition to [ReX(CO)$_5$].[78]

$$(7)$$

A review of transition metal fluoro complexes, including rhenium, has recently appeared.[79] More recently, trans-[ReF(CO)$_3$(PPh$_3$)$_2$] has been prepared in 86% yield by chloride abstraction from trans-[ReCl(CO)$_3$(PPh$_3$)$_2$] using Ag(SO$_3$CF$_3$), followed by addition of [PPN]F.[80]

Diiodine oxidation of $[Re_2I_2(CO)_6(THF)_2]$ at 25 °C in heptane generates the mixed valence complex $[Re_4I_8(CO)_6]$, in which the two central rhenium(III) atoms are joined by a metal–metal triple bond (0.227 9(1) nm). When $[Re_2I_2(CO)_8]$ and $I_2$ are heated in heptane a trinuclear complex, $[Re_3I_6(CO)_6]$, is formed in 53% yield.[81]

The substitution reactions of $[ReX(CO)_5]$ (X = Cl, Br) with group 15 donor ligands are catalyzed by PdO, as was observed for the substitution reactions of $[Re_2(CO)_{10}]$.[82,83] The reaction of $[ReBr(CO)_5]$ with phosphinimines ($Ph_3P=NR$) gives the isocyanide complexes $[ReBr(CO)_4(CNR)]$ (R = $Pr^i$, Ph) in >96% yield, and the monosubstitution product reacts with additional phosphinimine to give diisocyanide complexes in 73–76% yields.[84]

Substituted rhenium(I) carbonyl complexes of formula $[ReI(CO)_{5-n}(L)_n]$, (where $n = 1$, L = $PBu^n_3$; $n = 2$, L = $PEt_3$, $PBu^n_3$, $PPh_3$), and $[ReI(CO)_3(LL)]$ (where LL = diphos, dmpe, and TMEDA), exist in equilibrium with the triiodide adducts in the presence of $I_2$. $[Re(I_3)(CO)_3(dmpe)]$ was structurally characterized and found to have a linear $I_3$ ligand (I–I–I 177.85(4)°) with a ReI–$I_2$ distance of 0.316 2(1) nm and a ReI$_2$–I distance of 0.275 7(2) nm.[85]

Protonation of alkylmetal complexes with HX is also a useful route toward metal halide complexes and these reactions are discussed in Section 9.4.

The bis(acetonitrile) cation $[Re(CO)_2(NCMe)_2(PPh_3)_2][ClO_4]$ reacts with $NaNO_2$ to give a 91% yield of the bidentate nitrite complex $[Re(\eta^2-O_2N)(CO_2)(PPh_3)_2]$ (**26**). Sodium borohydride reduction of (**26**) in the presence of $PPh_3$ provides a high yield (98%) route to $[ReH(CO)_2(PPh_3)_3]$. Addition of CO to (**26**) gives a 1:1 mixture of the oxygen- and nitrogen-bound isomers, *mer*-$[Re(\eta^1-NO_2)(CO)_3(PPh_3)_2]$ and *mer*-$[Re(\eta^1-ONO)(CO)_3(PPh_3)_2]$ (Equation (8)). Treatment of (**26**) with HCl gives a 1:1 ratio of $[ReCl(CO)_3(PPh_3)_2]$ and $[ReCl_2(CO)(NO)(PPh_3)_2]$.[86]

$$\text{(8)}$$

(**26**)

## 9.4 RHENIUM LEWIS ACIDS

### 9.4.1 $[Re(CO)_5(FBF_3)]$

Complexes of rhenium with weakly coordinating anions (e.g., $OSO_2CF_3$, $OClO_3$, etc.) or organohalide ligands (e.g., $CH_2Cl_2$, RX) have been extensively studied as organometallic Lewis acids. Beck and co-workers have developed the chemistry of $[Re(CO)_5(FBF_3)]$ (**27**), and much of this work prior to 1988 has been summarized in an excellent comprehensive review on metal complexes of weakly coordinating anions.[87] The $BF_4$ ligand in (**27**) is readily substituted by a number of anionic and neutral species, representative examples of which are shown in Scheme 3.[88–95]

### 9.4.2 $[Re(\eta^1-ClCH_2Cl)(NO)(PPh_3)(Cp)][BF_4]$

The chemistry of the chiral Lewis acid equivalent $[Re(\eta^1-ClCH_2Cl)(NO)(PPh_3)(Cp)][BF_4]$ (**28**) has been extensively developed by Gladysz and co-workers. Optically active (**28**) is generated in 98–99% *ee* from protonation of the optically active methyl complex (**29**), which in turn is prepared from racemic $[Re(CO_2Me)(NO)(PPh_3)(Cp)]$, as shown in Scheme 4.[96,97] The presence of coordinated $ClCH_2Cl$ in (**28**) was confirmed by low-temperature $^{13}C$ NMR spectroscopic data ($-85$ °C, 78.3 ppm, doublet, $J_{PC} = 3.8$ Hz, Re–$ClCH_2Cl$).[96] The pentamethylcyclopentadienyl analogues are accessible by a similar route.[98] The absolute configurations at rhenium are assigned by considering the cyclopentadienyl ligand to be a pseudoatom of atomic number 30 and, in complexes with more than one chiral center, the rhenium configuration is specified first. A useful discussion of the practical aspects of optical rotation measurements for organometallic compounds is available.[99]

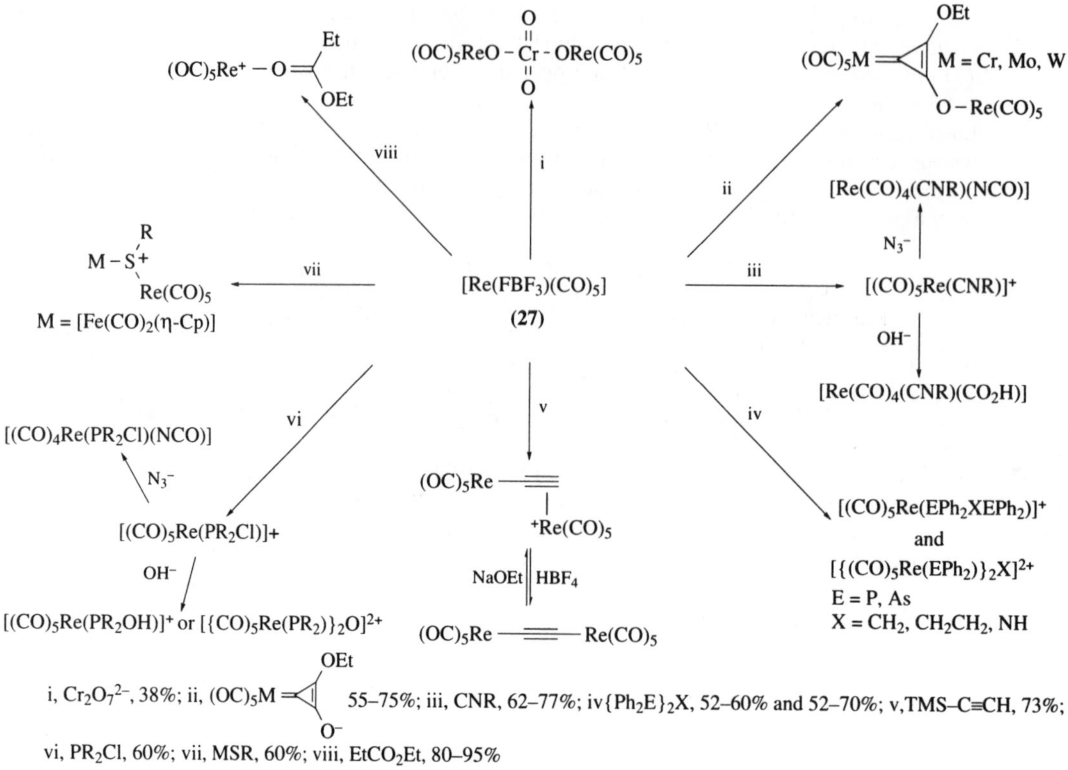

**Scheme 3**

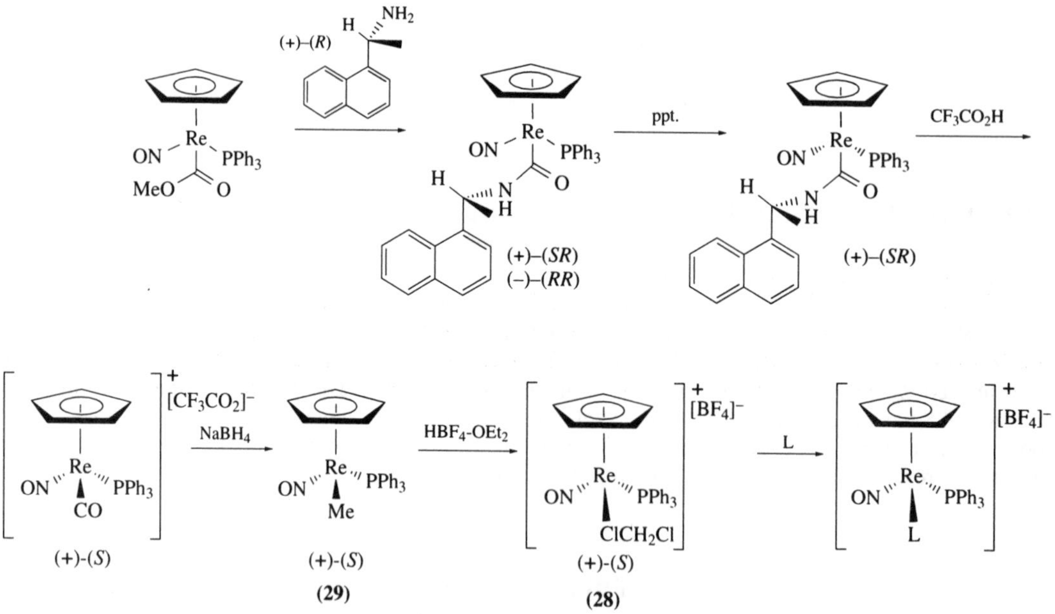

**Scheme 4**

Adducts of the chiral rhenium Lewis acid fragment $[Re(NO)(PPh_3)(Cp)]^+$ have been isolated for a wide array of organic and inorganic molecules including $R^1I$, $R^1OR^2$, $R^1SR^2$, $R^1S(=O)R^2$, $R^1R^2R^3N$, $R^1C\equiv N$, $R^1R^2C=NR^3$, $R^1R^2C=CR^3R^4$, $R^1C\equiv CR^2$, and $R^1R^2C=O$. These studies have resulted in systematic evaluations of binding trends and chiral recognition behavior which will be discussed under the various ligand sections. A key to understanding the properties and reactivity in these complexes is the coordination geometry (**30**) and the orientation of the $[Re(NO)(PPh_3)(Cp)]^+$ fragment HOMO (**31**), which interacts strongly with π-acceptor ligands in the fourth ligand site (L), thereby influencing the M–L conformational preferences.

(30)                (31)

Methyl complex (**29**) undergoes stereospecific reactions with protic acids HX at 0 °C to give products (**32**) with retention of configuration at rhenium (Equation (9)).[96] In contrast, reactions with halogens (Cl$_2$, Br$_2$, I$_2$) result in racemic halide complexes (6–11% *ee*). The former reaction proceeds via attack of the protic electrophile at the HOMO of (**29**) to give a square-pyramidal intermediate (**33**) followed by reductive elimination of methane and trapping by the nucleophile. Intermediate (**33**) is observable at low temperature by $^1$H NMR spectroscopy. Treatment of (+)-(*S*)-(**29**) with HPF$_6$·Et$_2$O at −78 °C in CH$_2$Cl$_2$ followed by addition of [PPN]Br gave the optically active bromide complex (+)-(*R*)-[Re(Br)(NO)(PPh$_3$)(Cp)] with overall retention of configuration in greater than 99% *ee*.[100]

(29)          (33)                        (32)     (9)

Protonation of [Re(H)(NO)(PPh$_3$)(Cp)] with HBF$_4$ at −78 °C leads to an observable dihydride, *cis*-[Re(H)$_2$(NO)(PPh$_3$)(Cp)]$^+$, which isomerizes to the *trans* isomer at −60 °C. At 22 °C, a 40:60 *cis:trans* ratio is established. The benzyl complex [Re(CH$_2$Ph)(NO)(PPh$_3$)(Cp)] reacts with HBF$_4$ in CH$_2$Cl$_2$ at −78 °C to give a benzyl hydride complex which slowly decomposes with liberation of toluene at ≥25 °C.[96]

Over the course of 30 h in CH$_2$Cl$_2$ solvent at 25 °C, the formate, trifluoroacetate, and iodide complexes (**32**) (X = O$_2$CH, O$_2$CCF$_3$, I) are configurationally stable. The bromide, chloride, and triflate racemize with $t_{1/2}$ = 3.1, 2.1, and 1.9 h, respectively. In benzene, the chloride and bromide complexes are configurationally stable over the course of 30 h at 25 °C.[101]

In CH$_2$Cl$_2$ solution at −20 °C, the PF$_6$ salt of (**28**) decomposes to uncharacterized products within 15 min, whereas the BF$_4$ salt of (**28**) preferentially binds one enantiomer of [Re(Cl)(NO)(PPh$_3$)(Cp)] to give [{Re(NO)(PPh$_3$)(Cp)}$_2$(μ-Cl)] in 64% yield.[100] The selectivity is due to unfavorable steric interactions between the cyclopentadienyl ligands in the *meso* diastereomer of the bridging halide complex. The prevailing steric interactions are enforced by a P–Re–X–Re torsion angle of 180° at both rhenium centers.[102] The pentamethylcyclopentadienyl complex [Re(Me)(NO)(PPh$_3$)(η-Cp*)] also gives a dichloromethane adduct upon protonation at −78 °C in CH$_2$Cl$_2$; however, subsequent oxidative addition ensues at −35 °C to produce [ReCl(CH$_2$Cl)(NO)(PPh$_3$)(η-Cp*)][BF$_4$]. Halide nucleophiles (X$^-$) attack the carbon atom of the coordinated ClCH$_2$Cl ligand to give XCH$_2$Cl and [ReCl(NO)(PPh$_3$)(η-Cp*)].[103]

Treatment of (**28**) with primary alkyl and acyl iodides generates the organoiodo complexes (**34**) in 64–86% isolated yields (Equation (10)). The reactivity order of RX toward (**28**) correlates with gas phase basicities, with RI > RBr > RCl. The bromides and chlorides are less readily formed than the iodides and could not be isolated. Coordination of the alkyl iodides activates them toward nucleophilic attack (Equation (10)). For example, coordination of iodoethane to rhenium leads to a rate acceleration of $3.3 \times 10^5$ for nucleophilic attack at the carbon–iodine bond by PPh$_3$ at 298 K. Coordinated aryl iodides failed to undergo nucleophilic substitution.[104] A theoretical study on [Re(IMe)(NO)(PH$_3$)(Cp)]$^+$ indicates that the LUMO is localized on iodine, which implies a reversible association of the nucleophile at iodine.[105] However, it has been experimentally determined that carbon–iodine bond cleavage occurs with inversion of configuration at carbon.[106] Metal-substituted alkyl halides, [M(CH$_2$I)(CO)$_5$], and (**28**) form bimetallic bridging haloalkyl complexes of formula [Re(NO)(PPh$_3$)(Cp){I(CH$_2$)M(CO)$_5$}][BF$_4$] in 78% yield.[107]

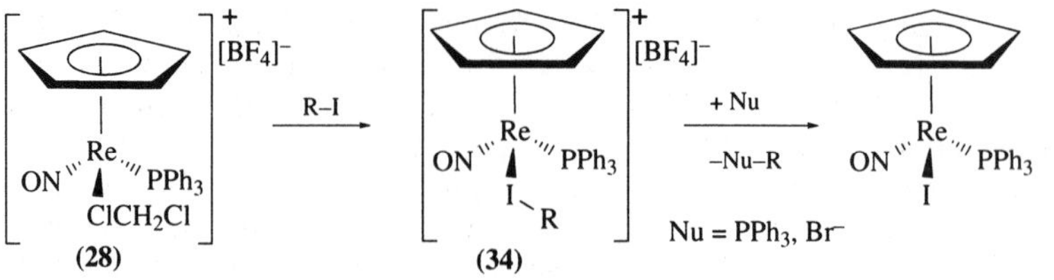

$$\begin{bmatrix} \overset{\ominus}{\underset{ON^{\text{\tiny I\!I\!I}}\underset{\displaystyle|}{\overset{\displaystyle|}{Re}}^{\text{\tiny I\!I\!I}}PPh_3}{ClCH_2Cl}} \end{bmatrix}^{+} [BF_4]^{-} \xrightarrow{R-I} \begin{bmatrix} \overset{\ominus}{\underset{ON^{\text{\tiny I\!I\!I}}\underset{\displaystyle I}{\overset{\displaystyle|}{Re}}^{\text{\tiny I\!I\!I}}PPh_3}{\underset{R}{}}} \end{bmatrix}^{+} [BF_4]^{-} \xrightarrow[-Nu-R]{+\,Nu} \overset{\ominus}{\underset{ON^{\text{\tiny I\!I\!I}}\underset{\displaystyle I}{\overset{\displaystyle|}{Re}}^{\text{\tiny I\!I\!I}}PPh_3}{}} \qquad (10)$$

(28)                                (34)                      Nu = PPh₃, Br⁻

## 9.5 RHENIUM HYDRIDES

### 9.5.1 Synthesis

The properties and reactivity of rhenium hydrides has been a very active research area during the 1980s and early 1990s, in part as a result of expanding interest in carbon–hydrogen bond activation chemistry and the discovery of $\eta^2$-dihydride ligands. A number of new or improved routes toward mononuclear rhenium hydrides have appeared since 1981.

The inorganic hydride $[ReH_7L_2]$ (L = triarylphosphines) is a useful precursor to organometallic hydrides through C–H bond activation routes as shown in Scheme 5 (TBE = tetrabutylethene).[108–14]

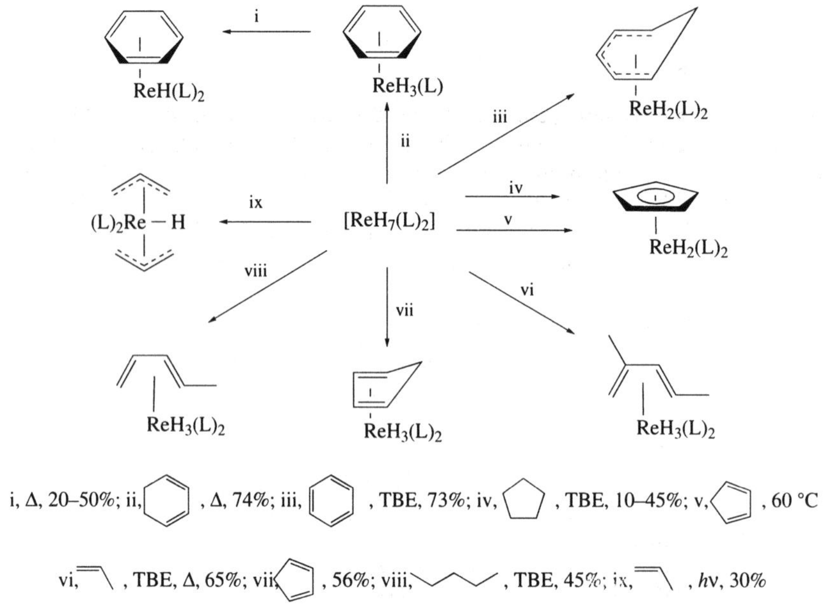

i, Δ, 20–50%; ii, ⬡ , Δ, 74%; iii, ⬡ , TBE, 73%; iv, ⬠ , TBE, 10–45%; v, ⬠ , 60 °C

vi, ⟍ , TBE, Δ, 65%; vii ⬠ , 56%; viii, ⟍⟋⟍ , TBE, 45%; ix, ⟍ , *hν*, 30%

**Scheme 5**

The cationic carbonyl complex $[Re(CO)_2(NO)(Cp)]^+$ is readily converted to $[ReH(CO)(NO)(Cp)]$ in 90% yield, via decarboxylation (NEt₃, H₂O, 50 °C, 1 h) of a hydroxycarbonyl intermediate, $[Re\{\eta^1\text{-}C(=O)OH\}(CO)(NO)(Cp)]$ (isolable in 71% yield).[115]

$[ReH(Cp)_2]$ is now available from $[ReCl_5]$ in 40% yield,[116] and $[ReH(\eta\text{-}Cp^*)_2]$ is accessible from rhenium atom vapor and Cp*H, albeit in 11% yield.[117] Both of these hydrides are converted to 17-electron rhenocene radicals upon photolysis.[118]

A number of routes toward the dihydride, *trans*-$[Re(H)_2(CO)_2(Cp)]$, *trans*-(35), have been developed and its chemistry extensively explored. Dihydride (35) is available from photolysis of $[Re(CO)_3(Cp)]$ in supercritical xenon pressurized with H₂,[119] and photolysis of $[Re(CO)_3(Cp)]$ and Et₂SiH₂ in heptane leads to (35) in 49% yield.[120] A more convenient preparation of (35) is from Zn/HOAc reduction of a mixture of *cis*- and *trans*-$[ReBr_2(CO)_2(Cp)]$.[121]

The monohydride $[ReH(CO)_2(Cp)]^-$ (36) is formed upon deprotonation of *trans*-(35) with KOH, but no reaction is observed upon attempted deprotonation of monohydride (36) with $LiNPr_2^i$ in either glyme or MeCN.[121] Protonation of (36) with CF₃CO₂H at −78 °C gives a 65:35 mixture of *cis*-(35) (IR; THF, −78 °C; 1987, 1900 cm⁻¹) and *trans*-(35) (IR; THF, −78 °C; 1996, 1920 cm⁻¹), which, upon warming to 24 °C, gives a 2:98 *cis*:*trans* mixture.[122] Photolysis of *trans*-(35) in a methylcyclohexane glass at 10 K leads to a 40:60 *cis*:*trans* photostationary state.

A nonobvious but high yield route to *cis*-(**35**) is from reaction of [(Cp)(CO)$_2$Re($\mu$-H)PtH(PPh$_3$)$_2$] and diphenylethyne, which gives an 84:16 *cis*:*trans* ratio at −9 °C. Multiple pathways have been proposed for isomerization of *cis*-(**35**) to *trans*-(**35**), depending on the mode of generation of *cis*-(**35**) and the presence of added bases or oxidants.[122]

The unusual dinuclear dihydride [Re$_2$($\mu$-H)$_2$(CO)$_4$($\eta$-Cp*)$_2$] (**37**) is formed in 90% yield upon exposure of [Re(CO)$_2$($\eta$-Cp*)]$_2$ (**38**) to H$_2$ gas (Equation (11)).[123] The rhenium–rhenium distance of 0.313 2(1) nm in (**37**) is appropriate for a net bond order of one, resulting from two three-center two-electron Re($\mu$-H)Re bonds and a rhenium–rhenium antibond. Dimer (**38**) contains two semibridging carbonyls and exhibits a Re=Re bond length of 0.272 3(1) nm. Complex (**38**) is formed in 98% yield from [Re(CO)$_2$(THF)($\eta$-Cp*)] via a surprising solid-state reaction which presumably involves a bridging carbonyl intermediate and not a dimerization of two [Re(CO)$_2$($\eta$-Cp*)] 16-electron fragments.

$$(11)$$

(**38**)            (**37**)

## 9.5.2 Physical Properties

The few available thermodynamic acidities of rhenium hydrides are listed in Table 4,[124] and Table 5 provides representative examples of rhenium hydride complexes along with $^1$H NMR spectroscopic data for the hydride ligands. The rhenium–hydrogen bond dissociation energy for [ReH(CO)$_5$] is reported at 314 kJ mol$^{-1}$, 29 kJ mol$^{-1}$ higher than the manganese–hydrogen value for [MnH(CO)$_5$].[124] Vibrational spectroscopy on matrix-isolated carbonyl hydrides [MH(CO)$_n$] (M = Mn, Re, Fe, Co) indicates a negatively polarized hydrogen with little variation in magnitude as M is varied.[125]

**Table 4** Thermodynamic acidities of rhenium hydrides.[124]

| Compound | p$K_a$ (Solvent) |
| --- | --- |
| [ReH(CO)$_5$] | 21.1 (MeCN), 13.6 (H$_2$O) |
| [MnH(CO)$_5$] | 14.2 (MeCN), 6.7 (H$_2$O) |
| [ReH$_2$(CO)$_2$(Cp)] | 23.0 (MeCN), 15.5 (H$_2$O) |
| [ReH$_2$(Cp)$_2$]$^+$ | 5.5 (60% dioxane/water), 26–30 (H$_2$O) |
| [ReH(NO)(PPh$_3$)(Cp)] | 26–30 (H$_2$O) |
| [ReH$_4$(PMe$_2$Ph)$_4$]$^+$ | 25.3 (MeCN) |
| [ReH$_4$(CO)(PMe$_2$Ph)$_3$]$^+$ | <19 (MeCN) |
| [ReH$_3$(CO)$_{12}$] | 3 (H$_2$O) |
| [Re$_3$H$_2$(CO)$_{12}$]$^-$ | 10 (H$_2$O) |
| [Re$_3$H(CO)$_{12}$]$^{2-}$ | 25 (H$_2$O) |

NMR spectroscopy has been employed to obtain M–H and H–H bond distances in hydride and dihydrogen complexes. Spin–lattice ($T_1$) relaxation times were proposed as a criterion to distinguish classical and nonclassical ($\eta^2$-H$_2$) hydride complexes based on the assumption that dipole–dipole interactions between the hydride ligands dominate the relaxation of the hydride NMR signals.[126] Dipole–dipole interactions between the hydride(s) and other nuclei in a complex may also contribute significantly to relaxation and must be considered in the interpretation of $T_{1(min)}$ values.[127–9] The relative contributions of proton–proton and proton–metal dipole–dipole interactions toward $T_{1(min)}$ have been determined from experiments on complexes with perdeuterated phosphine ligands.[127,128] More recently Berke and co-workers have demonstrated an "isotropic motion approach," which uses the observed selective ($T_{1s}$) and biselective ($T_{1bs}$) relaxation times for determining Re–H and $\eta^2$-H–H distances in nondeuterium-enriched complexes.[130]

For certain rhenium di- and polyhydride complexes, an equilibrium exists between dihydride and dihydrogen isomers. Protonation of the classical trihydride [Re(H)$_3$(CO)(PMe$_2$Ph)$_3$] at low temperature generates an equilibrium mixture of the tetrahydride [Re(H)$_4$(CO)(PMe$_2$Ph)$_3$][BF$_4$] (**39a**) and the pentagonal bipyramidal dihydrogendihydride [ReH$_2$($\eta^2$-H$_2$)(CO)(PMe$_2$Ph)$_3$][BF$_4$] (**39b**), with (**39a**) as the major species at 193 K (Equation (12)).[131] The equilibrium shifts toward the dihydrogen complex

**Table 5** Selected rhenium hydrides and dihydrogen complexes.

| Compound | $^1H$ NMR hydride shifts | Ref. |
|---|---|---|
| [ReH(CO)(NO)(Cp)] | $-8.5$ (br)[a] | 115 |
| [Re(H)(PPh$_3$)$_2$($\eta$-C$_6$H$_6$)] | $-7.2$(t, $J = 36$)[c] | 113 |
| [Re(H)(Cp*)$_2$] | $-13.24$[c] | 118 |
| [Re(H)(CO)$_2$(Cp)][NEt$_4$] | $-12.35$(s)[g] | 121 |
| [Re(H)(CO)$_2$(Cp)][K] | $-12.38$(s)[g] | 121 |
| [Re(H)(Me)(CO)$_2$(Cp)] | $-10.12$(q, $J = 6.5$)[h] | 121 |
| cis-[Re(H)$_2$(CO)$_2$(Cp*)] | $-8.72$(s) $-78$ °C[e] | 122 |
| trans-[Re(H)$_2$(CO)$_2$(Cp*)] | $-9.65$(s) $-78$ °C[e] | 122 |
| cis-[Re(H)$_2$(CO)$_2$(Cp)] | $-9.27$ $-78$ °C[e] | 122 |
| trans-[Re(H)$_2$(CO)$_2$(Cp)] | $-9.90$ $-78$ °C[e] | 122 |
| cis-[Re(H)(D)(CO)$_2$(Cp)] | $-9.296$(t, $J = 5.3$) $-34$ °C[e] | 122 |
| trans-[Re(H)(D)(CO)$_2$(Cp)] | $-9.959$(t, $J_{HD} = 2.0$) $-34$ °C[e] | 122 |
| [(Cp*)(CO)$_2$Re($\mu$-H)$_2$Re(CO)$_2$(Cp*)] | $-6.19$(s)[e] | 123 |
| [ReH($\mu$-H)$_2$(CO)$_8$] | $-9.38$[b] | 128 |
| [ReH(CO)$_3$\{P(OPr$^i$)$_3$\}$_2$] | $-6.17$(t, $J_{PH} = 22.7$) 298 K[c] | 129 |
| [ReH$_3$(CO)(PMe$_3$)$_3$] | $-6.28$(q, $J_{PH} = 23.0$) 298 K[e] | 130 |
| trans-[ReH(CO)(PMe$_3$)$_4$] | $-6.75$(m) 168 K[a] | 130 |
| [Re(H)$_4$(CO)(PMe$_2$Ph)$_3$][BF$_4$] | $-3.94$(q, $J = 17.7$) 188 K[a] | 131 |
| [Re(H)$_2$($\eta^2$-H$_2$)(CO)(PMe$_2$Ph)$_3$][BF$_4$] | $-4.86$(m), $-5.36$(br) 188 K[a] | 131 |
| [Re(H)$_2$(CO)$_3$(PMe$_3$)$_2$]$^+$ | $-4.70$(br,s, $\Delta = 94$ Hz) 181 K[a] | 132 |
| [Re(H)$_2$(CO)$_2$(PMe$_3$)$_3$][CF$_3$CO$_2$] | $-4.73$(br,s, $\Delta = 58$ Hz) 193 K[a] | 132 |
| [Re(H$_2$)(CO)(PMe$_3$)$_4$][CF$_3$CO$_2$] | $-6.45$(br,s, $\Delta = 35$ Hz) 213 K[a] | 132 |
| [Re(H)$_2$(CO)(PMe$_3$)$_4$][CF$_3$CO$_2$] | $-5.89$(m) 233 K[a] | 132 |
| [Re(H)$_4$(H$_2$)(CO)(PMe$_3$)$_3$][CF$_3$CO$_2$] | $-5.53$(app.t, $J_{HP} = 46$); $-6.0$(vbr,s) 168 K[a] | 132 |
| [Re(H)$_4$(CO)(PMe$_3$)$_3$][CF$_3$CO$_2$] | $-4.79$(q, $J_{HP} = 19.3$) 168 K[a] | 132 |
| [ReH(H$_2$)(CO)(NO)(PPr$^i$)$_2$][CF$_3$CO$_2$] | $-2.46$(t, $J_{HP} = 24.4$) 158 K[a] | 132 |
| [Re(H)(NO)(PPh$_3$)($\eta$-C$_5$H$_4$Me)] | $-8.79$(d, $J = 29.1$)[c] | 133 |
| [Re(H)(NO)(PPh$_3$)($\eta$-Cp)] | $-9.15$(d, $J = 29.7$)[c] | 133 |
| [Re(H)(NO)(PPh$_3$)($\eta$-Cp*)] | $-7.80$(d, $J = 31.8$)[c] | 133 |
| [ReH(CO)(PMe$_3$)$_4$] | $-7.04$(8 lines, tdd, $J_{PH} = 25.2$, $J_{PH} = 25.1$, $J_{PH} = 7.4$)[c] | 134 |
| [Re(H)($\eta^2$-CSO)(PMe$_3$)$_4$] | $-3.10$(qnt, $^2J_{PH} = 7.6$)[c] | 134 |
| [Re(H$_2$)(CO)(PMe$_3$)$_4$][BF$_4$] | $-5.75$(2nd order qnt)[d] | 134 |
| mer,trans-[ReH(CO)$_3$(PPh$_3$)$_2$] | $-5.12$(t, $^2J_{PH} = 18.2$)[a] | 134 |
| [Re(H)$_7$(PPh$_3$)$_2$] | $-4.20$(t, $J = 18$)[c] | 135 |
| [Re(H)$_5$(PPh$_3$)$_3$] | $-4.67$(q $J = 19.0$)[c] | 135 |
| [Re(H)$_3$(PPh$_3$)$_2$($\eta^4$-CpH)] | $-2.358$(t, $J = 38.8$); $-7.087$(t, $J = 19.4$) $-50$ °C[b] | 135 |
| [Re(H)$_2$\{P($p$-C$_6$H$_4$F)$_3$\}$_2$(Cp)] | $-10.30$(t, $J = 40.2$)[c] | 135 |
| [Re(H)$_2$\{P($p$-C$_6$H$_4$OMe)$_3$\}$_2$(Cp)] | $-9.888$(t, $J = 40.2$)[c] | 135 |
| trans-[Re(H)$_2$(PPh$_3$)$_2$(Cp)] | $-9.952$(t, $J = 40.1$)[c] | 136 |
| Re(H)$_2$(PMe$_3$)$_2$(Cp)] | $-12.13$(t, $J = 43.6$)[c] | 136 |
| [Re(H)$_2$\{P($p$-Tol)$_3$\}$_2$Cp] | $-9.901$(t, $J = 40.2$)[c] | 136 |
| [Re(H)$_2$\{P(Tol)$_3$\}(PPh$_3$)(Cp)] | $-9.923$(t, $J = 40.1$)[c] | 136 |
| [Re(H)$_2$(PMe$_3$)(PPh$_3$)(Cp)] | $-11.186$(dd, $J = 44.7,40.7$)[c] | 136 |
| [ReH(CNEt)$_3$(PPh$_3$)$_2$] | $-5.071$(t, $J = 19.6$)[c] | 137 |
| [Re(H)(CNMe)$_3$(PPh$_3$)$_2$] | $-5.114$(t, $J = 18.9$)[c] | 137 |
| [Re(H)(CO)$_3$(PPh$_3$)$_2$] | $-4.457$(t, $J = 18.2$)[c] | 137 |
| [Re(H)(Ph)(PMe$_3$)$_2$(Cp)] | $-12.98$(t, $J = 52.5$)[c] | 138 |
| [Re(H)(c-C$_3$H$_5$)(PMe$_3$)$_2$(Cp)] | $-13.29$ (td, $J = 49.7, 4.9$)[b] | 138 |
| [Re(H)$_2$(PMe$_3$)$_2$(Cp)] | $-12.14$(t, $J = 43.6$)[c] | 138 |
| [Re(H)($\eta$-C$_6$H$_{13}$)(PMe$_3$)$_2$(Cp)] | $-12.53$ (tt, $J = 50.1, 3.5$) $-20$ °C[b] | 138 |
| [Re(H)(c-C$_5$H$_{11}$)(PMe$_3$)$_2$(Cp)] | $-12.41$ (td, $J = 51.3, 5.4$) $-20$ °C[b] | 138 |
| [Re(H)($\eta^1$-CH$_2$PMe$_2$)(PMe$_3$)$_2$(Cp)] | $-12.77$(tm, $J = 50.8$) $-20$ °C[i] | 138 |
| [Re(H)(Me)(PMe$_3$)$_2$(Cp)] | $-12.34$ (tq, $J = 48.8, 2.6$) $-15$ °C[b] | 138 |
| [Re(H)(CH=CH$_2$)(PMe$_3$)$_2$(Cp)] | $-12.57$ (dt, $J = 4.2, 50.6$)[b] | 138 |
| trans-(H,Ph)[Re(H)(Ph)(CO)(PMe$_3$)(Cp*)] | $-9.18$(d, $J = 66.7$)[c] | 139 |
| [Re(H)$_2$(PPh$_3$)$_2$($\eta^5$-C$_4$H$_4$N)] | $-10.07$(t, $J = 41$)[a] | 140 |
| [Re(H)$_2$(PPh$_3$)$_2$($\eta^5$-C$_4$H$_4$NMe)]$^+$ | $-8.00$(s)[f] | 140 |
| [Re(H)(I)(PPh$_3$)$_2$($\eta^5$-C$_4$H$_4$N)] | $-10.23$(t, $J = 48$)[a] | 140 |
| [Re(H)$_2$(PPh$_3$)$_2$($\eta^4$-C$_4$H$_5$NMe)] | $-2.0$ and $-10.8$(br)[f] | 140 |
| endo-[Re(H)(CO)($\eta^3$-C$_3$H$_5$)(Cp*)] | $-11.65$ | 141 |
| exo-[Re(H)(CO)($\eta^3$-C$_3$H$_5$)(Cp*)] | $-9.23$ | 141 |
| cis-[Re(H)(SiPh$_3$)(CO)$_2$(Cp)] | $-9.01$(s) $-25$ °C[e] | 142 |
| trans-[Re(H)(SiPh$_3$)(CO)$_2$(Cp)] | $-9.71$(s) $-25$ °C[e] | 142 |
| [Re(H)(C$_5$H$_4$SiPh$_3$)(CO)$_2$][Li] | $-12.43$(s) 25 °C[e] | 142 |
| [ReH(MeOH)(NO)(PPh$_3$)$_3$]$^+$ | 1.6 ("featureless lump")[a] | 142 |
| [ReH(OMe)(NO)(PPh$_3$)$_2$] | 3.82 (t, $J = 22.9$)[a] | 143 |
| [ReH(X)(NO)(PPh$_3$)$_3$] (X = Cl, Br, N$_3$) | 0.47–0.79 (dt, $J_{PH} \simeq 27$, 68)[a] | 143 |
| [ReH(X)(C$_7$H$_7$NC)(NO)(PPh$_3$)$_2$] (X = OMe, F, Cl, Br, N$_3$, NCO, I, SCN) | 1.02–3.78 ($J_{PH} = 19.3$–22.2)[a] | 143 |

[a] CD$_2$Cl$_2$. [b] Toluene-d$_8$. [c] C$_6$D$_6$. [d] CD$_3$OD. [e] THF-d$_8$. [f] Acetone-d$_6$. [g] CD$_3$CN. [h] CDCl$_3$. [i] cyclohexane-d$_{12}$.

(39b) with increasing temperature, and above 283 K irreversible loss of $H_2$ occurs. The proposed nonclassical structure for (39b) is supported by the observation of the well-defined 1:1:1 triplet with $^1J_{HD} = 34$ Hz ($\delta$ −5.48) in the $^1H\ \{^{31}P\}$NMR spectrum of $[ReD_2(\eta^2\text{-HD})(CO)(PMe_2Ph)_3]^+$. Deuterium enrichment shifts the equilibrium toward the dihydrogen complex due to a greater vibrational zero-point energy difference between $Re(\eta^2\text{-HD})$ and $Re(\eta^2\text{-DD})$ relative to that between $ReH_2$ and $ReD_2$. The observation that hydride site exchange in (39b) is much faster than the (39a)⇌(39b) interconversion rate, and the assumption that the rearrangement barrier in an eight-coordinate system is very low (ruling out a tetrahydride intermediate that is different from (39a)), led to the proposal of a trihydrogen intermediate for the hydrogen site exchange within (39b) (Equation (13)). A bis(dihydrogen) intermediate was ruled out by the observation of diastereotopic methyl resonances for the equatorial $PMe_2Ph$ ligands in the $^1H\{^{31}P\}$NMR spectrum under conditions of rapid hydride–dihydrogen exchange.

$$ (12) $$

(39a)     P = PMe$_2$Ph     (39b)

$$ (13) $$

Protonation of $[ReH_3(CO)(PMe_3)_3]$ with $CF_3CO_2H$ at −95 °C also gives an equilibrium mixture of classical and nonclassical hydrides. In the dihydrogendihydride complex, rapid hydrogen scrambling was suggested to proceed through a symmetric tetrahydride intermediate (40), rather than a trihydrogen species (Equation (14)).[132]

$$ (14) $$

(40)

P = PMe$_3$, CO ligand not shown

More electron-rich metal centers favor dihydride ligands in preference to the dihydrogen isomers. Both $[Re(H)_4(PMe_2Ph)_4]^+$ and $[Re(H)_2(CO)(PMe_2Ph)_4]^+$ exist as classical hydrides, whereas $[Re(H)_2(CO)_2(PMe_2Ph)_3]^+$ exists in equilibrium with $[Re(H_2)(CO)_2(PMe_2Ph)_3]^+$.[131] At low temperature $[Re(H_2)(CO)(PMe_3)_4]^+$ exists solely as the dihydrogen complex, but warming above −30 °C results in irreversible isomerization to the classical hydride $[Re(H)_2(CO)(PMe_3)_4]^+$.[132]

### 9.5.3 Reactivity

Rhenium hydrides exhibit both acidic and nucleophilic character. Bursten and Gatter used molecular orbital studies to relate hydridic behavior to three-legged piano stool structures, $[MH(L)_2(Cp)]$, and protic behavior to four-legged piano stool hydride structures, such as (35).[144]

Hydride abstraction from $[ReH(CO)(NO)(Cp)]$ by $Ph_3CPF_6$ in the presence of donor ligands yields the cationic adducts $[Re(CO)(NO)(L)(Cp)][PF_6]$ (L = MeCN, THF, Me$_2$CO) (Scheme 6).[115]

The distinction between thermodynamic and kinetic acidity may be important when other protic ligands are present in a complex. For example, kinetic deprotonation of (41) by Bu$^n$Li at −78 °C occurs at the cyclopentadienyl ligand to give (42). Alkylation of (42) at −78 °C generates $[ReH(NO)(PPh_3)(\eta C_5H_4Me)]$, whereas warming solutions of (42) to −32 °C results in a proton shift from rhenium to the $C_5H_4Li$ ligand and subsequent alkylation gives $[Re(Me)(NO)(PPh_3)(Cp)]$ (Scheme 6). Formation of the kinetic product may be attributed to the rehybridization and negative charge delocalization involved in metal anion formation, compared to cyclopentadienyl deprotonation which entails neither rehybridization nor charge delocalization.[133]

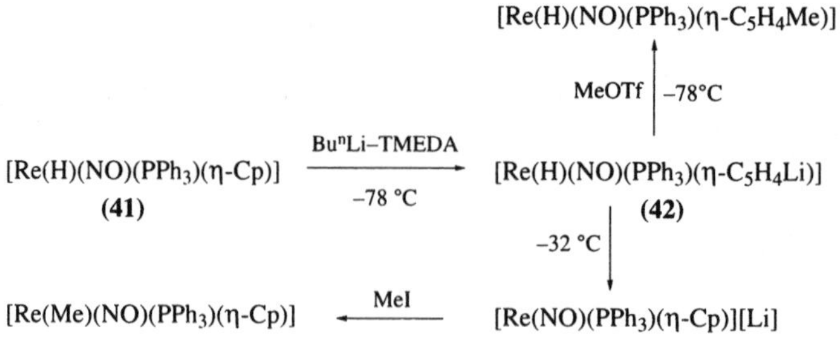

$[Re(H)(NO)(PPh_3)(\eta\text{-}C_5H_4Me)]$

$\uparrow$

MeOTf $\bigg|$ $-78$°C

$[Re(H)(NO)(PPh_3)(\eta\text{-}Cp)]$ $\xrightarrow[\text{-78 °C}]{\text{Bu}^n\text{Li-TMEDA}}$ $[Re(H)(NO)(PPh_3)(\eta\text{-}C_5H_4Li)]$

(41)                                                                (42)

$-32$ °C $\bigg|$

$[Re(Me)(NO)(PPh_3)(\eta\text{-}Cp)]$ $\xleftarrow{\text{MeI}}$ $[Re(NO)(PPh_3)(\eta\text{-}Cp)][Li]$

**Scheme 6**

Norton and co-workers reported the first quantitative measure of relative nucleophilicity for a series of transition metal hydrides based on the rates of $[Re\{C(=O)CH_2Me\}(CO)_4(NCMe)]$ (43) reduction by $[MH(CO)_n(L)]$ to give $[(CO)_4(MeCN)Re–M(CO)_n(L)]$ and propanal.[145] The reaction is first order in both (43) and $[MH(CO)_n(L)]$ and appears to involve nucleophilic attack at an unsaturated rhenium acyl complex by the pair of electrons in the H–Re bond to generate a species with a three-center, two-electron bond. On a relative scale, the kinetic nucleophilicites follow the order $[ReH(CO)_5](139) > [MnH(CO)_5](100) \gg [CrH(CO)_3(Cp)](1)$, which is the reverse of the kinetic acidity order.

Activated alkenes and alkynes insert into the metal–hydrogen bond of $[ReH(CO)(NO)(Cp)]$ to give $\sigma$-alkyl and $\sigma$-alkenyl complexes in low yields.[146] Whereas $[ReH(Cp)_2]$ does not react with acetone or trifluoroacetone[147] at 78 °C, it does react with activated alkynes to give $\sigma$-alkenyl complexes in moderate to excellent yields.[148] Reaction of $[ReH(Cp)_2]$ with the $\eta^2$-acyl ligand in $[Zr(Me)\{\eta^2\text{-}C(=O)Me\}(Cp)_2]$ at 25 °C gives $[(Cp)_2(Me)Zr(\mu\text{-}\eta^1,\eta^2\text{-}OCHMe)Re(Cp)_2]$.[149]

Oxidation of $[ReH(Cp)_2]$ with $CuCl_2$ or $AlX_3$ generates $[Re(Cp)_2][CuCl_2]$ and $[Re(HAlX_3)(Cp)_2]$, respectively.[150,151]

Rhenium hydrides also react with carbon dioxide to give metalloformato complexes, $[Re\{\eta^1\text{-}O(CH=O)\}(L)_n]$, in good yields.[134,152] In the case of *fac*-$[ReH(bipy)(CO)_3]$, kinetic studies are consistent with a direct associative hydride transfer to the carbon of $CO_2$ via a polar, charge-separated activated complex.[152] In contrast, the reaction of $[ReH(CO)_2(PPh_3)_3]$ with carbon suboxide occurs by dissociation of $PPh_3$, $C_3O_2$ coordination, and insertion to give (44) (Equation (15)).[153]

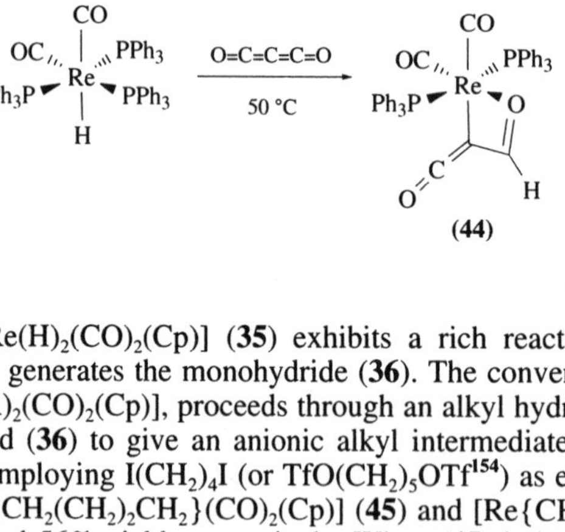

$$
\begin{array}{c}
\text{CO} \\
\text{OC}_{\prime\prime,}\,|\,_{,\backslash\backslash}\text{PPh}_3 \\
\text{Ph}_3\text{P}\blacktriangleright\!\!\underset{|}{\text{Re}}\!\!\blacktriangleleft\text{PPh}_3 \\
\text{H}
\end{array}
\xrightarrow[\text{50 °C}]{\text{O=C=C=C=O}}
\begin{array}{c}
\text{CO} \\
\text{OC}_{\prime\prime,}\,|\,_{,\backslash\backslash}\text{PPh}_3 \\
\text{Ph}_3\text{P}\blacktriangleright\!\!\text{Re}\!\!\blacktriangleleft\text{O} \\
\underset{\text{O}}{\overset{}{\underset{}{C}}}{=}\!\!\!\diagdown\,_{\text{H}} \\
\end{array}
\qquad (15)
$$

(44)

The dihydride complex $[Re(H)_2(CO)_2(Cp)]$ (35) exhibits a rich reactivity involving the hydride ligands. Deprotonation of (35) generates the monohydride (36). The conversion of $[ReH(CO)_2(Cp)][K]$ (36) to dialkyl products, $[Re(R)_2(CO)_2(Cp)]$, proceeds through an alkyl hydride intermediate which is in turn deprotonated by unreacted (36) to give an anionic alkyl intermediate which undergoes a second alkylation (Scheme 7).[121] By employing $I(CH_2)_4I$ (or $TfO(CH_2)_5OTf$[154]) as electrophiles, in the presence of dbu, the metallacycles $[Re\{CH_2(CH_2)_2CH_2\}(CO)_2(Cp)]$ (45) and $[Re\{CH_2(CH_2)_3CH_2\}(CO)_2(Cp)]$ are generated from (35) in 53% and 56% yield, respectively. When (45) is heated in benzene at 100 °C a 40% yield of methylcyclopropane is formed by the mechanism shown in Scheme 7.[121] Thermolysis of $[Re\{CH_2(CH_2)_3CH_2\}(CO)_2(Cp)]$ gives 1-pentene and cyclopentane.[154]

Casey *et al.* have examined the reactivity of dihydride (35) toward low-valent late metal complexes. For example, reaction of (35) and Vaska's complex in the presence of $PPh_3$ gives $[Re(CO)_3(Cp)]$ and $[Ir(H)_2(Cl)(L)(PPh_3)_2]$ (L = CO, 39%; L = $PPh_3$, 39%). The dihydride transfer was proposed to proceed by insertion of $[Ir(Cl)(CO)(PPh_3)_2]$ into a Re–H bond of (35) to form a bimetallic dihydride intermediate (46), with the second hydrogen transferred from rhenium to iridium as CO is simultaneously transferred from iridium to rhenium (Equation (16)).[155]

When (35) undergoes reaction with a low-valent metal lacking a CO for back transfer to rhenium only the first hydride of (35) is transferred. Thus, (35) and $[Pt(PPh_3)_2(CH_2=CH_2)]$ reversibly gives ethene and the rhenium–platinum dihydride (47) (Equation (17)).[156] The hydride ligands of (47) undergo rapid exchange, even at $-110$ °C on the NMR timescale. In addition, the phosphines undergo a slower self-exchange process.

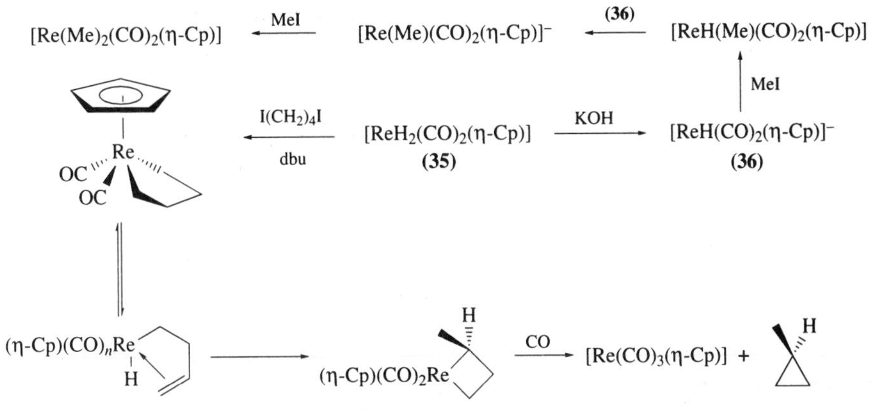

$[Re(Me)_2(CO)_2(\eta\text{-Cp})]$ $\xleftarrow{\text{MeI}}$ $[Re(Me)(CO)_2(\eta\text{-Cp})]^-$ $\xleftarrow{\textbf{(36)}}$ $[ReH(Me)(CO)_2(\eta\text{-Cp})]$

$\Big\uparrow$ MeI

$\xleftarrow[\text{dbu}]{\text{I(CH}_2)_4\text{I}}$ $[ReH_2(CO)_2(\eta\text{-Cp})]$ $\xrightarrow{\text{KOH}}$ $[ReH(CO)_2(\eta\text{-Cp})]^-$

**(35)** **(36)**

**Scheme 7**

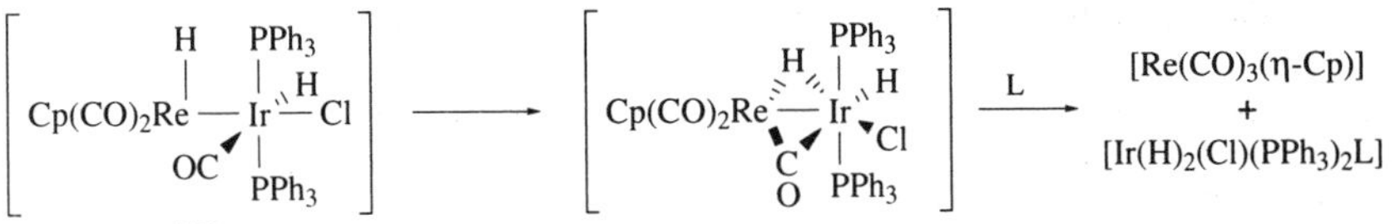

$$[Re(CO)_3(\eta\text{-Cp})] + [Ir(H)_2(Cl)(PPh_3)_2L] \tag{16}$$

**(46)**

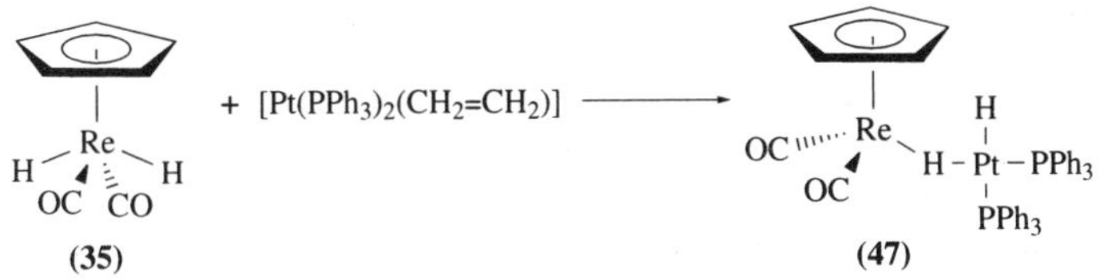

$$\textbf{(35)} + [Pt(PPh_3)_2(CH_2=CH_2)] \longrightarrow \tag{17}$$

**(47)**

Heterobimetallic dihydride (**47**) slowly catalyzes the hydrogenation of ethene at 25 °C but catalyst decomposition limits the system to four turnovers.[157] Reaction of (**47**) and 2 equiv. propyne gives the propene complex $[Re(CO)_2(\eta^2\text{-CH}_2=CHMe)(Cp)]$ in 99% yield. Significantly, no reaction is observed between (**35**) and propyne in the absence of platinum compounds. The reaction represents the first stoichiometric hydrogenation of an alkyne to a metal–alkene complex and is a useful preparation of pure cyclopentadienylrhenium–alkene complexes in high yield. The mechanism for reduction of 2-butyne is shown in Scheme 8. Again, a key feature is transfer of hydrogen from rhenium to platinum as the vinyl group coordinates to rhenium, thereby avoiding an unsaturated rhenium(I) intermediate. Consistent with the proposed mechanism is the observation that $[Re(H)(CO)_2(Cp)]K$ and $[Pt(\eta^1\text{-CMe}=CHMe)(PPh_3)_2][CF_3SO_3]$ also gives (**48**).[158]

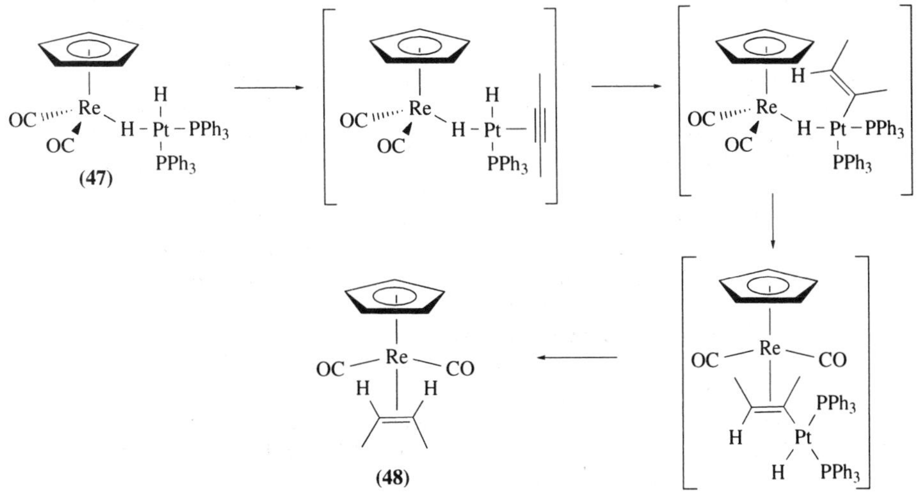

**(47)**

**(48)**

**Scheme 8**

The catalytic hydrogenation of alkynes in this system is prevented by the remarkable stability of the [Re(CO)$_2$(alkene)(Cp)] complexes. Substitution of the alkene ligand in [Re(CO)$_2$-($\eta^2$-CH$_2$=CHCH$_2$Bu$^t$)(Cp)] requires heating for days at 200 °C in the presence of PPh$_3$.[148]

Jones *et al.* have prepared the trihydride cyclopentadiene complex [ReH$_3$(PPh$_3$)$_2$($\eta^4$-CpH)] **(49)** in 56% yield from the reaction of [ReH$_7$(PPh$_3$)$_2$] and CpH at 25 °C.[159] In the solid state, **(49)** exists as a distorted pentagonal bipyramid with the three hydrides and two CpH double bonds all in the pentagonal plane. In solution, **(49)** exhibits complicated fluxional behavior including rotation of the CpH ring, which equilibrates the phosphine ligands, pairwise hydride exchange, and equilibration of the hydride and *endo*-$\eta^4$-CpH hydrogen by a reversible metal-to-ring hydride migration.[135] A similar migration was proposed for the cyclohexadiene complex [ReH$_3$(PPh$_3$)$_2$($\eta^4$-C$_6$H$_8$)] on the basis of spin saturation transfer and temperature dependent line shape analysis.[160] In both systems, the barriers for hydrogen migration were in the range of $\Delta G^{\ddagger} \simeq 63$ kJ mol$^{-1}$. A facile exchange between the ethene hydrogens and hydride ligands of [ReH$_3$\{PPh(Pr$^i$)$_2$\}$_2$(C$_2$H$_4$)$_2$] also occurs at 70 °C,[112] whereas [ReH$_3$(PMe$_2$Ph)$_2$($\eta^4$-*c*-C$_8$H$_{10}$)] undergoes an irreversible metal to ring hydride migration.[161]

At 60 °C, cyclopentadiene and [ReH$_7$(PPh$_3$)$_2$] generate [ReH$_2$(PPh$_3$)$_2$(Cp)] by a process in which the initially formed **(49)** reacts with [ReH$_5$(PPh$_3$)$_2$].[159] A dihydrogen transfer also occurs from **(49)** to [IrBr(CO)(dppe)] with formation of [ReH$_2$(PPh$_3$)$_2$(Cp)] and [IrH$_2$(Br)(CO)(dppe)]. A dihydride-bridged binuclear intermediate or transition state **(49)** was proposed for this dihydrogen transfer reaction.[159]

**(49)**

Complex **(49)** also undergoes a photochemical reaction to give the tetrahydride [ReH$_4$(PPh$_3$)(Cp)],[162] which has been structurally characterized.[135]

Dihydride [ReH$_2$(PPh$_3$)$_2$(Cp)] and its triarylphosphine derivatives undergo chemical and electrochemical oxidation to form long-lived 17-electron radical cations [ReH$_2$(PAr$_3$)$_2$(Cp)]$^{\cdot+}$. Further oxidation in acetonitrile solvent catalyzes the rapid formation of [ReH$_3$(PAr$_3$)$_2$(Cp)]$^+$ and [ReH(NCMe)(PAr$_3$)$_2$(Cp)]$^+$.[163]

## 9.6 CARBON–HYDROGEN BOND ACTIVATION

A number of rhenium hydride complexes serve as precursors to reactive organometallic species capable of inserting into carbon–hydrogen bonds. Jones and Maguire found that [ReH$_2$(PPh$_3$)$_2$(Cp)] is a catalyst precursor for exchange of deuterium between C$_6$D$_6$ and alkanes with turnover numbers ranging from 2.5 (secondary hydrogen exchange in propane) to ~1500 (α- and β-hydrogen exchange in THF).[136] The preference for primary over secondary exchange in propane was 20:1. In THF-d$_8$ solvent, methane is readily deuterated. Remarkably, even after 1000 turnovers the hydride ligands in [Re(H)$_2$(PPh$_3$)$_2$(Cp)] do not exchange with deuterium. Migration of both hydride ligands to the cyclopentadienyl ligand has been invoked to explain this latter result (Scheme 9).[164] Photolysis in the presence of other phosphines (e.g., PMe$_3$) leads to sequential ligand substitution and formation of [Re(H)$_2$(L)$_2$(Cp)] in quantitative yield.

Cyclopentadiene complex **(50)** serves as a useful precursor to monohydride complexes (Equation (18)).[137] The electron-rich hydride [ReH(PMe$_3$)$_5$] reversibly activates carbon–hydrogen bonds in both benzene and PMe$_3$, as indicated by deuteration of PMe$_3$ using C$_6$D$_6$.

$$[Re(H)(PMe_3)_5] \xrightarrow[93\%]{PMe_3} \mathbf{(50)} \xrightarrow{L} [Re(H)(PPh_3)_2(L)_3] \qquad (18)$$

L = CO, 90%
L = CNR, R = alkyl, 64–93%
L = CNR, R = 2,6-Xyl, 99%

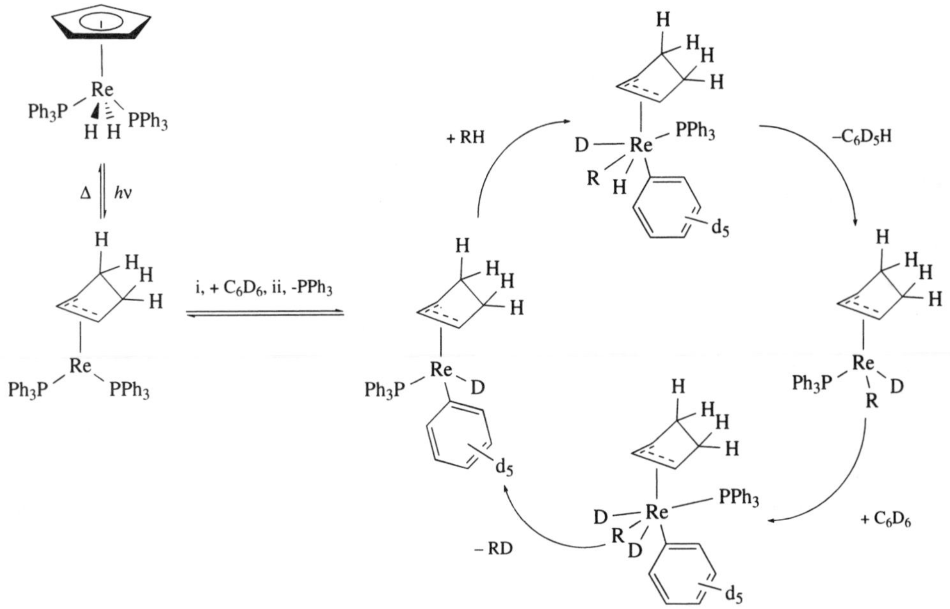

**Scheme 9**

The arene complex [ReH(PPh$_3$)$_2$($\eta$-C$_6$H$_6$)] catalyzes hydrogen–deuterium exchange between benzene-d$_6$ and other arenes by a mechanism which presumably involves Re$^I$–Re$^{III}$ oxidation state cycles.[165]

Wenzel and Bergman have thoroughly investigated the C–H activation chemistry of [Re(PMe$_3$)$_3$(Cp)] (**51**) (Scheme 10).[138] The highly air sensitive (**51**) was prepared from [Re(Cl)$_3$(PMe$_3$)$_3$] and Na/Hg, CpH, K$_2$CO$_3$, and purified in 38% yield by filtration through alumina(III) at −100 °C. Photolysis of (**51**) in cyclopropane (5–10 °C) or hexane (−30 °C) gives the isolable hydridoalkyl complexes (**52**) and (**53**) respectively. No products of insertion into a secondary C–H bond were observed with hexane substrate. Upon warming to room temperature in benzene, both alkyl complexes decompose to hydridophenyl complex (**54**).

Photolysis of (**51**) in cyclopentane at −50 °C resulted in only low conversion to insertion product (**55**), and rapid elimination of cyclopentane occured when the solution was warmed to room temperature ($T_{1/2}$ < 30 s). Photolysis in cyclohexane at higher temperatures (5–10 °C) gives only cyclometallated product (**56**) and a minor amount of the ($\eta^1$-dimethylphosphino)methyl complex (**57**). The fact that cyclohexane-derived products are not observed at this temperature made the use of cyclohexane as a solvent in the carbon–hydrogen bond activation of methane and ethene substrates possible. Irradiation of (**51**) in cyclohexane under 25 atm of methane gives (**58**) as the major product. Once again, the (hydrido)alkyl complex is thermally unstable at room temperature, and in benzene reductively eliminates methane to form (**54**). Photolysis of (**51**) in cyclohexane under 25 atm of ethene generates the vinyl complex (**59**) as the kinetic product. Upon warming to room temperature (**59**) isomerizes to the $\eta^2$-ethene complex (**60**). Irradiation of (**60**) in toluene-d$_8$ regenerates (**59**) as a 6:1 mixture of (**59**):(**60**).

These results indicate a greater selectivity for insertion into primary over secondary carbon–hydrogen bonds, but also a greater tendency to cyclometallate with the sterically crowded coordination sphere in (**51**) as compared to related, but sterically less congested, coordination environments of iridium and rhodium systems.

The phenyl hydride complex (**61**) and cyclometallated complex (**62**) are both converted back to [Re(CO)(N$_2$)(PMe$_3$)($\eta$-C$_5$Me$_5$)] under 14 MPa of N$_2$ (Scheme 11).[139] Even in C$_6$H$_6$ irradiation of [Re(CO)(N$_2$)(PCy$_3$)($\eta$-Cp*)] gives only cyclometallated product.

A major, largely unrealized, goal of C–H bond activation chemistry is efficient functionalization of the activated carbon center. An important lead in this regard is the [ReH$_7$(PAr$_3$)$_2$]-catalyzed dehydrogenation of cycloalkanes to cycloalkenes in the presence of 3,3-dimethylbut-1-ene as a hydrogen acceptor. For the case of methylcyclohexane, where Ar = $p$-FC$_6$H$_4$, three turnovers are observed in the formation of isomeric alkenes: 3-methylcyclohexene (29%), 4-methylcyclohexene (65%), and methylenecyclohexane (6%).[166] $n$-Alkanes (C$_6$–C$_8$) also react with [Re(H)$_7$(PPh$_3$)$_2$] at 79 °C in the presence of 3,3-dimethylbutene to give mixtures of $\eta^4$-diene trihydride complexes.[167] Scheme 12 provides a mechanistic rationale which invokes reactive 14-electron intermediates to account for the observed dehydrogenation chemistry.[168] The maximum number of turnovers was observed in the case of

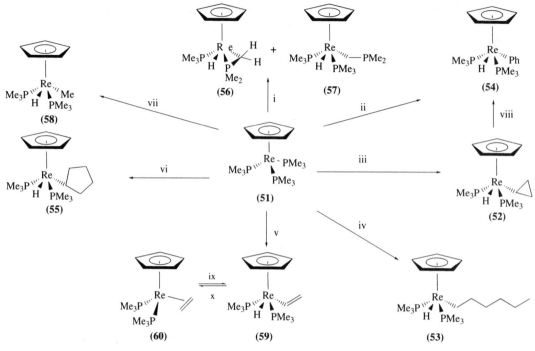

i, Cy, *h*v, 5–10 °C; ii, C₆H₆, *h*v, –10 °C; iii, △, *h*v, 5–10 °C; iv, *n*-C₆H₁₄, *h*v, –30 °C; v, CH₂=CH₂, Cy, *h*v, 5–10 °C;

vi, ⬠ , *h*v, –50 °C; vii, CH₄, Cy, *h*v, 5–10 °C; viii, C₆H₆, *h*v, ix, 20 °C; x, *h*v, 5–10 °C

**Scheme 10**

$$[Re(H)_3(L)_2]$$

cyclooctane and L = P(*p*-FC₆H₄)₃. Deactivation of the catalyst is thought to involve oligomerization or dehydrogenation of the 14-electron intermediate [Re(H)₃(L)₂].

The reactions of [Re(H)₇(L)₂] with pyrrole[140] and furan,[169] in the presence of 3,3-dimethylbutene, lead to the η-pyrrolyl dihydride (**63**) and 1-oxapentadienyl complex (**64**), respectively (Scheme 13). In the case of (**63**), conversion to an (η-pyrrolyl)iodohydridorhenium complex and addition of organolithium reagents results in nucleophilic substitution at the C-2 position of the pyrrolyl ring.

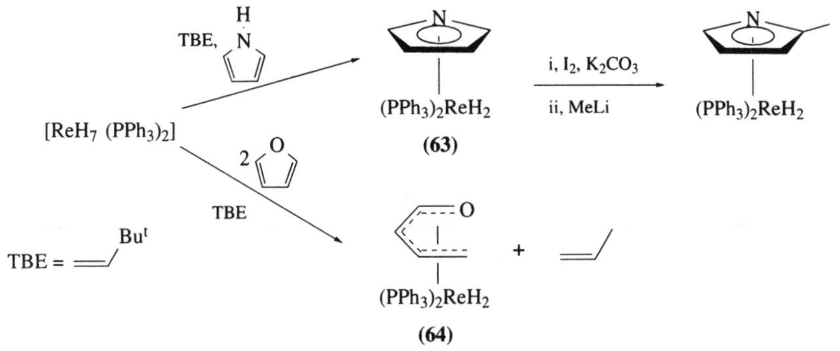

**Scheme 13**

## 9.7 RHENIUM ALKYLS

A number of useful routes toward rhenium alkyl complexes are available, including the alkylation of rhenates, reduction of rhenium acyls, and reduction of carbon monoxide.

### 9.7.1 Alkylation of [Re(CO)$_5$]$^-$

The most widely used route toward rhenium alkyl and acyl complexes involves reaction of anionic rhenium complexes with carbon electrophiles. Scheme 14 provides representative alkylations involving the strongly nucleophilic [Re$_2$(CO)$_5$][Na], which is readily available from Na/Hg reduction of [Re$_2$(CO)$_{10}$].[107,170-4] Both one- and two-electron processes may occur in the reactions of [Re(CO)$_5$]$^-$ with electrophiles. Kochi and co-workers found that reaction with *N*-methylacridinium cation at −30 °C gave both an alkylation adduct as well as 5% of [Re$_2$(CO)$_{10}$] and the homodimer *N,N*'-dimethyl-9,9-biacridanyl.[175]

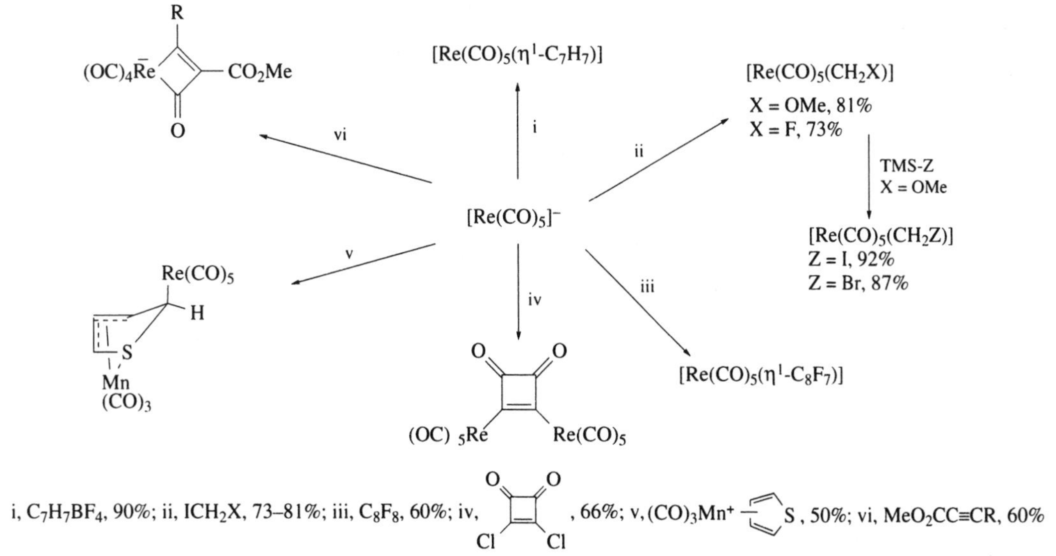

i, C$_7$H$_7$BF$_4$, 90%; ii, ICH$_2$X, 73–81%; iii, C$_8$F$_8$, 60%; iv, [diagram], 66%; v, (CO)$_3$Mn$^+$ [diagram] S , 50%; vi, MeO$_2$CC≡CR, 60%

**Scheme 14**

### 9.7.2 Rhenium Enolates

In the reaction of $[Re(CO)_5][Na]$ with $\alpha$-halocarbonyl compounds, the choice of halide is critical. For example, reaction with $\alpha$-bromo ketones or esters gives $[ReBr(CO)_5]$ from facile metal–halogen exchange, whereas reaction with $ClCH_2COR$ (R = OEt, Me, Ph) gives reasonable yields (55–63%) of the primary rhenium enolates $[Re(\eta^1\text{-}CH_2COR)(CO)_5]$.[176] The synthesis of secondary enolates is also problematic due to metal–halogen exchange; however, use of ethyl 2-((methylsulfonyl)oxy)propanoate gives a 47% yield of $[Re(CHMeCO_2Et)(CO)_5]$.[177] The monophosphine derivative (65) is available as a 97:3 ratio of *cis:trans* isomers in 86% yield from addition of a NaNp/THF solution to a THF solution of $[ReBr(CO)_4(PPh_3)]$ at $-78\,^\circ C$, followed by addition of excess ethyl chloroacetate (Scheme 15). Heating (65) with $PPh_3$ and MeCN gives the acetonitrile insertion product (66) in 81% yield, whereas abstraction of CO from (65) with $Me_3NO$ in MeCN gives the acetonitrile adduct (67). Amide enolates are accessible by similar routes and upon loss of a *cis*-CO ligand the amide oxygen coordinates to the metal.

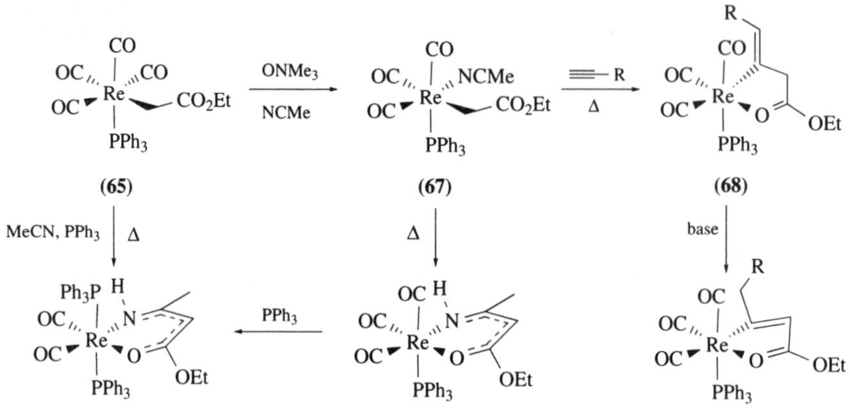

**Scheme 15**

Enolate (67) reacts with terminal alkynes to give metallacycle (68), which isomerizes to the endocyclic alkene (69) upon treatment with a Lewis base.[178] The base-catalyzed isomerization process involves an intramolecular 1,3-hydrogen shift, which is stereospecific with respect to the rhenium center (*anti* to the $PPh_3$ ligand). Decomplexation results from treatment with acid to give the unsaturated organic esters. Metallacycle formation is proposed to occur by oxidative addition of the alkyne C–H bond to give a seven-coordinate rhenium acetylide hydride, which rearranges to a vinylidene intermediate (70) (Equation (19)). Rearrangement of the carbon-bound enolate to an oxygen-bound enolate and attack at the electrophilic vinylidene $\alpha$-carbon then generates the metallacycle.

The acetonitrile enolate (67) also undergoes aldol condensation reactions with aldehydes. Treatment with benzaldehyde for 14 h at 55 °C produces $PhC(OH)HCH_2CO_2Et$ in 45% yield. When the reaction is run in the presence of TMS-Cl a 94% yield of $PhC(O\text{-}TMS)HCH_2CO_2Et$ is formed after only 30 min.[179]

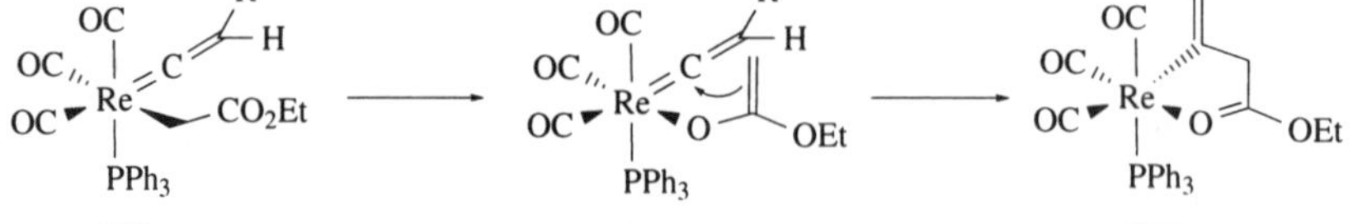

$$(19)$$

### 9.7.3 Alkylation of $[Re(NO)(PPh_3)(Cp)]^-$

Deprotonation of the hydride $[ReH(NO)(PPh_3)(Cp)]$ at $-70\,^\circ C$ and warming to $-30\,^\circ C$ provides $[Re(NO)(PPh_3)(Cp)]^-$, which is quenched with carbon electrophiles to access a variety of alkyl and acyl complexes: $[Re(R)(NO)(PPh_3)(Cp)]$ where R = Me (76%), $Bu^n$ (66%), allyl (90%), $CH_2C(O)Ph$ (65%), $TfO(CH_2)_nOTf$ ($n$ = 3, 4, 5, 8; 80–9%),[180] and C(=O)Ph (68%).[133]

Due to inhomogeneity problems, Grignard reactions at the coordinated CO ligand in $[Re(NO)(PPh_3)(CO)(Cp)]^+$ gave only poor yields of acyl complexes. However, acyl complexes are readily prepared (66–99% yields) by Grignard reaction at the methyl ester ligand in $[Re(CO_2Me)(NO)(PPh_3)(Cp)]$, and the resultant acyls are converted to alkyls in high yield by $BH_3 \cdot THF$ reduction.[181]

### 9.7.4 Deprotonation at the α-Carbon

The polarity of rhenium–carbon σ-bonds as well as the strong π-donor character of the HOMO leads to destabilization of carbanions at the α-C of L in $[Re(L)(NO)(PPh_3)(Cp)]$ complexes.[182] Treatment of $[Re(Me)(NO)(PPh_3)(Cp)]$ with $Bu^nLi/TMEDA$ gives only the lithiocyclopentadienyl complex $[Re(Me)(NO)(PPh_3)(\eta\text{-}C_5H_4Li)]$, with no evidence for deprotonation at the methyl group. By using a cyano substituent to stabilize an α-C carbanion, Gladysz and co-workers have succeeded in generating the first transition metal substituted carbanions (Scheme 16). Treatment of the neutral, coordinatively saturated cyanomethyl complex (71) with $Bu^nLi/TMEDA$ at –98 °C, gives a 2:1 ratio of carbanions (72) and (73). At –78 °C, (72) isomerizes to (73). Addition of $MeOSO_2CF_3$ to the mixture of anions gives the methylated complex (SR,RS)-(74) as the exclusive product. Deprotonation of (74) and methylation gives only the cyclopentadienyl-ring alkylation product.

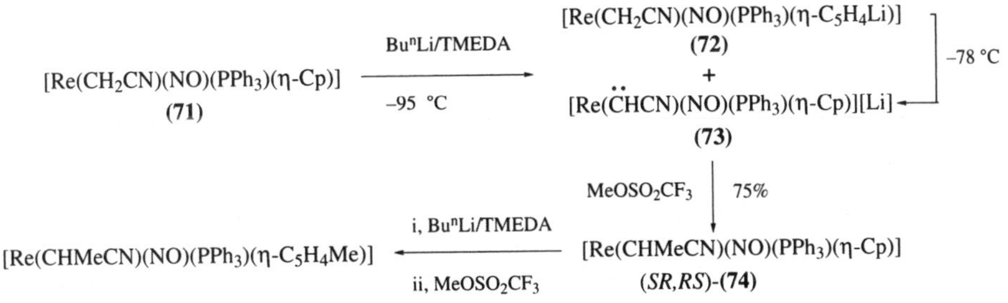

**Scheme 16**

The corresponding cationic ylide (75) is also deprotonated at the α-carbon to give a neutral ylide (76), which is alkylated at carbon to give (SR,RS)-(77) in 83% yield (Equation (20)). Notably, the carbanion (73) and ylide (76) alkylate with opposite stereochemistry.

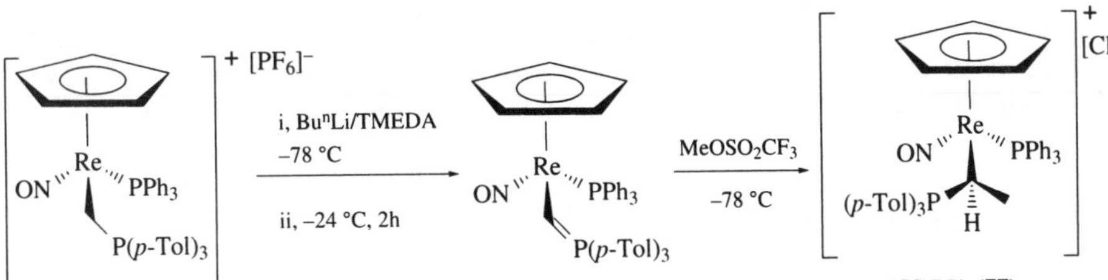

(20)

### 9.7.5 Hydride Abstraction

The orientation of the HOMO and the chiral environment at rhenium exert a profound effect on the stereochemistry of reactions at ligands coordinated to the $[Re(NO)(PPh_3)(Cp)]^+$ fragment. π-Acceptor ligands such as alkylidenes exhibit favored conformations, which maximize overlap of the rhenium HOMO with the empty p-orbital of the carbene fragment. This phenomenon is the basis for the development of synthetic routes toward diastereomerically and enantiomerically pure alkylrhenium complexes $[Re(CHDR)(NO)(PPh_3)(Cp)]$, where R = aryl or alkyl.[183] Hydride abstraction from the benzyl complex $[Re(CH_2Ph)(NO)(PPh_3)(Cp)]$ (78) at –78 °C gives a kinetic benzylidene isomer (79k) which isomerizes at 25 °C to the thermodynamic conformer (79t) (Scheme 17). The rhenium–carbon double bond conformations of these geometric carbene isomers allow for maximum overlap of the carbene p-orbital with the rhenium HOMO (d-orbital). Irradiation of a >99:1 mixture of (79t):(79k) at –78 °C results in a 55:45 (t:k) photostationary state. The propylidene complex exhibits similar photochemical behavior.[184] Nucleophiles (LiEt₃BD, MeLi, EtMgBr, PhCH₂MgCl, PMe₃, and MeONa) attack the carbene carbon *anti* to the bulky PPh₃ ligand with a high degree of stereoselectivity (generally 92–95:8–5 diastereomer ratios).[185]

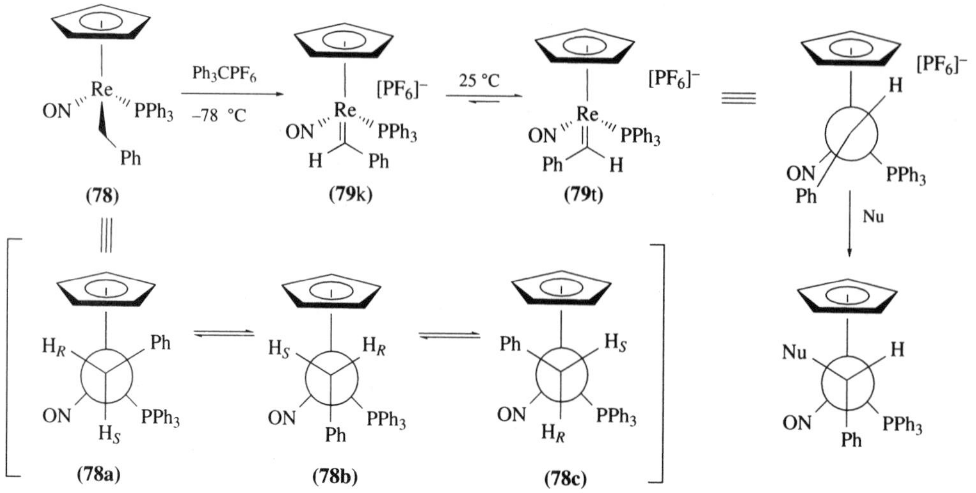

**Scheme 17**

Deuterium-labeling studies demonstrate that only the pro-R hydride of (**78**) is abstracted by Ph₃CPF₆ and abstraction occurs from a direction *anti* to the PPh₃ ligand in rotamer (**78a**) (Scheme 17).[186] Of the three staggered Re–C$_\alpha$ rotamers (**78a**), (**78b**) and (**78c**), the most stable is the one with the R group between the medium-sized Cp and small NO ligand. An extended Hückel analysis of the two conformers with a C$_\alpha$-hydrogen *anti* to the large PPh₃ ligand indicates that (**78a**) is more stable than (**78b**). The key to rationalizing this relative stability is the smaller ON–Re–PPh₃ angle of 90° compared to the 135° ON–Re–(Cp centroid) and (Cp centroid)–Re–PPh₃ angles.[187,188]

Electrochemical studies indicate that α-hydride abstraction in these systems involves an electron transfer preequilibrium followed by rate-determining hydrogen atom transfer.[189] Remarkably, even alkyl complexes of the type [Re(CH₂CH₂R)(NO)(PPh₃)(Cp)] undergo α-hydride abstraction and it is only with the more-substituted alkyls ReCH₂CHR¹R² and ReCHR¹CH₂R² that the expected β-hydride abstraction predominates.[189] Treatment of the methyl complex [Re(Me)(NO)(PPh₃)(Cp)] in MeCN with 1 equiv. ferricinium ion results in simple electron transfer and ultimately formation of a 1:3 ratio of [Re(CH₂=CH₂)(NO)(PPh₃)(Cp)]⁺ and [Re(NCMe)(NO)(PPh₃)(Cp)]⁺.[190]

A clever synthesis of chiral methyl groups takes advantage of the stereoselective hydride abstraction and nucleophilic addition reactions at chiral rhenium alkyl and benzylidene complexes.[191] The stereochemical outcome requires retention of configuration at carbon in the HBr cleavage step.

### 9.7.6 Carbon Monoxide Reduction

The reduction of coordinated CO by hydride reagents was intensively studied in the late 1970s and early 1980s as part of efforts directed at understanding the metal-catalyzed reduction of CO by H₂. By judicious choice of reducing reagents and reaction conditions [Re(CO)(NO)(L)(Cp)]⁺ (L = CO, PPh₃) is selectively reduced to neutral formyl [Re(CHO)(NO)(L)(Cp)], hydroxymethyl [Re(CH₂OH)-(NO)(L)(Cp)] (**80**), and methyl [Re(Me)(NO)(L)(Cp)]⁺ (L = CO (**81**); L = PPh₃ (**29**)) compounds.[191-3] Treatment of (**29**) with Ph₃CBF₄ gives the isolable cationic alkylidene [Re(=CH₂)(NO)(PPh₃)(Cp)]⁺, whereas (**81**) reacts with Ph₃CBF₄ to give a cationic ethene complex [Re(η²-CH₂=CH₂)(CO)(NO)(Cp)]⁺ via the unisolable methylene [Re(=CH₂)(CO)(NO)(Cp)]⁺.[192] Treatment of hydroxymethyl (**80**) with HBF₄ gives a 2:1 mixture of (**81**) and [Re(CO)₂(NO)(Cp)]⁺.[192]

The stepwise reduction of the CO ligand in [Re(NO)(CO)(L)(Cp)]⁺ has been examined by a molecular orbital analysis which suggests the initial site of hydride attack is at the nitrosyl ligand, followed by rearrangement to a formyl complex.[194]

Substitution of the η-Cp ligand by the pure σ-donor 1,4,7-triazocyclononane ligand has been used to stabilize more highly charged analogues of [Re(CO)(L)(NO)(Cp)]⁺. Thus, hydride reduction of dication (**82**) generates cationic formyl and methyl (**83**) complexes (Scheme 18).[195,196] The enantiomers of (**83**) were resolved as (1S)-3-bromocamphorsulfonate salts and treatment of optically pure (**83**) with aqueous acid gave methane and the halide complex with retention of configuration at rhenium. Acids with nonnucleophilic counterions such as 1 M $HClO_4$ or $H_2SO_4$ do not react with (**83**). The protonation reaction has been proposed to involve dissociation and protonation of a secondary amine ligand followed by loss of methane and halide attack.

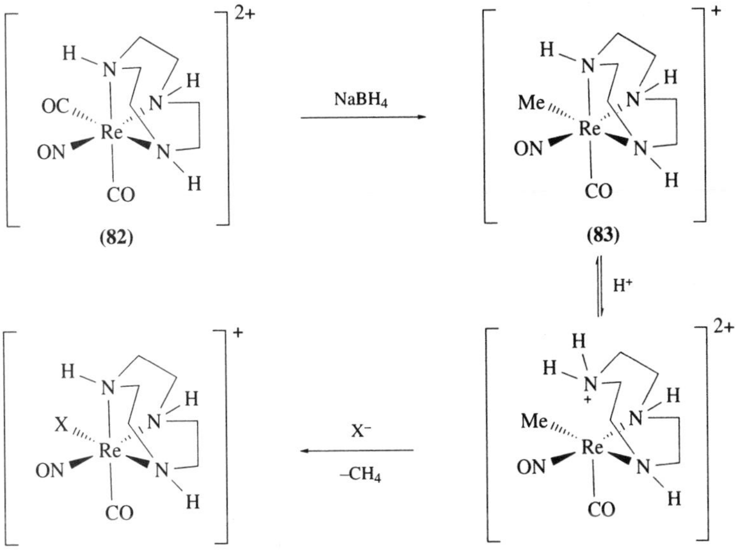

**Scheme 18**

Attempts at generating a stable pentacarbonyl rhenium–hydroxyalkyl complex from decarbonylation of hydroxyacyl precursors [Re{C(=O)CHPhOH}(CO)$_5$] and protodesilylation of α-(silyloxy)alkyl complexes [Re{CHPh(O-TMS)}(CO)$_5$] gave only [ReH(CO)$_5$] and benzaldehyde, presumably from decomposition of the desired hydroxyalkyl.[197] An alternative successful route toward hydroxymethyl complexes of the type [(CO)$_4$Re{PPh$_2$(o-C$_6$H$_4$CHOH)}] takes advantage of entropic stabilization provided by the chelate structure.[198]

The octahedral formyl- and acylrhenium complexes [Re{C(=O)R}(CO)$_2${P(OEt)$_3$}(L)] (R = H, Me; L = CO, P(OEt)$_3$) are available in very good yields (70–91%) from NaBH$_4$ or LiCuMe$_2$ attack at coordinated CO of [Re(CO)$_3${P(OEt)$_3$}(L)][BF$_4$].[199]

Kochi and co-workers were the first to report a radical chain mechanism for metal formyl to metal hydride conversion. Whereas *cis*-[(CO)$_5$ReRe(CO)$_4${C(=O)H}][NBu$_4$] (**84**) is stable at 25 °C in the dark under N$_2$, the addition of AIBN leads to rapid formation of *cis*-[(CO)$_5$ReRe(CO)$_4$H][NBu$_4$] in 90% yield.[200] Anionic formyl (**84**) acts as a hydride donor toward aldehydes, alkyl halides, and other metal carbonyls. In reactions with K(Bu$^s$)$_3$BH, formaldehyde (20%) and [Re$_2$(CO)$_9$]K$_2$ (91%) are formed.[201]

### 9.7.7 Acetylides and Carbides

The η$^1$-acetylide complexes [Re(η$^1$-C≡CR)(NO)(PPh$_3$)(L)], (where L is Cp, η-C$_5$H$_4$Me, η-Cp* and R = H, alkyl, TMS) are available from deprotonation of η$^2$-alkyne precursors. Deprotonation of the terminal acetylide complexes generates the lithiocarbide [Re(η$^1$-C≡CLi)(NO)(PPh$_3$)(η-Cp*)], (**85**), which reacts with electrophiles to give the derivatives [Re(η$^1$-C≡CR)(NO)(PPh$_3$)(η-Cp*)] (**88**) (where R = D, TMS, Ph$_3$Sn, Me) (Scheme 19).[202] In the case of the η-Cp acetylide complex, a mixture of the mono- and dilithio intermediates [(Re(η$^1$-C≡CLi)(NO)(PPh$_3$)(η-C$_5$H$_4$X)] (X = H, Li) is generated. When transition metal halides are employed as electrophiles dinuclear dicarbides result.[203] Reaction of the lithiocarbide complex with metal carbonyls such as [Mn(CO)$_3$(Cp)], followed by *O*-alkylation with Me$_3$OBF$_4$, gives methoxy carbene complexes. Subsequent reaction with BF$_3$ generates tricarbide (**86**).[204] Two resonance structures for (**86**) are the propadienylidene (**86a**) and acetylide-substituted carbyne (**86b**), the latter being the major contributor to the ground state structure based on $^{13}$C NMR spectroscopic and x-ray data. Four-carbon chains (**87**) are accessible through oxidative coupling of the ethynyl complex (**88**) (R = H). Oxidation of (**87**) with AgPF$_6$ gives the dicationic butatrienediylidene complexes (**89**).[205]

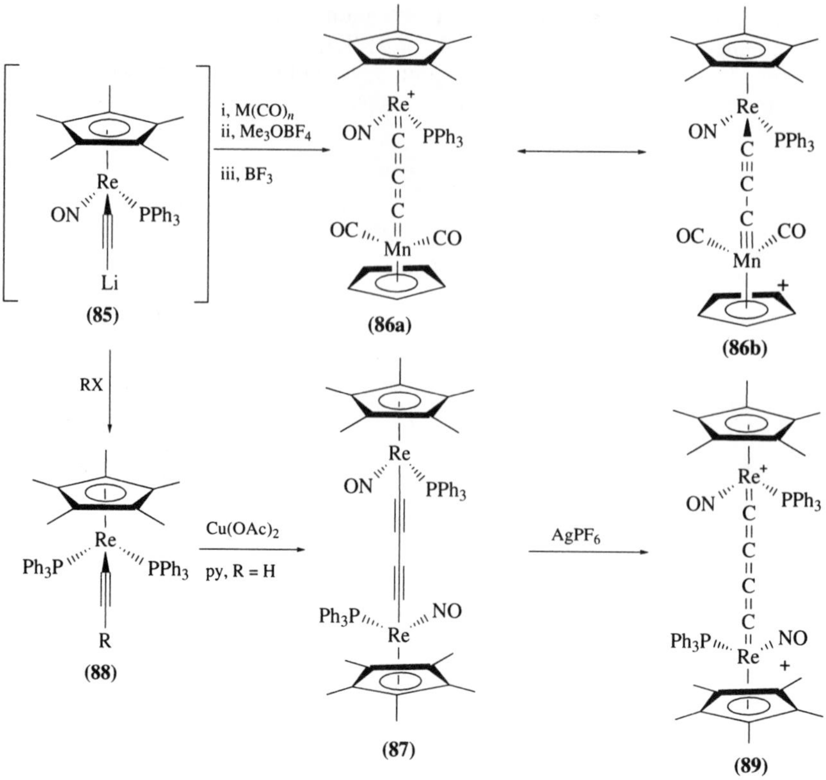

**Scheme 19**

Dicarbide–carbonylrhenium complexes [(CO)$_5$Re–C≡C–Re(CO)$_5$] are not accessible from [M$_2$C$_2$] (M = Na, Li) and [Re(CO)$_5$(FBF$_3$)], and diiodoethyne and [Re(CO)$_5$]$^-$ gives only [ReI(CO)$_5$] and [Re$_2$I$_2$(CO)$_8$]. However, [Re(CO)$_5$(FBF$_3$)] and HC≡C-TMS forms the σ,π-ethynide complex [(CO)$_5$Re-(μ-η$^1$,η$^2$-C≡CH)Re(CO)$_5$][BF$_4$] (73%), which is deprotonated reversibly with NaOEt to give an 88% yield of [(CO)$_5$Re–C≡C–Re(CO)$_5$].[92]

Beck *et al.* also prepared a μ$_2$-carbido complex [(TPP)Fe=C=Re(CO)$_4$Re(CO)$_5$] in 51% yield from [(TPP)Fe=CCl$_2$] and 2 equiv. [Re(CO)$_5$]$^-$, where TPP = tetraphenylporphyrinate.[206]

## 9.7.8 Vinyl Ligands

Chiral vinylrhenium complexes of the type [Re(CX=CHR)(NO)(PPh$_3$)(Cp)] (90) (X = H, OMe) are available in 72–86% yield from deprotonation of the corresponding alkylidene complexes (Scheme 20). For R = Me (X = H) and R = Pr$^n$ (X = H) the (E)/(Z) C=C geometric isomers equilibrate over the course of hours at 25 °C to give 84:16 and 92:8 (E)/(Z) mixtures, respectively. The diminished π-acidity of the vinyl ligand compared to acyl and alkylidene ligands results in a Re–C$_\alpha$ conformation, which is influenced significantly by the steric bulk of the PPh$_3$ ligand. This conformational preference is reinforced by electronic influences of frontier orbital overlap and minimum Re–C$_\alpha$ π-repulsive interactions. The crystal structure of (Z)-(90) (R = Bz, X = OMe) indicates an ON–Re–C$_\alpha$–C$_\beta$ torsion angle (θ) of 47°, whereas the corresponding angle for (E)-(90) (R = Bz, X = H) is 175.5°. Difference $^1$H NOE data indicate the Re–C$_\alpha$ conformation (θ) is near 0° in solution, and protonation studies support the view that conformations with θ = 0 are more reactive toward protonation. Alkylation of (Z)-(90) (R = Me, X = OMe) with CD$_3$I gives the isobutyroyl complex (91) in 88% yield as a 96:4 ratio of diastereomers. Thus, the electrophile attacks the alkene face opposite to the PPh$_3$ ligand with a high degree of 1,3-asymmetric induction from rhenium to carbon.[207]

Proulx and Bergman discovered a carbon monoxide to vinyl ligand conversion in the reaction of [Re(R)(CO)$_5$] (where R = Me, Ph) with the tantalum alkylidene complex [Ta(=CH$_2$)(Me)(Cp)$_2$] to give vinylrhenium complexes (Scheme 21).[208] This unusual transformation is important from the fundamental perspective of CO deoxygenation and coupling reactions. Isotope labeling and low-temperature NMR spectroscopy studies support the mechanism shown in Scheme 21. The zwitterionic intermediate (92) is actually isolable when the fluorinated alkyl [Re(CF$_2$CF$_2$CF$_3$)(CO)$_5$] is employed.

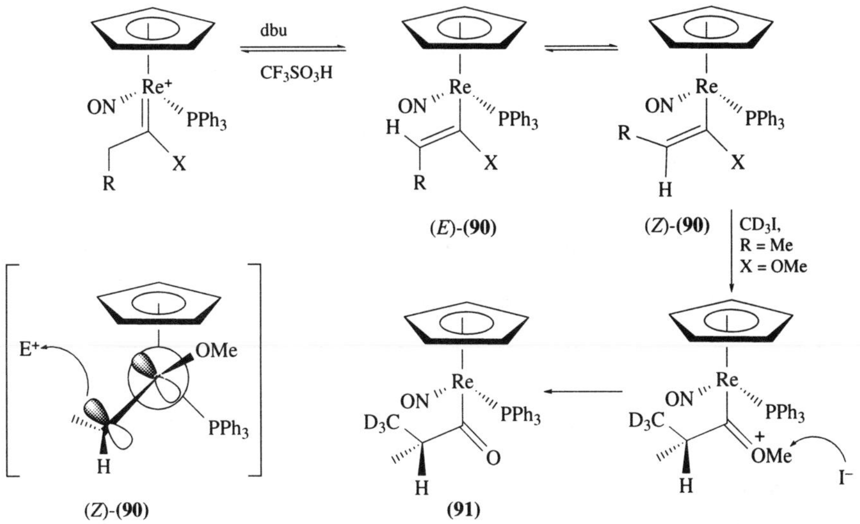

dbu

CF$_3$SO$_3$H

(*E*)-(**90**)

(*Z*)-(**90**)

CD$_3$I,
R = Me
X = OMe

E$^+$

OMe

PPh$_3$

H

(*Z*)-(**90**)

(**91**)

I$^-$

**Scheme 20**

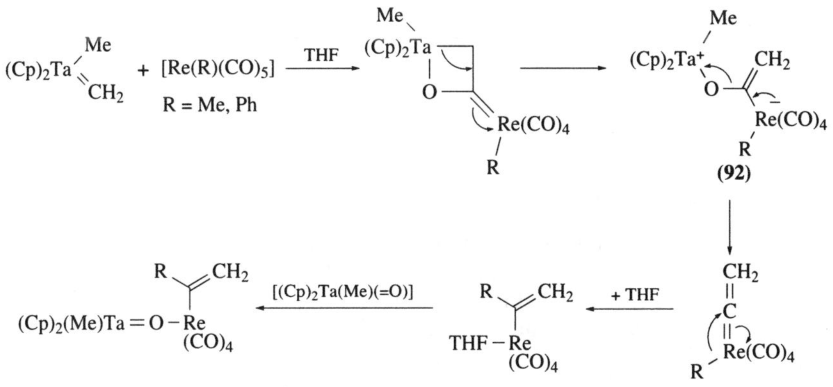

(Cp)$_2$Ta + [Re(R)(CO)$_5$]

CH$_2$

R = Me, Ph

THF

(Cp)$_2$Ta

O

Re(CO)$_4$

R

(Cp)$_2$Ta$^+$

CH$_2$

O

Re(CO)$_4$

R

(**92**)

CH$_2$

C

Re(CO)$_4$

R

(Cp)$_2$(Me)Ta=O−Re
(CO)$_4$

[(Cp)$_2$Ta(Me)(=O)]

R

CH$_2$

THF−Re
(CO)$_4$

+ THF

R

CH$_2$

**Scheme 21**

Phosphaalkenyl complexes [Re{C(O-TMS)=PR}(CO)(NO)(η-Cp*)] (R = TMS, Ph, Bu$^t$) are accessible in low to moderate yield from reaction of [LiP(TMS)(R)]·2THF with the metal carbonyl cation [Re(CO)$_2$(NO)(η-Cp*)]$^+$.[209] Table 6 shows a selection of vinyl complexes.

**Table 6** Selected vinyl complexes.

| Complex | Ref. |
|---|---|
| [Cp$_2$Re(η$^1$-CH=CHR)] (R = CO$_2$Me, CN, CF$_3$, COMe) | 148 |
| [Cp$_2$Re{η$^1$-CH=C(R$^2$)CH=CR$^1$H}](R$^1$, R$^2$ = CO$_2$Me; R$^1$,R$^2$ = COMe; R$^1$ = CO$_2$Me, R$^2$ = COMe; R$^1$ = COMe, R$^2$ = CO$_2$Me) | 148 |
| [Re(η$^1$-CH=CHR)(NO)(PPh$_3$)(η-Cp)] (R = H, Me, Pr$^n$, Bz, CPh$_3$) | 207 |
| [Re(η$^1$-CR$^1$=CHR$^2$)(NO)(PPh$_3$)(η-Cp)] (R$^1$ = OMe, R$^2$ = H, Me, Bz, Ph) | 207 |
| [Re{η$^1$-C(R)=CH$_2$}{η$^1$-O=TaMe(Cp)$_2$}(CO)$_4$] | 208 |
| [Re(η$^1$-CH=CHCH$_2$Bu$^t$)(η$^1$-CH$_2$CH=CH$_2$)(CO)$_2$(Cp)] | 210 |
| [Re(η$^1$-CH=CHCH$_2$Bu$^t$)(Me)(CO)$_2$(η-Cp)] | 210 |
| [Re(η$^1$-CH=CHCH$_2$Bu$^t$)(CH≡CH)(CO)$_2$(η-Cp)] | 211 |
| [Re(η$^1$-CH=CHCH$_2$Bu$^t$)(CO)$_2$(η-Cp)][K] | 212 |
| [Re(η$^1$-CH=CHCH$_2$Bu$^t$)(Me)(CO)$_2$(η-Cp)] | 212 |
| [Re{η$^1$-(1-cyclopentene)}(NO)(PPh$_3$)(η-Cp)] | 213 |
| [Re(η$^1$-CH=CHBr)(Br)(CO)$_2$(η-Cp*)] | 214 |
| [Re(η$^1$-CH=CH$_2$)(PMe$_3$)$_2$(η-C$_6$H$_6$)] | 215 |
| [Re(η$^1$-CH=CHCHMe$_2$)(CO)$_3$(dppe)] | 216 |

## 9.8 RHENIUM ACYLS

### 9.8.1 Decarbonylation

The interconversion of acylrhenium and alkylrhenium complexes has been examined in detail for the diacylrhenate complex *cis*-[Re{C(=O)Me}{C(=O)Ph}(CO)$_4$][NMe$_4$] (**93**) (Scheme 22). Three selectively $^{13}$C-labeled complexes of (**93**) were used to elucidate the stereochemistry of their conversion to decarbonylation products. The phenyl migration product (**94**) is thermodynamically favored over the more rapidly formed methyl migration product (**95**). Acetyl (**94**) equilibrates with benzoyl (**95**) via a five-coordinate intermediate and CO scrambling in this intermediate is slow relative to methyl migration, but comparable to the rate of phenyl migration.[217]

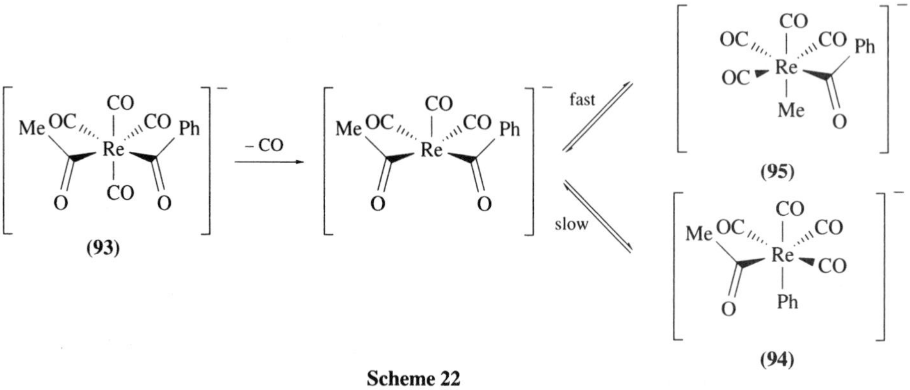

**Scheme 22**

### 9.8.2 Acyl Ligand Coupling Reactions

Goldberg and Bergman were the first to report the direct observation of alkylation at the metal center of an acylmetallate.[218] Treatment of the acylrhenate [Re(COMe)(CO)$_2$(Cp)][Li] (**96**) with the hard alkylating agent [(Me)$_3$O][PF$_6$] favors a Fischer carbene complex, whereas alkylation with the softer electrophile MeI favors the alkylacylrhenium complex (**97**) (Equation (21)). Protic solvents such as H$_2$O favor formation of (**97**) due to solvation of the electronegative metallate oxygen. Reaction of (**96**) with MeI in THF gives (**97**) quantitatively.

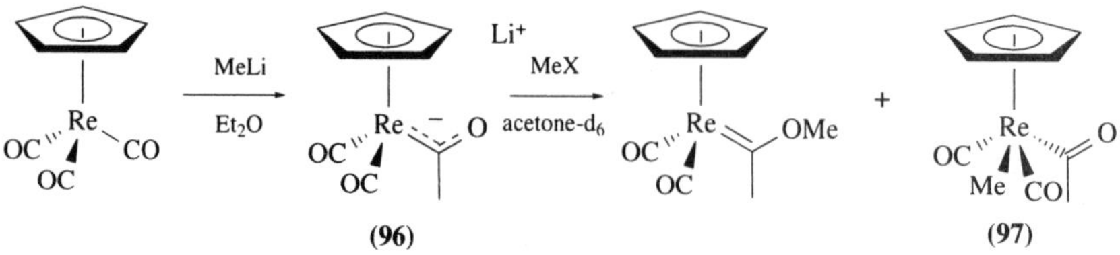

|  | | |
|---|---|---|
| (Me)$_3$OPF$_6$ | 78 : | 22 |
| MeI | 2 : | 98 |

Thermolysis of (**97**) in the presence of PPh$_3$ quantitatively generates [Re(CO)$_2$(PPh$_3$)($\eta$-Cp)] and acetone, whereas photolysis of (**97**), or the dialkyl complex [Re(Me)$_2$(CO)$_2$($\eta$-Cp)], under 20 atm of CO gives 2,3-butanedione (78%) and [Re(CO)$_3$(Cp)] (100%). Radical-trapping experiments as well as labeling studies with $^{13}$CO and mixtures of acetyl- and propionylrhenium complexes are consistent with a mechanism involving homolytic cleavage of the metal–alkyl and metal–acyl bonds.

Reaction of [ReCl(CO)$_3$(L)$_2$] (L = P(OMe)$_3$, P(OPr$^i$)$_3$, P(OPh)$_3$, PMe$_3$, PEt$_3$) with Bu$^t$Li and PhC≡CH generates $\eta^2$-$\alpha$-acylvinyl complexes (**98**) in moderate to good yields (Scheme 23).[219] The reaction of Bu$^t$Li without subsequent addition of PhC≡CH results in an $\eta^2$-acyl complex ($^{13}$C NMR 305.6 ppm) of structure (**99a**) or (**99b**). A reasonable pathway from (**99**) to (**98**) proceeds through the vinylidene intermediate [Re(COBu$^t$)(=C=CPhH)(CO)$_2$(L)$_2$], and involves vinylidene–acyl ligand coupling.

Casey and co-workers have systematically developed the concept of concerted organometallic reactions in a series of extensive studies on CO insertion and acyl–allyl coupling reactions at rhenium. Concerted organometallic reactions involve the combination of two elementary processes in order to avoid the intervention of a high-energy coordinatively unsaturated intermediate.[1]

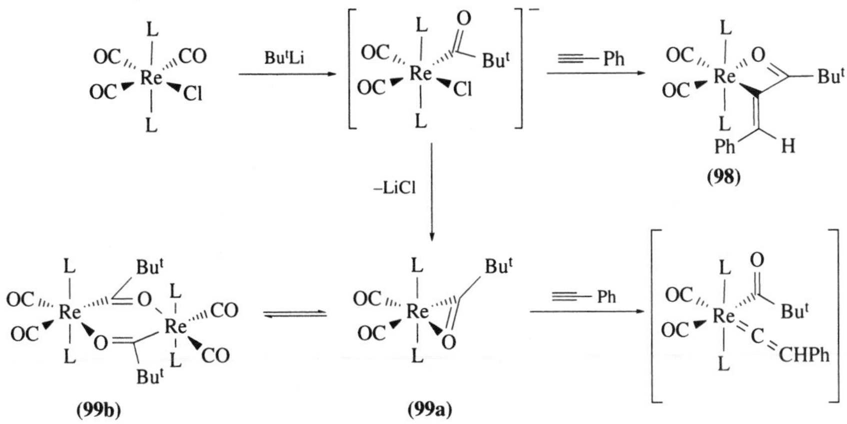

**Scheme 23**

The anionic vinylrhenium complex [Re(CH=CHCH$_2$Bu$^t$)(CO)$_2$(Cp)]$^-$ reacts with allyl bromide at the metal to give an allyl vinyl rhenium complex (**100**), which is unstable at 25 °C toward allyl vinyl ketone (**101**) (Scheme 24).[210] Insight into this remarkably facile CO insertion reaction is found in the observation that the analogous methyl vinyl complex (**102**) is stable at 90 °C for hours, and $\eta^1$-allyl alkyl complex (**103**) rearranges to $\eta^3$-allyl acyl (**104**) at 25 °C. The less congested allyl methyl complex analogue of (**103**) exists as a 7:1 equilibrium mixture with an $\eta^3$-allyl acetyl complex (**104**).[220] These results support a mechanism for the conversion of (**100**) to (**101**), in which two consecutive concerted organometallic rearrangements occur. The vinyl migration to CO is assisted by $\eta^1$- to $\eta^3$-allyl rearrangement, and coupling of the allyl and acyl ligands is facilitated by complexation of the vinyl double bond. In both cases, high-energy unsaturated rhenium intermediates are avoided. The net inversion of allyl regiochemistry, as determined by deuterium labeling, is consistent with the allyl double bond swinging toward the incipient acyl ligand in (**100**).

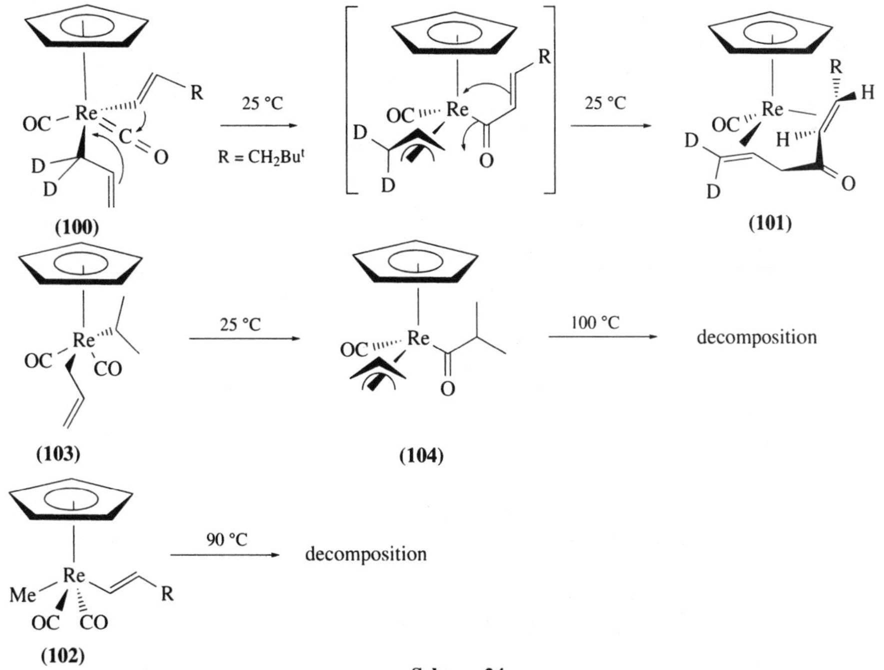

**Scheme 24**

The bis(alkene) geometry in (**101**) is "crossed" with the terminal alkene parallel to the plane of the cyclopentadienyl ligand and the internal double bond perpendicular to that plane. In the formation of (**101**), complexation of the vinyl double bond occurs most readily from the *s-trans* configuration of the α,β-unsaturated acyl ligand, which leads directly to the observed "crossed" alkene geometry. Complex (**104**), which lacks a vinyl double bond, does not undergo acyl–$\eta^3$-allyl coupling up to 100 °C, at which point decomposition occurs.

Propargyl σ- to π-rearrangements also assist the migration of alkyl groups to CO.[211] Thus, vinyl σ-propargyl complex (**105**) rearranges rapidly at 23 °C to give the observable allenyl vinyl ketone complex (**106**), which further isomerizes at 23 °C to allenyl vinyl ketone complex (**107**) and *exo,syn*-π-allyl-σ-vinyl complex (**108**) (1:1.3 ratio) (Scheme 25). At 105 °C, (**107**) rearranges to the *endo,syn* isomer (**109**). Allenyl complexes (**106**) and (**107**) are diastereomers which differ in the face of the enone coordinated to rhenium. The conversion of (**107**) to (**109**) is explained by a [1,5]-hydrogen migration from the alkene to the bottom face of the allenyl central carbon. It is estimated that the $\eta^1-\eta^3$ propargyl rearrangement in (**105**) lowers the activation barrier for CO insertion by $\geq 29$ kJ mol$^{-1}$. The geometry requirements for [1,5]-hydrogen shifts to give *exo*-allyl (**108**) and *endo*-allyl (**109**) have been analyzed in detail.

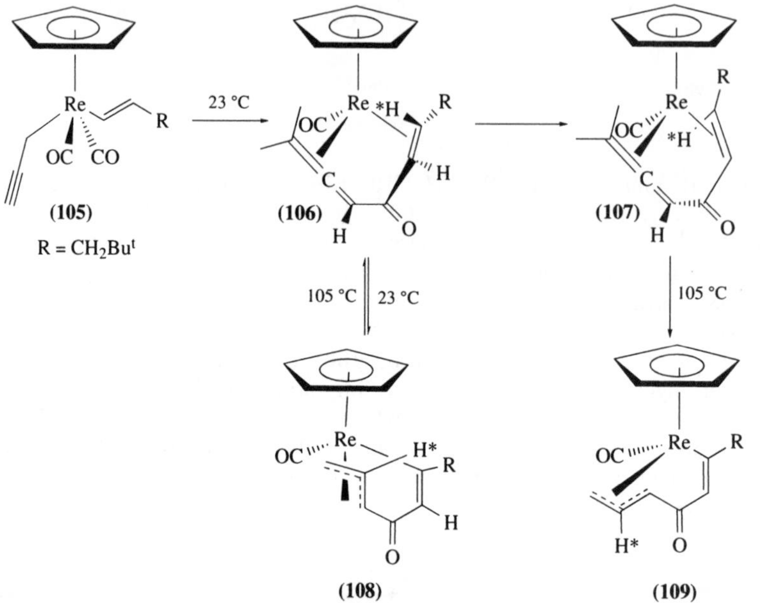

**Scheme 25**

Hughes and co-workers have reported a fascinating concerted decarbonylation reaction in which cyclopropene acyl (**110**) is converted to cyclopropenyl (**111**) with allylic rearrangement of the migrating group (Equation (22)). Of particular significance is the observation that the cyclopropyl analogue of (**110**) is thermally and photochemically inert toward decarbonylation. Compound (**111**) does not undergo [1,3]-sigmatropic shifts of the [Re(CO)$_5$] moiety at 63 °C.[221] The related [1,7]-sigmatropic shift is also a high-energy process in [Re($\eta^1$-C$_7$H$_7$)(CO)(NO)(Cp)] and is not observed at 130 °C on the $^1$H NMR timescale.[222]

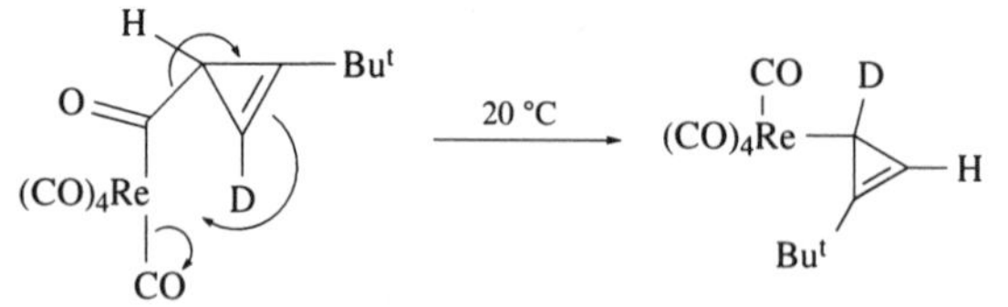

(22)

Berke and co-workers prepared the cyclopropenylrhenium complex (**112**) in 70% yield from Na[Re(CO)$_5$] and triphenylcyclopropenylium cation. Thermolysis of (**112**) in *n*-hexane gives $\eta^2$-propenetriyl complex (**113**) and oxocyclobutenyl complex (**114**) (Scheme 26). This latter product appears to be the result of an initial CO insertion followed by ring expansion since (**113**) does not give (**114**) in the presence of CO.[223]

Rhenium carbonyl acyls with bulky substituents, such as *t*-butyl and cyclohexyl, are converted to acyl-bridged compounds (**115**) in 67–95% yield upon heating.[224] The metalla-β-diketonate complexes of the type [Re(COMe)$_2$(CO)$_4$][NMe$_4$] undergo reaction with organometallic Lewis acids to give bis(acyl)-bridged bimetallic complexes such as (**116**).[225] The rhenaacetylacetone complex [Re(CMeOHO-CMe)(CO)$_4$] undergoes condensation reactions with a variety of amino acid esters to give rhena-β-ketoimine complexes (**117**).[226]

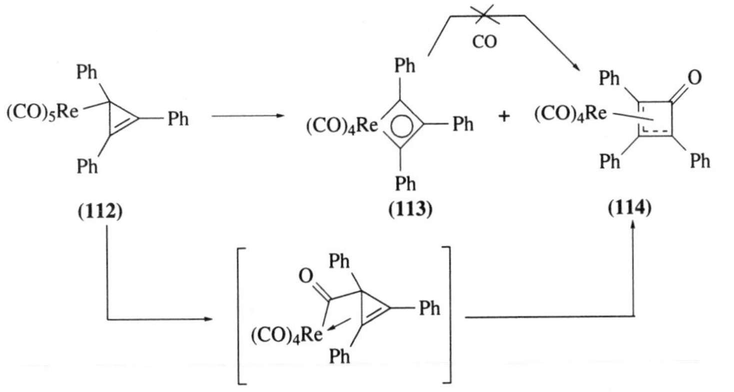

**Scheme 26**

Treatment of the (metalla-β-diketonato)difluoroboron complex [Re{CRO(BF$_2$)OCMe}(CO)$_4$] with base gives η$^3$-allyl products (**118**) from coupling of the adjacent acyl ligands.[227] A related coupling reaction occurs with (triacylrhenato)boron complexes.[228]

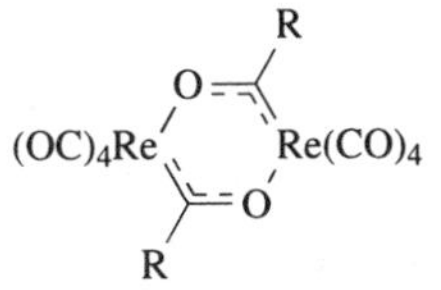

(**115**)

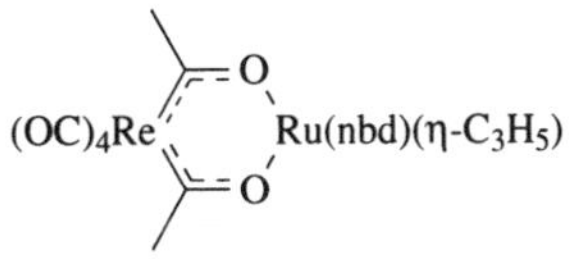

(**116**)

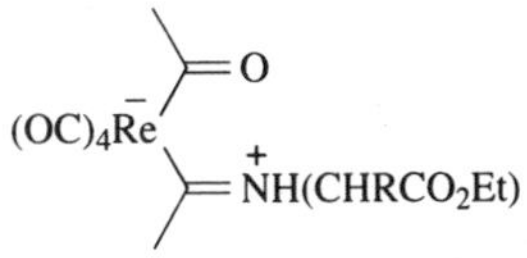

(**117**)

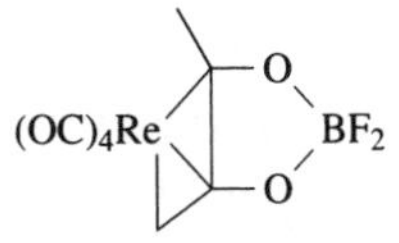

(**118**)

### 9.8.3 Deprotonation of Acyl Ligands

In the reactions of [Re{C(=O)CH$_2$R}(NO)(PPh$_3$)(Cp)] with strong bases, either enolate formation or η-Cp deprotonation can occur. Addition of lithium diisopropylamide (LDA) to the optically active acyl [Re{C(=O)Me}(NO)(PPh$_3$)(Cp)] in THF at −78 °C followed by MeI gives the methyl complex [Re(Me)(NO)(PPh$_3$){η$^5$-C$_5$H$_4$(COMe)}] in 79–80% yield, with retention of configuration at rhenium (≥90% *ee*).[229] When the addition of MeI is delayed until 12 min after deprotonation, only 20% *ee* is observed for the alkylation product. Thus, the metal-based anion has limited configurational stability at −78 °C. Acyl migration to the ring results in charge stabilization by conversion of a carbanion to the more stabilized metal-based anion. Use of the weaker base LiN(TMS)$_2$ with [Re{C(=O)CH$_2$Ph}(NO)(PPh$_3$)(Cp)] at 0 °C and alkylation with MeOSO$_2$CF$_3$, also at 0 °C, gives [Re{C(=O)CHMePh}(NO)(PPh$_3$)(Cp)] as the major product in 53% yield. The intermediate enolate anion was spectroscopically observed and found to be stable in THF solution at 25 °C for nearly 2 d.

The pentamethylcyclopentadienyl analogue [Re{C(=O)CH$_2$Ph}(NO)(PPh$_3$)(η-Cp*)] is converted to an enolate anion with Bu$^n$Li at −70 °C and subsequent quenching with MeI gives the α-phenylpropionyl complex [Re{C(=O)CHPhMe}(NO)(PPh$_3$)(η-Cp*)] in 72% yield as a 96:4 ratio of diastereomers. The stereochemistry of alkylation is consistent with a reactive Re–C$_α$ conformation in which the C=C bond of the enolate is conjugated with the HOMO of the [Re(NO)(PPh$_3$)(η-Cp*)] fragment. Electrophiles attack the enolate from a direction anti to the bulky PPh$_3$ ligand.[229]

Enolate (**119**) undergoes reaction with metal carbonyl electrophiles to give μ-malonyl complexes (e.g., (**120**), (**121**)), or a μ-ketene complex (**122**), depending on the specific nature of the electrophile (Scheme 27).[230,231] The malonyl complex (**120**) represents the first metal–acyl complex to exist in equilibrium with an observable enol tautomer. Deuterium-labeling studies indicate that the enolization is a stereoselective one, with the methylene hydrogen *exo* to the PPh$_3$ ligand enolizing more rapidly than the *endo* hydrogen.[232] In the reaction of (**119**) with [ReBr(CO)$_4$(PPh$_3$)] to give bridging ketene (**122**), a malonyl intermediate is observed by NMR spectroscopy.

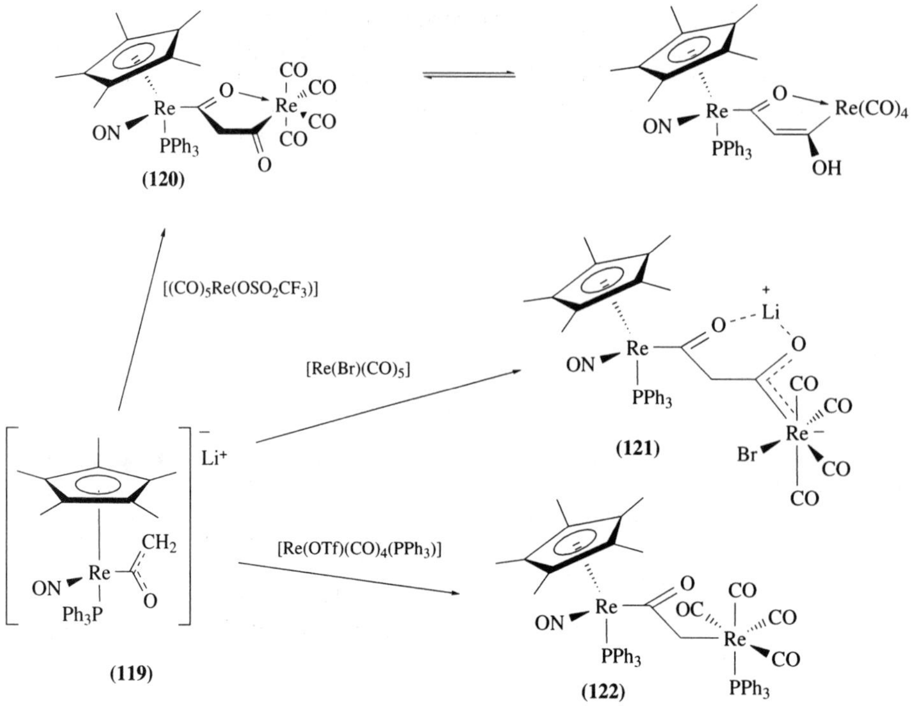

**Scheme 27**

### 9.8.4 Metal Carboxylic Acids

Deprotonation of [Re(CO₂H)(NO)(PPh₃)(Cp)] and subsequent reaction with main-group halides (Ph₃GeBr 52% and Ph₃SnCl 86%) gives the bridging carboxylate complexes [Re(CO₂GePh₃)(NO)(PPh₃)(Cp)] and [Re(CO₂SnPh₃)(NO)(PPh₃)(Cp)].[233] [Re(CO₂H)(CO)(NO)(Cp)] also reacts with [ZrCl(Me)(Cp)₂] to give carboxylate (**123**) in 82% yield.[234] Reaction of [Re(CO₂H)(CO)(NO)(Cp)] with 2 equiv. [ZrH(Cl)(Cp)₂] gives [Re(CO)(NO)(Cp)(μ-η¹,η¹-CH₂O)ZrCl(Cp)₂] in 61% yield in addition to the μ-oxo complex [{ZrCl(Cp)₂}₂O].

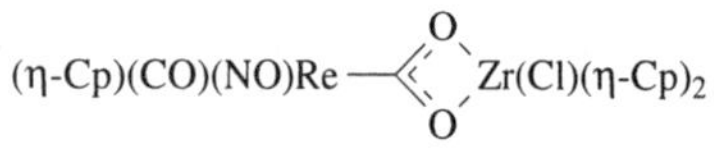

(**123**) 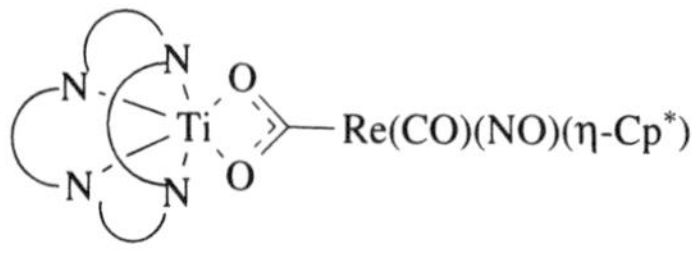                                                        (**124**)

A cycloaddition reaction between [Ti(=O)(tmtaa)] and [Re(CO)₂(NO)(η-Cp*)][BF₄] gives (**124**) in 90% yield (tmtaa = dianion of 7,16-dihydro-6,8,15,17-tetramethyldibenzo[*b,i*][1,4,8,11]tetra-azacyclotetradecine).[235] Geoffroy and co-workers have also prepared μ₂,η³-CO₂ complexes by net [2 + 2]-cycloaddition of the metal–oxo bond in [M(=O)(Cp)₂] (M = W, Mo) across the carbon–oxygen bond of the carbonyl ligand in [Re(NO)(CO)₂(η-Cp*)]⁺.[236]

The iron carboxylate [Fe(CO₂)(CO)(PPh₃)(Cp)][K] reacts with [Re(CO)₅(FBF₃)] to give [(CO)₅Re(μ₂-η²-CO₂)Fe(CO)(PPh₃)(Cp)], and subsequent thermally induced decarbonylation gives [(CO)₄Re(μ₂-η³-CO₂)Fe(CO)(PPh₃)(Cp)].[237]

## 9.9 RHENIUM–CARBON MULTIPLE BONDS

### 9.9.1 Carbene and Vinylidene

Traditionally, metal–carbene complexes have been classified as either electrophilic or nucleophilic, with low oxidation state rhenium carbenes in the former category. For example, the rhenium–methylidene complex [Re(=CH₂)(NO)(PPh₃)(Cp)]⁺ is attacked by dimethylsulfide to give the sulfonium salt [Re(CH₂SMe₂⁺)(NO)(PPh₃)(Cp)] in 90–95% yield.[238]

In 1989, Casey and Nagashima reported the synthesis of the neutral alkyl-substituted carbene complexes (125) in 50–60% yields.[239] The carbene α-hydrogen of (125) is observed at δ 16.11 in the [1]H NMR spectrum and the carbene carbon at 292.3 (d, $J_{CH}$ = 134 Hz) in the [13]C NMR spectrum. These carbenes exhibit a reactivity which is unique for low-valent alkyl-substituted carbene complexes undergoing reaction at the carbene carbon with both electrophiles and nucleophiles.

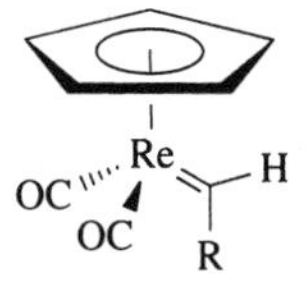

(125a) R = Me
(125b) R = CH$_2$CH$_2$Bu$^t$

Deuterated carbene (125b) undergoes stereospecific addition of HCl across the rhenium–carbon double bond to generate a single diastereomer of the alkylrhenium complex *cis*-(126), which rearranges at −13 °C to give a 1:1 equilibrium mixture of the two diastereomers (Scheme 28).[240] The stereochemistry of the HCl addition was not determined. At 17 °C, (126) decomposes to an η$^2$-alkene product with deuterium equally distributed between the *cis* and *trans* positions at C-1. Reaction of (125) with excess HCl leads to reduction and chain extension of the carbene ligand to form the hydroxy carbene (127) by a process in which the deuterated alkyl group in *cis*-(126) reversibly migrates to CO with retention of stereochemistry.

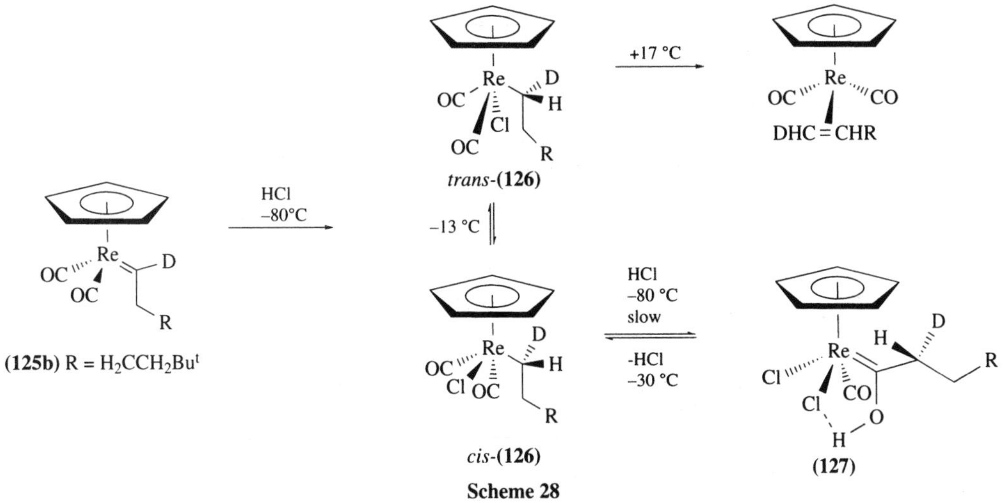

**Scheme 28**

Dimethyl cuprate attacks (125) at the carbene carbon to give [Re(CHMeCH$_2$CH$_2$Bu$^t$)(CO)$_2$(Cp)]$^−$ and Bu$^t$OK deprotonates (125) at the α-carbon to give the anionic vinyl complex [Re(CH=CHCH$_2$Bu$^t$)-(CO)$_2$(Cp)]$^−$.[212]

The rhenium–carbon bond length in the parent electrophilic methylidene complex [Re(=CH$_2$)(NO)(PPh$_3$)(η-Cp*)]$^+$ is 0.189 8(18) nm, and $\Delta G^{\ddagger}$(114 °C) ≥79 kJ mol$^{-1}$ for rotation about the double bond.[241] In the IR spectrum the antisymmetric C–H stretches are split into doublets at 2987 cm$^{-1}$ and 2976 cm$^{-1}$. In solution [Re(=CH$_2$)(NO)(PPh$_3$)(Cp)]$^+$ decomposes at room temperature to the η$^2$-ethene complex [Re(CH$_2$=CH$_2$)(NO)(PPh$_3$)(Cp)]$^+$.[242] The coupling process exhibits remarkable enantiomer self-recognition with the (R,R) and (S,S) transition states preferred over the (R,S) and (S,R) transition states. The mechanism appears to involve mutual front-side attack of the two Re–CH$_2$ moieties in the rate-determining step. The methylidene complex is unreactive toward [Re(Me)(NO)(PPh$_3$)(Cp)] except for degenerate hydride transfer.

Related rhenium–vinylidene complexes are available from acyl precursors as shown in Scheme 29. For rotation about the metal–carbon double bond in (128) (R = H) $\Delta G^{\ddagger}$(110 °C) >77.8 kJ mol$^{-1}$, compared to $\Delta G^{\ddagger}$(27 °C) >62.8 kJ mol$^{-1}$ for the methylidene analogue.[243] Although the steric bulk is further removed from the metal in the case of the vinylidene ligand, it is also a better π-acceptor ligand, which increases the metal–carbon double bond character. The metal-bound vinylidene carbon is observed in the range 327–336 ppm in the [13]C NMR spectra. For R = Ph (128) exists as a 1:3 ratio of kinetic (anticlinal) ac-(**128**) to thermodynamic (synclinal) sc-(**128**) isomers. This compares to a 99:1 ratio for the analogous carbene complex. Deprotonation of (**128**) (R = Ph) gives the acetylide [Re(C≡CPh)(NO)(PPh$_3$)(Cp)], which is alkylated at 0 °C to give a 99:1 ratio of anticlinal to synclinal [Re(C=C=CPhMe)(NO)(PPh$_3$)(Cp)]$^+$. At 25 °C, an equilibrium ratio of 25:75 ac:sc is established. Irradiation of the sample results in a 50:50 photostationary state of sc:ac isomers, which returns to thermal equilibrium in the dark. The vinylidene ligand undergoes stereospecific nucleophilic attack by PMe$_3$ at the α-carbon.

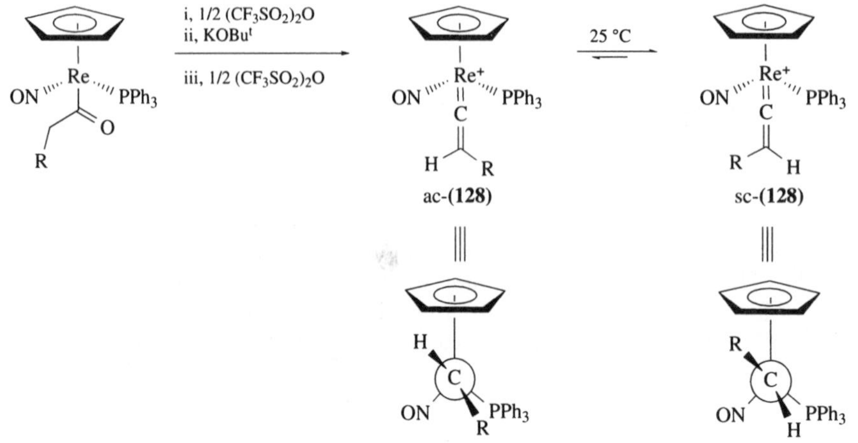

**Scheme 29**

Angelici and co-workers have found that [Re(CO)$_5$X] reacts with aziridines and epoxides in the presence of [BrCH$_2$CH$_2$NH$_3$]$^+$[Br]$^-$ to form cyclic aminooxy- and dioxycarbene complexes (Equation (23)).[244] The mechanism involves nucleophilic attack of aziridine (or BrCH$_2$CH$_2$O$^-$ or oxirane) at a CO ligand.

$$[ReX(CO)_5] + \overline{YCH_2CH_2} \xrightarrow[25\,°C]{Br^-} cis\text{-}[ReX(CO)_4(=\overline{COCH_2CH_2Y})] \qquad (23)$$

$$X = Cl, Br, I \quad Y = NH \text{ or } O$$

The alkyl–carbene complex *cis*-[Re(Me)(CO)$_4$(=$\overline{COCH_2CH_2NMe}$)] was prepared in the hope of observing an alkyl-to-carbene migration, but thermolysis in the presence of PMe$_3$ or PPh$_3$ gave only low yields of uncharacterized substitution products. Reaction of *cis*-[ReBr(CO)$_4$(=$\overline{COCH_2CH_2NH}$)] with KHB(pz)$_3$ gives *fac*-[Re(CO)$_3${η$^2$-HB(pz)$_3$}(=$\overline{COCH_2CH_2NH}$)], which loses CO upon photolysis to give the η$^3$-HB(pz)$_3$ dicarbonyl complex.[245] Treatment of [ReBr(CO)$_4$($\overline{COCH_2CH_2O}$)] with dimethylthiocarbamate ion regenerates [ReBr(CO)$_5$] by a process which involves initial nucleophilic attack at the *sp*$^3$ carbon of the cyclic carbene.[246]

Alkylation of the anion generated from [Re$_2$(CO)$_{10}$] and LiSiPh$_3$ gives ketenyl complex (**129**) (10% yield), in addition to the mono- and bis(carbene) complexes [(CO)$_9$Re$_2${=C(OEt)SiPh$_3$}] and [(CO)$_8$Re$_2${=C(OEt)SiPh$_3$}$_2$].[247]

Reaction of [Re$_2$(CO)$_{10}$] with CH$_2$=CHCH$_2$CH$_2$MgBr gives a terminal carbene complex [(CO)$_5$Re–Re(CO)$_4${=C(OEt)CH$_2$CH$_2$CH=CH$_2$}], which converts to the alkene–carbene complex (**130**) upon abstraction of CO with Me$_3$NO.[248] The reaction of [Re$_2$(CO)$_{10}$] and [Ti(η$^2$-C$_2$H$_4$)(η-Cp*)$_2$] at −20 °C gives the titanoxycarbene complex (**131**) in 57% yield.[249]

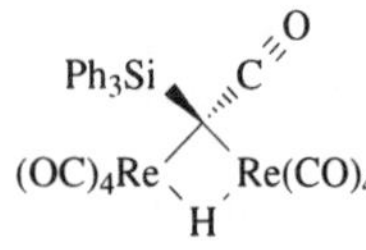

(**129**) 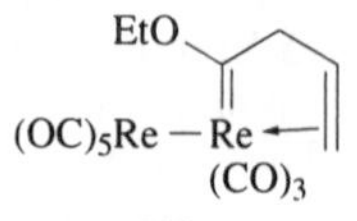                              (**130**) 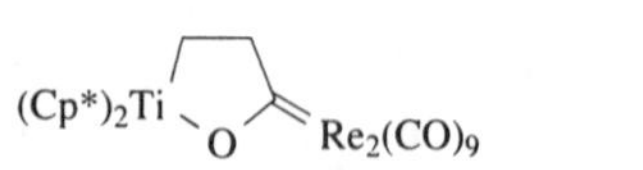                              (**131**)

Both hydroxy- and methoxycarbene complexes are accessible from protonation or alkylation of the formyl group in *mer,trans*-[Re(CHO)(CO)$_3$(PPh$_3$)$_2$].[250] Aminolysis of the methoxycarbene gives aminocarbenes, also in excellent yield. The aminocarbenes are deprotonated to give iminomethyl or formidoyl complexes and are reduced with [Et$_4$N][BH$_4$] to aminomethyl complexes.[251] The dinitrogen complex *trans*-[ReCl(N$_2$)(dppe)$_2$] serves as a precursor to isocyanide,[252] allene,[253] and vinylidene complexes.[254] Protonation of the isocyanide ligand in [ReCl(CNR)(dppe)$_2$] gives *trans*-[ReCl($\equiv$CNHR)(dppe)$_2$]$^+$,[255] whereas protonation of the allene complex [ReCl($\eta^2$-CH$_2$=C=CHPh)(dppe)$_2$] gives a cationic metallacyclopropene.[256,257] The benzoyldiazenido complex [ReCl$_2$(NNCOPh)(PPh$_3$)$_2$] reacts with P$\equiv$CBu$^t$ to give [ReCl$_2$(NNCOPh)(P$\equiv$CBu$^t$)(PPh$_3$)$_2$], which undergoes a phosphaalkyne–hydrazido coupling to form a phosphidocarbene complex.[258]

### 9.9.2 Carbyne

Pombeiro and co-workers have prepared aminocarbyne complexes *trans*-[ReCl(CNH$_2$)(dppe)$_2$]$^+$ from the trimethylsilyl isocyanide complex *trans*-[Re(Cl)CN-TMS(dppe)$_2$] and HBF$_4$.[259] The related reaction of *trans*-[ReCl(CNR)(dppe)$_2$] (R = Me, Bu$^t$) and [CoCl$_2$(THF)$_{1.5}$], [ReCl$_3$(O)(PPh$_3$)$_2$], or [WCl$_4$(PPh$_3$)] gives the dinuclear adducts [ReCl{CNR(M)}(dppe)$_2$], where M = CoCl$_2$(THF), ReCl$_3$(O)(PPh$_3$), or WCl$_4$(PPh$_3$).[261] These dinuclear complexes exhibit IR bands in the range 1550–1600 cm$^{-1}$ as is observed for the related amino–carbyne complexes.[255] The alkyl-substituted carbyne complexes are formed upon protonation of the vinylidene complexes *trans*-[ReCl(=C=CHR)(dppe)$_2$].[261] The redox properties of *trans*-[ReCl(L)(dppe)$_2$] where L is a vinylidene, isocyanide, carbyne, or aminocarbyne ligand have been reported.[262] A convenient single-pot synthesis of *trans*-[ReF($\equiv$CCH$_2$R)(dppe)$_2$][BF$_4$] in 40–65% yields is shown in Equation (24).[263]

$$[ReCl(N_2)(dppe)_2][BF_4] \xrightarrow[\substack{TlBF_4,\ NH_4BF_4 \\ THF}]{HC\equiv CR} \textit{trans-}[ReF(\equiv CCH_2R)(dppe)_2][BF_4] \qquad (24)$$

Reduction of the seven-coordinate cations [ReCl$_2$(CNR)$_3$(PMePh$_2$)$_2$]$^+$ with zinc or aluminum in wet THF generates the alkylamino–carbyne complexes [ReCl($\equiv$CNHR)(CNMe)$_2$(PMePh$_2$)$_2$][SbF$_6$], where R = Me or Bu$^t$.[264,265]

Geoffroy and co-workers have extensively developed the chemistry of cyclopentadienylrhenium–carbyne complexes. The reaction of [Re($\equiv$CR)(CO)$_2$(Cp)][BCl$_4$] (**130**)-BCl$_4$ and [PPN]NO$_2$ at 0 °C generates the acyl complex (**131**) in 74% yield (Scheme 30).[266] Formally, the reaction involves the addition of O$^-$ to the cationic carbyne ligand and substitution of NO for CO. The reaction may involve nitrite attack at the carbyne carbon and migration of NO from oxygen to rhenium and CO loss.

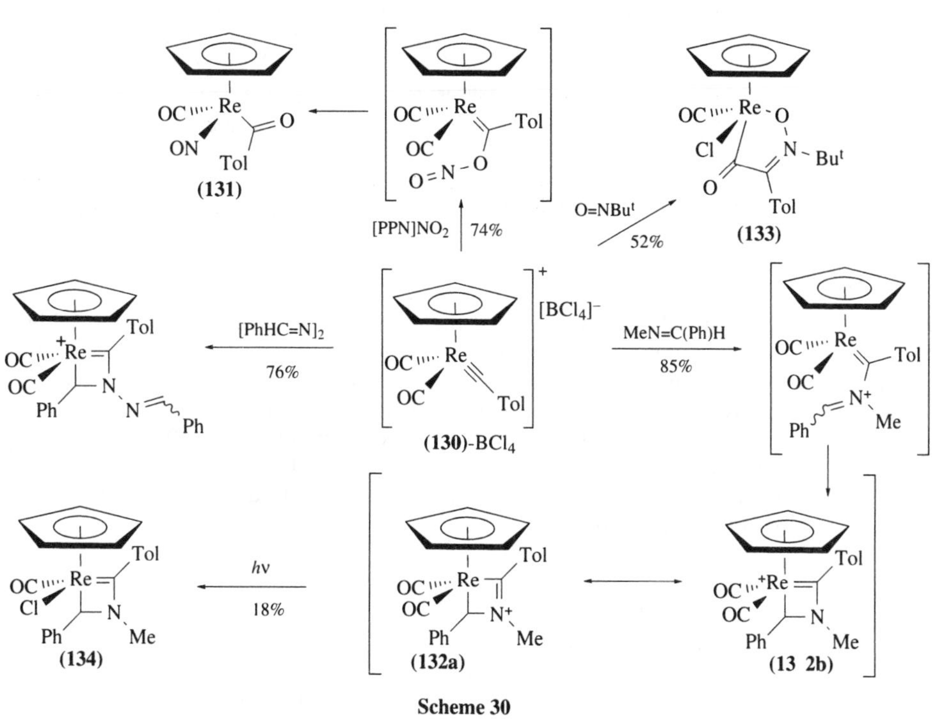

**Scheme 30**

Imines and Bu$^t$N=O also attack the carbyne carbon of (130) to give the metallaazetine complex (132) (85%) and (133) (52%), respectively (Scheme 30). Structural characterization indicates that (132a) is the major resonance contributor for the metallaazetine ring.[267] Photolysis of (132) leads to loss of CO and Cl$^-$ abstraction from BCl$_4^-$ to give (134). Treatment of (132) with NaOH fragments the metallacycle and forms benzaldehyde and the aminocarbene [Re{=CNHMe(Tol)}(CO)$_2$(Cp)] in 49% yield. By changing from the reactive BCl$_4^-$ counterion to BPh$_4^-$, the reaction of (130) and Bu$^t$N=O gave a 70% yield of [Re{ON(Bu$^t$)C(Tol)}(CO)$_2$(Cp)], which is converted to (133) in 90% yield upon addition of [PPN]Cl. The reaction is reversed by halide abstraction with AgBF$_4$.[268]

More recently (130)-BPh$_4$ has been shown to undergo addition and cycloaddition reactions with *cis*-azoarenes ((135), 90%), epoxides ((136), 36%), 3,3-dimethyloxetane ((137), 40%), 2-methylaziridine ((138), 91%), propene sulfide ((139), 82%), and benzophenonehydrazone ((140), 71%) (Scheme 31).[269] Simple reaction of (130) with Cl$^-$ gives the carbene complex [Re{=C(Cl)Tol}(CO)$_2$(Cp)] in 23% yield.[270] Similar reaction with metal carbonyl anions generates dinuclear µ-carbene complexes.[271]

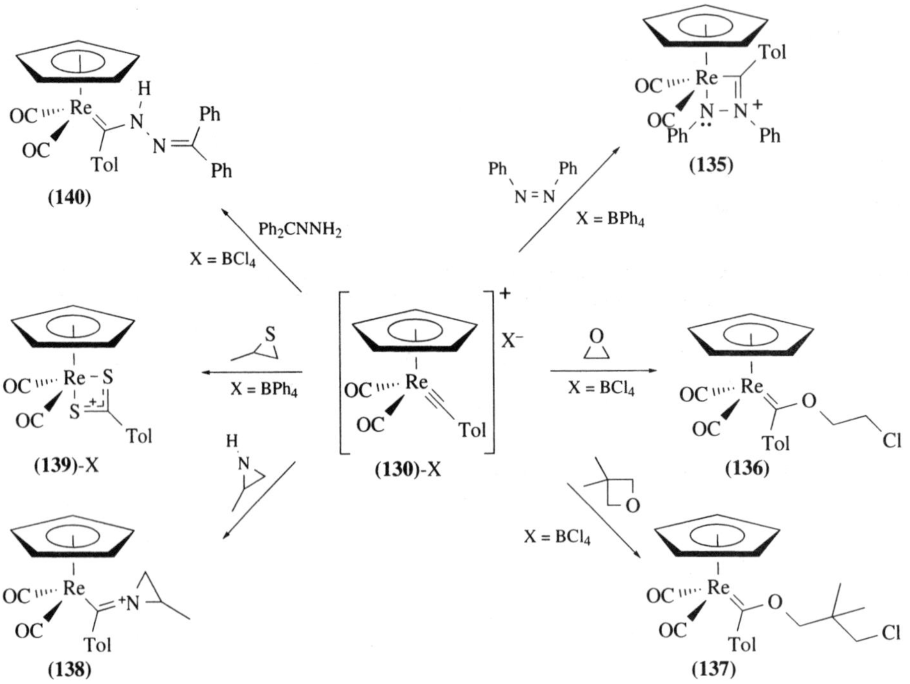

**Scheme 31**

# 9.10 π-COMPLEXES OF HYDROCARBONS

## 9.10.1 Alkene Complexes

Synthetic routes toward η$^2$-alkene complexes of rhenium include obvious ones, such as direct ligand substitution and β-hydrogen elimination from rhenium alkyls, as well as less obvious routes as described below. Table 7 provides a listing of selected η$^2$-alkene complexes of rhenium.

An example of the β-hydrogen elimination route is found in the reaction of [Re(Cl)$_3$(PMe$_2$Ph)$_3$] with excess ethyllithium to give a 54% yield of the bis(ethene) complex [ReH(PMe$_2$Ph)$_3$(η$^2$-CH$_2$=CH$_2$)$_2$] (Equation (25)).[272] At ambient temperatures one of the two ethene ligands rotates rapidly on the NMR timescale ($\Delta G^{\ddagger} = 51$ kJ mol$^{-1}$ at 0 °C). The ethene ligands are readily displaced by dihydrogen to give trihydrido- or pentahydridorhenium complexes.

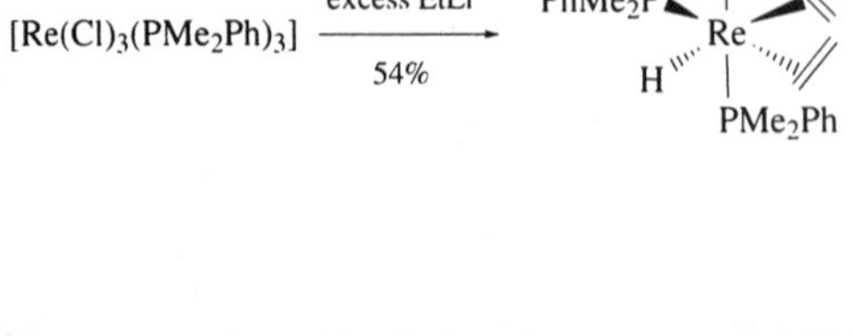

(25)

**Table 7** Selected alkene complexes.

| Complex | Ref. |
|---|---|
| $[Re(\eta^2\text{-CHR}^1=\text{CHR}^2)(CO)_2(Cp)]$ ($R^1 = H$, $R^2 = Bu^t$; $R^1 = R^2 = Me$; $R^1 = TMS$, $R^2 = C\equiv C\text{-TMS}$) | 156 |
| $[Re(\eta^2\text{-CH}_2=\text{CHMe})(CO)_2(Cp)]$ | 135 |
| $[Re\{\eta^4\text{-CHR}=\text{CHC}(=O)\text{CH}_2\text{CH}=\text{CH}_2\}(CO)(Cp)]$ ($R = CH_2Bu^t$) | 220 |
| $[Re\{\eta^4\text{-CHR}=\text{CHC}(=O)\text{CH}=\text{C}=\text{CH}_2\}(CO)(Cp)]$ ($R = CH_2Bu^t$) | 211 |
| $[Re(\eta^2\text{-CH}_2=\text{CH}_2)(NO)(PPh_3)(Cp)]^+$ | 242 |
| $[Re(\eta^2\text{-CH}_2=\text{CHR})(CO)_5]^+$ ($R = H$, Me) | 273 |
| $[\{Re(CO)_5\}_2(\eta^2\text{-}\eta^2\text{-diene})]$ (diene = 1,5-hexadiene, 1,4,-cyclohexadiene) | 274 |
| $[Re(\eta^2\text{-CH}_2=\text{CHR})(CO)_2(Cp^*)]$ ($R = H$, Me, $CH_2OMe$, $C_6H_{13}$) | 141 |
| $[Re(\eta^2\text{-CHR}^1=\text{CHR}^2)(CO)_2(Cp^*)]$ ($R^1 = Me$, $R^2 = Et$; $R^1 = Me$, $R^2 = C_5H_{11}$) | 141 |
| $[Re(\eta^2\text{-}\overline{\text{CH}=\text{CHCHRCH}_2\text{CH}_2})(CO)_2(Cp^*)]$ ($R = H$, OMe) | 141 |
| $[Re(\eta^2\text{-CMe}_2=\text{CHCOMe})(CO)_2(Cp^*)]$ | 275 |
| $[Re(\eta^2\text{-alkene})(CO)_2(Cp^*)]$ (alkene = 4-methylcyclohexene, cyclooctene, 1,3-cyclohexadiene, 1,4-cyclohexadiene, allene) | 275 |
| $[Re(\eta^2\text{-CH}_2=\text{CHMe})(CO)(PMe_3)(\eta\text{-Cp}^*)]$ | 275 |
| $[Re(\eta^2\text{-CH}_2=\text{CHCH}_2R)(CO)_2(Cp)]$ ($R = Me$, $CH(CO_2Et)_2$) | 276 |
| $[Re(\eta^2\text{-CH}_2=\text{CHR})(NO)(PPh_3)(Cp)]^+$ ($R = CHMe_2$, $Bu^t$, TMS) | 277 |
| $[Re(\eta^2\text{-cycloalkene})(NO)(PPh_3)(\eta\text{-C}_5H_4Me)]^+$ (cycloalkene = cyclopentene, cyclohexene, cycloheptene, cyclooctene) | 213 |
| $[Re(\eta^2\text{-CH}_2=\text{CHR})(NO)(PPh_3)(Cp)]^+$ ($R = Me$, Et, $Pr^n$, Ph, Bz, $CH_2CPh_3$) | 278 |
| $[Re(\eta^2\text{-CH}_2=\text{CMe}_2)(NO)(PPh_3)(Cp)]^+$ | 278 |
| $[Re(\eta^2\text{-CHPh}=\text{CHPh})(CO)_2(Cp)]$ | 279 |
| $[Re(\eta^2\text{-CHPh}=\text{CHMe})(CO)_2(Cp)]$ | 279 |
| $[Re(\eta^2\text{-cycloheptatriene})(CO)(NO)(Cp)]^+$ | 280 |
| $[Re(\eta^2\text{-CH}_2=\text{CH}_2)(PMe_3)_2(\eta\text{-C}_6H_6)]^+$ | 215 |

The synthesis of $\eta^2$-alkene and $\eta^2$-alkyne rhenium complexes by simple ligand substitution chemistry has met with variable degrees of success. The reaction of $[Re(CO)_5(FBF_3)]$ and TMS-CH=CH$_2$ gives an 80% yield of $[Re(CO)_5(\mu\text{-}\eta^1\text{:}\eta^2\text{-CH=CH}_2)Re(CO)_5]^+[BF_4]^-$ which in turn reacts with $I^-$ to give $[Re(\eta^2\text{-CH}_2\text{CH}_2)(CO)_5]^+$ in 45% yield.[273] Reaction with butadiene gives a 96% yield of $[Re(\eta^2\text{-CH}_2\text{CHCHCH}_2)(CO)_5]^+$. The IR stretch of the coordinated alkene is shifted to an energy 100 cm$^{-1}$ lower than that of the free alkene.[274]

Photolysis of $[Re(CO)_3(\eta\text{-Cp}^*)]$ or $[Re(CO)_2(N_2)(\eta\text{-Cp}^*)]$ in the presence of alkenes generates the corresponding $\eta^2$-alkene complexes in low to moderate yield (6–67%). In some cases, allyl hydrido by-products are also formed from carbon–hydrogen bond activation chemistry.[141,275]

The best high-yield synthesis of pure rhenium–alkene complexes in the cyclopentadienyl dicarbonyl series is from the reaction of the heterobimetallic dihydride $[(Cp)(CO)_2Re(\mu\text{-H})PtH(PPh_3)_2]$ (**47**) and alkynes (Equation (26)).[156] The mechanism of this reaction was discussed in Section 9.5.3 and the alkene complexes, $[Re(CO)_2(cis\text{-}\eta^2\text{-CR}^1H=\text{CHR}^2)(Cp)]$ ($R^1 = H$, $R^2 = Bu^t$; $R^1 = R^2 = Me$; $R^1 = TMS$, $R^2 = C\equiv C\text{-TMS}$) are generated in 67–91% yield. With diphenylethyne a 1.5:1 mixture of *cis*- and *trans*-stilbene complexes (**141**) is obtained. Isolated *cis*-(**141**) isomerizes to *trans*-(**141**) over the course of days at 127 °C, but no isomerization was seen under the reaction conditions. One mechanistic possibility for the *trans*-hydrogenation of the alkyne involves an intramolecular, reversible hydride transfer from rhenium to a vinyl ligand on platinum (Scheme 32).

$$(26)$$

The (Z)-alkene complex *cis*-(**142**) undergoes an acid-catalyzed isomerization to an equilibrium mixture of *cis*-(**142**) and *trans*-(**142**) (45:55) (Scheme 33).[281] Treatment of $[Re(CO)_2\text{-}(\eta^2\text{-CH}_2=\text{CHCH}_2Me)(Cp)]$ (**143**) with $CF_3CO_2D$ leads to deuterium exchange at the vinyl hydrogen sites with no isomerization to the $\eta^2$-2-butene complex (**142**). In a similar fashion, treatment of (**142**) with $CF_3CO_2D$ led to (**142**)-$d_1$ with no formation of the $\eta^2$-1-butene complex (**143**). The observed incorporation of deuterium into the vinyl sites and not the methyl sites is explained by stereospecific protonation at the vinyl carbon, rotation about the carbon–carbon bond, and stereospecific deprotonation. The failure of (**142**) and (**143**) to undergo acid-catalyzed interconversion requires a strong agostic interaction and an "in place rotation" of the agostic methyl group as shown in Scheme 33. It is possible to elaborate the coordinated $\eta^2$-alkene ligand in these systems by a hydride abstraction to give an $\eta^3$-allyl complex, followed by nucleophilic attack at the allyl ligand.[276]

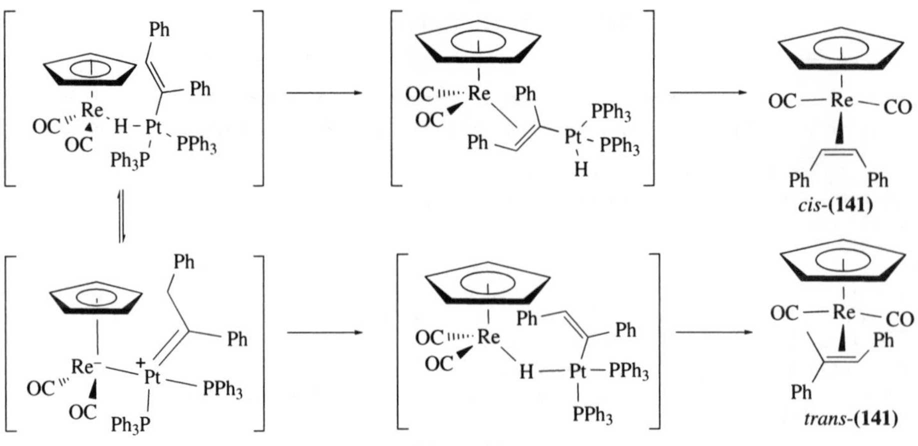

Scheme 32

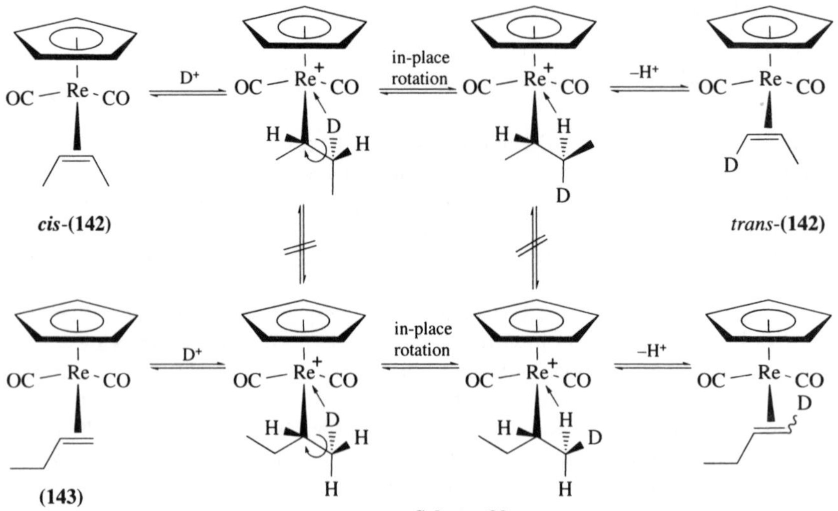

Scheme 33

The substitution-labile cationic chiral complexes [Re(NO)(PPh$_3$)(L)(η-Cp*)][BF$_4$] (L = ClCH$_2$Cl, ClPh) bind monosubstituted alkenes to give high yields (89–91%) of the η$^2$-alkene complexes as ~2:1 mixtures of the diastereomers (RS,SR)-(145) and (RR,SS)-(145) (Scheme 34).[277] There is no evidence that these substitution reactions proceed by a dissociative mechanism. The observation that each diastereomer of the η$^2$-styrene complex (145) (R = Ph) is generated in optically active form is consistent with configurational stability at rhenium under the conditions of their synthesis. Nuclear Overhauser enhancement (NOE) experiments support the indicated conformations for both diastereomers. Although the kinetic diastereoselectivities are modest, when the (RS,SR)-(145)/(RR,SS)-(145) mixtures are heated in PhCl at 100 °C for 24 h the (RR,SS)-diastereomer converts to the (RS,SR)-isomer, with a ~95:5 ratio of isomers established at equilibrium. Deuterium-labeling experiments prove that the isomerization process is intramolecular, which requires that the metal exchange alkene enantiofaces without dissociation and without rotation of the =CHR terminus. Two mechanisms are consistent with the deuterium labeling, kinetic, and stereochemical studies. Both involve the movement of rhenium through the nodal plane of the alkene, one process via a carbon–hydrogen σ-bond complex and the other via a vinyl hydride oxidative addition–reductive elimination sequence (Scheme 35).

Complexes of cyclic alkenes have also been reported in this system and subsequent treatment with base leads to a remarkable vinylic deprotonation and formation of η$^1$-vinyl complexes (Scheme 36).[213] The observed kinetic preference for vinylic deprotonation over allylic deprotonation is without precedent. The reactions of the η$^1$-allyl complexes (E)-[Re(η$^1$-CH$_2$CH=CHR$^1$)(NO)(PPh$_3$)(Cp)] (R$^1$ = H, Ph) with electrophiles (H$^+$, D$^+$, R$^+$) also gives alkene complexes (79–92% yield) as (75–60):(25–40) mixtures of diastereomers.[278]

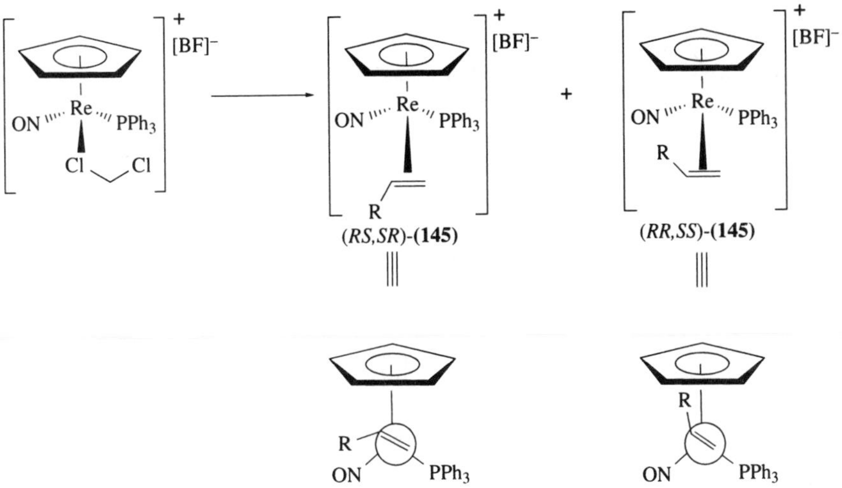

**Scheme 34**

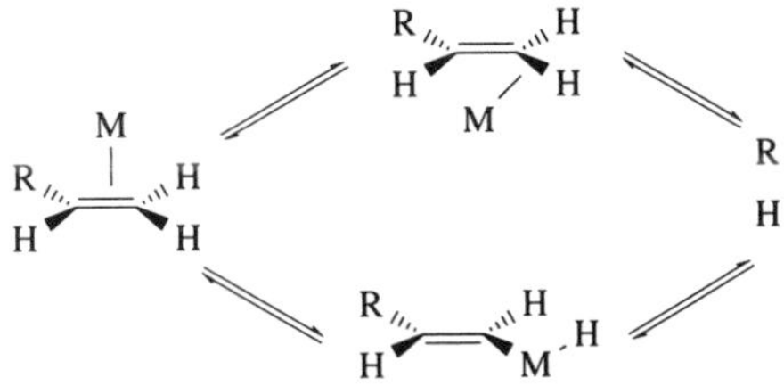

**Scheme 35**

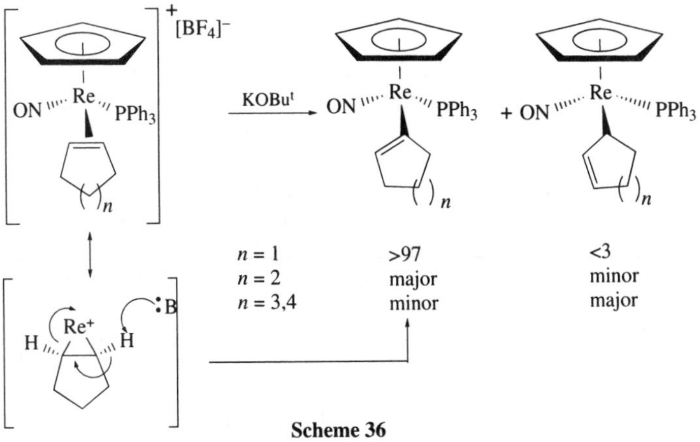

| | | |
|---|---|---|
| $n = 1$ | >97 | <3 |
| $n = 2$ | major | minor |
| $n = 3,4$ | minor | major |

**Scheme 36**

### 9.10.2 Alkyne Complexes

Low yields (~30–38%) of $\eta^2$-alkyne complexes are available from photolysis of [Re(CO)$_3$(Cp')] (Cp' = $\eta$-Cp, $\eta$-Cp*, $\eta$-C$_5$H$_4$Me) and R$^1$C≡CR$^2$ (e.g., R$^1$ = H, Me, Ph and R$^2$ = H; R$^1$ = R$^2$ = Me).[279,282] At 25 °C [Re(CO)$_2$(THF)($\eta$-Cp*)] and 5 equiv. 2-butyne give an 85% yield of the $\eta^2$-butyne complex.[283] Photolysis of indenyl(tricarbonyl)rhenium in the presence of diphenylethyne gives a dicarbonyl $\eta^2$-alkyne complex, whereas thermolysis in the presence of phenylethyne gives the octahedral acetylide [Re(C≡CPh)(CO)$_3$(py)$_2$].[284]

Hydride abstraction from the $\eta^2$-alkyne complex [Re($\eta^2$-MeC≡CMe)(CO)$_2$($\eta$-Cp*)] generates a cationic ($\eta^3$-propargyl) complex (**144**), which is attacked by nucleophiles at the central carbon of the propargyl unit to give the first rhenacyclobutene complexes (Scheme 37).[283] In the $^{13}$C NMR spectrum of the metallacycles the quaternary propargyl carbons are observed at δ 76.6 (s) and 56.7 (s).

An attempt to extend this hydride abstraction reaction to the terminal $\eta^2$-alkyne complex [Re(CO)$_2$($\eta^2$-HC≡CMe)($\eta$-Cp*)] leads, not to a propargyl complex, but rather to an allyl complex,

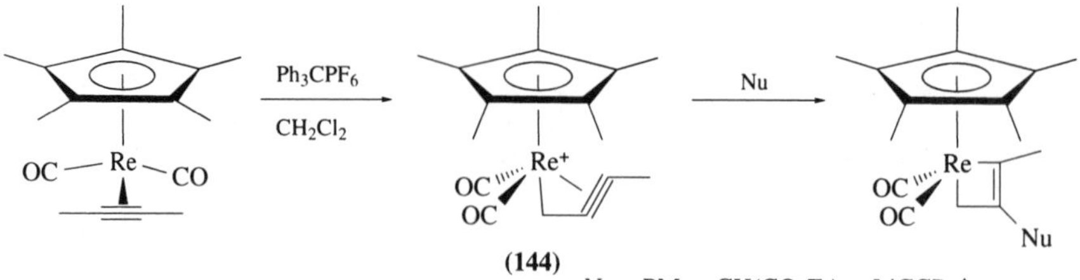

**(144)**

Nu = PMe$_3$, CH(CO$_2$Et)$_2^-$, LiCCBu$^t$

**Scheme 37**

$[Re(CO)_2\{\eta^3\text{-Ph}_2CC(Ph)CHMe\}(\eta\text{-Cp*})]^+$. The reaction involves addition of trityl cation to the terminal alkyne carbon followed by successive [1,2]-hydrogen and [1,2]-phenyl shifts.[285]

Addition of Br$_2$ to the $\eta^2$-ethyne complex $[Re(CO)_2(\eta^2\text{-HC}\equiv\text{CH})(\eta\text{-Cp*})]$ gives a 30% yield of the $\eta^1$-vinyl bromide complex $[ReBr(\eta^1\text{-CH}=\text{CHBr})(CO)_2(\eta\text{-Cp*})]$.[214]

The first synthesis of a rhenium–benzyne complex was discovered upon addition of 2 equiv. PMe$_3$ or PMe$_2$Ph to $[Re(2\text{-MeC}_6H_4)_4]$ at $-40\,°C$. The paramagnetic benzyne complex $[Re(2\text{-MeC}_6H_4)_2(\eta^2C_6H_3Me)(PMe_2R)_2]$ is formed in nearly quantitative yield, and the solid-state data are consistent with a delocalized aromatic benzyne ring (Equation (27)).[286]

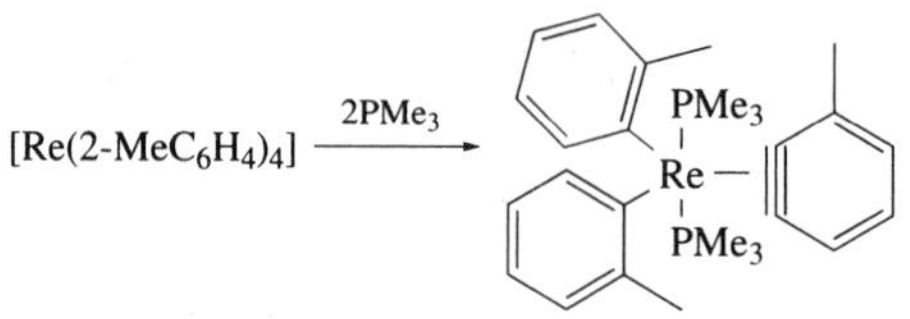

$$[Re(2\text{-MeC}_6H_4)_4] \xrightarrow{2PMe_3} \qquad\qquad\qquad (27)$$

Alkyne complexes of the type $[Re(NO)(PPh_3)(R^1C\equiv CR^2)(Cp)]^+$ are also accessible from alkyne and $[Re(NO)(PPh_3)(ClPh)(Cp)]^+$ in good to excellent yields (73–97%). Deprotonation of the terminal $\eta^2$-alkyne ligands gives $\eta^1$-acetylide complexes and thermolysis of the $\eta^2$-alkyne complexes generates vinylidene ligands.

The Re-$(\eta^2\text{-HC}\equiv\text{CR})$ rotational barrier for R = Me or Et is much higher ($>92$ kJ mol$^{-1}$, $180\,°C$) than the related barrier in the alkene complex $[Re(NO)(PPh_3)(\eta^2\text{-CH}_2CH_2)(Cp)]$ (68.6 kJ mol$^{-1}$, $96\,°C$).[287] For comparison $\Delta G^{\ddagger} = 34.7$ kJ mol$^{-1}$ for rotation of the $\eta^2$-ethene ligand in $[Re(CO)_2(\eta^2\text{-CH}_2CH_2)(Cp)]$.[220]

The rhenium(III) $\eta^2$-alkyne complexes $[ReCl_2(\eta^2\text{-R}^1C\equiv CR^2)(\eta\text{-Cp*})]$ ($R^1 = R^2 = $ Ph, Me, Et; $R^1 = $ Ph, $R^2 = $ Me) **(146)** exhibit a rich reactivity including alkyne–allyl[288] and alkyne–alkene[289] coupling reactions as shown in Scheme 38. Treatment of **(146)** with CH$_2$=CHCH$_2$MgCl at $-78\,°C$ followed by thermolysis in toluene gives the 2,4-pentadienylrhenium(III) complex **(147)** in 70–80% yield. The transformation arises from alkyne insertion into an $\eta^1$-allyl–rhenium bond and a 1,3-hydride rearrangement to give the $\pi$-conjugated ligand.

The dihalo($\eta^2$-alkyne) complexes are surprisingly inert toward metallacycle formation; however, treatment of **(146)** with a catalytic amount of HBF$_4$ in the presence of excess ethene generates diene complex **(148)** in 85% yield. This latter reaction presumably involves metallacyclopentene intermediates. The role of the acid appears to be in facilitating the loss of the chloro ligand as HCl, which generates the requisite open coordination site for the alkene reactant.[284]

Additional alkyne complexes are described in Section 9.13.3, which discusses rhenium–oxo complexes. Table 8 provides a listing of selected $\eta^2$-alkyne complexes.

### 9.10.3 Allyl Complexes

The propene complex $[Re(\eta^2\text{-CH}_2=\text{CHMe})(CO)_2(\eta\text{-Cp*})]$ gives a 6.4:1 ratio of *endo* to *exo* $\eta^3$-allyl complexes upon reaction with Ph$_3$CBF$_4$. The *exo–endo* interconversion occurs without scrambling of the *syn*- and *anti*-protons, which is consistent with a rotation mechanism.[291]

The allyl complexes $[Re(\eta^3\text{-H}_2CCR^1CR^2R^3)(CO)_4]$ ($R^1 = R^2 = R^3 = $ H; $R^1 = $ Me, $R^2 = R^3 = $ H; $R^1 = R^2 = $ H, $R^3 = $ Me; and $R^1 = $ H, $R^2 = R^3 = $ Me) are generated in 52–68% yield by an allyl transfer reaction between $[Re(CO)_5]Na$ and $[PdCl(\eta^3\text{-allyl})]_2$.[292] Treatment of $[Re(\eta^3\text{-C}_3H_5)(CO)_4]$ with HBF$_4\cdot$Et$_2$O in CH$_2$Cl$_2$ generates $[Re(FBF_3)(OEt_2)(CO)_4]$, which serves as a source of the 14-electron $[Re(CO)_4]^+$ fragment. THF displaces both labile ligands to give a 1:1 mixture of *cis*- and *trans*-$[Re(CO)_4(THF)_2][BF_4]$.

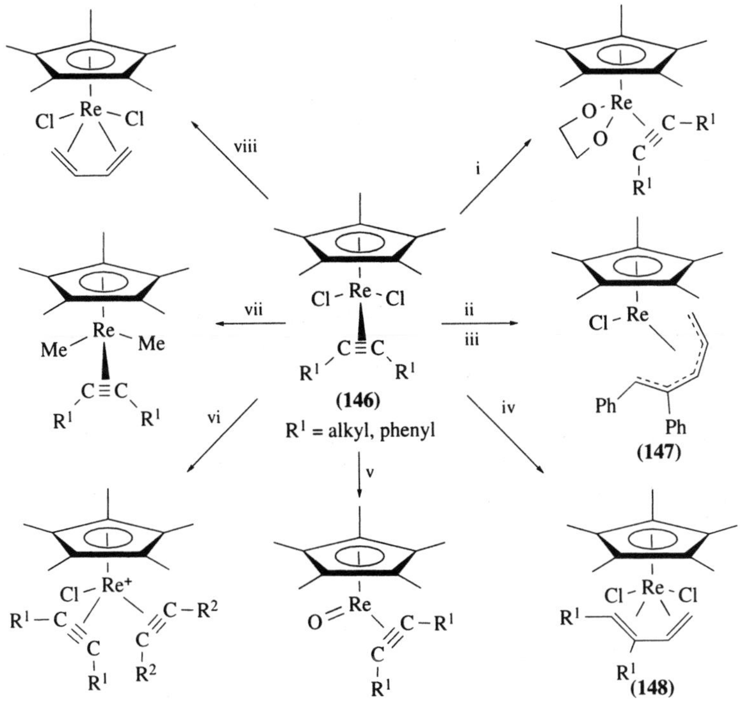

i, NaO⌒ONa ; ii, C₃H₅MgBr; iii, Δ, PhMe; iv, 3 CH₂=CH₂, cat. HBF₄; v, LiOH; vi, R²C≡CR², AgSbF₆;

vii, 2MeMgX; viii, ⌒⌒, cat. HBF₄

**Scheme 38**

**Table 8** Selected alkyne complexes.

| Complex | Ref. |
|---|---|
| $[Re(\eta^2\text{-}RC\equiv CH)(NO)(PPh_3)(\eta\text{-}Cp)]^+$ (R = Me, H) | 202 |
| $[Re(\eta^2\text{-}MeC\equiv CH)(NO)(PPh_3)(\eta\text{-}C_5H_4Me)]^+$ | 202 |
| $[Re(\eta^2\text{-}RC\equiv CH)(NO)(PPh_3)(\eta\text{-}Cp^*)]^+$ (R = H, Me) | 202 |
| $[Re(\eta^2\text{-}RC\equiv CR)(NO)(PPh_3)(\eta\text{-}Cp)]^+$ (R = Me, Et, Ph) | 278 |
| $[Re(\eta^2\text{-}HC\equiv CBu^t)(NO)(PPh_3)(\eta\text{-}Cp)]^+$ | 278 |
| $[Re(\eta^2\text{-}HC\equiv CR)(CO)_2(\eta\text{-}Cp)]$ (R = H, Me) | 282 |
| $[Re(\eta^2\text{-}MeC\equiv CMe)(CO)_2(\eta\text{-}Cp)]$ | 282 |
| $[Re(\eta^2\text{-}CH\equiv CR)(CO)_2(\eta\text{-}Cp^*)]$ (R = H, Me, Ph) | 282 |
| $[Re(\eta^2\text{-}MeC\equiv CMe)(CO)_2(\eta\text{-}Cp^*)]$ | 282 |
| $[ReCl_2(\eta^2\text{-}R^1C\equiv CR^2)(\eta\text{-}Cp^*)]^+$ (R¹ = R² = Me, Et, Ph; (R¹ = Me, R² = Ph) | 289 |
| $[ReX_2(\eta^2\text{-}MeC\equiv CMe)(\eta\text{-}Cp^*)]$ (X = Br, I) | 289 |
| $[Re(\eta^2\text{-}PhC\equiv CPh)(CO)(NO)(L)]$ (L = P(OPr$^i$)$_3$, PMe$_3$, PEt$_3$, PPr$^i_3$, PCy$_3$) | 290 |

Sutton and co-workers have observed an allylic carbon–hydrogen activation in rhenium $\eta^2$-propene complexes with formation of an $\eta^3$-allyl hydride product.[141]

Treatment of $[Re(\eta\text{-}C_3H_4R)(CO)_4]$ (R = H, Me), with NaN(TMS)₂ at 70–85 °C gives the anionic cyanide complex $[Re(CN)(\eta\text{-}C_3H_4R)(CO)_3][Na]$ and O(TMS)₂, and subsequent alkylation at nitrogen with Me₃OBF₄ or MeI gives a 94% yield of the allyl isonitrile complex $[Re(\eta^3\text{-}CH_2CRCH_2)(CO)_3(CNMe)]$.[293] Photolysis of $[Re(CO)_3(Cp)]$ and allylic alcohol or dienes in the presence of HBF₄ provides access to cationic allyl complexes as shown in Scheme 39.[294] In this series, the parent allyl $[Re(CO)_2(\eta\text{-}C_3H_5)(Cp)]^+$ exists as an equilibrium mixture of *exo-* and *endo-*conformers with the *endo-*isomer prevailing and a 51 kJ mol$^{-1}$ barrier for interconversion.[295]

The $\eta^5$-pentadienyl ligand serves as a precursor to $\eta^1$-*cis* and $\eta^1$-*trans*-pentadienyl ligands. Addition of 2 equiv. PR₃ (R = Me or Et) to $[Re(\eta^5\text{-}C_5H_7)(CO)_3]$ in refluxing Et₂O gives *fac*-[Re-($\eta^1\text{-}C_5H_7)(CO)_3(PR_3)_2]$ (R = Et, 85% yield). Over the course of weeks at 25 °C, the pentadienyl ligand isomerizes to the *trans*-isomer.[296] Reaction with Et₂PCH₂CH₂PEt₂ (depe) gives $[Re(\eta^1\text{-}cis\text{-}C_5H_7)(CO)_3(depe)]$, which is converted to the first structurally characterized $\eta^1$-3-pentadienyl complex $[Re(\eta^1\text{-}3\text{-}C_5H_7)(CO)_3(depe)]$ upon photolysis at 0 °C. This complex in turn rearranges to the *trans*-isomer via a phosphine dissociation–reassociation mechanism.[297]

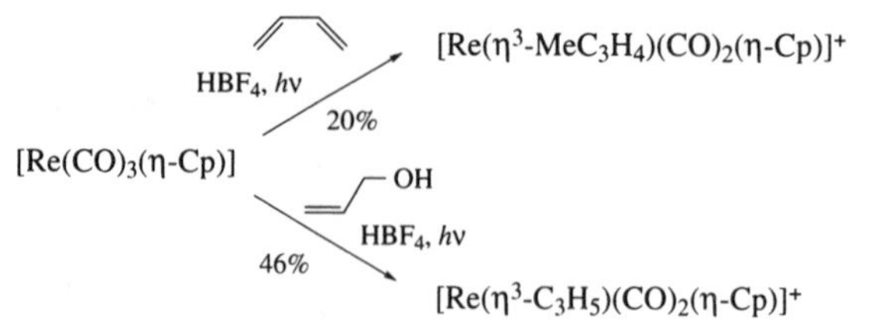

$$[Re(CO)_3(\eta\text{-}Cp)] \xrightarrow[\quad 20\% \quad]{HBF_4,\ h\nu} [Re(\eta^3\text{-}MeC_3H_4)(CO)_2(\eta\text{-}Cp)]^+$$

$$\xrightarrow[\quad 46\% \quad]{HBF_4,\ h\nu} [Re(\eta^3\text{-}C_3H_5)(CO)_2(\eta\text{-}Cp)]^+$$

**Scheme 39**

### 9.10.4 Arene Complexes

Mononuclear arene complexes of rhenium are accessible by heating [ReBr(CO)$_5$], AlCl$_3$ and the arene at 150 °C. In this manner [Re(CO)$_3$(η-C$_6$HMe$_5$)][PF$_6$] is prepared in 71% yield. The parent compound [Re(CO)$_3$(η-C$_6$H$_6$)]$^+$ is prepared as its AlCl$_3$Br$^-$ salt, which is stable in CH$_2$Cl$_2$ solution for weeks at −10 °C. As expected, the methylated derivatives (C$_6$Me$_6$ and C$_6$HMe$_5$) are more stable than the parent ligand toward arene displacement by oxygen donor solvents. Grignard reagents and PBu$^n_3$ react at the arene ligand to give substituted cyclohexadienyl complexes in low to moderate yields.[298,299] Treatment of the neutral cyclohexadienyl complexes with NOPF$_6$ provides the cationic nitrosyl complexes (e.g., [Re(CO)$_2$(NO)(η$^5$-C$_6$Me$_7$)][PF$_6$], 52% yield). The (η$^5$-C$_6$Me$_5$PhH) analogue exhibits 44.8 kJ mol$^{-1}$ barrier to rotation of the Re(CO)$_2$(NO) tripod.[300] Hydride adds to the hexadienyl ring to give cyclohexadiene complexes with the hydride *endo* to the metal, whereas other nucleophiles (CN$^-$, PBu$_3$) simply substitute for a CO ligand.[298]

An η$^2$-arene complex is accessible by reversible protonation of the aryl complex [Re-(η$^1$C$_6$H$_4$R)(CO)(NO)(Cp)] (R = H, Me (*o, m, p*), CF$_3$).[301] The complex (R = H) is fluxional by a process in which rhenium migrates from one η$^2$-position on the ring to another via an η$^1$-arenium intermediate.

The hydride complex [ReH(CO)(NO)(Cp)] reacts with Ph$_3$CPF$_6$ at −78 °C to give an η$^2$-arene [Re(3,4-η$^2$-PhCHPh$_2$)(CO)(NO)(Cp)][PF$_6$], which is deprotonated by NEt$_3$ at −78 °C to give *para*- and *meta*-isomers of [Re(η$^1$-C$_6$H$_4$CHPh$_2$)(CO)(NO)(Cp)].[280]

Irradiation of [Re(CO)$_3$(η-Cp*)] in benzene gives the doubly coordinated benzene complex [(Cp*)(OC)$_2$Re(μ-η$^2$,η$^2$-C$_6$H$_6$)Re(CO)$_2$(Cp*)] (149), which on prolonged photolysis generates [Re(η-Cp*)(η-C$_6$H$_6$)].[302]

Green and co-workers have prepared a variety of mono- and dinuclear η$^6$-rhenium arenes by cocondensation (−196 °C) of rhenium atoms with hydrocarbons.[303–9] Cocondensation of benzene, PMe$_3$, and rhenium atoms gives the dinuclear arene complex [Re(PMe$_3$)$_2$(η-C$_6$H$_6$)]$_2$ in 20% isolated yield based on rhenium evaporated from the furnace. Reduction of this dimer at a potassium metal film provides the mononuclear anion [Re(PMe$_3$)$_2$(η-C$_6$H$_6$)][K] in >90% yield and treatment with MeI gives a 62% yield of [ReI(PMe$_3$)$_2$(η-C$_6$H$_6$)], which in turn serves as a precursor to a variety of organometallic derivatives.[215]

### 9.10.5 Cyclopentadienyl Ligand Chemistry

#### 9.10.5.1 *Synthesis and reactivity*

A number of substituted cyclopentadienyl(tricarbonyl) rhenium derivatives of the type [Re(CO)$_3$-(η-Cp')] are accessible from [ReBr(CO)$_5$] and the corresponding thallium reagents, Cp'Tl, including Cp' = C$_5$H$_4$PPh$_2$,[310] C$_5$H$_4$Bz,[311] C$_5$H$_4$Ph,[311] C$_5$H$_4$CO$_2$Et,[312] C$_5$H$_4$COH,[312] and C$_5$H$_4$COMe.[312] The pentabenzylcyclopentadienyl complex [Re(CO)$_3${η-C$_5$Bz$_5$}] is accessible in low yield (13%) from [Re$_2$(CO)$_{10}$].[313] Fulvalenedithallium, [TlC$_5$H$_4$C$_5$H$_4$Tl], and [ReBr(CO)$_5$] give the fulvalene complex [Re(CO)$_3${η$^5$,η$^5$-H$_4$C$_5$C$_5$H$_4$}Re(CO)$_3$] in 94% yield.[314] The tris(cyclopentadienyl) complex [Re(Cp)$_2$-(η$^1$-Cp)] is prepared in >95% yield from [ReCl(Cp)$_2$] and KCp.[315]

The dichalcogenides [{Re(CO)$_3$(C$_5$H$_4$E)}$_2$] (150) (E = S, Se, Te) are formed from air oxidation of the chalcogenates [Re(CO)$_3$(η-C$_5$H$_4$ELi)], which are generated *in situ* from {Re(CO)$_3$(η-C$_5$H$_4$Li)] and the chalcogens (Scheme 40).[316] Protonation gives the corresponding chalcogenols, which are also oxidized to (150). The tellurol was more labile than the thiols and selenols and not characterized as a pure compound. The chalcogenate also reacts with CH$_2$I$_2$ to give [{Re(CO)$_3$C$_5$H$_4$E}$_2$CH$_2$].[317]

Reaction of [ReBr(CO)$_4$(PPh$_3$)] and (*E,E*)-1,4-dilithio-1,4-diphenyl-1,3-butadiene followed by protonation or alkylation gives [Re(CO)$_3${η-C$_5$H$_2$Ph$_2$(OH)}] (R = H, Me).[318]

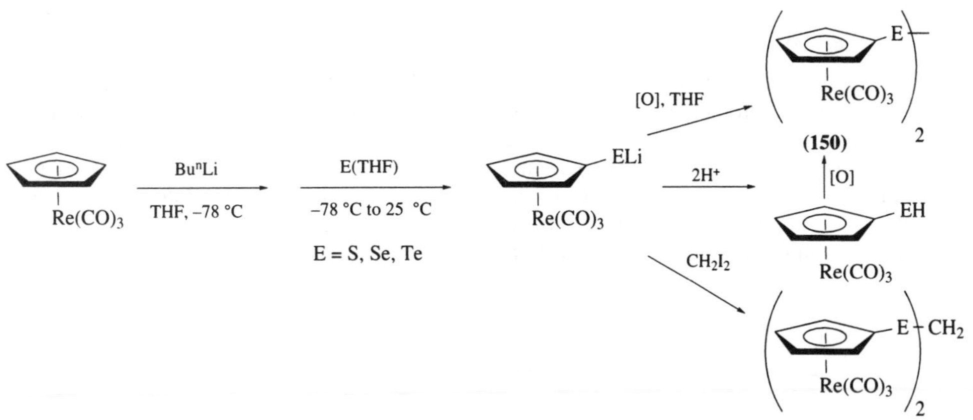

**Scheme 40**

The carbinol complexes $[\{(CO)_3Re(C_5H_4)\}_2C(OH)R]$ (R = H, Me, Ph, $CF_3$) are metallated at only one of the two $\alpha$-positions of the cyclopentadienyl ring when treated with excess $Bu^nLi$.[319] For the tertiary carbinols, intramolecular hydrogen bonding to rhenium has been proposed on the basis of IR spectroscopic evidence and molecular mechanics calculations.[320]

As discussed elsewhere in this review, Gladysz and co-workers have discovered a series of metal-to-$\eta$-$C_5H_4Li$ ligand migrations upon deprotonation of $[Re(R^1)(CO)(NO)(Cp)]$, where $R^1 = CHO$,[321] H,[134] and $CR^2O$ ($R^2 = Me$, Ph, $CH_2Ph$).[322] In contrast, where $R^1 = $ alkyl (e.g., Me, Bz) migration is not observed.[229] Fenske–Hall calculations on the formyl migration favor a mechanism involving an $\eta^5$- to $\eta^1$-$C_5H_4Li$ ring slip followed by CHO migration.[323] Deprotonation of $[ReCl(NO)(PPh_3)(Cp)]$ at $-78\,°C$ gives $[ReCl(NO)(PPh_3)(\eta$-$C_5H_4Li)]$, which is converted to the dianion $[ReCl(NO)(PPh_3)(\eta$-$C_5H_3Li_2)]$ upon reaction with $Bu^nLi/TMEDA$ at $0\,°C$. Quenching with MeI gives $[ReCl(NO)(PPh_3)(\eta$-$C_5H_3Me_2)]$ in 63% yield.[324] Sequential reaction of $[Re(PPh_2)(NO)(PPh_3)(Cp)]$ with $Bu^nLi$ and $ClPPh_2$ at $-78\,°C$ gives $[Re(PPh_2)(NO)(PPh_3)(\eta$-$C_5H_4PPh_2)]$. The optically active bis(phosphido) complex forms a heterobimetallic asymmetric hydrogenation catalyst upon reaction with $[Rh(nbd)Cl]_2$ and $AgPF_6$. Enamides are hydrogenated to $\alpha$-amino acids and esters with 100–300 turnovers per hour.[325]

Deprotonation of *cis*-$[ReH(SiPh_3)(CO)_2(Cp)]$ at $-78\,°C$ gives $[ReH(CO)_2(\eta$-$C_5H_4SiPh_3)][Li]$ from rapid migration of the $SiPh_3$ ligand in preference to the hydride ligand.[142] Protonation of the anion at $-78\,°C$ gives the *cis*-dihydride, which isomerizes to the *trans*-dihydride at $25\,°C$.

Photolysis of $[Re(CO)_3(Cp)]$ and its $\eta$-Cp* analogue in the presence of phosphines and phosphites[326,327] gives rise to the corresponding $[Re(CO)_2(L)(Cp)]$ complexes (where L = $PMe_3$, $PMe_2Ph$, $PMePh_2$, and $PPh_3$) in yields of 19–70%. When $[Re(CO)_3(\eta$-Cp*)] is photolyzed in cyclohexane at $20\,°C$ $[Re_2(CO)_3(\eta$-Cp*)_2] and $[Re_2(CO)_5(\eta$-Cp*)_2] are formed with the former generated in 30% yield based on unreacted starting material.[328]

The electrocatalysis of ligand exchange has been extended to the conversion of $[Re(CO)_2(py)(Cp)]$ and $PPh_3$ to $[Re(CO)_2(PPh_3)(Cp)]$ (86% yield).[329] This chain reaction involves an associative substitution process at a 17-electron radical-cation intermediate. Reduction-catalyzed substitutions in $[Re(phen)(CO)_3(NCMe)]^+$ have also been described.[330]

Photolysis of $[Re(CO)_3(\eta$-Cp*)] in neat $C_6F_6$ gives a 60% yield of the air-stable complex $[Re(C_6F_5)(CO)_2(\eta^6$-$C_5Me_4CH_2)]$ (**151**). Reaction of (**151**) with $PMe_3$ and HCl give products resulting from nucleophilic and electrophilic attack, respectively, at the $\eta^6$-$C_5Me_4CH_2$ ligand (Equation (28)).[331]

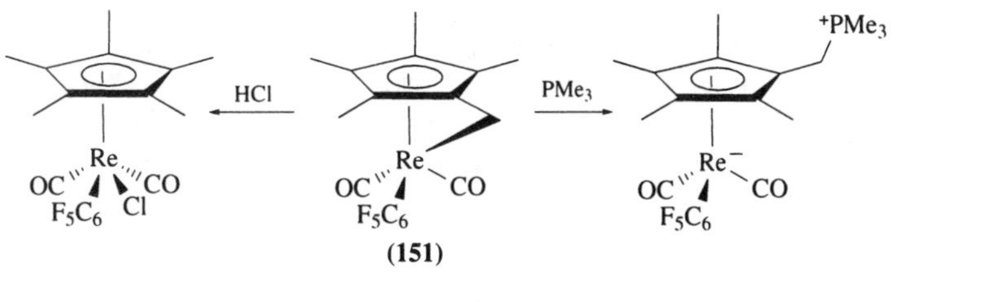

(28)

$[Re(CO)_3(\eta$-Cp*)] and $X_2$ (X = Br, I) in $THF/H_2O$ gives a mixture of *cis*- and *trans*-$[ReX_2(CO)_2(\eta$-Cp*)], whereas reaction of $[ReCl(CO)_3(\eta$-Cp*)][SbCl_6]$ in $H_2O$ gives the *cis*-dichloro complex.[332] Photolysis of *cis*-$[Re(CO)_2(Me)_2(\eta$-Cp*)] also leads to the *trans*-isomer.[333]

### 9.10.5.2 *Ring slippage*

Casey *et al.* have extensively investigated rhenium–cyclopentadienyl ring slippage reactions whereby $\eta^5$, $\eta^3$, $\eta^1$, and "$\eta^0$" cyclopentadienyl complexes interconvert. Reaction of [Re(Me)(NO)(CO)(Cp)] (**152**) with 2 equiv. PMe$_3$ at 25 °C generates the bis(phosphine) $\eta^1$-cyclopentadienyl complex [Re(Me)(NO)(CO)(PMe$_3$)$_2\eta^1$-Cp] (**153**) (Scheme 41). At 90 °C the equilibrium mixture of (**152**) and (**153**) is converted to acyl complex (**154**) and methyl complex (**155**). Rate studies indicate a monophosphine intermediate, presumably with an $\eta^3$-cyclopentadienyl ligand.[334] When conversion to (**154**) and (**155**) is inhibited by high concentrations of PMe$_3$, the cyclopentadienyl ketene complex (**156**) is formed.

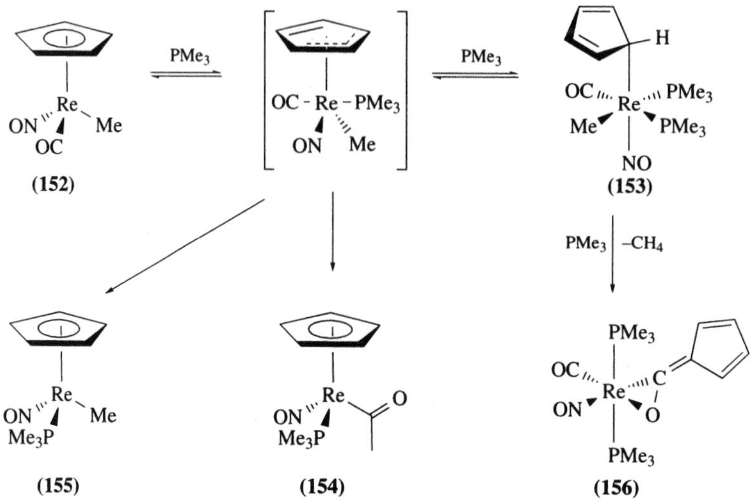

**Scheme 41**

Reaction of [Re(Me)(NO)(PMe$_3$)(Cp)] with excess PMe$_3$ cleanly generates the $\eta^1$-cyclopentadienyl complex (**157**), and at 50 °C the cyclopentadienyl ligand is heterolytically cleaved from rhenium in a reversible reaction to give the ionic [Cp]$^-$ complex (**158**) in addition to the $\eta^1$-(1,3-cyclopentadiene) complexes (**159**) and (**160**) (Scheme 42).[335]

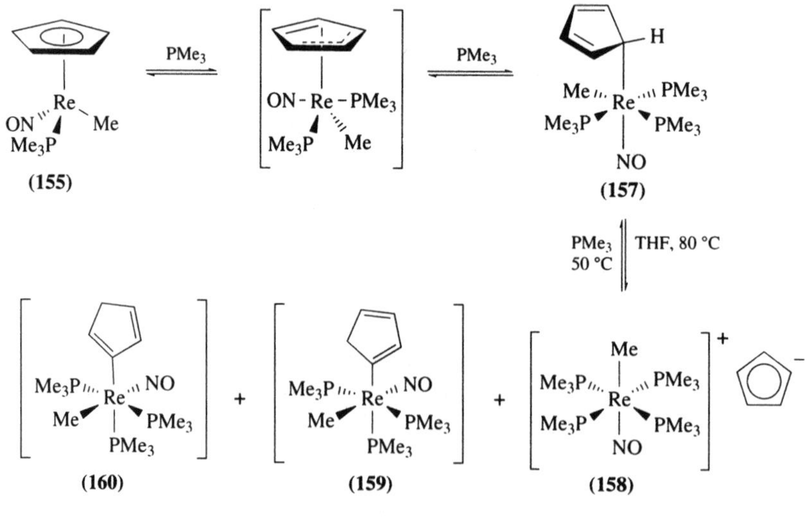

**Scheme 42**

The $\eta^5$–$\eta^3$ ring slip is more difficult to induce than the $\eta^3$–$\eta^1$ transformation, leading to the possibility that two different incoming ligands may be incorporated into one $\eta^5$–$\eta^1$ isomerization reaction. Phosphine-induced cyclopentadienyl ring slippage has been employed to catalyze CO insertion into a methylrhenium complex.[336] The carbonylation of [Re(Me)(NO)(PMe$_3$)(Cp)] (**155**) to form [Re(COMe)(NO)(PMe$_3$)(Cp)] (**154**) under 5 atm of CO at 90 °C is accelerated by a factor of >25 by added PMe$_3$. Kinetic and phosphine exchange studies indicate a PMe$_3$-induced cyclopentadienyl ring slippage mechanism.

The nonnitrosyl-containing complex [Re(CO)$_3$(Cp)] reacts reversibly with PMe$_3$ at 64 °C to give [Re(CO)$_3$(PMe$_3$)$_2$($\eta^1$-Cp)].[337] Again the reaction is first order in both [PMe$_3$] and [Re], indicative of an $\eta^3$-Cp monophosphine intermediate. At 100 °C the phosphine substitution product [Re(CO)$_2$(PMe$_3$)(Cp)] is formed. The indenyl complex [Re(CO)$_3$($\eta^5$-C$_9$H$_7$)] is converted to an $\eta^1$-indenyl bis(phosphine) complex much more quickly than the cyclopentadienyl analogue ($\Delta\Delta G^{\ddagger}$ ≈58.6 kJ mol$^{-1}$), possibly due to formation of a fully aromatic benzene ring in the $\eta^3$-indenyl intermediate.[338]

Thermolysis of [Re(CO)$_5$($\eta^1$-Cp)] and [Re(CO)$_5$($\eta^1$-fluorenyl)] gives CO loss and ring slippage products. In contrast, long-wavelength excitation gives rhenium–carbon bond homolysis and formation of [Re$_2$(CO)$_{10}$] and C$_{10}$H$_{10}$ or 9,9'-bifluorene, whereas high-energy irradiation leads to CO loss. A comparison of thermal reaction rates for [Re(CO)$_5$(R)] where R is methyl, cyclopentadienyl, and 9-fluorenyl suggests concerted ring slippage and CO loss.[339]

Photolysis of [ReH(Cp)$_2$] in a CO matrix reversibly generates [ReH(CO)(Cp)($\eta^3$-Cp)] and [ReH(CO)$_2$(Cp)($\eta^1$-Cp)].[340]

Rest and co-workers have observed the photochemical conversion of [Re(CO)$_5$($\eta^1$-C$_7$H$_7$)] to [Re(CO)$_4$($\eta^3$-C$_7$H$_7$)], [Re(CO)$_3$($\eta^5$-C$_7$H$_7$)], and [Re(CO)$_2$($\eta^7$-C$_7$H$_7$)] in frozen gas matrices at 12 K. No 16-electron species were observed prior to ring slippage, once again pointing to CO loss concerted with the ring slip.[341]

## 9.11 KETONE AND ALDEHYDE COMPLEXES

Reaction of [Re(=CH$_2$)(PPh$_3$)(Cp)][PF$_6$] with PhIO, S=PPh$_3$, or cyclohexene sulfide, and Se=PPh$_3$ or KSeCN generates the corresponding $\eta^2$-formaldehyde,[342] $\eta^2$-thioformaldehyde,[343,344] and $\eta^2$-seleno-formaldehyde[345] complexes [Re($\eta^2$-CH$_2$=X)(NO)(PPh$_3$)(Cp)]$^+$ (X = O, 83%; X = S 32% or 95%; X = Se, 39% or 45%). The solid-state structures have been determined for each member of the series, and a detailed comparison of the bonding geometries provided.[344] The preferred conformations maximize overlap of the ligand $\pi^*$-acceptor orbital and the rhenium donor orbital (HOMO). Optically active [Re(NO)(PPh$_3$)(ClCH$_2$Cl)(Cp)]$^+$ reacts with aldehyde and methyl ketone substrates to give $\eta^2$-aldehyde (**161**)[346] and $\eta^1$-ketone (**162**)[347] complexes in high chemical (77–91%) and optical ($\geq$95% *ee*) yields. Both spectroscopic and crystallographic data indicate that one aldehyde enantioface binds to rhenium with high selectivity. Retention of configuration is observed at rhenium. In the ketone complexes rhenium is *cis* to the smaller carbonyl substituent.

(161)        (162)

The $\sigma$-ketone coordination mode is characterized by a medium-strong $\nu$(C=O) band at 1554–1625 cm$^{-1}$ in the IR spectra and a $^{13}$C NMR resonance at 216–240 ppm. The $\pi$-aldehyde coordination mode does not exhibit a $\nu$(C=O) stretch in the IR spectra and the aldehydic carbon is observed at 79–89 ppm in the $^{13}$C NMR spectra. In addition, the aldehyde hydrogen is shifted upfield to $\delta$ 5.2–6.8 in the $^1$H NMR spectra upon coordination.

Although there is no evidence for $\pi$-isomers in the case of methyl ketone complexes, for aldehyde complexes the spectroscopic data suggest small equilibrium concentrations of $\sigma$-isomers. Aromatic aldehydes exist as varying mixtures of $\pi$–$\sigma$-isomers.[348] The nitrosyl ligand stretch in the IR spectra of $\sigma$-ketone complexes is observed at 1697–1680 cm$^{-1}$, whereas for aliphatic $\pi$-aldehydes the nitrosyl stretch is observed at 1740–1729 cm$^{-1}$. The $\sigma$-ketone complexes are thus stronger $\sigma$-donors and weaker $\pi$-acceptors than the $\pi$-aldehyde complexes. Aromatic aldehydes with electron-donating substituents give greater ratios of $\sigma$–$\pi$-isomers. Aryl ring–carbonyl conjugation appears to be enhanced in the $\sigma$-binding mode.

In the case of the σ-acetone complex, $\Delta G^{\ddagger}_{133\,K} = 25$ kJ mol$^{-1}$ for methyl group interconversion. The process is intramolecular and appears to involve a linear Re–O=C(Me)$_2$ species. Electron-withdrawing substituents on the ketone carbonyl favor π-bonding to such an extent that for (XCH$_2$)$_2$CO (X = Cl, F) and pentafluoroacetophenone the π-isomer is observed.[349]

Both the η$^2$-aldehyde and η$^1$-ketone complexes undergo highly diastereoselective (and enantioselective) reductions to give alkoxide complexes (75–99% *de* for ketones and 68–98% *de* for aldehydes).[346,347,350] The dominant product stereochemistry is consistent with hydride attack on the σ-ketone ligand from a direction *anti* to the PPh$_3$ ligand. In the case of aldehyde reduction, the ambiguity of σ- vs. π-coordination arises. In an elegant study, Klein and Gladysz used a detailed kinetic analysis to determine that the σ-isomers of [Re{C$_6$F$_5$C(=O)H}(NO)(PPh$_3$)(Cp)] (**163**) are more reactive toward CN$^-$ than the π-isomers (Scheme 43).[351] In CH$_2$Cl$_2$ solution at 26 °C (**163**) exists as a 98:2 mixture of two configurational diastereomers (**163**)-π and (**163**)-π′, which slowly interconvert via the σ-isomers (**163**)-σ and (**163**)-σ′ ($\Delta G^{\ddagger}_{300K} > 62.8$ kJ mol$^{-1}$). Reaction of (**163**) with cyanide generates the cyanohydrin alkoxide complex (**164**) as a 96–97:4–3 mixture of diastereomers. The kinetic data requires an intermediate in the reaction of (**163**) with CN$^-$ and permits the relative reactivity of (**163**)-σ and (**163**)-π toward CN$^-$ to be set at >10$^5$ in CDCl$_2$F at −83 °C. Furthermore, when a series of aromatic aldehydes with electron-releasing substituents is examined it is found that the better donor substituents give more rapid rates of CN$^-$ attack! The alkoxide products react with (−)-(*R*)-C$_6$H$_5$CH(OAc)-(CO$_2$H)/DCC/4-(dimethylamino)pyridine to give organic ester products as comparable mixtures of diastereomers along with the rhenium carboxylate complex [Re{O(C=O)CH(OAc)-C$_6$H$_5$}(NO)(PPh$_3$)(Cp)] in >99% *ee*.[346]

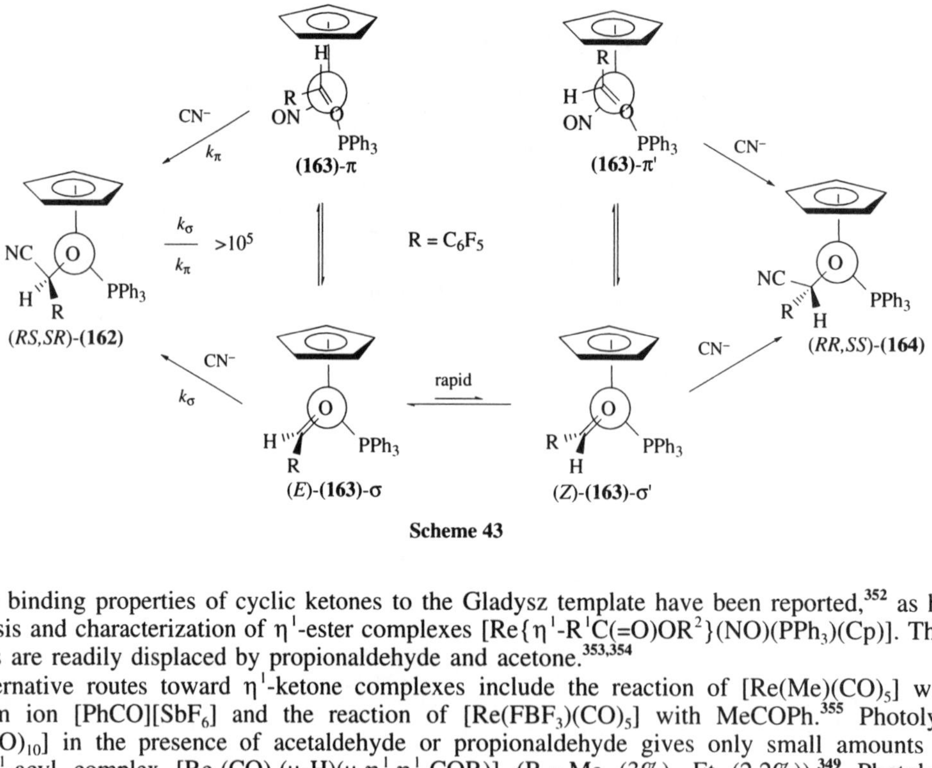

**Scheme 43**

The binding properties of cyclic ketones to the Gladysz template have been reported,[352] as has the synthesis and characterization of η$^1$-ester complexes [Re{η$^1$-R$^1$C(=O)OR$^2$}(NO)(PPh$_3$)(Cp)]. The ester ligands are readily displaced by propionaldehyde and acetone.[353,354]

Alternative routes toward η$^1$-ketone complexes include the reaction of [Re(Me)(CO)$_5$] with the acylium ion [PhCO][SbF$_6$] and the reaction of [Re(FBF$_3$)(CO)$_5$] with MeCOPh.[355] Photolysis of [Re$_2$(CO)$_{10}$] in the presence of acetaldehyde or propionaldehyde gives only small amounts of the μ-η$^1$,η$^1$-acyl complex [Re$_2$(CO)$_8$(μ-H)(μ,η$^1$,η$^1$-COR)] (R = Me (3%), Et (2.2%)).[349] Photolysis of [Re(CO)$_3$(η-Cp*)] and aldehydes in Et$_2$O provides access to the η$^2$-aldehyde complexes [Re-{η$^2$-RC(=O)H}(CO)$_2$(η-Cp*)] in low yields.[356]

## 9.12 NITROGEN, PHOSPHORUS, AND ARSENIC LIGANDS

The chiral rhenium amine complexes [Re(NR$^1$$_2$R$^2$)(NO)(PPh$_3$)(Cp)][OTf] are prepared in high yield by sequential reaction of [Re(Me)(NO)(PPh$_3$)(Cp)] with HOTf and NR$^1$$_2$R$^2$.[361] Substitution of cyanide and azide for the amine ligand occurs with >98% *ee*.[357] Complexes of aromatic nitrogen heterocycles are prepared in a similar fashion.[358] Deprotonation of primary and secondary amine complexes with Bu$^n$Li gives amido complexes in quantitative yields. The crystal structure of [Re(NMe$_2$)(NO)(PPh$_3$)(Cp)][OTf] exhibits a pyramidal amido nitrogen with the Ph$_3$P–Re–N–(lone pair) torsion angle of 59.3°, similar to that found in the corresponding neutral phosphido and cationic sulfide complexes.[359] Reaction of the amido complexes with TfOH and TfOR[2] indicate that the basicity and nucleophilicity of the amido nitrogen is greater than that of organic amines.[360] The amido complexes (SR)- and (SS)-[Re(NHCHMePh)(NO)(PPh$_3$)(Cp)] epimerize at rhenium by a process involving anchimeric assistance to PPh$_3$ loss and formation of a planar trigonal rhenium intermediate.[361]

Sutton and co-workers have extensively investigated the chemistry of rhenium aryldiazenido complexes [Re(CO)$_2$(N$_2$R)(Cp)][BF$_4$] (165), prepared from the corresponding arenediazonium salt and [Re(CO)$_2$(THF)(Cp)]. Reaction of (165) with NaBH$_4$ gives aryldiazene complexes, reaction with MeLi gives hydrazido derivatives, and reaction with MeLi, PhLi, and Bu$^n$Li gives monocarbonyl compounds formulated as acyl or benzoyl derivatives.[362,363] Metal carboxylic acid and hydrido derivatives are accessible upon treatment of (165) with hydroxide.[364] Related reactivity toward nucleophiles give monocarbonyl aryldiazenido complexes of the type [Re(L)(CO)(N$_2$Ar)(Cp)] and [Re(L)(CO)(N$_2$Ar)-(η-Cp*)], where L = Cl$^-$, Br$^-$, I$^-$, NCO, CO$_2$R, CO$_2$NH$_2$, OC(=O)H, CONHMe, and CONMe$_2$.[365,366]

Azobenzene complexes [Re(trans-N$_2$R$_2$)(CO)$_2$(Cp)] are prepared from [Re(CO)$_2$(THF)(Cp)] and the trans-azobenzenes in hexane. In the solid state, the azobenzene is bound "side-on" with unequal Re–N bond lengths. In solution the complex exists in equilibrium with an η$^1$-isomer. The η$^2$-isomer exhibits additional fluxional behavior presumed to involve nitrogen inversion.[367]

Two recent reviews cover the extensive photophysical studies reported for diimine (tricarbonyl)rhenium complexes, and this chemistry will not be discussed here.[368,369]

Clark and Kilner have prepared amidino complexes of the type (166) which contain a delocalized bidentate N,N'-chelated amidino ligand.[370] Reaction of [{ReBr(CO)$_4$}$_2$] or [ReX(CO)$_5$] with amidines give the o-metallated complexes (167). Related chemistry provides access to N,N'-diarylamidines R$^2$NHC(R$^1$)NR$^2$ (R$^1$ = H, Me, Ph; R$^2$ = Ph, Tol) and carbamoyl derivatives.[371] The redox properties of [Re(CO)$_2$(PPh$_3$)$_2$(ArNXNAr)] where X = N or CH have been reported.[372] The rhenium carbonyl halides [ReBr(CO)$_5$] react with amino(imino)thiophosphoranes (R$_3$C)(R$_3$Si)NP(S)=NCR$_3$ or R(S)P(NCR$_3$)SiR$_2$NCR$_3$ to give (168) in 37% and 68% yield, respectively.[373] The tricyclic rhenium complex (169) (R = TMS) is formed in 35% yield from [ReBr(CO)$_3$(THF)]$_2$ and (TMS)$_2$N–P(=N-TMS)$_2$.[374] The chemistry of [Re(CO)$_4$Re{N(R)PCl(NR$_2$)NR}] provides access to a number of exotic (tricarbonyl)rhenium complexes such as dinuclear complex (170).[375] The air-stable hydrazido complex [ReCl$_2$(NHNHCOPh)(NNHCOPh)(PPh$_3$)$_2$] is formed in 55% yield from benzoylhydrazine and [ReCl$_2$(N$_2$COPh)(PPh$_3$)$_2$]. The linear (ReNN 175°) hydrazido (2 –) ligand (NNHCOPh) functions as a four-electron donor.[376] An improved synthesis of [ReCl$_2$(NPh)(NH$_3$)(PMe$_2$Ph)$_2$] (70% yield) involves reaction of [ReCl$_4$(PMe$_2$Ph)$_2$] with PhNHNH$_2$.[377] Diazonium salts displace a chelating phosphine ligand in trans-[ReCl(CO){Ph$_2$PC(=CH$_2$)PPh$_2$}$_2$] to give mer-[ReCl(CO)(N$_2$Ar){η$^1$-Ph$_2$PC(=CH$_2$)PPh$_2$}-{η$^2$-Ph$_2$PC(=CH$_2$)PPh$_2$}] in which one of the diphosphine ligands is bidentate and the other monodentate.[378] Mixed Schiff base complexes of rhenium(I) such as (171) have been prepared and structurally characterized.[379]

Rhenium complexes with tridentate nitrogen ligands have also been reported including the 1,4,7-triazacyclononane ligand (172),[380] the hydrotris(1-pyrazolyl)borate ligand (173),[381] and tridentate cyano ligand (174).[382]

(172)                        [Re{HB(pz)₃}(CO)₂(THF)]                        (174)

Rhenium dinitrogen complexes are now well known and only a few recent examples are given here. Sutton and co-workers have isolated [Re(CO)₂(N₂)(η-Cp*)] from both photolysis of [Re(CO)₃(η-Cp*)] under 10.3 MPa of N₂ (53% yield) and reduction of [Re(CO)₂(N₂Ar)(η-Cp*)][BF₄] with NaBH₄ (44% yield). Treatment of the dinitrogen complex with HX gives the dihalides *cis*-[Re(CO)₂(X)₂(η-Cp*)], which are converted to the *trans*-isomers upon photolysis.[383] The diazenido route has been used to prepare a series of phosphine derivatives [Re(CO)(N₂)(PR₃)(η-Cp*)] in 39–51% yield.[384] The rate of linkage isomerism of [Re(CO)₂(¹⁵N≡¹⁴N)(η-Cp*)] to [Re(CO)₂(¹⁴N≡¹⁵N)(η-Cp*)] was determined from ¹⁵N NMR intensities ($\Delta G^{\ddagger} = 90$ kJ mol⁻¹). Crossover experiments indicate that the isomerization is intramolecular with a first-order rate constant at 287 K is $1.45 \times 10^{-2}$ min⁻¹. No isomerization was observed for the dinitrogen ligand in [Re(CO)(N₂)(PMe₃)(η-Cp*)].[385]

The tris(dinitrogen) and bis(dinitrogen) complexes [Re(N₂)ₙ(CO)₃₋ₙ(Cp)] (n = 3, 2) have been observed in supercritical xenon (sc Xe) at 25 °C and in N₂ matrices at 20 K. In scXe solution both of these complexes are surprisingly stable toward N₂ substitutions by CO. The bis(dinitrogen) complex is stable toward 1 atm of CO and even survives exposure to air for two weeks at room temperature![386]

The nitrosyl complexes *trans*-[ReCl(NO)(dppe)₂][X]₂ where X = NO₃ and BF₄ is formed from the reaction of either NOBF₄ or NO with *trans*-[ReCl(N₂)(dppe)₂]. The nitrate counterion is formed from oxidation of NO.[387]

The explosive perchlorate salt of [ReH(MeOH)(NO)(PPh₃)₃]⁺ reacts with coordinating anions to give [ReH(X)(NO)(PPh₃)₃] (X = OMe, F, Cl, Br, I, N₃, NCO, and SCN).[143]

Heating [ReCl(CO)₅] and NEt₄Cl at reflux in Bu₂O gives [Re₂Cl₂(μ-Cl)₂(CO)₆], which is nitrosylated with [NO][BF₄] to give a 90% yield of [Re₂Cl₂(μ-Cl)₂(CO)₄(NO)₂]. Heating the nitrosyl complex and L in MeCN leads to the mononuclear nitrosyls [ReCl₂(CO)(NO)(L)₂] (L = phosphines or phosphites) in 70–83% yields. The dichloride is a convenient starting material for a number of trigonal-bipyramidal and octahedral organometallic nitrosyl derivatives.[290]

The mononuclear rhenium phosphide complex [Re(NO)(PPh₂)(PPh₃)(Cp)] is formed in 97% yield by deprotonation of [Re(NO)(PPh₂H)(PPh₃)(Cp)]⁺ with BuᵗOK. The PPh₂ phosphorus is pyramidal but the bond angles suggest increased *p*-character in the PPh₂ lone pair. The Re–PPh₂ bond length (0.2461 (3) nm) compares with the Re–PPh₃ bond length of 0.2358 (3) nm. The torsion angle between the PPh₂ lone pair and the rhenium fragment HOMO is 59.7°.[359] The phosphorus inversion barrier in the PTol₂ analogue has $\Delta G^{\ddagger} = 54.4$ kJ mol⁻¹ compared to trialkyl phosphine inversion barriers of >125 kJ mol⁻¹.[359,388] The labile rhenium diphosphene complex [Re(η¹-P=PAr)(CO)(NO)(η-Cp*)] is stable in solution and has been trapped in 31% yield by [Cr(CO)₅(C₈H₁₄)] as the isolable μ-PPAr dinuclear complex [Re(CO)(NO)(η-C₅Me₅)(μ,η¹,η¹-P=PAr)Cr(CO)₅].[389]

The diphosphido-bridged early/late heterobimetallic complex [(Cp)₂Zr(μ-PPh₂)₂ReH(CO)₃] is prepared in 17% yield from [Re₂(CO)₁₀] and [Zr(PPh₂)₂(Cp)₂] at 25 °C in THF. The structural data suggest at most a very weak dative bond between the d⁰ zirconium and d⁶ rhenium centers.[390]

A series of E₂Ph₄ complexes of the type [Re₂X₂(CO)₆(E₂Ph₄)] (175), where E = As, Sb, P and X = halide, have been prepared from [Re₂Br₂(CO)₆(THF)₂] or [Re₂I₂(CO)₈] and E₂Ph₄. The stabilities qualitatively decreased from phosphorus to antimony to arsenic based on the observed tendency of THF to displace E₂Ph₄.[391]

Rheingold and co-workers have prepared and structurally characterized the tris(homocubane) complex *cyclo*-[{(AsMe)₇(As)Re(CO)₄}Re₂(CO)₆] (176) in 26% yield from [Re₂(CO)₁₀] and pentamethylcyclopentaarsine, *cyclo*-[(AsMe)₅], at 70 °C. Cluster (176) contains a nine-membered As₈Re ring coordinated to two Re(CO)₃ groups. Demethylation converts one arsenic from a two-electron donor to a one-electron donor bearing a nonbonded lone pair.[392] When [Re₂(CO)₁₀] and [(C₆H₅As)₆] are heated at 130 °C the dinuclear complex [Re₂{μ-As(Ph)₂}₂(CO)₈] is isolated in 8% yield along with [Re₃(CO)₁₀(AsPh)₇(As)]. It thus appears that a 2ReAs to Re₂As + As disproportionation is involved in the arsenic dealkylation.[393]

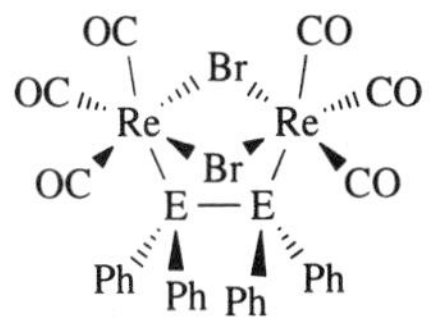

(175)

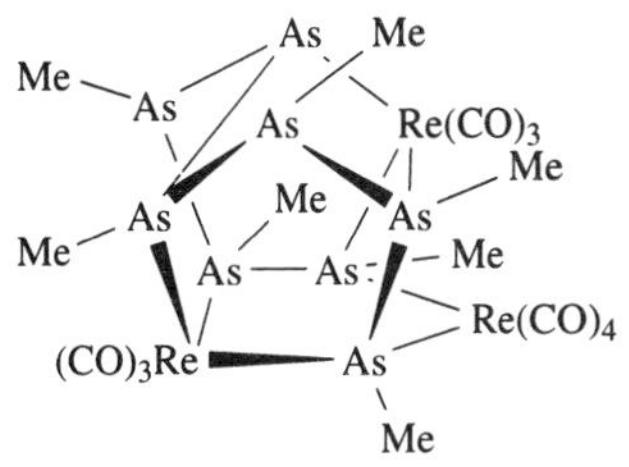

(176)

## 9.13 OXYGEN, SULFUR, AND SELENIUM LIGANDS

### 9.13.1 Alkoxide and Alcohol Complexes

Complexes containing rhenium–oxygen single bonds are of interest due to the importance of metal-mediated alcohol, alkoxide, and carbonyl reactivity. Bergman and Simpson have accessed rhenium(I) aryloxide complexes [Re(OAr)(CO)$_3$(L)$_2$] from the reaction of *fac-cis*-[Re(Me)(CO)$_3$(L)$_2$] (L$_2$ = 2PPh$_3$, 2PMe$_3$, diars) and *p*-cresol. The alkoxide complexes are available in 75–86% yield from [Re(CO)$_3$(L)$_2$(OTf)] and NaOR (R = Me, Et). The presence of *cis*-phosphine or arsine ligands prohibits the undesired formation of clusters containing bridging alkoxide ligands. The rhenium–oxygen bond is cleaved by the Brønsted and Lewis acids HX, H$_2$NR, R$_2$PH, RPH$_2$, HSR, H$_2$S, [WH(CO)$_3$(Cp)], and ROH. Exchanges with first-row substrates (ROH, R$_2$NH) are reversible, while exchanges with second-row substrates (HCl, RSH, R$_2$PH) are irreversible. The alkoxides do not react cleanly with carbon acids but conversion to alkyl- and alkenylrhenium complexes is possible using alkyl- and alkenylboranes.[216]

The reactions of Brønsted acids with the alkoxide complexes is affected by both the p$K_a$ of the acid and the tendency of the acid to associate to the alkoxide ligand oxygen. The hydrogen-bonded proton was located in the x-ray structure of [Re{(MeO)···HOC$_6$H$_4$Me}(CO)$_3$(depe)].[394]

The exchange of added alcohol with coordinated alkoxide ligands is more rapid than that observed between coordinated aryloxide ligand and added phenols. The conversion of [Re{(MeO)···HOC$_6$H$_4$X}(CO)$_3$(diars)] to [Re(OC$_6$H$_4$X)(CO)$_3$(diars)] exhibits an increase in rate when the X substituent lowers the O–H bond dissociation energy or increases the acidity of the phenol. The transition state therefore has both proton and hydrogen atom transfer character (Scheme 44).

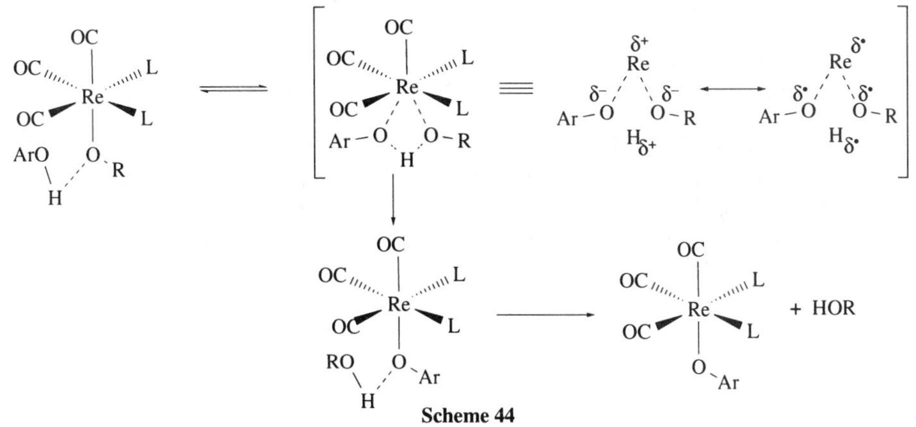

**Scheme 44**

Hydrogen bonding has been observed between the alcohol hydrogen and perchlorate anion in [ReH(CO)(NO)(PPh$_3$)$_2$(ROH)][ClO$_4$].[395] Orchin and co-workers have observed the reversible absorption of atmospheric CO$_2$ by rhenium(I) alkoxide complexes such as *fac*-[Re(OMe)(CO)$_3$(dppe)]. The carbonato products react with CS$_2$ to give xanthates and with PhNCO to give urethanes in nearly quantitative yield.[396]

The first well-defined low-valent transition metal catalyst for epimerization of secondary alcohols is based on [Re(OMe)(NO)(PPh$_3$)(Cp)] (177).[397] In the presence of 10 mol.% (177) diastereomerically pure secondary alcohols epimerize to mixtures of diastereomers at 65–90 °C. Initial alkoxide ligand exchange between (177) and HOCR$^1$R$^2$H is followed by epimerization at rhenium (~35 °C) and then at carbon (65 °C). The alkoxide oxygen lone pairs provide anchimeric assistance for PPh$_3$ dissociation. The trigonal-planar rhenium either recoordinates $PPh_3$ to give epimerization at rhenium or undergoes β-hydride elimination to reversibly form a ketone hydride complex with epimerization at carbon (Scheme 45). Added $PPh_3$ strongly inhibits the rate of epimerization at carbon but not at rhenium.

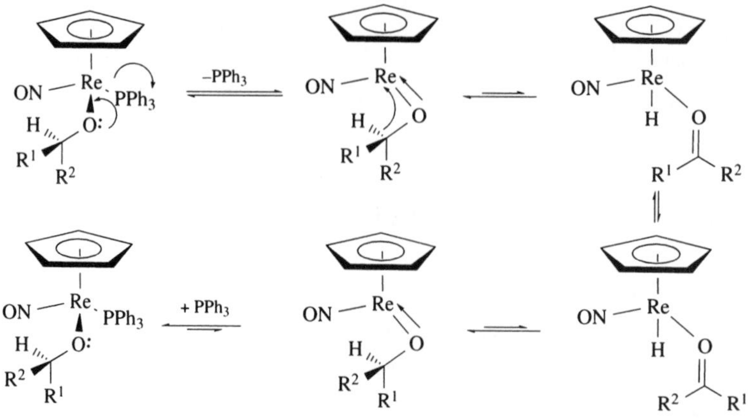

**Scheme 45**

Epimerization of *exo*-borneol gave 33% *exo*-borneol, 62% *endo*-borneol, and 5% camphor. The pentamethylcyclopentadienyl analogue proved to be 2–4 times more active than (**177**). A conformational model provides a rationale for the observed epimerization reactions based on the relative stabilities of the diastereomeric alkoxide complexes.[397]

The alcohol complexes $[Re(NO)(ROH)(PPh_3)(Cp)]^+$ (R = alkyl) are accessible in 91–95% yields from the alcohol and $[Re(NO)(CH_2Cl_2)(PPh_3)(Cp)][BF_4]$ at −50 °C. Deprotonation gives the corresponding alkoxide complexes. The alcohol ligand is displaced by aldehydes and the Lewis acid binding affinities toward the Gladysz template have been determined to follow the trend O=CHR > ROH > ROR > $CH_2Cl_2$ > $\eta^2$-$C_6H_6$.[398] Protonation of the methylrhenium complex $[Re(Me)(NO)(PPh_3)(Cp)]$ with formic acid gives the formate complex $[Re\{\eta^1\text{-}OC(=O)\text{-}H\}(NO)(PPh_3)(Cp)]$, which undergoes decarboxylation at 70–130 °C to the rhenium hydride. Possible mechanisms include formate dissociation to a configurationally stable tight ion pair followed by hydride transfer to rhenium, and a concerted decarboxylation from a four-centered transition state.[399]

### 9.13.2 Sulfur and Selenium Complexes

Interest in hydrodesulfurization of coal has stimulated studies into the organometallic chemistry of thiophene and other sulfur-containing compounds. The S-bound dihydrothiophene complex $[Re(\overline{SCH=CHCH_2CH_2})(CO)_5][SO_3CF_3]$ is prepared in 84% yield from $[Re(OSO_2CF_3)(CO)_5]$ but decomposes upon warming or exposure to air.[400] The related tetrahydrothiophene, 1,4-dithiane, 1,3-dithiane, and 1,3,5-trithiane pentacarbonylrhenium complexes have been prepared in 80–92% yield and the barriers for inversion at the coordinated sulfur atoms are 57 kJ mol$^{-1}$ (278 K) and 62 kJ mol$^{-1}$ (308 K) for the 1,3-dithiane and 1,3,5-trithiane ligands, respectively.[401]

Choi and Angelici have extensively developed the thiophene coordination chemistry of pentamethylcyclopentadienyl(dicarbonyl)rhenium. Photolysis of $[Re(CO)_3(\eta\text{-}Cp^*)]$ in THF generates $[Re(CO)_2(THF)(\eta\text{-}Cp^*)]$, which reacts with thiophene at 25 °C to give the air stable thiophene complex (**178**) in 38% yield (Scheme 46). Whereas free thiophene does not react with $[Fe_2(CO)_9]$ under mild conditions, reaction with the coordinated thiophene ligand of (**178**) occurs at −40 °C to 25 °C in THF to generate the dinuclear complex (**179**) in 59% yield.[402] A series of sulfur-bound thiophene complexes $[Re(CO)_2(Th)(Cp')]$, where Cp' is Cp or Cp* and Th is 2-$MeC_4H_3S$, 2,5-$Me_2C_4H_2S$, $Me_4C_4S$, and dibenzothiophene (DBT) have been prepared in 15–45% yield by the same route and similar reactivity toward $[Fe_2(CO)_9]$ is observed.[403] The thiophene ligands are substituted by $PPh_3$ by a dissociative mechanism for which the rates follow the order $C_4H_4S$ (3000) > 3-$MeC_4H_3S$ (1200) > 2-$MeC_4H_3S$ (91) > 2,5-$Me_2C_4H_2S$ (13) > $Me_4C_4S$ (2.7). Increasing the number of methyl substituents makes sulfur a better σ-donor and a more tightly bound ligand despite the increase in steric bulk.[404] The benzo[*b*]thiophene (BT) complexes (**180**) exist as an equilibrium mixture of $\eta^1$(S)- and 2,3-$\eta^2$-isomers with the equilibrium shifted more to the $\eta^2$-form for the η-Cp* than the Cp complex. The isomerization process is intramolecular. The 2- and 3-methylbenzo[*b*]thiophenes are stronger donor ligands than BT and exist solely as the $\eta^1$(S)-isomers. For the BT complex, alkylation with $Me_3O^+$ occurs at sulfur with generation of an $\eta^2$-S-MeBT complex.[405] Selenophenes form similar complexes with qualitative differences in behavior compared to thiophene complexes. The selenophene complexes $[Re(CO)_2(Sel)(Cp')]$, where Cp' is $\eta$-Cp or $\eta$-Cp* and Sel is $C_4H_4Se$ or $2$-$MeC_4H_3Se$, exist as equilibrium mixtures of $\eta^1$- and $\eta^2$-isomers. The presence of methyl substituents shift the equilibrium to the $\eta^1$-isomer and the $2,5$-$Me_2C_4H_2Se$ complex exists solely as the $\eta^1$-isomer. The $\eta^2$-coordination mode shifts the CO stretching frequency in the IR spectrum $\sim 35$ cm$^{-1}$ higher than that in the $\eta^1$(Se) mode. Spin saturation transfer experiments indicate that the $\eta^2$-$C_4H_4Se$ complex is fluxional by a process in which rhenium migrates between the two double bonds. Reaction of $[Re(CO)_2(SeC_4H_4)(\eta$-$Cp^*)]$ with $[Fe_2(CO)_9]$ gives three iron carbonyl adducts including the $\eta^4$-coordination of $Fe(CO)_3$ to the diene unit of coordinated $C_4H_4Se$ as well as a selenium-bound $Fe(CO)_4$ dinuclear product (181). The $^{77}$Se chemical shifts of the $\eta^1$(Se) isomers occur $100$–$150$ ppm upfield of those for the $\eta^2$-isomers.[406]

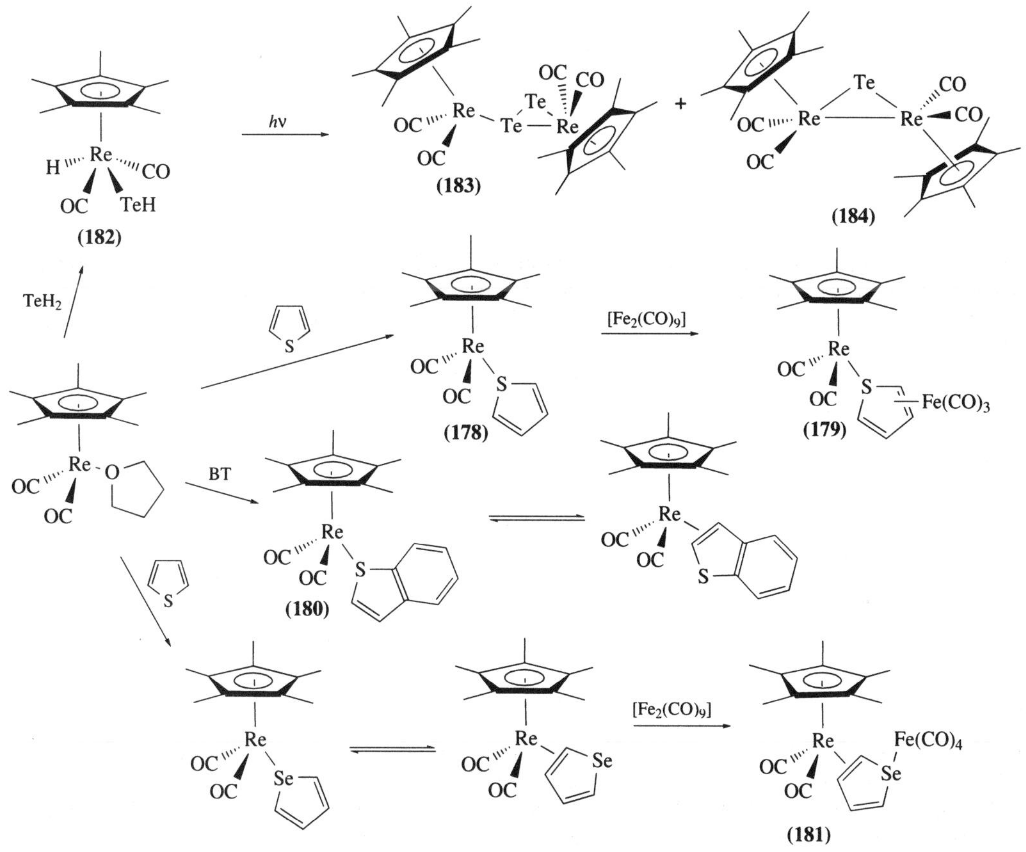

Scheme 46

$[Re(CO)_3(\eta$-$Cp^*)]$ oxidatively adds $H_2Te$ at $25\,°C$ to give (182) (10% yield), (183) (9%), and (184) (24%) (Scheme 46). Photolysis of (182) gives (183), whereas (182) and $[Re(CO)_3(\eta$-$Cp^*)]$ combine to give (184) (70% yield). The bridging tellurium atom in (184) is alkylated upon treatment with $CF_3SO_3Me$.[407] Both $^{125}$Te and $^{77}$Se NMR spectroscopy have been used to distinguish the bonding modes of tellurium and selenium.[408]

Rosini and Jones have examined the reactivity of thiophene toward complexes containing ligands which can directly assist in hydrogenation and carbon–sulfur bond cleavage chemistry. Thus, reaction of $[ReH_7(PPh_3)_2]$ and excess thiophene in the presence of the hydrogen-acceptor, *t*-butylethene, gives the $\eta^4$-$C_4H_5S$ complex (185) in 74% isolated yield.[409] The *exo*-ring hydrogen is observed at 5.13 ppm and the *endo*-hydrogen at 3.68 ppm in the $^1$H NMR spectrum. Deuterium-labeling studies support the mechanism shown in Scheme 47. When benzene solutions of (185) are heated at $60\,°C$ in the presence of $PMe_3$, uncoordinated tetrahydrothiophene is formed as well as an 87% yield of the orthometallated complex (186).

Photolysis of (185) gives five rhenium complexes: (*E*)- and (*Z*)-(187), *syn*- and *anti*-(188), and (189) by the mechanism shown in Scheme 48. The key photochemical step appears to be dissociation of the

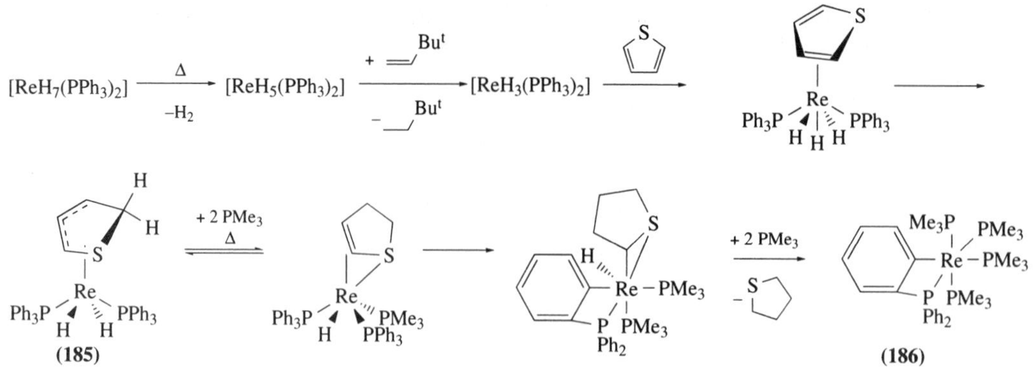

Scheme 47

alkene to form a sulfur-bound 2,3-dihydrothiophene complex, thereby preventing further hydrogenation of the alkene. At 25 °C complexes (**187**) and (**188**) are rapidly converted to a butene–thiolate complex, $[ReH_2(SCH=CHEt)(PMe_3)_4]$, which at 70 °C under 1 atm of $H_2$ undergoes quantitative conversion to the butanethiolate complex $[ReH_2(SBu^n)(PMe_3)_4]$. These studies represent the first homogeneous hydrogenolysis reactions of thiophene.[409]

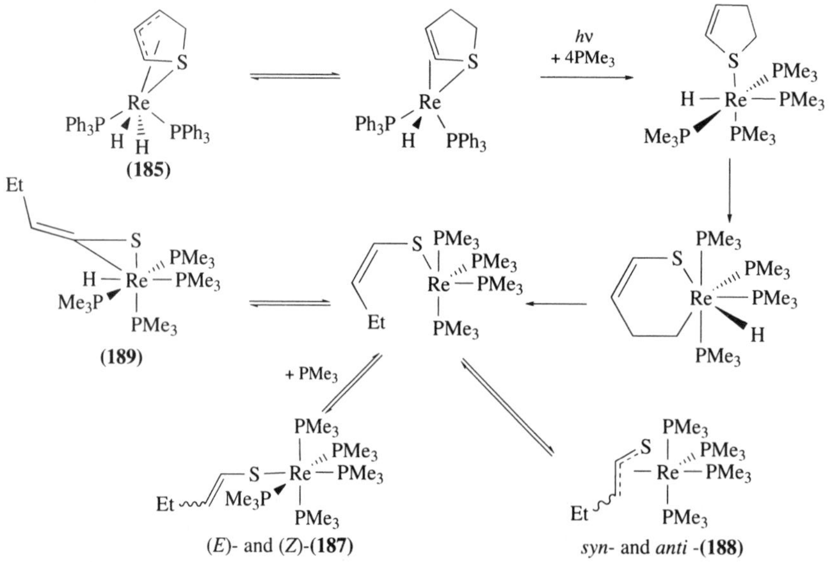

Scheme 48

Herberhold *et al.* have isolated the disulfide complex $[Re(CO)_2(\eta^2\text{-}S_2)(Cp)]$ (38% yield) along with the dinuclear complexes $[Re_2(CO)_4(\mu\text{-}S)(Cp)_2]$ and (**190**) from $[Re(CO)_2(THF)(Cp)]$ and $S_8$ or COS.[410] In the solid-state structure of the disulfide complex, the S–S bond distance of 0.199 6(5) nm is between that for single and double bonds. The CO stretching absorptions in the IR spectrum appear at 2014(s) cm and 1945(s) cm$^{-1}$, indicating that $S_2$ is a better acceptor ligand than CO.[411]

The pentamethylcyclopentadienyl analogue (**191**) is prepared in 47% yield and undergoes 3-chloroperbenzoic acid oxidation to (**192**) in 50% yield. Further oxidation generates (**193**) which was not isolated in pure form. The π-acceptor capacity of this ligand series follows the expected order $S_2 < S_2O < S_2O_2$.[412] The μ-S dinuclear complex $[Re_2(CO)_4(\mu\text{-}S)(\eta\text{-}Cp^*)_2]$ is oxidized to the μ-SO$_2$ analogue in 48% yield.[413,414]

The doubly bridged tetrasulfido dimer $[Re(CO)_3(\eta^2\text{-}S_4)]_2^{2-}$ (**194**) is formed in 75% yield from $[Re(CO)_5]^-$ and sulfur.[415]

Octahedral mononuclear rhenium carbonyl complexes with a variety of chelating sulfur ligands have been prepared ((**195**)–(**199**)).[416-21] Selenium-chelating ligands of the type MeSeCH=CHSeMe have also been ligated to rhenium and the energy barriers for inversion at selenium and sulfur atoms determined.[422] Decacarbonylrhenium reacts with $K_2Se_4$ at 100 °C in DMF in the presence of $Ph_4PBr$ to give $[Re_2(Se_4)_2(CO)_6][PPh_4]_2$ in 25% isolated yield. Each rhenium is coordinated to an $Se_4^{2-}$ chain with the α-Se of each chain bridging to a second rhenium.[423] An interesting synthesis of $[Re_2(\mu\text{-}SBu^i)_2(CO)_8]$ involves a ligand exchange reaction between $[ReBr(CO)_5]$ and $[Fe_2(\mu\text{-}SR)(\mu\text{-}CO)(CO)_6]^-$.[424] The reaction of $[Re_2X_2(CO)_6(THF)_2]$ and $S_8$ in $CS_2$ gives the labile $S_8$ dinuclear complex $[Re_2X_2(CO)_6(S_8)]$ which liberates $S_8$ in coordinating solvents.[425]

(190)  (191)  (192)  (193)

(194)  (195)  (196)

(197)  (198)  (199)

### 9.13.3 Oxo Complexes

In 1984, Mayer *et al.* reported the synthesis of $[Re(O)(I)(\eta^2\text{-}MeC{\equiv}CMe)_2]$ (**200**), arguably the first low-valent oxo complex.[426] The original synthetic procedure involved reaction of $[Re(O)_2(I)(AsPh_3)_2]$ with 2 equiv. 2-butyne. However, an improved synthesis of (**200**) (80–95% yields) employs the reaction of $[Re(O)(I)_3(PPh_3)_2]$ and alkyne.[427] A variety of alkynes, $R^1C{\equiv}CR^2$, can be used, including $R^1$, $R^2$ = Et; $R^1 = Bu^t$, $R^2$ = H; and $R^1$ = Ph, $R^2$ = Me. For (**200**), the rhenium–oxygen distance of 0.169 7(3) nm as well as the Re–O stretching frequency in the infrared spectrum (980 cm⁻¹) are consistent with a rhenium(V) oxo formulation. In terms of chemical reactivity, the complexes exhibit properties also consistent with a rhenium(V) oxidation state: they are relatively resistant to further oxidation by $I_2$ and $O_2$, they do not undergo an oxidative conversion to a metallacyclopentadiene complex, and there is no oxidation wave in the cyclic voltammagram to the onset of THF solvent oxidation. On the other hand, the alkyne carbon–carbon bond length of 0.133 0(7) nm, the relatively low ethyne stretching frequencies (1700–1800 cm⁻¹) and the relatively low field chemical shifts for the alkyne carbons in the ¹³C NMR spectrum all point to a rhenium(III) formulation with the two alkyne ligands donating a total of six electrons to the metal. Furthermore, extended Hückel calculations rationalize the observed reactivity and structural properties in terms of a rhenium(III) $d^4$ oxo formulation.[426]

Electrochemical or chemical reduction ($Bu^tZnCl$, $NaC_{10}H_8$, $CoCp_2$) of (**200**) gives a mixture of two rhenium dimers: a $Re^{III}$–$Re^I$ mixed-valence isomer (**201**) and the $Re^{II}$–$Re^{II}$ symmetric dimer (**202**) (Scheme 49). Oxidation of the rhenium(I) anions $[Re(O)(RC{\equiv}CR)_2][Na]$ (**203**) also generates (**201**) and (**202**). Mechanistically, the symmetric dimers appear to form by loss of $I^-$ from (**200**) and dimerization of the resultant radicals $[Re(O)(RC{\equiv}CR)_2]$. The μ-alkyne ligand in (**201**) has an unusual twist and the $\eta^2$-alkyne ligands at the rhenium(I) center are fluxional on the NMR timescale. Upon heating, (**201**) is converted to (**202**) with no scrambling of the alkyne ligands between the rhenium centers.[428]

The rhenium(I), $d^6$, oxo anion (**203**) is prepared by reduction of (**200**) with 2 equiv. reducing agent. The structure is trigonal planar with the rhenium 0.012 nm out of the plane defined by the oxo and alkyne midpoints. Anion (**203**) is a strong base and reducing agent, and reacts with a variety of electrophiles to give the derivatives shown in Scheme 49.[429] Reactions with primary and secondary alkyl halides occurs at least in part by an electron-transfer pathway.[430] Dialkylzinc reagents also react with (**205**) to give oxoalkyl complexes in 75–90% yields.

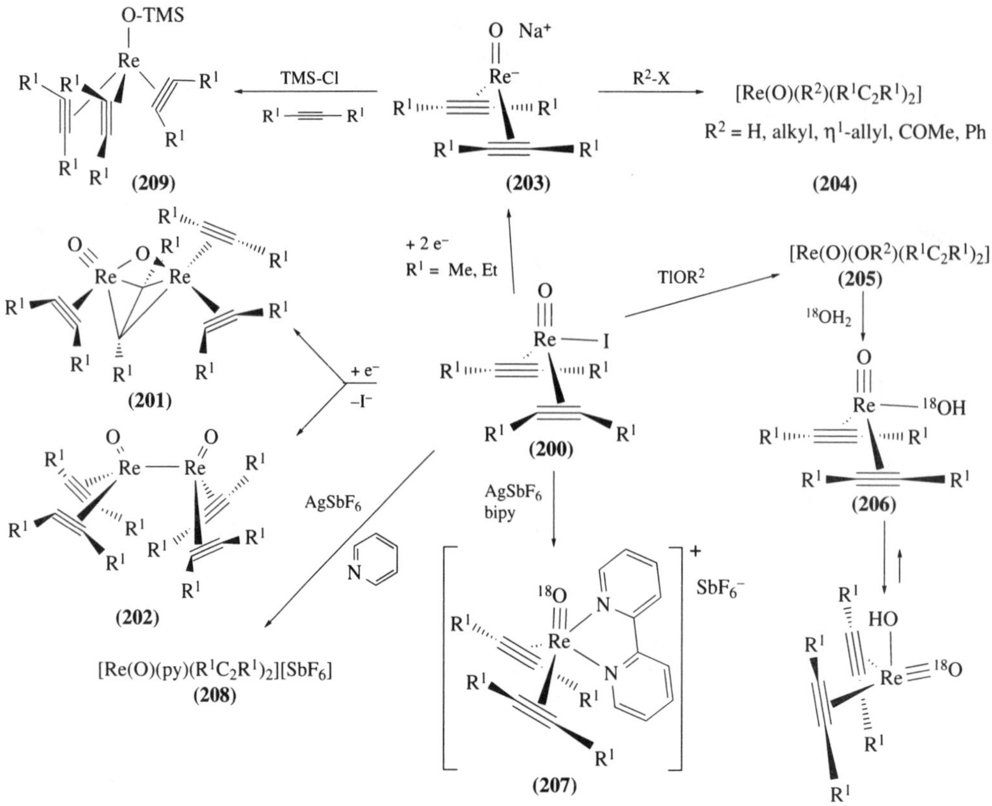

**Scheme 49**

Large kinetic barriers toward CO insertion–deinsertion and β-hydrogen elimination–hydride migration are attributed to the absence of an available empty orbital on rhenium. However, protonation of the oxo group in the isopropyl complex (**204**)-Pr ($R^2 = Pr^i$) reduces the Re–O π-interaction and leads to formation of the oxo hydride (**204**)-H ($R^2 = H$), as a result of β-hydrogen elimination. Upon protonation at oxygen the alkyne ligands also become fluxional due to the ability to π-donate to rhenium in any orientation without competition from the strong Re–O π-interaction.[431]

In contrast to the stability of the oxoalkyl complexes, the oxoalkoxides (**205**) are much more reactive. Metathesis of thallium alkoxides and (**200**) generate $[Re(O)(OR^2)(R^1C\equiv CR^1)_2]$ (**205**) in 60–80% yields. At 100 °C the alkoxide complexes decompose by a variety of pathways including β-hydrogen elimination. Carbon monoxide insertion into the Re–OR bond gives alkoxy carbonyl complexes and ligand exchange occurs with ROH, $NHR_2$, $H_2S$, HCl, and $H_2O$. The exchange mechanism involves an associative front-side attack at the tetrahedral rhenium, possibly mediated by a hydrogen bond between the exchanging heteroatoms.[432] Reaction of the ethoxide or methylamido complexes [Re(O)-(X)(MeC≡CMe)₂], where X = OEt, NHMe with 1 equiv. water gives the oxohydroxide complex [Re(O)(OH)(MeC≡CMe)₂], (**206**). Oxygen-18 labeling experiments indicate that (**206**) undergoes a surprisingly slow tautomerization, which exchanges the two oxygen sites. The unusually slow nature of the proton transfer is attributed to the large change in structure and bonding which must accompany the interconversion of the rhenium oxygen multiple and single bonds. Since the alkyne ligands are oriented perpendicular to the Re≡O bond, and alkyne rotation has a high barrier, the tautomerization requires significant alkyne ligand movement.[433]

A large number of pyridine derivatives have been prepared by reaction of (**200**) with $AgSbF_6$ and nitrogen heterocycles. A rapid ligand exchange reaction occurs between coordinated and free pyridine with retention at the metal. The structurally characterized bipyridine adduct (**207**) probably models the transition state for these ligand exchange reactions.[434] In the structure of (**207**), the Re–O bond is coplanar with the bipyridine ligand. The Re–N distances of 0.224 6(3) nm and 0.233 7(4) nm are longer than that observed for the pyridine adduct (**208**) (0.211 8(3) nm).[434]

The tris(alkyne) complexes $[Re(O-TMS)(RC\equiv CR)_3]$ (**209**) are available in 15–40% isolated yields from reaction of the rhenium(I) anion $[Re(O)(RC\equiv CR)_2][Na]$ (**203**) and excess alkyne in the presence of TMS-Cl. Reaction of (**209**) with HX gives $[ReX(RC\equiv CR)_3]$, where X = OC(=O)Me, OTf (**210**), OPh, $OCH(CF_3)_2$, Cl. Triflate (**210**) is also accessible from TMS-OTf and (**209**), AgOTf and $[ReI(RC\equiv CR)_3]$, or (**203**) and 2 equiv. TMS-OTf. The triflate ligand is displaced by a variety of other neutral ligands such as $PMe_3$, $OPMe_3$, MeCN, py, and $C_5H_5NO$. Reaction of (**210**) with NaOMe gives the methoxide complex, [Re(OMe)(RC≡CR)$_3$], which undergoes alkyne exchange reactions by a dissociative pathway. The rates of alkyne exchange for a series of complexes [ReX(RC≡CR)$_3$] vary with the basicity of X (OMe ≈ O-TMS, OH > OPh ≫ OAc, OTf).[435]

The rhenium hydroxide complex [Re(OH)(EtC≡CEt)$_3$] (**211**) is accessible by deprotonation of the aquo complex [Re(OH$_2$)(EtC≡CEt)$_3$][OTf] with KOH (Equation (29)). In benzene solution (**211**) rearranges to an oxo–hydride complex [Re(O)H(EtC≡CEt)$_2$] and EtC≡CEt. The hydrogen migration occurs either synchronously with or prior to alkyne loss.[436]

## 9.14 REFERENCES

1. C. P. Casey, *Science*, 1993, **259**, 1552.
2. W. A. Herrmann, D. Mihalios, K. Öfele, P. Kiprof and F. Belmedjahed, *Chem. Ber.*, 1992, **125**, 1795.
3. A. M. Stolzenberg and E. L. Muetterties, *J. Am. Chem. Soc.*, 1983, **105**, 822.
4. M. Marchionna, M. Lami, A. M. R. Galletti and G. Braca, *Gazz. Chim. Ital.*, 1993, **123**, 107.
5. F. Calderazzo, U. Mazzi, G. Pampaloni, R. Poli, F. Tisato and P. F. Zanazzi, *Gazz. Chim Ital.*, 1989, **119**, 241.
6. L. S. Crocker, G. L. Gould and D. M. Heinekey, *J. Organomet. Chem.*, 1988, **342**, 243.
7. Y. Zhen and J. D. Atwood, *J. Am. Chem. Soc.*, 1989, **111**, 1506.
8. K. Yasufuku, H. Noda, J. Iwai, H. Ohtani, M. Hoshino and T. Kobayashi, *Organometallics*, 1985, **4**, 2174.
9. D. G. Leopold and V. Vaida, *J. Am. Chem. Soc.*, 1984, **106**, 3720.
10. W. K. Meckstroth, R. T. Walters, W. L. Waltz, A. Wojcicki and L. M. Dorfman, *J. Am. Chem. Soc.*, 1982, **104**, 1842.
11. A. Fox, J. Malito and A. Pöe, *J. Chem. Soc., Chem. Commun.*, 1981, 1052.
12. J. M. Hanckel, K.-W. Lee, P. Rushman and T. L. Brown, *Inorg. Chem.*, 1986, **25**, 1852.
13. K.-W. Lee and T. L. Brown, *J. Am. Chem. Soc.*, 1987, **109**, 3269.
14. H. W. Walker, G. B. Rattinger, R. L. Belford and T. L. Brown, *Organometallics*, 1983, **2**, 775.
15. L. S. Crocker, B. M. Mattson, D. M. Heinekey and G. K. Schulte, *Inorg. Chem.*, 1988, **27**, 3722.
16. L. S. Crocker, B. M. Mattson and D. M. Heinekey, *Organometallics*, 1990, **9**, 1011.
17. L. S. Crocker, D. M. Heinekey and G. K. Schulte, *J. Am. Chem. Soc.*, 1989, **111**, 405.
18. J. A. Armstead, D. J. Cox and R. Davis, *J. Organomet. Chem.*, 1982, **236**, 213.
19. R. R. Andréa, W. G. J. de Lange, T. van der Graaf, M. Rijkhoff, D. J. Stufkens and A. Oskam, *Organometallics*, 1988, **7**, 1100.
20. I. V. Karsanov, Y. P. Ivakhnenko, V. S. Khandkarova, A. I. Prokof'ev, A. Z. Rubezhov and M. I. Kabachnik, *J. Organomet. Chem.*, 1989, **379**, 1.
21. C. P. Cheng, H. S. Chen and S. R. Wang, *J. Organomet. Chem.*, 1989, **359**, 71.
22. A. Alberti and M. Benaglia, *J. Organomet. Chem.*, 1992, **434**, 151.
23. K. A. M. Creber, T.-I. Ho, M. C. Depew, D. Weir and J. K. S. Wan, *Can. J. Chem.*, 1982, **60**, 1504.
24. D. R. Gard and T. L. Brown, *J. Am. Chem. Soc.*, 1982, **104**, 6340.
25. K.-W. Lee, W. T. Pennington, A. W. Cordes and T. L. Brown, *Organometallics*, 1984, **3**, 404.
26. J. D. Atwood, *Inorg. Chem.*, 1981, **20**, 4031.
27. G. W. Harris, J. C. A. Boeyens and N. J. Coville, *J. Chem. Soc., Dalton Trans.*, 1985, 2277.
28. W. L. Ingham and N. J. Coville, *J. Organomet. Chem.*, 1992, **423**, 51.
29. G. W. Harris, J. C. A. Boeyens and N. J. Coville, *Organometallics*, 1985, **4**, 908, 914.
30. K.-W. Lee and T. L. Brown, *Organometallics*, 1985, **4**, 1025.
31. B. F. G. Johnson, *J. Organomet. Chem.*, 1991, **415**, 109.
32. M. O. Albers, J. C. A. Boeyens, N. J. Coville and G. W. Harris, *J. Organomet. Chem.*, 1984, **260**, 99.
33. M. O. Albers, N. J. Coville and E. Singleton, *J. Organomet. Chem.*, 1987, **326**, 229.
34. S.-J. Wang and R. J. Angelici, *Inorg. Chem.*, 1988, **27**, 3233.
35. S. P. Schmidt, F. Basolo and W. C. Trogler, *Inorg. Chim. Acta*, 1987, **131**, 181.
36. P. O. Nubel and T. L. Brown, *J. Am. Chem. Soc.*, 1984, **106**, 644.
37. K.-H. Franzreb and C. G. Kreiter, *J. Organomet. Chem.*, 1983, **246**, 189.
38. K.-W. Lee and T. L. Brown, *Organometallics*, 1985, **4**, 1030.
39. P. O. Nubel and T. L. Brown, *J. Am. Chem. Soc.*, 1984, **106**, 3474.
40. E. Guggolz, F. Oberdorfer and M. L. Ziegler, *Z. Naturforsch., Teil B*, 1981, **36**, 1060.
41. C. G. Kreiter, K. H. Franzreb and W. S. Sheldrick, *Z. Naturforsch., Teil B*, 1986, **41**, 904.
42. M. Green, A. G. Orpen, C. J. Schaverien and I. D. Williams, *J. Chem. Soc., Chem. Commun.*, 1983, 1399.
43. S. Zhang, T. L. Brown, Y. Du and J. R. Shapley, *J. Am. Chem. Soc.*, 1993, **115**, 6705.
44. M. J. Mays, D. W. Prest and P. R. Raithby, *J. Chem. Soc., Dalton Trans.*, 1981, 771.
45. D. B. Pourreau, R. R. Whittle and G. L. Geoffroy, *J. Organomet. Chem.*, 1984, **273**, 333.

46. P. O. Nubel and T. L. Brown, *Organometallics*, 1984, **3**, 29.
47. K. H. Franzreb and C. G. Kreiter, *Z. Naturforsch., Teil B*, 1984, **39**, 81.
48. A. D. Shaposhnikova, R. A. Stadnichenko, V. K. Bel'skii and A. A. Pasynskii, *Organomet. Chem. USSR*, 1988, **1**, 522.
49. S. Top, M. Gunn, G. Jaouen, J. Vaissermann, J.-C. Daran and M. J. McGlinchey, *Organometallics*, 1992, **11**, 1201.
50. K.-W. Lee, W. T. Pennington, A. W. Cordes and T. L. Brown, *J. Am. Chem. Soc.*, 1985, **107**, 631.
51. R. D. Adams, G. Chen and J. Yin, *Organometallics*, 1991, **10**, 1278.
52. R. D. Adams, G. Chen and J. Yin, *Organometallics*, 1991, **10**, 2087.
53. R. D. Adams, G. Chen, Y. Chi, W. Wu and J. Yin, *Organometallics*, 1992, **11**, 1480.
54. R. D. Adams, L. Chen and W. Wu, *Organometallics*, 1993, **12**, 1257.
55. K.-Y. Shih, P. E. Fanwick and R. A. Walton, *Organometallics*, 1993, **12**, 347.
56. K.-Y. Shih, P. E. Fanwick and R. A. Walton, *J. Am. Chem. Soc.*, 1993, **115**, 9319.
57. T. J. Henly, *Coord. Chem. Rev.*, 1989, **93**, 269.
58. R. D. Adams, J. E. Cortopassi and S. B. Falloon, *Organometallics*, 1992, **11**, 3794.
59. T. Beringhelli, G. D'Alfonso, M. Freni, G. Ciani, M. Moret and A. Sironi, *J. Organomet. Chem.*, 1988, **339**, 323.
60. T. Beringhelli, G. D'Alfonso, M. Freni, G. Ciani, M. Moret and A. Sironi, *J. Organomet. Chem.*, 1991, **412**, C4.
61. T. Beringhelli, G. D'Alfonso, A. Minoja, G. Ciani, M. Moret and A. Sironi, *Organometallics*, 1991, **10**, 3131.
62. T. Beringhelli, G. D'Alfonso, M. Freni, G. Ciani, M. Moret and A. Sironi, *J. Organomet. Chem.*, 1990, **399**, 291.
63. T. Beringhelli, G. Ciani, G. D'Alfonso, H. Molinari, A. Sironi and M. Freni, *J. Chem. Soc., Chem. Commun.*, 1984, 1327.
64. H.-J. Haupt, P. Balsaa and U. Flörke, *Angew. Chem., Int. Ed. Engl.*, 1988, **27**, 263.
65. S. R. Wang, S.-L. Wang, C. P. Cheng and C. S. Yang, *J. Organomet. Chem.*, 1992, **431**, 215.
66. G. Ciani, G. D'Alfonso, M. Freni, P. Romiti and A. Sironi, *J. Chem. Soc., Chem. Commun.*, 1982, 339.
67. G. Ciani, G. D'Alfonso, M. Freni, P. Romiti and A. Sironi, *J. Chem. Soc., Chem. Commun.*, 1982, 705.
68. T. Beringhelli, G. D'Alfonso, M. De Angelis, G. Ciani and A. Sironi, *J. Organomet. Chem.*, 1987, **322**, C21.
69. G. Hsu, S. R. Wilson and J. R. Shapley, *Inorg. Chem.*, 1991, **30**, 3881.
70. S. W. Simerly, S. R. Wilson and J. R. Shapley, *Inorg. Chem.*, 1992, **31**, 5146.
71. T. J. Henly, J. R. Shapley and A. L. Rheingold, *J. Organomet. Chem.*, 1986, **310**, 55.
72. T. Beringhelli, G. D'Alfonso, G. Ciani, A. Sironi and H. Molinari, *J. Chem. Soc., Dalton Trans.*, 1988, 1281.
73. T. Beringhelli, G. D'Alfonso, G. Ciani, A. Sironi and H. Molinari, *J. Chem. Soc., Dalton Trans.*, 1990, 1901.
74. H.-J. Haupt, U. Flörke and P. Balsaa, *Acta Crystallogr., Part C*, 1988, **44**, 61.
75. T. Beringhelli, G. D'Alfonso, A. Minoja, G. Ciani and D. M. Proserpio, *Inorg. Chem.*, 1993, **32**, 803.
76. A. C. C. Wong, P. G. Edwards, G. Wilkinson, M. Motevalli and M. B. Hursthouse, *J. Chem. Soc., Dalton Trans.*, 1988, 219.
77. S. P. Schmidt, W. C. Trogler and F. Basolo, *J. Am. Chem. Soc.*, 1984, **106**, 1308.
78. W. L. Ingham and N. J. Coville, *Organometallics*, 1992, **11**, 2551.
79. N. M. Doherty and N. W. Hoffman, *Chem. Rev.*, 1991, **91**, 553.
80. N. W. Hoffman, N. Prokopuk, M. J. Robbins, C. M. Jones and N. M. Doherty, *Inorg. Chem.*, 1991, **30**, 4177.
81. F. Calderazzo, F. Marchetti, R. Poli, D. Vitali and P. F. Zanazzi, *J. Chem. Soc., Dalton Trans.*, 1982, 1665.
82. A. E. Leins and N. J. Coville, *J. Organomet. Chem.*, 1991, **407**, 359.
83. N. J. Coville, P. Johnston, A. E. Leins and A. J. Markwell, *J. Organomet. Chem.*, 1989, **378**, 401.
84. L.-C. Chen, M.-Y. Chen, J.-H. Chen, Y.-S. Wen and K.-L. Lu, *J. Organomet. Chem.*, 1992, **425**, 99.
85. R. Ambrosetti, W. Baratta, D. B. Dell'Amico, F. Calderazzo and F. Marchetti, *Gazz. Chim. Ital.*, 1990, **120**, 511.
86. K. L. Leighton, K. R. Grundy and K. N. Robertson, *J. Organomet. Chem.*, 1989, **371**, 321.
87. W. Beck and K. Sünkel, *Chem. Rev.*, 1988, **88**, 1405.
88. E. Fritsch, J. Heidrich, K. Polborn and W. Beck, *J. Organomet. Chem.*, 1992, **441**, 203.
89. P. M. Fritz, J. Breimair, B. Wagner and W. Beck, *J. Organomet. Chem.*, 1992, **426**, 343.
90. P. Steil, U. Nagel and W. Beck, *J. Organomet. Chem.*, 1988, **339**, 111.
91. P. Steil, U. Nagel and W. Beck, *J. Organomet. Chem.*, 1989, **366**, 313.
92. P. Steil, W. Sacher, P. M. Fritz and W. Beck, *J. Organomet. Chem.*, 1989, **362**, 363.
93. P. M. Fritz *et al.*, *Z. Naturforsch., Teil B*, 1988, **43**, 665.
94. J. Milke, C. Missling, K. Sünkel and W. Beck, *J. Organomet. Chem.*, 1993, **445**, 219.
95. M. Appel, J. Heidrich and W. Beck, *Chem. Ber.*, 1987, **120**, 1087.
96. J. M. Fernández and J. A. Gladysz, *Organometallics*, 1989, **8**, 207.
97. J. H. Merrifield, C. E. Strouse and J. A. Gladysz, *Organometallics*, 1982, **1**, 1204.
98. Y.-H. Huang, F. Niedercorn, A. M. Arif and J. A. Gladysz, *J. Organomet. Chem.*, 1990, **383**, 213.
99. M. A. Dewey and J. A. Gladysz, *Organometallics*, 1993, **12**, 2390.
100. J. M. Fernández and J. A. Gladysz, *Inorg. Chem.*, 1986, **25**, 2672.
101. J. H. Merrifield, J. M. Fernández, W. E. Buhro and J. A. Gladysz, *Inorg. Chem.*, 1984, **23**, 4022.
102. C. H. Winter, A. M. Arif and J. A. Gladysz, *Organometallics*, 1989, **8**, 219.
103. C. H. Winter and J. A. Gladysz, *J. Organomet. Chem.*, 1988, **354**, C33.
104. C. H. Winter, W. R. Veal, C. M. Garner, A. M. Arif and J. A. Gladysz, *J. Am. Chem. Soc.*, 1989, **111**, 4766.
105. P. T. Czech, J. A. Gladysz and R. F. Fenske, *Organometallics*, 1989, **8**, 1806.
106. A. Igau and J. A. Gladysz, *Organometallics*, 1991, **10**, 2327.
107. Y. Zhou and J. A. Gladysz, *Organometallics*, 1993, **12**, 1073.
108. D. Baudry, J.-M. Cormier, M. Ephritikhine and H. Felkin, *J. Organomet. Chem.*, 1984, **277**, 99.
109. D. Baudry, M. Ephritikhine, H. Felkin and J. Zakrzewski, *J. Chem. Soc., Chem. Commun.*, 1982, 1235.
110. D. Baudry, M. Ephritikhine and H. Felkin, *J. Chem. Soc., Chem. Commun.*, 1982, 606.
111. D. Baudry, P. Boydell and M. Ephritikhine, *J. Chem. Soc., Dalton Trans.*, 1986, 525.
112. N. J. Hazel, J. A. K. Howard and J. L. Spencer, *J. Chem. Soc., Chem. Commun.*, 1984, 1663.
113. D. Baudry, M. Ephritikhine and H. Felkin, *J. Organomet. Chem.*, 1982, **224**, 363.
114. W. D. Jones and J. A. Maguire, *Organometallics*, 1985, **4**, 951.
115. J. R. Sweet and W. A. G. Graham, *Organometallics*, 1982, **1**, 982.
116. D. M. Heinekey and G. L. Gould, *Organometallics*, 1991, **10**, 2977.
117. F. G. N. Cloke, J. P. Day, J. C. Green, C. P. Morley and A. C. Swain, *J. Chem. Soc., Dalton Trans.*, 1991, 789.

118. J. A. Bandy *et al.*, *J. Am. Chem. Soc.*, 1988, **110**, 5039.
119. S. M. Howdle and M. Poliakoff, *J. Chem. Soc., Chem. Commun.*, 1989, 1099.
120. J. K. Hoyano and W. A. G. Graham, *Organometallics*, 1982, **1**, 783.
121. G. K. Yang and R. G. Bergman, *J. Am. Chem. Soc.*, 1983, **105**, 6500.
122. C. P. Casey, R. S. Tanke, P. N. Hazin, C. R. Kemnitz and R. J. McMahon, *Inorg. Chem.*, 1992, **31**, 5474.
123. C. P. Casey, H. Sakaba, P. N. Hazin and D. R. Powell, *J. Am. Chem. Soc.*, 1991, **113**, 8165.
124. S. S. Kristjánsdóttir and J. R. Norton, in 'Transition Metal Hydrides', ed. A. Dedieu, VCH, New York, 1992, p. 309.
125. R. L. Sweany and J. W. Owens, *J. Organomet. Chem.*, 1983, **255**, 327.
126. D. G. Hamilton and R. H. Crabtree, *J. Am. Chem. Soc.*, 1988, **110**, 4126.
127. P. J. Desrosiers, L. Cai, Z. Lin, R. Richards and J. Halpern, *J. Am. Chem. Soc.*, 1991, **113**, 4173.
128. X.-L. Luo, H. Liu and R. H. Crabtree, *Inorg. Chem.*, 1991, **30**, 4740.
129. T. Beringhelli, G. D'Alfonso, M. Freni and A. P. Minoja, *Inorg. Chem.*, 1992, **31**, 848.
130. D. G. Gusev, D. Nietlispach, A. B. Vymenits, V. I. Bakhmutov and H. Berke, *Inorg. Chem.*, 1993, **32**, 3270.
131. X.-L. Luo and R. H. Crabtree, *J. Am. Chem. Soc.*, 1990, **112**, 6912.
132. D. G. Gusev, D. Nietlispach, I. L. Eremenko and H. Berke, *Inorg. Chem.*, 1993, **32**, 3628.
133. G. L. Crocco and J. A. Gladysz, *J. Am. Chem. Soc.*, 1988, **110**, 6110.
134. D. L. Allen, M. L. H. Green and J. A. Bandy, *J. Chem. Soc., Dalton Trans.*, 1990, 541.
135. W. D. Jones and J. A. Maguire, *Organometallics*, 1987, **6**, 1301.
136. W. D. Jones and J. A. Maguire, *Organometallics*, 1986, **5**, 590.
137. W. D. Jones and J. A. Maguire, *Organometallics*, 1987, **6**, 1728.
138. T. T. Wenzel and R. G. Bergman, *J. Am. Chem. Soc.*, 1986, **108**, 4856.
139. J. M. Aramini, F. W. B. Einstein, R. H. Jones, A. H. Klahn-Oliva and D. Sutton, *J. Organomet. Chem.*, 1990, **385**, 73.
140. H. Felkin and J. Zakrzewski, *J. Am. Chem. Soc.*, 1985, **107**, 3374.
141. J.-M. Zhuang and D. Sutton, *Organometallics*, 1991, **10**, 1516.
142. P. Pasman and J. J. M. Snel, *J. Organomet. Chem.*, 1986, **301**, 329.
143. J. Y. Chen, K. R. Grundy and K. N. Robertson, *Can. J. Chem.*, 1989, **67**, 1187.
144. B. E. Bursten and M. G. Gatter, *Organometallics*, 1984, **3**, 941.
145. B. D. Martin, K. E. Warner and J. R. Norton, *J. Am. Chem. Soc.*, 1986, **108**, 33.
146. I. A. Lobanova, V. I. Zdanovich, N. E. Kolobova and V. N. Kalinin, *Organomet. Chem. USSR*, 1990, **3**, 469.
147. J. A. Labinger and K. H. Komadina, *J. Organomet. Chem.*, 1978, **155**, C25.
148. G. E. Herberich and W. Barlage, *J. Organomet. Chem.*, 1987, **331**, 63.
149. J. A. Marsella and K. G. Caulton, *J. Am. Chem. Soc.*, 1980, **102**, 1747.
150. V. M. Ishchenko, B. M. Bulychev, G. L. Soloveichik, V. K. Bel'sky and O. G. Ellert, *Polyhedron*, 1984, **3**, 771.
151. V. M. Ishchenko, G. L. Soloveichik, B. M. Bulychev and T. A. Sokolova, *Russ. J. Inorg. Chem. (Engl. Transl.)*, 1984, **29**, 66.
152. B. P. Sullivan and T. J. Meyer, *Organometallics*, 1986, **5**, 1500.
153. G. L. Hillhouse, *J. Am. Chem. Soc.*, 1985, **107**, 7772.
154. E. Lindner and W. Wassing, *Organometallics*, 1991, **10**, 1640.
155. C. P. Casey, E. W. Rutter, Jr. and K. J. Haller, *J. Am. Chem. Soc.*, 1987, **109**, 6886.
156. C. P. Casey *et al.*, in 'Organic Synthesis via Organometallics', eds. K. H. Dotz and R. W. Hoffmann, Vieweg, Braunschweig, 1991, p. 187.
157. C. P. Casey and E. W. Rutter, Jr., *J. Am. Chem. Soc.*, 1989, **111**, 8917.
158. C. P. Casey and Y. Wang, *Organometallics*, 1992, **11**, 13.
159. W. D. Jones and J. A. Maguire, *J. Am. Chem. Soc.*, 1985, **107**, 4544.
160. D. Baudry, M. Ephritikhine, H. Felkin and J. Zakrzewski, *J. Organomet. Chem.*, 1984, **272**, 391.
161. M. C. L. Trimarchi, M. A. Green, J. C. Huffman and K. G. Caulton, *Organometallics*, 1985, **4**, 514.
162. N. J. Hazel, J. A. K. Howard and J. L. Spencer, *J. Chem. Soc., Chem. Commun.*, 1984, 1663.
163. M. R. Detty and W. D. Jones, *J. Am. Chem. Soc.*, 1987, **109**, 5666.
164. G. P. Rosini, J. A. Maguire and W. Jones, American Chemical Society National Meeting, San Diego, CA, March 13–17, 1994, Division of Inorganic Chemistry, Paper 22. American Chemical Society, Washington, DC.
165. W. D. Jones and M. Fan, *Organometallics*, 1986, **5**, 1057.
166. D. Baudry, M. Ephritikhine, H. Felkin and R. Holmes-Smith, *J. Chem. Soc., Chem. Commun.*, 1983, 788.
167. D. Baudry, M. Ephritikhine, H. Felkin and J. Zakrzewski, *Tetrahedron Lett.*, 1984, **25**, 1283.
168. H. Felkin, T. Fillebeen-Khan, Y. Gault, R. Holmes-Smith and J. Zakrzewski, *Tetrahedron Lett.*, 1984, **25**, 1279.
169. D. Baudry *et al.*, *J. Chem. Soc., Chem. Commun.*, 1983, 813.
170. D. M. Heinekey and W. A. G. Graham, *J. Organomet. Chem.*, 1982, **232**, 335.
171. S. J. Doig, R. P. Hughes, S. L. Patt, D. E. Samkoff and W. L. Smith, *J. Organomet. Chem.*, 1983, **250**, C1.
172. W. Beck, M. J. Schweiger and G. Müller, *Chem. Ber.*, 1987, **120**, 889.
173. B. Niemer, T. Weidmann and W. Beck, *Z. Naturforsch., Teil B*, 1992, **47**, 509.
174. L. L. Padolik, J. Gallucci and A. Wojcicki, *J. Organomet. Chem.*, 1990, **383**, C1.
175. R. E. Lehmann, T. M. Bockman and J. K. Kochi, *J. Am. Chem. Soc.*, 1990, **112**, 458.
176. J. G. Stack, J. J. Doney, R. G. Bergman and C. H. Heathcock, *Organometallics*, 1990, **9**, 453.
177. J. J. Doney, R. G. Bergman and C. H. Heathcock, *J. Am. Chem. Soc.*, 1985, **107**, 3724.
178. J. G. Stack, R. D. Simpson, F. J. Hollander, R. G. Bergman and C. H. Heathcock, *J. Am. Chem. Soc.*, 1990, **112**, 2716.
179. E. R. Burkhardt, J. J. Doney, J. G. Stack, C. H. Heathcock and R. G. Bergman, *J. Mol. Catal.*, 1987, **41**, 41.
180. C. Roger, T.-S. Peng and J. A. Gladysz, *J. Organomet. Chem.*, 1992, **439**, 163.
181. W. E. Buhro, A. Wong, J. H. Merrifield, G.-Y. Lin, A. C. Constable and J. A. Gladysz, *Organometallics*, 1983, **2**, 1852.
182. G. L. Crocco, K. E. Lee and J. A. Gladysz, *Organometallics*, 1990, **9**, 2819.
183. W. A. Kiel, G.-Y. Lin, G. S. Bodner and J. A. Gladysz, *J. Am. Chem. Soc.*, 1983, **105**, 4958.
184. F. B. McCormick, W. A. Kiel and J. A. Gladysz, *Organometallics*, 1982, **1**, 405.
185. W. A. Kiel *et al.*, *J. Am. Chem. Soc.*, 1982, **104**, 4865.
186. W. A. Kiel, W. E. Buhro and J. A. Gladysz, *Organometallics*, 1984, **3**, 879.
187. S. Georgiou and J. A. Gladysz, *Tetrahedron*, 1986, **42**, 1109.

188. J. I. Seeman and S. G. Davies, *J. Am. Chem. Soc.*, 1985, **107**, 6522.
189. G. S. Bodner, J. A. Gladysz, M. F. Nielsen and V. D. Parker, *J. Am. Chem. Soc.*, 1987, **109**, 1757.
190. M. Tilset, G. S. Bodner, D. R. Senn, J. A. Gladysz and V. D. Parker, *J. Am. Chem. Soc.*, 1987, **109**, 7551.
191. E. J. O'Connor, M. Kobayashi, H. G. Floss and J. A. Gladysz, *J. Am. Chem. Soc.*, 1987, **109**, 4837.
192. C. P. Casey, M. A. Andrews, D. R. McAlister, W. D. Jones and S. G. Harsy, *J. Mol. Catal.*, 1981, **13**, 43.
193. W. Tam, G.-Y. Lin, W.-K. Wong, W. A. Kiel, V. K. Wong and J. A. Gladysz, *J. Am. Chem. Soc.*, 1982, **104**, 141.
194. R. F. Fenske, M. C. Milletti and M. Arndt, *Organometallics*, 1986, **5**, 2316.
195. C. Pomp and K. Wieghardt, *Inorg. Chem.*, 1988, **27**, 3796.
196. K. Wieghardt, C. Pomp, B. Nuber and J. Weiss, *Inorg. Chem.*, 1986, **25**, 1659.
197. J. C. Selover, G. D. Vaughn, C. E. Strouse and J. A. Gladysz, *J. Am. Chem. Soc.*, 1986, **108**, 1455.
198. G. D. Vaughn, C. E. Strouse and J. A. Gladysz, *J. Am. Chem. Soc.*, 1986, **108**, 1462.
199. C. Sontag, O. Orama and H. Berke, *Chem. Ber.*, 1987, **120**, 559.
200. B. A. Narayanan, C. Amatore, C. P. Casey and J. K. Kochi, *J. Am. Chem. Soc.*, 1983, **105**, 6351.
201. W. Tam, M. Marsi and J. A. Gladysz, *Inorg. Chem.*, 1983, **22**, 1413.
202. J. A. Ramsden, W. Weng and J. A. Gladysz, *Organometallics*, 1992, **11**, 3635.
203. J. A. Ramsden, W. Weng, A. M. Arif and J. A. Gladysz, *J. Am. Chem. Soc.*, 1992, **114**, 5890.
204. W. Weng, J. A. Ramsden, A. M. Arif and J. A. Gladysz, *J. Am. Chem. Soc.*, 1993, **115**, 3824.
205. Y. Zhou, J. W. Seyler, W. Weng, A. M. Arif and J. A. Gladysz, *J. Am. Chem. Soc.*, 1993, **115**, 8509.
206. W. Beck, W. Knauer and C. Robl, *Angew. Chem., Int. Ed. Engl.*, 1990, **29**, 318.
207. G. S. Bodner *et al.*, *J. Am. Chem. Soc.*, 1987, **109**, 7688.
208. G. Proulx and R. G. Bergman, *J. Am. Chem. Soc.*, 1993, **115**, 9802.
209. L. Weber, K. Reizig, K. Boese and M. Polk, *Organometallics*, 1986, **5**, 1098.
210. C. P. Casey, P. C. Vosejpka and J. A. Gavney, Jr., *J. Am. Chem. Soc.*, 1990, **112**, 4083.
211. C. P. Casey, T. L. Underiner, P. C. Vosejpka, J. A. Gavney, Jr. and P. Kiprof, *J. Am. Chem. Soc.*, 1992, **114**, 10 826.
212. C. P. Casey, P. C. Vosejpka and F. R. Askham, *J. Am. Chem. Soc.*, 1990, **112**, 3713.
213. J. J. Kowalczyk, A. M. Arif and J. A. Gladysz, *Chem. Ber.*, 1991, **124**, 729.
214. H. G. Alt and H. E. Engelhardt, *J. Organomet. Chem.*, 1988, **346**, 211.
215. M. L. H. Green, D. O'Hare and J. M. Wallis, *Polyhedron*, 1986, **5**, 1363.
216. R. D. Simpson and R. G. Bergman, *Organometallics*, 1992, **11**, 3980.
217. C. P. Casey and L. M. Baltusis, *J. Am. Chem. Soc.*, 1982, **104**, 6347.
218. K. I. Goldberg and R. G. Bergman, *J. Am. Chem. Soc.*, 1989, **111**, 1285.
219. S. Feracin, H.-U. Hund, H. W. Bosch, E. Lippmann, W. Beck and H. Berke, *Helv. Chim. Acta*, 1992, **75**, 1305.
220. C. P. Casey, P. C. Vosejpka, T. L. Underiner, G. A. Slough and J. A. Gavney, Jr., *J. Am. Chem. Soc.*, 1993, **115**, 6680.
221. D. M. DeSimone, P. J. Desrosiers and R. P. Hughes, *J. Am. Chem. Soc.*, 1982, **104**, 4842.
222. J. R. Sweet and W. A. G. Graham, *J. Organomet. Chem.*, 1981, **217**, C37.
223. C. Löwe, V. Shklover and H. Berke, *Organometallics*, 1991, **10**, 3396.
224. M. J. Schweiger, U. Nagel and W. Beck, *J. Organomet. Chem.*, 1988, **355**, 289.
225. E. Lippmann, C. Robl, H. Berke, H. D. Kaesz and W. Beck, *Chem. Ber.*, 1993, **126**, 933.
226. P. G. Lenhert, C. M. Lukehart and K. Srinivasan, *Inorg. Chem.*, 1984, **23**, 438.
227. P. G. Lenhert, C. M. Lukehart and K. Srinivasan, *J. Am. Chem. Soc.*, 1984, **106**, 124.
228. C. M. Lukehart and W. L. Magnuson, *J. Am. Chem. Soc.*, 1984, **106**, 1333.
229. P. C. Heah, A. T. Patton and J. A. Gladysz, *J. Am. Chem. Soc.*, 1986, **108**, 1185.
230. J. M. O'Connor, R. Uhrhammer, A. L. Rheingold and D. L. Staley, *J. Am. Chem. Soc.*, 1989, **111**, 7633.
231. J. M. O'Connor, R. Uhrhammer, A. L. Rheingold and D. M. Roddick, *J. Am. Chem. Soc.*, 1991, **113**, 4530.
232. J. M. O'Connor, R. Uhrhammer and R. K. Chadha, *Polyhedron*, 1993, **12**, 527.
233. D. R. Senn, J. A. Gladysz, K. Emerson and R. D. Larsen, *Inorg. Chem.*, 1987, **26**, 2737.
234. C. T. Tso and A. R. Cutler, *J. Am. Chem. Soc.*, 1986, **108**, 6069.
235. C. E. Housmekerides *et al*, *Inorg. Chem.*, 1992, **31**, 4453.
236. R. S. Pilato, G. L. Geoffroy and A. L. Rheingold, *J. Chem. Soc., Chem. Commun.*, 1989, 1287.
237. D. H. Gibson, M. Ye and J. F. Richardson, *J. Am. Chem. Soc.*, 1992, **114**, 9716.
238. F. B. McCormick, W. B. Gleason, X. Zhao, P. C. Heah and J. A. Gladysz, *Organometallics*, 1986, **5**, 1778.
239. C. P. Casey and H. Nagashima, *J. Am. Chem. Soc.*, 1989, **111**, 2352.
240. C. P. Casey, H. Sakaba and T. L. Underiner, *J. Am. Chem. Soc.*, 1991, **113**, 6673.
241. A. T. Patton, C. E. Strouse, C. B. Knobler and J. A. Gladysz, *J. Am. Chem. Soc.*, 1983, **105**, 5804.
242. J. H. Merrifield, G.-Y. Lin, W. A. Kiel and J. A. Gladysz, *J. Am. Chem. Soc.*, 1983, **105**, 5811.
243. D. R. Senn, A. Wong, A. T. Patton, M. Marsi, C. E. Strouse and J. A. Gladysz, *J. Am. Chem. Soc.*, 1988, **110**, 6096.
244. M. M. Singh and R. J. Angelici, *Inorg. Chem.*, 1984, **23**, 2699.
245. S.-J. Wang and R. J. Angelici, *J. Organomet. Chem.*, 1988, **352**, 157.
246. G. L. Miessler, S. Kim, R. A. Jacobson and R. J. Angelici, *Inorg. Chem.*, 1987, **26**, 1690.
247. E. O. Fischer, P. Rustemeyer, O. Orama, D. Neugebauer and U. Schubert, *J. Organomet. Chem.*, 1983, **247**, 7.
248. C. Alvarez-Toledano *et al.*, *J. Organomet. Chem.*, 1987, **328**, 357.
249. K. Mashima, K. Jyodoi, A. Ohyoshi and H. Takaya, *Bull. Chem. Soc. Jpn.*, 1991, **64**, 2065.
250. D. H. Gibson, S. K. Mandal, K. Owens and J. F. Richardson, *Organometallics*, 1990, **9**, 1936.
251. D. H. Gibson and K. Owens, *Organometallics*, 1991, **10**, 1216.
252. A. J. L. Pombeiro, C. J. Pickett and R. L. Richards, *J. Organomet. Chem.*, 1982, **224**, 285.
253. D. L. Hughes, A. J. L. Pombeiro, C. J. Pickett and R. L. Richards, *J. Chem. Soc., Chem. Commun.*, 1984, 992.
254. A. J. L. Pombeiro, S. S. P. R. Almeida, M. F. C. G. Silva, J. C. Jeffrey and R. L. Richards, *J. Chem. Soc., Dalton Trans.*, 1989, 2381.
255. A. J. L. Pombeiro, M. F. N. N. Carvalho, P. B. Hitchcock and R. L. Richards, *J. Chem. Soc., Dalton Trans.*, 1981, 1629.
256. A. J. L. Pombeiro, D. L. Hughes, R. L. Richards, J. Silvestre and R. Hoffmann, *J. Chem. Soc., Chem. Commun.*, 1986, 1125.
257. M. F. N. N. Carvalho, R. A. Henderson, A. J. L. Pombeiro and R. L. Richards, *J. Chem. Soc., Chem. Commun.*, 1989, 1796.
258. P. B. Hitchcock, M. F. Meidine, J. F. Nixon and A. J. L. Pombeiro, *J. Chem. Soc., Chem. Commun.*, 1991, 1031.

259. A. J. L. Pombeiro, D. L. Hughes, C. J. Pickett and R. L. Richards, *J. Chem. Soc., Chem. Commun.*, 1986, 246.
260. M. Fernanda, N. N. Carvalho, A. J. L. Pombeiro, E. G. Bakalbassis and C. A. Tsipis, *J. Organomet. Chem.*, 1989, **371**, C26.
261. A. J. L. Pombeiro, A. Hills, D. L. Hughes and R. L. Richards, *J. Organomet. Chem.*, 1988, **352**, C5.
262. M. Amélia, N. D. A. Lemos and A. J. L. Pombeiro, *J. Organomet. Chem.*, 1988, **356**, C79.
263. S. S. P. R. Almeida, J. J. R. F. Da Silva and A. J. L. Pombeiro, *J. Organomet. Chem.*, 1993, **450**, C7.
264. S. Warner and S. J. Lippard, *Organometallics*, 1989, **8**, 228.
265. R. N. Vrtis, C. P. Rao, S. Warner and S. J. Lippard, *J. Am. Chem. Soc.*, 1988, **110**, 2669.
266. J. B. Sheridan, G. L. Geoffroy and A. L. Rheingold, *J. Am. Chem. Soc.*, 1987, **109**, 1584.
267. B. M. Handwerker, K. E. Garrett, G. L. Geoffroy and A. L. Rheingold, *J. Am. Chem. Soc.*, 1989, **111**, 369.
268. B. M. Handwerker, K. E. Garrett, K. L. Nagle, G. L. Geoffroy and A. L. Rheingold, *Organometallics*, 1990, **9**, 1562.
269. L. A. Mercando, B. M. Handwerker, H. J. MacMillan, G. L. Geoffroy, A. L. Rheingold and B. E. Owens-Waltermire, *Organometallics*, 1993, **12**, 1559.
270. E. O. Fisher, J. Chen and K. Scherzer, *J. Organomet. Chem.*, 1983, **253**, 231.
271. J. Chen, Y. Yu, K. Liu, G. Wu and P. Zheng, *Organometallics*, 1993, **12**, 1213.
272. S. Komiya and A. Baba, *Organometallics*, 1991, **10**, 3105.
273. K. Raab, U. Nagel and W. Beck, *Z. Naturforsch., Teil B*, 1983, **38**, 1466.
274. W. Beck, K. Raab, U. Nagel and W. Sacher, *Angew. Chem., Int. Ed. Engl.*, 1985, **24**, 505.
275. F. W. B. Einstein, R. H. Jones, A. H. Klahn-Oliva and D. Sutton, *Organometallics*, 1986, **5**, 2476.
276. C. P. Casey and C. S. Yi, *Organometallics*, 1990, **9**, 2413.
277. T.-S. Peng and J. A. Gladysz, *J. Am. Chem. Soc.*, 1992, **114**, 4174.
278. G. S. Bodner, T.-S. Peng, A. M. Arif and J. A. Gladysz, *Organometallics*, 1990, **9**, 1191.
279. F. W. B. Einstein, K. G. Tyers and D. Sutton, *Organometallics*, 1985, **4**, 489.
280. J. R. Sweet and W. A. G. Graham, *Organometallics*, 1983, **2**, 135.
281. C. P. Casey and C. S. Yi, *Organometallics*, 1991, **10**, 33.
282. H. G. Alt and H. E. Engelhardt, *J. Organomet. Chem.*, 1988, **342**, 235.
283. C. P. Casey and C. S. Yi, *J. Am. Chem. Soc.*, 1992, **114**, 6597.
284. A. S. Botsanov, Y. T. Struchkov, V. I. Zhdanovich, P. V. Petrovskii and N. E. Kolobova, *Organomet. Chem. USSR*, 1989, **2**, 548.
285. C. P. Casey, C. S. Yi and J. A. Gavney, Jr., *J. Organomet. Chem.*, 1993, **443**, 111.
286. J. Arnold, G. Wilkinson, B. Hussain and M. B. Hursthouse, *J. Chem. Soc., Chem. Commun.*, 1988, 704.
287. J. J. Kowalczyk, A. M. Arif and J. A. Gladysz, *Organometallics*, 1991, **10**, 1079.
288. R. A. Fischer and W. A. Herrmann, *J. Organomet. Chem.*, 1989, **377**, 275.
289. W. A. Herrmann, R. A. Fischer and E. Herdtweck, *Organometallics*, 1989, **8**, 2821.
290. H.-U. Hund, U. Ruppli and H. Berke, *Helv. Chim. Acta*, 1993, **76**, 963.
291. R. J. Batchelor, F. W. B. Einstein, J.-M. Zhuang and D. Sutton, *J. Organomet. Chem.*, 1990, **397**, 69.
292. R. Krämer, E. Lippmann, K. Noisternig, M. Steimann, U. Nagel and W. Beck, *Chem. Ber.*, 1993, **126**, 927.
293. M. Moll, H. Behrens, H.-J. Seibold and P. Merbach, *J. Organomet. Chem.*, 1983, **248**, 329.
294. V. V. Krivykh, O. V. Gusev and M. I. Rybinskaya, *J. Organomet. Chem.*, 1989, **362**, 351.
295. V. V. Krivykh, O. V. Gusev, P. V. Petrovskii and M. I. Rybinskaya, *J. Organomet. Chem.*, 1989, **366**, 129.
296. J. R. Bleeke, D. J. Rauscher and D. A. Moore, *Organometallics*, 1987, **6**, 2614.
297. J. R. Bleeke and P. L. Earl, *Organometallics*, 1989, **8**, 2735.
298. R. D. Pike, T. J. Alavosus, C. A. Camaioni-Neto, J. C. Williams, Jr. and D. A. Sweigart, *Organometallics*, 1989, **8**, 2631.
299. Y. K. Chung, E. D. Honig and D. A. Sweigart, *J. Organomet. Chem.*, 1983, **256**, 277.
300. R. D. Pike *et al.*, *Organometallics*, 1992, **11**, 2841.
301. J. R. Sweet and W. A. G. Graham, *J. Am. Chem. Soc.*, 1983, **105**, 305.
302. H. van der Heijden, A. G. Orpen and P. Pasman, *J. Chem. Soc., Chem. Commun.*, 1985, 1576.
303. M. L. H. Green and D. O'Hare, *J. Chem. Soc., Chem. Commun.*, 1985, 332.
304. M. L. H. Green, D. O'Hare and G. Parkin, *J. Chem. Soc., Chem. Commun.*, 1985, 356.
305. M. L. H. Green, N. D. Lowe and D. O'Hare, *J. Chem. Soc., Chem. Commun.*, 1986, 1547.
306. M. L. H. Green and D. O'Hare, *J. Chem. Soc., Dalton Trans.*, 1987, 403.
307. A. E. Derome, M. L. H. Green and D. O'Hare, *J. Chem. Soc., Dalton Trans.*, 1986, 343.
308. F. G. N. Cloke, A. E. Derome, M. L. H. Green and D. O'Hare, *J. Chem. Soc., Chem. Commun.*, 1983, 1312.
309. J. A. Bandy, F. G. N. Cloke, M. L. H. Green, D. O'Hare and K. Prout, *J. Chem. Soc., Chem. Commun.*, 1984, 240.
310. M. D. Rausch and W. C. Spink, *Synth. React. Inorg. Metal. Org. Chem.*, 1989, **19**, 1093.
311. P. Singh, M. D. Rausch and T. E. Bitterwolf, *J. Organomet. Chem.*, 1988, **352**, 273.
312. S. S. Jones, M. D. Rausch and T. E. Bitterwolf, *J. Organomet. Chem.*, 1990, **396**, 279.
313. M. D. Rausch, W.-M. Tsai, J. W. Chambers, R. D. Rogers and H. G. Alt, *Organometallics*, 1989, **8**, 816.
314. M. D. Rausch, W. C. Spink, B. G. Conway, R. D. Rogers and J. L. Atwood, *J. Organomet. Chem.*, 1990, **383**, 227.
315. C. Apostolidis, B. Kanellakopulos, R. Maier, J. Rebizant and M. L. Ziegler, *J. Organomet. Chem.*, 1991, **409**, 243.
316. M. Herberhold and M. Biersack, *J. Organomet. Chem.*, 1990, **381**, 379.
317. M. Herberhold and M. Biersack, *J. Organomet. Chem.*, 1993, **443**, 1.
318. R. Ferede, J. F. Hinton, W. A. Korfmacher, J. P. Freeman and N. T. Allison, *Organometallics*, 1985, **4**, 614.
319. N. M. Loim, A. G. Ginzburg and M. V. Galakhov, *Organomet. Chem. USSR*, 1991, **4**, 471.
320. E. S. Shubina *et al.*, *J. Organomet. Chem.*, 1992, **434**, 329.
321. P. C. Heah and J. A. Gladysz, *J. Mol. Catal.*, 1985, **31**, 207.
322. P. C. Heah and J. A. Gladysz, *J. Am. Chem. Soc.*, 1984, **106**, 7636.
323. M. C. Milletti and R. F. Fenske, *Organometallics*, 1989, **8**, 420.
324. G. L. Crocco and J. A. Gladysz, *Chem. Ber.*, 1988, **121**, 375.
325. B. D. Zwick, A. M. Arif, A. T. Patton and J. A. Gladysz, *Angew. Chem., Int. Ed. Engl.*, 1987, **26**, 910.
326. R. J. Angelici, G. Facchin and M. M. Singh, *Synth. React. Inorg. Metal. Org. Chem.*, 1990, **20**, 275.
327. A. H. Klahn, C. Leiva, K. Mossert and X. Zhang, *Polyhedron*, 1991, **10**, 1873.
328. J. K. Hoyano and W. A. G. Graham, *J. Chem. Soc., Chem. Commun.*, 1982, 27.
329. J. W. Hershberger, C. Amatore and J. K. Kochi, *J. Organomet. Chem.*, 1983, **250**, 345.

330. D. P. Summers, J. C. Luong and M. S. Wrighton, *J. Am. Chem. Soc.*, 1981, **103**, 5238.
331. A. H. Klahn, M. H. Moore and R. N. Perutz, *J. Chem. Soc., Chem. Commun.*, 1992, 1699.
332. G. Díaz, A. H. Klahn and C. Manzur, *Polyhedron*, 1988, **7**, 2743.
333. R. H. Hill and B. J. Palmer, *Organometallics*, 1989, **8**, 1651.
334. C. P. Casey and W. D. Jones, *J. Am. Chem. Soc.*, 1980, **102**, 6154.
335. C. P. Casey, J. M. O'Connor and K. J. Haller, *J. Am. Chem. Soc.*, 1985, **107**, 1241.
336. C. P. Casey, R. A. Widenhoefer and J. M. O'Connor, *J. Organomet. Chem.*, 1992, **428**, 99.
337. C. P. Casey, J. M. O'Connor, W. D. Jones and K. J. Haller, *Organometallics*, 1983, **2**, 535.
338. C. P. Casey and J. M. O'Connor, *Organometallics*, 1985, **4**, 384.
339. K. M. Young, T. M. Miller and M. S. Wrighton, *J. Am. Chem. Soc.*, 1990, **112**, 1529.
340. J. Chetwynd-Talbot, P. Grebenik, R. N. Perutz and M. H. A. Powell, *Inorg. Chem.*, 1983, **22**, 1675.
341. A. K. Campen, R. Narayanaswamy and A. J. Rest, *J. Chem. Soc., Dalton Trans.*, 1990, 823.
342. W. E. Buhro, S. Georgiou, J. M. Fernández, A. T. Patton, C. E. Strouse and J. A. Gladysz, *Organometallics*, 1986, **5**, 956.
343. W. E. Buhro, M. C. Etter, S. Georgiou, J. A. Gladysz and F. B. McCormick, *Organometallics*, 1987, **6**, 1150.
344. W. E. Buhro, A. T. Patton, C. E. Strouse, J. A. Gladysz, F. B. McCormick and M. C. Etter, *J. Am. Chem. Soc.*, 1983, **105**, 1056.
345. F. B. McCormick, *Organometallics*, 1984, **3**, 1924.
346. C. M. Garner *et al.*, *J. Am. Chem. Soc.*, 1990, **112**, 5146.
347. D. M. Dalton, J. M. Fernández, K. Emerson, R. D. Larsen, A. M. Arif and J. A. Gladysz, *J. Am. Chem. Soc.*, 1990, **112**, 9198.
348. N. Q. Méndez, J. W. Seyler, A. M. Arif and J. A. Gladysz, *J. Am. Chem. Soc.*, 1993, **115**, 2323.
349. D. P. Klein, D. M. Dalton, N. Q. Méndez, A. M. Arif and J. A. Gladysz, *J. Organomet. Chem.*, 1991, **412**, C7
350. F. Agbossou, J. A. Ramsden, Y.-H. Huang, A. M. Arif and J. A. Gladysz, *Organometallics*, 1992, **11**, 693.
351. D. P. Klein and J. A. Gladysz, *J. Am. Chem. Soc.*, 1992, **114**, 8710.
352. D. M. Dalton and J. A. Gladysz, *J. Chem. Soc., Dalton Trans.*, 1991, 2741.
353. I. Saura-Llamas, D. M. Dalton, A. M. Arif and J. A. Gladysz, *Organometallics*, 1992, **11**, 683.
354. G. Bringmann, O. Schupp, K. Peters, L. Walz and H. G. von Schnering, *J. Organomet. Chem.*, 1992, **438**, 117.
355. M. Appel, W. Sacher and W. Beck, *J. Organomet. Chem.*, 1987, **322**, 351.
356. R. Birk, H. Berke, H.-U. Hund, K. Evertz, G. Huttner and L. Zsolnai, *J. Organomet. Chem.*, 1988, **342**, 67.
357. M. A. Dewey, D. A. Knight, D. P. Klein, A. M. Arif and J. A. Gladysz, *Inorg. Chem.*, 1991, **30**, 4995.
358. M. A. Dewey, D. A. Knight, A. M. Arif and J. A. Gladysz, *Z. Naturforsch., Teil B*, 1992, **47**, 1175.
359. W. E. Buhro, B. D. Zwick, S. Georgiou, J. P. Hutchinson and J. A. Gladysz, *J. Am. Chem. Soc.*, 1988, **110**, 2427.
360. M. A. Dewey, D. A. Knight, A. Arif and J. A. Gladysz, *Chem. Ber.*, 1992, **125**, 815.
361. M. A. Dewey and J. A. Gladysz, *Organometallics*, 1990, **9**, 1351 and references therein.
362. C. F. Barrientos-Penna, F. W. B. Einstein, T. Jones and D. Sutton, *Inorg. Chem.*, 1982, **21**, 2578.
363. C. F. Barrientos-Penna, C. F. Campana, F. W. B. Einstein, T. Jones, D. Sutton and A. S. Tracey, *Inorg. Chem.*, 1984, **23**, 363.
364. C. F. Barrientos-Penna, A. B. Gilchrist and D. Sutton, *Organometallics*, 1983, **2**, 1265.
365. C. F. Barrientos-Penna, A. H. Klahn-Oliva and D. Sutton, *Organometallics*, 1985, **4**, 367.
366. C. F. Barrientos-Penna, A. B. Gilchrist, A. H. Klahn-Oliva, A. J. L. Hanlan and D. Sutton, *Organometallics*, 1985, **4**, 478.
367. F. W. B. Einstein, D. Sutton and K. G. Tyers, *Inorg. Chem.*, 1987, **26**, 111.
368. D. J. Stufkens, *Comments Inorg. Chem.*, 1992, **13**, 359.
369. K. S. Schanze, D. B. MacQueen, T. A. Perkins and L. A. Cabana, *Coord. Chem. Rev.*, 1993, **122**, 63.
370. J. A. Clark and M. Kilner, *J. Chem. Soc., Dalton Trans.*, 1983, 2613.
371. J. A. Clark and M. Kilner, *J. Chem. Soc., Dalton Trans.*, 1984, 389.
372. P. Zanello *et al.*, *Polyhedron*, 1988, **7**, 195.
373. O. J. Scherer and J. Kerth, *J. Organomet. Chem.*, 1983, **243**, C33.
374. O. J. Scherer, J. Kerth and M. L. Ziegler, *Angew. Chem., Int. Ed. Engl.*, 1983, **22**, 503.
375. O. J. Scherer, P. Quintus, J. Kaub and W. S. Sheldrick, *Chem. Ber.*, 1987, **120**, 1463.
376. J. R. Dilworth, R. A. Henderson, P. Dahlstrom, T. Nicholson and J. A. Zubieta, *J. Chem. Soc., Dalton Trans.*, 1987, 529.
377. H. M. Ali and G. J. Leigh, *J. Chem. Soc., Dalton Trans.*, 1986, 213.
378. S. W. Carr, X. L. R. Fontaine and B. L. Shaw, *J. Chem. Soc., Dalton Trans.*, 1987, 3067.
379. R. Rossi, A. Marchi, L. Magon, A. Duatti, U. Casellato and R. Graziani, *Inorg. Chim. Acta*, 1989, **160**, 23.
380. C. Pomp, H. Duddeck, K. Wieghardt, B. Nuber and J. Weiss, *Angew. Chem., Int. Ed. Engl.*, 1987, **26**, 924.
381. M. Angaroni, G. A. Ardizzoia, G. D'Alfonso, G. LaMonica, N. Masciocchi and M. Moret, *J. Chem. Soc., Dalton Trans.*, 1990, 1895.
382. D. T. Plummer, G. A. Kraus and R. J. Angelici, *Inorg. Chem.*, 1983, **22**, 3492.
383. F. W. B. Einstein, A. H. Klahn-Oliva, D. Sutton and K. G. Tyers, *Organometallics*, 1986, **5**, 53.
384. A. H. Klahn and D. Sutton, *Organometallics*, 1989, **8**, 198.
385. A. Cusanelli and D. Sutton, *J. Chem. Soc., Chem. Commun.*, 1989, 1719.
386. S. M. Howdle, P. Grebenik, R. N. Perutz and M. Poliakoff, *J. Chem. Soc., Chem. Commun.*, 1989, 1517.
387. Y. Wang, J. J. R. F. Da Silva, A. J. L. Pombeiro, M. A. Pellinghelli and A. Tiripicchio, *J. Organomet. Chem.*, 1992, **430**, C56.
388. B. D. Zwick, M. A. Dewey, D. A. Knight, W. E. Buhro, A. M. Arif and J. A. Gladysz, *Organometallics*, 1992, **11**, 2673.
389. L. Weber, G. Meine, R. Boese and D. Bläser, *Chem. Ber.*, 1988, **121**, 853.
390. P. Y. Zheng and D. W. Stephan, *Can. J. Chem.*, 1989, **67**, 1584.
391. I. Bernal, J. D. Korp, F. Calderazzo, R. Poli and D. Vitali, *J. Chem. Soc., Dalton Trans.*, 1984, 1945.
392. A.-J. DiMaio and A. L. Rheingold, *Organometallics*, 1987, **6**, 1138.
393. A.-J. DiMaio, S. J. Geib and A. L. Rheingold, *J. Organomet. Chem.*, 1987, **335**, 97.
394. R. D. Simpson and R. G. Bergman, *Organometallics*, 1993, **12**, 781.
395. K. R. Grundy and K. N. Robertson, *Inorg. Chem.*, 1985, **24**, 3898.
396. S. K. Mandal, D. M. Ho and M. Orchin, *Organometallics*, 1993, **12**, 1714.
397. I. Saura-Llamas and J. A. Gladysz, *J. Am. Chem. Soc.*, 1992, **114**, 2136.

398. S. K. Agbossou, W. W. Smith and J. A. Gladysz, *Chem. Ber.*, 1990, **123**, 1293.
399. J. H. Merrifield and J. A. Gladysz, *Organometallics*, 1983, **2**, 782.
400. N. N. Sauer and R. J. Angelici, *Inorg. Chem.*, 1987, **26**, 2160.
401. J. Heidrich and W. Beck, *J. Organomet. Chem.*, 1988, **354**, 91.
402. M.-G. Choi and R. J. Angelici, *J. Am. Chem. Soc.*, 1989, **111**, 8753.
403. M.-G. Choi and R. J. Angelici, *Organometallics*, 1991, **10**, 2436.
404. M.-G. Choi and R. J. Angelici, *Inorg. Chem.*, 1991, **30**, 1417.
405. M.-G. Choi and R. J. Angelici, *Organometallics*, 1992, **11**, 3328.
406. M.-G. Choi and R. J. Angelici, *J. Am. Chem. Soc.*, 1991, **113**, 5651.
407. W. A. Herrmann, C. Hecht, E. Herdtweck and H.-J. Kneuper, *Angew. Chem., Int. Ed. Engl.*, 1987, **26**, 132.
408. W. A. Herrmann and H.-J. Kneuper, *J. Organomet. Chem.*, 1988, **348**, 193.
409. G. P. Rosini and W. D. Jones, *J. Am. Chem. Soc.*, 1992, **114**, 10767.
410. M. Herberhold, D. Reiner, K. Ackermann, U. Thewalt and T. Debaerdemaeker, *Z. Naturforsch., Teil B*, 1984, **39**, 1199.
411. M. Herberhold, D. Reiner and U. Thewalt, *Angew. Chem., Int. Ed. Engl.*, 1983, **22**, 1000.
412. M. Herberhold and B. Schmidkonz, *J. Organomet. Chem.*, 1986, **308**, 35.
413. M. Herberhold, B. Schmidkonz, U. Thewalt, A. Razavi, H. Schöllhorn, W. A. Herrmann and C. Hecht, *J. Organomet. Chem.*, 1986, **299**, 213.
414. M. Herberhold and B. Schmidkonz, *J. Organomet. Chem.*, 1988, **358**, 301.
415. T. S. A. Hor, B. Wagner and W. Beck, *Organometallics*, 1990, **9**, 2183.
416. U. Abram and B. Lorenz, *Z. Naturforsch., Teil B*, 1993, **48**, 771.
417. R. Rossi, A. Marchi, A. Duatti, L. Magon, U. Casellato and R. Graziani, *J. Chem. Soc., Dalton Trans.*, 1987, 2299.
418. E. W. Abel, D. Ellis and K. G. Orrell, *J. Chem. Soc., Dalton Trans.*, 1992, 2243.
419. P. C. Servaas, D. J. Stufkens, A. Oskam, P. Vernooijs, E. J. Baerends, D. J. A. De Ridder and C. H. Stam, *Inorg. Chem.*, 1989, **28**, 4104.
420. U. Kunze and A. Bruns, *J. Organomet. Chem.*, 1985, **292**, 349.
421. A. J. Deeming, M. Karim, P. A. Bates and M. B. Hursthouse, *Polyhedron*, 1988, **7**, 1401.
422. E. W. Abel *et al.*, *J. Chem. Soc., Dalton Trans.*, 1982, 2065.
423. S. C. O'Neal, W. T. Pennington and J. W. Kolis, *Can. J. Chem.*, 1989, **67**, 1980.
424. A. A. Pasynskii *et al.*, *Organomet. Chem. USSR*, 1990, **3**, 226.
425. W. Baratta and F. Calderazzo, *Organometallics*, 1993, **12**, 1489.
426. J. M. Mayer, D. L. Thorn and T. H. Tulip, *J. Am. Chem. Soc.*, 1985, **107**, 7454.
427. A. B. Manion, T. K. G. Erikson, E. Spaltenstein and J. M. Mayer, *Organometallics*, 1989, **8**, 1871.
428. E. Spaltenstein and J. M. Mayer, *J. Am. Chem. Soc.*, 1991, **113**, 7744.
429. E. Spaltenstein, R. R. Conry, S. C. Critchlow and J. M. Mayer, *J. Am. Chem. Soc.*, 1989, **111**, 8741.
430. R. R. Conry and J. M. Mayer, *Organometallics*, 1991, **10**, 3160.
431. E. Spaltenstein, T. K. G. Erikson, S. C. Critchlow and J. M. Mayer, *J. Am. Chem. Soc.*, 1989, **111**, 617.
432. T. K. G. Erikson, J. C. Bryan and J. M. Mayer, *Organometallics*, 1988, **7**, 1930.
433. T. K. G. Erikson and J. M. Mayer, *Angew. Chem., Int. Ed. Engl.*, 1988, **27**, 1527.
434. J. M. Mayer, T. H. Tulip, J. C. Calabrese and E. Valencia, *J. Am. Chem. Soc.*, 1987, **109**, 157.
435. R. R. Conry and J. M. Mayer, *Organometallics*, 1993, **12**, 3179.
436. S. K. Tahmassebi, R. R. Conry and J. M. Mayer, *J. Am. Chem. Soc.*, 1993, **115**, 7553.

# 10
# High-valent Organorhenium Compounds

DAVID M. HOFFMAN
*University of Houston, TX, USA*

## 10.1 INTRODUCTION

In this chapter organorhenium compounds in oxidation states +4 to +7 are reviewed. Ligand charges are assigned using the "oxidation state formalism."[1] Alkynes are considered neutral ligands, although it is recognized there is an electronic ambiguity in many cases. For simplicity, Re–O and Re–NR are drawn as Re=O and Re=NR, although in all but a few select examples the oxo and imido ligand contributions to the metal are known to be more than four electrons and the bond orders are greater than 2. There are several organorhenium reviews that cover compounds with rhenium in high oxidation states,[2] including the valuable one by Boag and Kaesz.[3]

## 10.2 COMPLEXES WITHOUT ALKYLIDENE, ALKYLIDYNE, OR CONJUGATED POLYENE LIGANDS

### 10.2.1 Rhenium(IV)

[Re($o$-Tol)$_4$], a monomer with a slightly distorted tetrahedral geometry in the solid state, is prepared from [ReCl$_4$(THF)$_2$] and Grignard reagent in low yield.[4] Rhenium(IV) tetraalkyls were previously reported to be trimers.[3] Magnetic measurements indicate that [Re($o$-Tol)$_4$] has only one unpaired electron (low-spin). The addition of phosphines to [Re($o$-Tol)$_4$] gives rhenium(III) benzyne complexes[5] and the reaction with O$_2$ gives [ReO($o$-Tol)$_4$].[4]

The alkyne complex [ReCl$_4$(PhCCPh)(OPCl$_3$)] is prepared from ReCl$_5$, alkyne, and POCl$_3$,[6] and the THF and acetonitrile adducts are prepared by OPCl$_3$ ligand displacement.[7] [PPh$_4$][ReCl$_5$(PhCCPh)] is prepared from [PPh$_4$]Cl and [{Re($\mu$-Cl)Cl$_3$(PhCCPh)}$_2$],[8] a compound synthesized directly from ReCl$_5$, or obtained from CH$_2$Cl$_2$ solutions of [ReCl$_4$(PhCCPh)(OPCl$_3$)].[6,7] Diphenylacetylene is readily displaced from [ReCl$_4$(PhCCPh)(OPCl$_3$)] by 1,2-bis(diphenylphosphino)ethane (dppe).[9] The related aryl oxide derivatives [Re(O-Xyl)$_4$(RCCR)] (Xyl-2,6-Me$_2$C$_6$H$_3$; R = Me, Et, Ph) are prepared by reacting the alkynes with square-planar [Re(O-Xyl)$_4$],[10] but the square-planar complex with bulky 2,6-diisopropylphenoxide ligands does not form similar adducts with internal alkynes.

Polymer-supported triphenylphosphine reduces [ReO$_2$Me(RCCR)] (R = Me, Et) to give [{Re(O)Me(RCCR)}$_2$($\mu$-O)].[11] The Re–Re single bond distance in this complex (0.264 nm) and low C–C stretching frequency are consistent with a rhenacyclopropene bonding description.

Reaction of [Re$_2$($\mu$-H)$_4$H$_3$(PPh$_3$)$_4$(NCMe)$_2$]$^+$ with Bu$^t$NC gives diamagnetic Re$^{IV}$ [Re$_2$($\mu$-H)$_4$H$_3$(PPh$_3$)$_4$(CNBu$^t$)]$^+$.[12] Oxidation of [Re$_2$($\mu$-H)$_4$H$_3$(PPh$_3$)$_4$(CNBu$^t$)]$^+$ and Re$^{III}$ [Re$_2$($\mu$-H)$_3$H$_2$(PPh$_3$)$_4$(CNBu$^t$)$_2$]$^+$ with NO$^+$ yields paramagnetic Re$^V$–Re$^{IV}$ [Re$_2$($\mu$-H)$_4$H$_3$(PPh$_3$)$_4$(CNBu$^t$)]$^{2+}$ and Re$^{IV}$–Re$^{III}$ [Re$_2$($\mu$-H)$_3$H$_2$(PPh$_3$)$_4$(CNBu$^t$)$_2$]$^{2+}$, respectively. The oxidized complexes react much more rapidly with excess isocyanide than their diamagnetic precursors.

### 10.2.2 Rhenium(V)

#### 10.2.2.1 Oxo complexes

Rhenium(V) monooxo alkyl and aryl complexes exhibit square-pyramidal or octahedral coordination. They are diamagnetic with two electrons residing in a nonbonding $d_{x^2-y^2}$ orbital ($z$-axis parallel to the Re–O bond vector, $y$-axis between the ligands).

[ReOR$_3$L] complexes are the most common type of monooxo compound. The synthesis of (1) (Scheme 1) involves the use of Grignard reagent (Equation (1)), but the reaction fails for the Me and neopentyl derivatives.[13] The methyl derivative (2) (Scheme 1) is prepared by reacting ReOCl$_2$(OEt)py$_2$ with AlMe$_3$ in the presence of phosphine to give [ReOClMe$_2$(PMe$_2$R)$_2$], followed by treatment with ZnMe$_2$ at low temperature.[14] An alternative synthesis of [ReOR$^1_3$(PR$^2_3$)] compounds entails reduction of the rhenium(VI) dimers [{ReOR$_3$}$_2$($\mu$-O)] with phosphine (to give phosphine oxide),[13] but the preparation of [{ReOR$_3$}$_2$($\mu$-O)] is relatively tedious. [ReO(CH$_2$TMS)$_3$(PMe$_2$R)] is thermally stable, but [ReOMe$_3$(PMe$_2$R)] decomposes readily to give a methylidene complex.[15] [ReOEt$_3$(PPh$_3$)$_2$], prepared from [ReOCl$_3$(PPh$_3$)$_2$] and AlCl$_2$Et, is also reported to be thermally unstable, giving ethane and ethene on decomposition at room temperature.[16]

$$ReOCl_2(OEt)py_2 + 3\,MgCl(CH_2TMS) + PMe_2R \longrightarrow ReO(CH_2TMS)_3(PMe_2R) + 2\,py + 2\,MgCl_2 + MgCl(OEt) \quad (1)$$

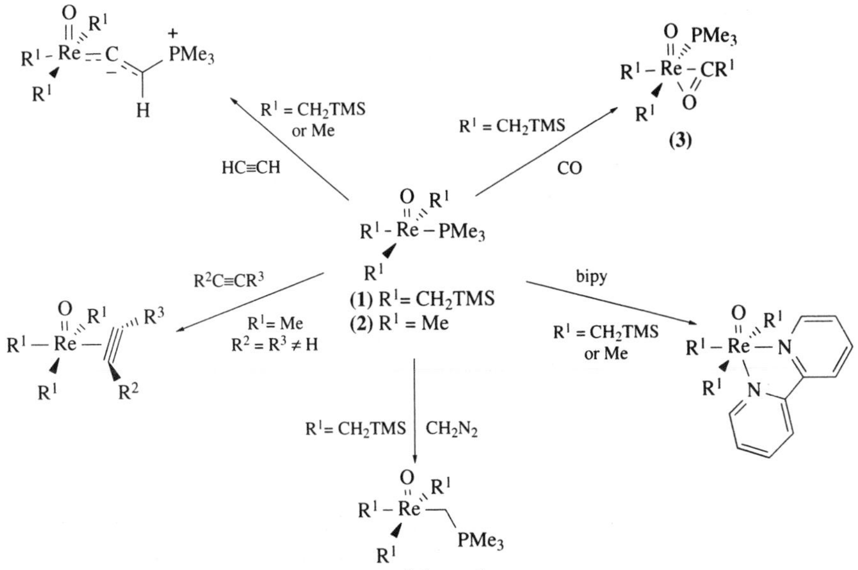

**Scheme 1**

Selected reactions of (1) and (2) are shown in Scheme 1,[13,14,17,18] including formal Re–P bond insertion reactions of $CH_2N_2$ with (1),[17] and ethyne with both (1) and (2).[18] In the case of ethyne insertion, the isolation of alkyne adducts of the type $[ReOMe_3(R^2CCR^3)]$ suggests that the ethyne "insertion" reactions actually proceed via external phosphine attack on an initially formed ethyne complex. $[ReOClMe_2(PMe_2R)_2]$ undergoes an analogous double ethyne insertion reaction to yield $[ReO\{C(H)C(H)PMe_3\}_2Me_2]Cl$.[18] Spectroscopic and structural data, as well as theoretical considerations, suggest that $[ReO\{C(H)C(H)PMe_3\}R_3]$ and $[ReO\{C(H)C(H)PMe_3\}_2Me_2]^+$ are organometallic analogues of resonance-stabilized ylides. The oxoacyl (3) (Scheme 1) is hydrolyzed to $[ReO\{C(O)Me\}(CH_2TMS)_2(PMe_3)]$ in the presence of a catalytic amount of base.[17]

Pseudooctahedral $[ReOCl_2Me(bipy)]$ is prepared by reaction of $[ReO_3Me(bipy)]$ with TMS-Cl, followed by addition of $PPh_3$. Substitution of the chlorides using Grignard reagents affords $[ReOR_2Me(bipy)]$ (Scheme 2).[19]

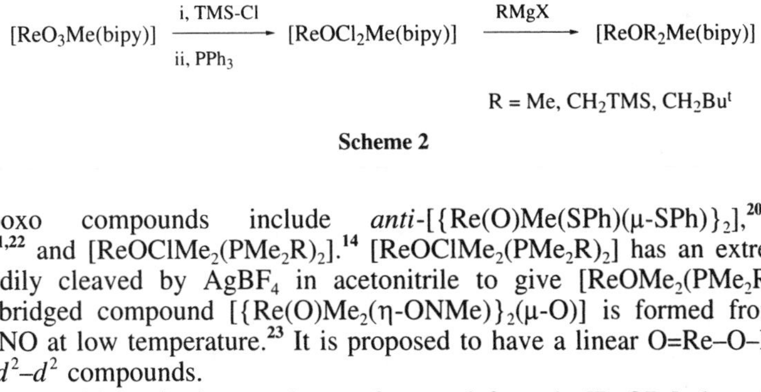

$$[ReO_3Me(bipy)] \xrightarrow[\text{ii, PPh}_3]{\text{i, TMS-Cl}} [ReOCl_2Me(bipy)] \xrightarrow{RMgX} [ReOR_2Me(bipy)]$$

$$R = Me, CH_2TMS, CH_2Bu^t$$

**Scheme 2**

Other monooxo compounds include *anti*-$[\{Re(O)Me(SPh)(\mu\text{-}SPh)\}_2]$,[20] $[Re(O)Me(1,2\text{-}(O)(NH)C_6H_4)_2]$,[21,22] and $[ReOClMe_2(PMe_2R)_2]$.[14] $[ReOClMe_2(PMe_2R)_2]$ has an extremely long Re–Cl bond that is readily cleaved by $AgBF_4$ in acetonitrile to give $[ReOMe_2(PMe_2R)_2(NCMe)]^+$. The rhenium(V) oxo-bridged compound $[\{Re(O)Me_2(\eta\text{-}ONMe)\}_2(\mu\text{-}O)]$ is formed from the reaction of $Li_2[ReMe_8]$ with NO at low temperature.[23] It is proposed to have a linear O=Re–O–Re=O group, as is found in related $d^2$–$d^2$ compounds.

Several square-pyramidal anionic complexes of general formula $[ReOR_4]^-$ have been prepared.[24,25] Equation (2) illustrates the synthesis of one compound. The syntheses of other $M[ReOR_4]_x$ compounds (M = $[Li(THF)_2]$, R = Me, $x = 1$; M = $[Mg(THF)_4]$, R = Me, $x = 2$; M = $[Mg(THF)_2]$, R = $CH_2TMS$, $x = 2$; M = $[Mg(THF)_2]$, R = $o$-$(CH_2)C_6H_4$, $x = 2$) are similar. $[Re_2O_7]$, $[ReOCl_4]$, and $[ReOCl_3(PPh_3)_2]$ are also commonly used as starting materials in the preparations. The anions are diamagnetic and can be readily oxidized to give rhenium(VI) complexes.

$$[Me_3NH][ReO_4] + 7\,MeMgCl\,(or\,MgMe_2) \xrightarrow{THF} [Mg(THF)_4][ReOMe_4]_2 \qquad (2)$$

There are two types of rhenium(V) dioxo complexes, $[ReO_2RL_n]$ and $[ReO_2R_2]^-$. Trapping reactions are used to prepare the $[ReO_2RL_n]$ compounds. As an example, $[ReO_3Me]$ reacts with polymer-bound triphenylphosphine to produce "$ReO_2Me$," which is then trapped by alkynes (Scheme 3).[11,26] Similarly, $Re^{VII}$ $[ReO_2(CH_2Bu^t)_2Ph]$ is photolyzed or thermolyzed to give neopentylbenzene and "$ReO_2(CH_2Bu^t)$," which is captured by pyridine to give *trans*-[ReO$_2$(CH$_2$CMe$_3$)(py)$_3$] or by alkyne to yield [ReO$_2$(CH$_2$Bu$^t$)(alkyne)].[27] [ReO$_2$RL] complexes where L is a π-acceptor ligand other than alkyne have not been prepared. In the [ReO$_2$R(alkyne)] complexes the alkyne plane bisects the O–Re–O angle.

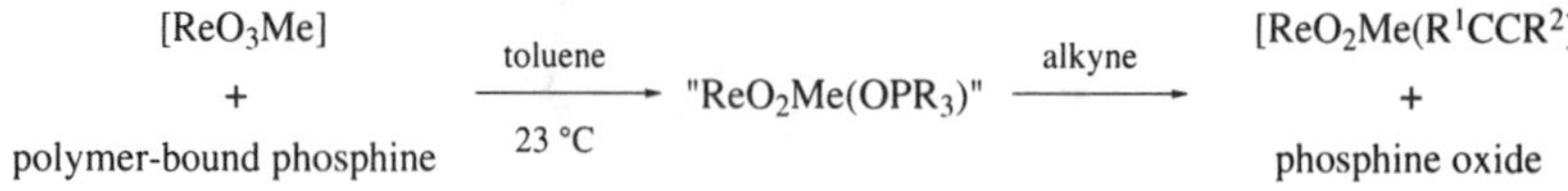

$$[\text{ReO}_3\text{Me}] \; + \; \text{polymer-bound phosphine} \xrightarrow[\;23\,°C\;]{\text{toluene}} \text{"ReO}_2\text{Me(OPR}_3\text{)"} \xrightarrow{\text{alkyne}} [\text{ReO}_2\text{Me(R}^1\text{CCR}^2\text{)}] \; + \; \text{phosphine oxide}$$

R$^1$ = R$^2$ = H, Me, Et, Ph, TMS; R$^1$ = H, R$^2$ = Et, Pr$^n$, Ph; R$^1$ = Me, R$^2$ = Bu$^t$, Ph

**Scheme 3**

Diamagnetic M[ReO$_2$(CH$_2$Bu$^t$)$_2$] (M = [Li(MeCN)$_2$], [Na(MeCN)]) compounds are prepared by reducing the rhenium(VI) dimer [{Re(μ-O)O(CH$_2$Bu$^t$)$_2$}$_2$] with alkali metal, and the [NEt$_4$]$^+$ salt is prepared by cation exchange.[28] [Mg(THF)$_2$][ReO$_2$(aryl)$_2$] (aryl = Mes or Xyl) is synthesized by reacting [ReO$_4$]$^-$ or [Re$_2$O$_7$] with 7 equiv. of MgBr(aryl).[29] The [ReO$_2$R$_2$]$^-$ compounds are readily oxidized, yielding [{Re(μ-O)O(CH$_2$Bu$^t$)$_2$}$_2$] and [ReO$_2$(Mes)$_2$]. The [ReO$_2$(CH$_2$Bu$^t$)$_2$]$^-$ anions exhibit highly distorted tetrahedral geometries (∠O–Re–O = 127–138° and ∠C–Re–C = 81–84°) that are attributed to electronic factors.[28]

Reduction of Re$^{VI}$ [{Re(μ-O)OMe$_2$}$_2$] quantitatively produces green oxygen-sensitive [{Re(μ-O)OMe$_2$}$_2$]$^-$, which upon oxidation with O$_2$ gives (4), [CoCp$_2$][ReO$_4$] and methane.[30] Compound (4) has an average oxidation state of $5\frac{2}{3}$ and Re–Re distances (0.265 nm) that are normal for single bonds. Formally, (4) is the product of [{Re(μ-O)OMe$_2$}$_2$] and [ReO$_2$Me$_2$]$^-$. Compound (5) (R = CH$_2$TMS), prepared serendipitously from the reaction of [Re(NPh)Cl$_3$(PMe$_3$)$_2$] with Mg(CH$_2$TMS)$_2$,[31] has a rare unsupported M–M double bond linking ReO$_2$R and [Re(PMe$_3$)$_4$]$^+$ fragments. The [Re(PMe$_3$)$_4$]$^+$ fragment is additionally complexed via a long Re–O bond to [ReO$_2$R$_2$]$^-$.

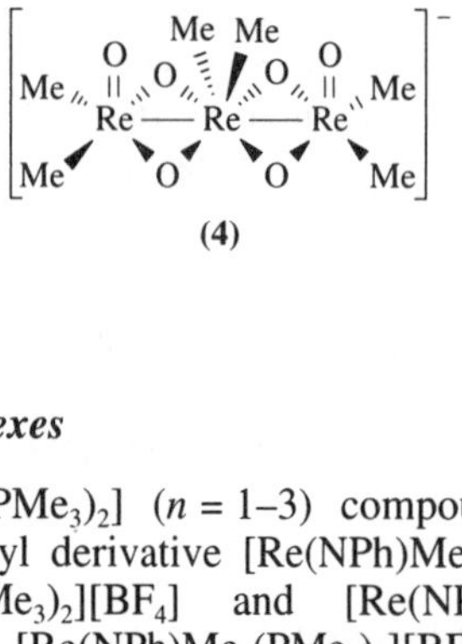

(4) 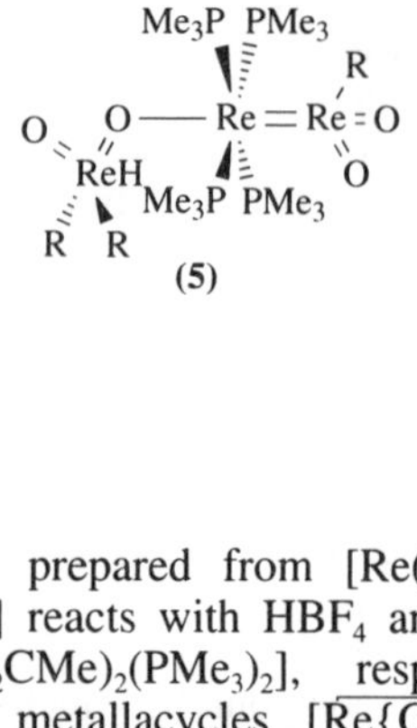                              (5)

### 10.2.2.2 Imido complexes

[Re(NPh)Cl$_n$Me$_{3-n}$(PMe$_3$)$_2$] (n = 1–3) compounds are prepared from [Re(NPh)Cl$_3$(PMe$_3$)$_2$] and MgMe$_2$.[31] The trimethyl derivative [Re(NPh)Me$_3$(PMe$_3$)$_2$] reacts with HBF$_4$ and acetic acid to give [Re{N(H)Ph}FMe$_2$(PMe$_3$)$_2$][BF$_4$] and [Re(NPh)Me(O$_2$CMe)$_2$(PMe$_3$)$_2$], respectively, and with [Ph$_3$C][BF$_4$] to give [Re(NPh)Me$_2$(PMe$_3$)$_2$][BF$_4$]. The metallacycles [$\overline{\text{Re}\{\text{C(Bu}^i\text{)=}}$C(Bu$^i$)C=C(H)-Pr$^i$}(Ndip){OCMe(CF$_3$)$_2$}(Bu$^i$CCBu$^i$)] and [$\overline{\text{Re}\{\text{C(R}^1\text{)=}}$C(R$^1$)C=C(H)R$^2$}(Ndip){OCH(CF$_3$)$_2$}py] (R$^1$ = Et, R$^2$ = Me; R$^1$ = Bu$^i$, R$^2$ = Pr$^i$) (dip = 2,6-diisopropylphenyl) have been obtained from metallacyclobutadiene compounds (see Section 10.3.3.2).

Compound (6) is the starting material for [Re(Ndip)$_2$(CH$_2$Bu$^t$)L] compounds (Scheme 4).[32,33] The structure of the acetone complex (7), which resembles the [ReO$_2$R(alkyne)] compounds mentioned in the previous section, is typical of the adducts. Compound (6) has also been prepared from the reaction of 0.5 equiv. of Zn(CH$_2$Bu$^t$)$_2$ with [Re(Ndip)$_2$Cl(py)$_2$], a compound that reacts as well with 2-butyne or Bu$^t$CH$_2$C≡CCH$_2$Bu$^t$ to give [Re(Ndip)$_2$Cl(alkyne)].[33,34] The [Re(Ndip)$_2$(CH$_2$Bu$^t$)L] compounds shown in Scheme 4 can also be synthesized from [Re(Ndip)$_2$(CH$_2$Bu$^t$)(PMe$_2$Ph)], which is made from (6) or [Re(Ndip)$_2$(CH$_2$Bu$^t$)Cl$_2$], PMe$_2$Ph, and zinc dust.[32,33] Equations (3) and (4) give products that are structurally analogous to (7).[34] An alternative electrophile in reaction (4) is H$^+$, delivered from [Et$_2$NH$_2$][OTf] (OTf = OSO$_2$CF$_3$).

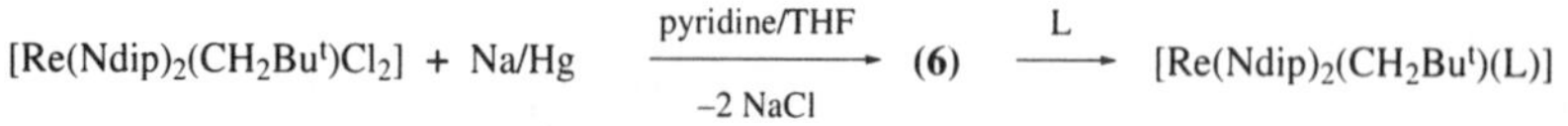

$$[\text{Re(Ndip)}_2(\text{CH}_2\text{Bu}^t)\text{Cl}_2] + \text{Na/Hg} \xrightarrow[-2\,\text{NaCl}]{\text{pyridine/THF}} (6) \xrightarrow{\text{L}} [\text{Re(Ndip)}_2(\text{CH}_2\text{Bu}^t)(\text{L})]$$

L = PMe$_2$Ph, MeC≡CMe, Me$_2$C=O, Bu$^t$(H)C=O, norbornene

**Scheme 4**

$$\text{(6)} \qquad \text{(7)} \qquad \text{(8)}$$

$$[\text{Re(Ndip)}_3\text{H}] + L \longrightarrow [\text{Re(Ndip)}_2\{\text{N(H)dip}\}L] \qquad (3)$$

$$L = HC{\equiv}CH,\ MeC{\equiv}CMe,\ Bu^t(H)C{=}O,\ H_2C{=}CH_2,\ norbornene$$

$$Na[\text{Re(Ndip)}_2(R^2C{\equiv}CR^2)] + R^1\text{-}X \xrightarrow{-NaX} [\text{Re(Ndip)}_2R^1(R^2C{\equiv}CR^2)] \qquad (4)$$

$$R^1\text{-}X = MeI\ or\ 2,4,6\text{-}Me_3H_3C_6CH_2Cl;\ R^2 = CH_2Bu^t$$

$[\text{Re(Ndip)}_2(\text{CH}_2\text{Bu}^t)(\text{PMe}_2\text{Ph})]$ adds the small $\pi$-acceptor ligands ethene, ethyne, and CO to give five-coordinate $[\text{Re(Ndip)}_2(\text{CH}_2\text{Bu}^t)(\text{PMe}_2\text{Ph})(L)]$ compounds[32] such as (8). The reaction with ethene is reversible, and the ethyne complex rearranges to give $[\text{Re(Ndip)}_2(\text{CH}_2\text{Bu}^t)\{\text{C(H)C(H)(PMe}_2\text{Ph})\}]$, a compound with a ylide ligand analogous to the one in $[\text{ReO}\{\text{C(H)C(H)PMe}_3\}R_3]$ shown in Scheme 1.

$[\text{Re(NBu}^t)_2(\text{Xyl})_2]^-$, an imido analogue of the $[\text{ReO}_2R_2]^-$ compounds, is formed on reduction of $[\text{Re(NBu}^t)_2(\text{Xyl})_2]$ with magnesium or Na/Hg.[35]

The isolobal relationship between bent $d^2$ $[\text{Re(NR)}_2]^+$ and $MCp_2$ fragments[32] provides a convenient predictive tool for the bisimido fragment reactivity and a rationalization for much of the structural chemistry.

### 10.2.2.3 Nitrido complexes

$[\text{Re(N)R}_2(\text{PPh}_3)_2]$ (R = Me, C$\equiv$CBu$^t$, Ph, $p$-Tol) compounds are formed from $[\text{Re(N)Cl}_2(\text{PPh}_3)_2]$ and 2 equiv. of LiR.[3,36] The crystal structure of $[\text{Re(N)Me}_2(\text{PPh}_3)_2]$ reveals a trigonal-pyramidal/square-pyramidal structure with the nitride ligand in the equatorial/apical position. Excitation ($\lambda > 400$ nm) of the solid compounds at 23 °C and 77 K gives red emission, and $[\text{Re(N)(C}\equiv\text{CBu}^t)_2(\text{PPh}_3)_2]$ in benzene exhibits room-temperature luminescence.

### 10.2.2.4 Complexes without oxo, imido, or nitrido ligands

$[\text{ReMe}_6]$ reacts with an excess of NO at low temperature in hydrocarbon solution to yield diamagnetic Re$^V$ $[\text{ReMe}_6(\text{NO})]$.[23]

The interesting rhenium(III) hydrido–ethene complex $[\text{ReH}_3(\text{C}_2\text{H}_4)_2(\text{PR}_3)]$ (PR$_3$ = PPhPr$^i_2$, P(cyclopentyl)$_3$) oxidatively adds H$_2$ to give $[\text{ReH}_5(\text{C}_2\text{H}_4)(\text{PR}_3)_2]$.[37] A related compound, $[\text{ReH}_5(\eta^2\text{-}2,3\text{-dimethylbutadiene})(\text{dppe})]$, is believed to be the precursor to the 18-electron allyl complex $[\text{ReH}_4(\eta\text{-}2\text{-isopropylallyl})(\text{dppe})]$.[38] $[\text{ReH}_4(\eta\text{-}2\text{-isopropylallyl})(\text{dppe})]$ has a pentagonal-bipyramidal structure with one phosphorus atom and four hydrides in the pentagonal plane.

Protonation of $[\text{ReH}_3(\text{CO})(\text{PMe}_2\text{Ph})_3]$ with HBF$_4$ at low temperature produces, according to an NMR study, an equilibrium mixture of the rhenium(III) dihydrogen complex $[\text{Re(H}_2)\text{H}_2(\text{CO})(\text{PMe}_2\text{Ph})_3]^+$ and Re$^V$ $[\text{ReH}_4(\text{CO})(\text{PMe}_2\text{Ph})_3]^+$.[39] The compounds are proposed to interconvert ($\Delta G^{\ddagger} = 51.5$ kJ mol$^{-1}$) by transfer of a proton from the $\eta^2$-H$_2$ ligand to an adjacent terminal hydride. At 0 °C dihydrogen is irreversibly and rapidly lost.

Reaction of ReCl$_5$ with 1 equiv. of PhC$\equiv$CPh in the presence of POCl$_3$ is reported to give $[\text{ReCl}_5(\text{PhCCPh})]$.[40]

## 10.2.3 Rhenium(VI)

### 10.2.3.1 Reactions of [ReMe$_6$]

Excess Bu$^t$NC reacts with $[\text{ReMe}_6]$ at −78 °C to give the triple insertion product $[\text{Re}\{\eta^1\text{-C(Me)=NBu}^t\}_3(\text{CNBu}^t)_2]$.[41] Reactions of NO with $[\text{ReMe}_6]$ and Li$_2[\text{ReMe}_8]$ have also been carried out, yielding diamagnetic rhenium(V) compounds.[23]

### 10.2.3.2 Oxo and sulfido complexes

All of the reported monoxoo compounds are of general formula [ReOR$_4$] and have square-pyramidal structures with an apical oxo ligand. Their synthesis generally involves oxidation of isolated or inferred [ReOR$_4$]$^-$ compounds with oxygen or H$_2$O$_2$.[24,25,29] New compounds synthesized by this method include R = Mes, o-Tol, and p-MeOC$_6$H$_4$ derivatives.[29] [ReO(o-Tol)$_4$] has also been prepared by adding O$_2$ to extremely oxygen-sensitive Re$^{IV}$ [Re(o-Tol)$_4$].[4]

Dioxo [ReO$_2$(aryl)$_2$] (aryl = Xyl, Mes) compounds have been prepared by oxidizing [ReO$_2$(aryl)$_2$]$^-$ and,[29,42] more conveniently, by treating [ReO$_3$(O-TMS)] with 2 equiv. of Grignard reagent.[42] The [ReO$_2$(aryl)$_2$] compounds have distorted tetrahedral geometries with large O–Re–O and small C–Re–C angles. [ReO$_2$(Mes)$_2$] reacts with NO to give rhenium(VII) [ReO$_3$(NMes$_2$)] via a proposed [ReO$_2$(Mes){η$^2$-ON(Mes)}] intermediate and with NO$_2$ to yield [ReO$_3$(Mes)].[43]

In contrast to the monomeric aryl derivatives, alkyl ReO$_2$R$_2$ compounds are all *syn* dimers with bridging oxo groups and a Re–Re single bond, (9). Derivatives where R = Me, Et, CH$_2$TMS, CH$_2$Bu$^t$, and CH$_2$CMe$_2$Ph have been synthesized.[28a,44-6] The syntheses generally involve interaction of an oxide of rhenium with AlR$_3$, ZnR$_2$, or other alkylating agents (e.g., Equation (5)).

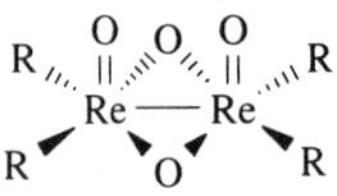

$$\text{(9)} \qquad \text{(10)} \qquad \text{(11)}$$

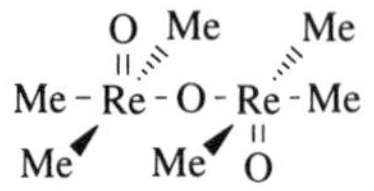

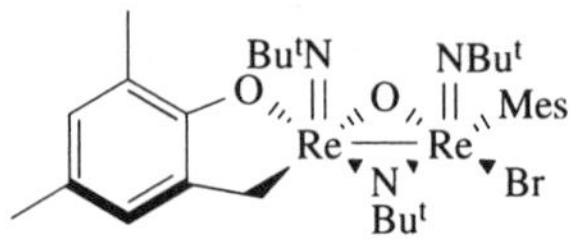

$$4\,[Re_2O_7] + 3\,ZnEt_2 \xrightarrow[-78\,°C]{THF/ether} [\{Re(\mu\text{-}O)(O)Et_2\}_2] + 3\,Zn(ReO_4)_2 + C_4H_{10} \qquad (5)$$

Compound (9) is a precursor to numerous other organorhenium compounds. Reduction of [{Re(μ-O)(O)Me$_2$}$_2$] produces [{Re(μ-O)(O)Me$_2$}$_2$]$^-$ quantitatively,[30] and reduction of [{Re(μ-O)O(CH$_2$Bu$^t$)$_2$}$_2$] gives [ReO$_2$(CH$_2$Bu$^t$)$_2$]$^-$.[28] Oxidation of [{Re(μ-O)O(CH$_2$Bu$^t$)$_2$}$_2$] with bromine cleaves the Re–Re bond to yield [ReO$_2$Br(CH$_2$Bu$^t$)$_2$],[47] and pyridine N-oxide formally inserts oxygen into the Re–Re bond of (9) to give rhenium(VII) compounds with one bridging oxo group (Scheme 5).[45,48] Reaction of [{Re(μ-O)O(CH$_2$Bu$^t$)$_2$}$_2$] with 1 equiv. of H$_2$S in pyridine gives *syn*-[(Bu$^t$CH$_2$)$_2$(O)Re(μ-O)(μ-S)Re(O)(CH$_2$Bu$^t$)$_2$] and with 2 equiv. of H$_2$S yields a mixture of *syn*- and *anti*-[{Re(μ-S)O(CH$_2$Bu$^t$)$_2$}$_2$], both of which have been structurally characterized.[49] The *syn* and *anti* isomers interconvert in pyridine solution. [{Re(μ-O)(O)Me$_2$}$_2$] catalyzes the ring-opening metathesis polymerization of cyclopentene in the presence of SnMe$_4$/AlCl$_3$.[44]

$$[\{ReO_2R_2(py)\}_2(\mu\text{-}O)] \xleftarrow[R = Me]{pyN\text{-}oxide} \text{(9)} \xrightarrow[R = CH_2Bu^t]{pyN\text{-}oxide} [\{ReO_2R_2\}_2(\mu\text{-}O)]$$

**Scheme 5**

The methyl compound (10) is prepared by reacting [{Re(μ-O)(O)Me$_2$}$_2$] or [{ReO(O-CMe$_2$CMe$_2$O)Me}$_2$(μ-O)] with ZnMe$_2$, or [ReO$_3$(O-TMS)] with AlMe$_3$,[44,45] and new preparations of [{ReO(CH$_2$TMS)$_3$}$_2$(μ-O)] have been reported.[24,28a] Potentially rich sources of other derivatives are [{ReOCl$_2$Me(OPPh$_3$)}$_2$(μ-O)], prepared from [ReO$_3$Me] and TMS-Cl/PPh$_3$,[26] and [{ReO(OCR$_2$CR$_2$O)MeL}$_2$(μ-O)] (R = Me, L = none; R = H, L = py), prepared by reductive condensation between [ReO$_3$Me] and HOCR$_2$CR$_2$OH in methanol.[50] All of the [{ReOR$_3$}$_2$(μ-O)] compounds have *anti* structures and are diamagnetic, the two "d" electrons residing in a b$_g$ MO that is the antisymmetric combination of rhenium d$_{yz}$ orbitals (y-axis along the Re–O–Re vector).

[{ReO(CH$_2$TMS)$_3$}$_2$(μ-O)] is readily oxidized by O$_2$, DMSO, or pyridine N-oxide to give [ReO$_2$(CH$_2$TMS)$_3$] and is reduced by phosphine to give [ReO(CH$_2$TMS)$_3$(PR$_3$)].[13]

### 10.2.3.3 Imido complexes

[Re(NBu$^t$)$_2$(aryl)$_2$] (aryl = Mes, Xyl) have been prepared by treating [Re(NBu$^t$)$_2$Cl$_3$] with excess Grignard reagent.[35] [Re(NBu$^t$)$_2$(Xyl)$_2$] is a distorted tetrahedron with large N–Re–N (122°) and small C–Re–C (102°) angles similar to the analogous d$^1$ and d$^2$ oxo compounds mentioned previously. [Re(NBu$^t$)$_2$(Xyl)$_2$] is reduced by Na/Hg or magnesium to give [Re(NBu$^t$)$_2$(Xyl)$_2$]$^-$ and is oxidized by

AgX to yield $[Re(NBu^t)_2(Xyl)_2]X$ (X = $PF_6$ or $CF_3SO_3$), a rare example of an organometallic rhenium(VII) cation. $[Re(NBu^t)_2(Xyl)_2]$ reacts with NO to form diamagnetic $[Re(NBu^t)_2(Xyl)_2(NO)]$.

Compound (**11**) is one product formed in the reaction of "aged" $[Re(NBu^t)_2Cl_3]$ or $[Re(NBu^t)_2Cl(OH)_2]$ with MgBr(Mes) (compare with compound (**9**)).[35]

## 10.2.4 Rhenium(VII)

### 10.2.4.1 Oxo complexes

Herrmann and co-workers have published several high-yield practical syntheses of the important compound (**12**) (Scheme 6) based on the use of $SnMe_4$ as a methylating agent;[3,19,44,51,52] the best one appears to be Equation (6), where the driving force is formation of Sn–O bonds.[52] Numerous derivatives (Table 1) have been prepared by reacting $[Re_2O_7]$ with dialkyl and diarylzinc reagents at low temperatures.[44,51-5] Unbranched, acyclic alkyl compounds are more thermally stable than branched derivatives, and the thermal stability decreases with increasing chain length. $[ReO_3Et]$, for example, undergoes only slow decomposition via radical paths (i.e., not β-hydride elimination) at 60 °C. Highly substituted aryls are thermally more stable than the phenyl compound, which decomposes slowly even at room temperature.

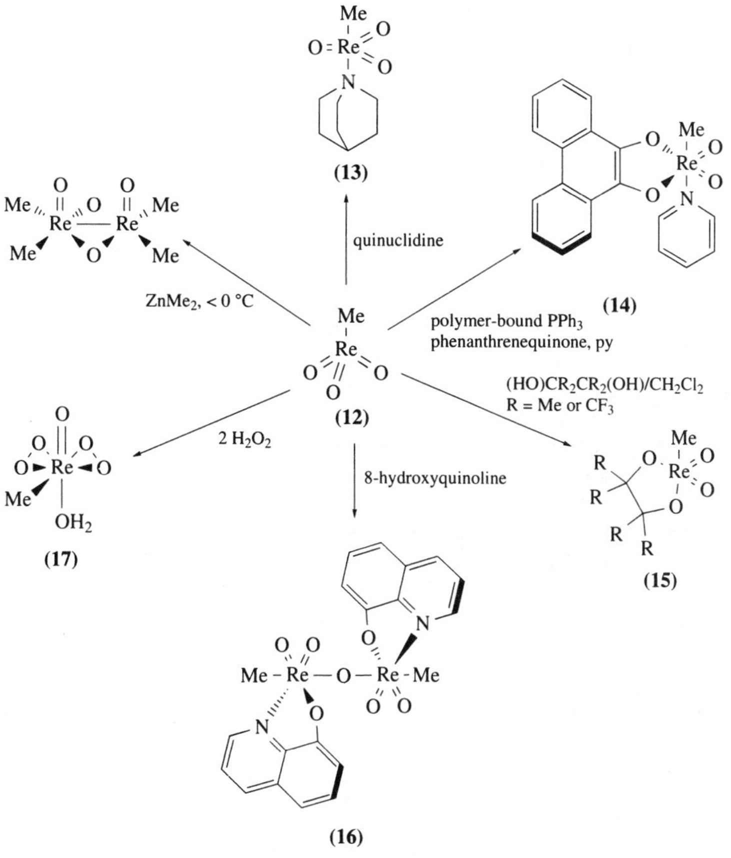

**Scheme 6**

$$[Re_2O_7] + 2\,SnMe_4 + \text{perfluoroglutaric anhydride} \xrightarrow{\ MeCN\ } 2\,(\mathbf{12}) + Me_3SnO(O=)C(CF_2)_3C(=O)OSnMe_3 \quad (6)$$

The $[ReO_3R]$ compounds have been described as Lewis-acidic 14-electron systems. This description does not take into account the ability of the oxo groups to contribute up to six electrons each to the metal, but the compounds do form Lewis-base adducts (Scheme 6 and Table 1).[19,51,53,55-7] The quinuclidine adduct (**13**) has a characteristic structure, but, for example, the aniline adduct of (**12**) crystallizes with two isomers in the unit cell, one similar to (**13**) and the other with the aniline *trans* to an oxo group. Octahedral $[ReO_3MeL_2]$ compounds are also known (Table 1).

**Table 1**  Selected [ReO₃R] compounds and ligand adducts.

| Compound | Ref. |
|---|---|
| [ReO₃R] | |
| R = Me, Et, Buⁱ, CH₂TMS, CH₂Buᵗ, Buᵗ, Prⁿ, *n*-pentyl, *n*-heptyl, isopentyl, 2-methyl(butyl), 2-ethyl(hexyl), cyclopropyl | 44,51,52,53 |
| R = Ph, Mes,ᵃ Xyl, *p*-Tol, C₆H₂-2,6-Me₂-4-(O-TMS),ᵃ C₆H₂-2,6-Me₂-4-(OH) | 54,55 |
| [ReO₃RL] | |
| R = Me, Et,ᵃ Buⁱ'ᵃ CH₂TMS,ᵃ CH₂Buᵗ,ᵃ Buᵗ, Prⁿ, *n*-pentyl, *n*-heptyl, isopentyl, 2-methyl(butyl), 2-ethyl(hexyl), cyclopropyl;ᵃ L = quinuclidine | 19,51,53,56 |
| R = Ph, C₆F₅, Mes, Xyl, *p*-Tol, C₆H₂-2,6-Me₂-4-(O-TMS),ᵃ C₆H₂-2,6-Me₂-4-(OH); L = quinuclidine | 55 |
| R = Ph; L = THFᵃ | 55 |
| R = Me; L = py, pyridine *N*-oxide, aniline,ᵃ 2-methoxyaniline, 3-methoxyaniline, 4-methoxyaniline, Mn(η-C₄H₄N)(CO)₃, Me₂NCH₂CN, Me₂NCH₂C(=O)Me | 19,51,56 |
| [ReO₃RL₂] | |
| R = Me; L₂ = (NH₃)₂, bipy, 4,4′-Buᵗbipy, 2-aminomethylpyridine, *N*-(2-aminoethyl)pyrrolidine, en, 1,4-piperazine | 19,51,56,57 |
| R = CH₂TMS, L₂ = en | 53 |
| [{ReO₃Me}₂(μ-L)] | 19,51,56 |
| L = 1,4-diazabicyclo[2.2.2]octane (dabco), *N,N*′-dimethyl-1,4-piperazine, hexamethyltetraamine | |
| M[ReO₃MeX] | 57 |
| M = [NBuⁿ₄]; X = Cl, Br | |
| M = [PPN]; X = Cl | |
| M₂[ReO₃MeX₂] | 57 |
| M = [NBuⁿ₄]; X = Cl, Br | |
| M = [PPN]; X = Cl | |

ᵃ Structurally characterized.

The oxo groups in (**12**) are converted to imido ligands with isocyanate (see Section 10.2.4.2), and photolysis of (**12**) leads to Re–Me bond homolysis via ligand-to-metal charge transfer.[58] Other reactions that result in reduction of (**12**) have been reported; for example, ZnMe₂ reacts to give [{Re(μ-O)(O)Me₂}₂][44,45] and polymer-bound triphenylphosphine removes an oxo ligand to form "ReO₂Me" *in situ*, which can then be captured by alkyne[11,26] (Scheme 3) or reoxidized with phenanthrenequinone to give (**14**).[20] Ethylene glycol and pinacol react with (**12**) in methanol to give rhenium(VI) oxo-bridged compounds but reactions with pinacol or perfluoropinacol in CH₂Cl₂ give rhenium(VII) (**15**).[50,59] A related water elimination reaction forms (**16**) in Scheme 6.[21,22]

Reactions of (**12**) with catechols and 2-aminothiophenol in the presence of pyridine or X⁻ (X⁻ = Br⁻, Cl⁻) causes elimination of water and formation of six-coordinate chelates that are akin to (**14**) in Scheme 6.[20,22] Thiols react with (**12**) under mild conditions, eliminating 2 equiv. of H₂O to yield [Re(O)Me(SPh)₄] and [Re(O)Me(S̄S̄)₂] (S̄S̄ = 1,2-(S)₂C₆H₄, 1,2-(S)₂C₆H₃-4-Me) and, similarly, 2-aminophenol gives [Re(O)Me(1,2-(O)(HN)C₆H₄)₂].[20-2] Rhenium(VII) compounds with only one oxo or imido ligand are unprecedented, suggesting there may be S–S interactions in the thiolates and an imido group in the amidophenolate. At 50 °C, [Re(O)Me(SPh)₄] releases PhSSPh to form [{Re(O)Me(SPh)(μ-SPh)}₂].

Over time, (**12**) dissolved in water gives a strongly acidic solution containing perrhenic acid, evolution of methane, and precipitation of a solid of formula C₀.₉₁H₃.₃ReO₃.₀.[60] The solid is proposed to be a polymeric sheet of (**12**) with oxo bridging groups and every tenth or so rhenium atom missing a methyl ligand. In the presence of H₂O the polymer is thermally or photochemically converted to [ReO₃]. Interestingly, (**12**) is released from the polymer under pressure, and the pyridine adduct of (**12**) is formed when the polymer is placed in pyridine.

Compound (**12**) has been used in catalytic alkene metathesis, aldehyde alkenation, and alkene epoxidation reactions.[2d] In solution, (**12**) catalyzes the ring-opening metathesis polymerization of norbornene and the metathesis of nonfunctionalized acyclic alkenes in the presence of RₙAlCl₃₋ₙ (R = Me, Et; *n* = 1, 2).[44,61] Supported on γ-alumina and silica–alumina, (**12**) efficiently metathesizes functionalized alkenes,[61,62] and on Nb₂O₅ it metathesizes acyclic alkenes.[63,64] When [ReO₃(CD₃)] is used in the former case, CD₂=CHR products are obtained but how the putative Re=CD₂ intermediate is formed is not clear. In the case of the Nb₂O₅ supported catalyst a Re=CH₂ species is apparently not formed from the methyl ligand of (**12**).[16,65]

Compound (**12**) catalyzes aldehyde alkenation according to Equation (7), including reactions involving diazoacetates and diazomalonates that give alkenes with terminal carboxyl groups.[66] The key steps in the catalytic cycle are proposed to be phosphine reduction of (**12**) to [ReO₂Me(OPR¹₃)]; oxidation of [ReO₂Me(OPR¹₃)] with N₂CR²R³ to give [ReO₂(=CR²R³)Me], phosphine oxide, and N₂;

addition of aldehyde to Re=CR$^2$R$^3$ to form a rhenaoxacyclobutane; and finally, release of alkene with reformation of (**12**).

$$R^4C(=O)H + R^2R^3C=N_2 + PR^1_3 \xrightarrow{\text{cat. (12)}} R^4CH=CR^2R^3 + N_2 + P(O)R^1_3 \qquad (7)$$

Compound (**12**) and other alkyl derivatives catalyze the epoxidation of alkenes by H$_2$O$_2$ under mild conditions, and direct the subsequent chemistry of the epoxides.[51,67,68] The active species in the epoxidation, (**17**), is formed quantitatively when H$_2$O$_2$ reacts with (**12**) (Scheme 6).[68] Alkenes, such as 2,3-dimethylbut-2-ene, react with (**17**) to form epoxide and a monoperoxo complex (**18**) (Equation (8)), which can also be prepared by adding just 1 equiv. of H$_2$O$_2$ to (**12**). Compound (**18**) does not catalyze epoxidation efficiently, but it reacts with H$_2$O$_2$ to reform (**17**), thus continuing the cycle. In the absence of excess H$_2$O$_2$, the epoxide and water (from (**17**)) react to give (**12**) and diol, which then react to form (**15**) (Scheme 6).

$$(17) + Me_2C=CMe_2 \longrightarrow [ReO_2(\eta^2\text{-}O_2)Me(H_2O)] + \text{epoxide} \qquad (8)$$

$$(18)$$

[ReO$_2$(CH$_2$Bu$^t$)$_2$X(py)] (X = Br or F) has been used to make rhenium(VII) alkyl, aryl, alkoxide and thiolate derivatives,[27,47,48] including [ReO$_2$(CH$_2$Bu$^t$)$_2$(SR)] (R = Me or Ph), [ReO$_2$(CH$_2$Bu$^t$)$_2$(OR)] (R = Me, CH$_2$Bu$^t$ or Bu$^t$),[48] and [ReO$_2$(CH$_2$Bu$^t$)$_2$R] (R = Me, CH$_2$Bu$^t$, CH$_2$TMS, Ph).[27,47,69] [ReO$_2$R$_3$] compounds where R = Me[3,45] or CH$_2$TMS[13] can be prepared by oxidation of [{ReOR$_3$}$_2$(μ-O)] with amine *N*-oxide. Thermolysis or photolysis of [ReO$_2$(CH$_2$Bu$^t$)$_3$] causes α-hydrogen abstraction, giving [ReO$_2${C(H)Bu$^t$}(CH$_2$Bu$^t$)],[70] whereas thermolysis or photolysis of [ReO$_2$(CH$_2$Bu$^t$)$_2$Ph] gives "ReO$_2$(CH$_2$Bu$^t$)" and neopentylbenzene,[27] and photolysis of [ReO$_2$(CH$_2$TMS)$_3$] yields (**9**).[13] The [ReO$_2$(CH$_2$Bu$^t$)$_2$(OR)] compounds decompose in room light to give a mixture of products including [{ReO$_2$(CH$_2$Bu$^t$)$_2$}$_2$(μ-O)], which is better prepared by stirring [ReO$_2$(CH$_2$Bu$^t$)$_2$Br(py)] with Ag$_2$O, hydrolyzing [ReO$_2$(CH$_2$Bu$^t$)$_2$(OBu$^t$)], or oxidizing (**9**) with pyridine *N*-oxide.[48] Pyridine *N*-oxide and (**9**) have also been used to synthesize the methyl derivative [{ReO$_2$Me$_2$(py)}$_2$(μ-O)],[45] and the preparation of a related compound (**16**) is shown in Scheme 6. The rhenium(VII) dioxo centers in all of these compounds have *cis*-oxo ligands.

### 10.2.4.2 Imido complexes

The triimido compounds [Re(NR$^2$)$_3$R$^1$], where R$^1$ = *o*-Tol, Xyl, Mes, Me, Et, C≡CPh, CH$_2$CH=CH$_2$, and CH$_2$Bu$^t$, are prepared from readily available [Re(NR$^2$)$_3$(O-TMS)] and Grignard or lithium reagents.[42,71,72] Also, (**12**) reacts with 3 equiv. of 2,6-dimethylphenylisocyanate in refluxing toluene to give [Re(NXyl)$_3$Me],[73] and trigonal-planar [Re(Ndip)$_3$]$^-$ reacts rapidly with alkyl halides to give [Re(Ndip)$_3$R], R = Me, CH$_2$Mes, CH$_2$Cl, and CHPh(Cl).[34,74] [Re(NXyl)$_3$Me] exists as a monomer in nonpolar solvents and as a dimer with two bridging imido ligands in polar solvents.[71] In contrast to analogous trioxo compounds, [Re(NR$^2$)$_3$R$^1$] complexes do not form adducts of the type [Re(NR$^2$)$_3$R$^1$(L)].

One of the imido ligands in the triimido compounds can be removed as an amine in protonation reactions; for example, [Re(NXyl)$_3$Me] reacts with 2 equiv. of PhSH to afford [Re(NXyl)$_2$Me(SPh)$_2$] and with catechol and pyridine to give [Re(NXyl)$_2$Me(O$_2$C$_6$H$_4$)(py)].[73] [Re(NBu$^t$)$_2$Me(O$_2$C$_6$H$_4$)] is synthesized in a similar way. [Re(NR$^2$)$_3$R$^1$] (R$^2$ = Bu$^t$, R$^1$ = *o*-Tol, Xyl, Mes;[35,42] R$^2$ = Xyl, 2,6-Cl$_2$C$_6$H$_3$, R$^1$ = CH$_2$Bu$^t$)[72] reacts with 3 equiv. of HCl to give diimido [Re(NR$^2$)$_2$Cl$_2$R$^1$] and [R$^2$NH$_3$]Cl, a reaction which illustrates the remarkable stability of the Re–R bond to acid. Similarly, HCl and py·HCl react with [Re(NXyl)$_3$Me] to form [Re(NXyl)$_2$Cl$_3$Me]$^-$ and [Re(NXyl)$_2$Cl$_2$Me(py)], respectively.[73]

A direct route to diimido alkyl derivatives involves the reactions of [Re(NR$^1$)$_2$Cl$_3$] with 1 equiv. of MgBr(aryl) (aryl = *o*-Tol, Xyl, Mes) to give [Re(NBu$^t$)$_2$Cl$_2$(aryl)],[42] and with 2 equiv. of MgBr(*o*-Tol) to yield [Re(NBu$^t$)$_2$Cl(*o*-Tol)$_2$].[35] Similarly, [Re(Ndip)$_2$Cl$_3$(py)] reacts with Zn(CH$_2$Bu$^t$)$_2$ to give [Re(Ndip)$_2$Cl$_2$(CH$_2$Bu$^t$)]. [Re(NBu$^t$)$_2$Cl(CH$_2$TMS)$_2$][72] and [Re(NBu$^t$)$_2$R$_3$] (R = Me, CH$_2$Ph, or CH$_2$TMS) compounds are synthesized using AlMe$_3$, MgCl(CH$_2$Ph), and Zn(CH$_2$TMS)$_2$.[75,76] The neopentyl derivative [Re(NBu$^t$)$_2$(CH$_2$Bu$^t$)$_3$] is not isolable; instead, it undergoes α-hydrogen abstraction with formation of [Re(NBu$^t$)$_2${C(H)Bu$^t$}(CH$_2$Bu$^t$)].[75,76] The mixed methyl–neopentyl derivative [Re(NXyl)$_2$(CH$_2$Bu$^t$)$_2$Me] is formed from Mg(CH$_2$Bu$^t$)Cl and [Re(NXyl)$_2$Cl$_2$Me(py)].[73] Rare examples of rhenium(VII) cations, [Re(NBu$^t$)$_2$(aryl)$_2$]$^+$ (aryl = Xyl, *o*-Tol), have been synthesized by reacting [Re(NBu$^t$)$_2$Cl(*o*-Tol)$_2$] with Ag[PF$_6$] and [Re(NBu$^t$)$_2$(aryl)$_2$] with AgX (X = PF$_6$ or CF$_3$SO$_3$).[35]

The structures of the diimido complexes are trigonal-bipyramidal or square-pyramidal with *cis*-imido ligands. The most important reactions of the diimido alkyl compounds are their conversion to alkylidene complexes (Section 10.3.1) and their reduction in the presence of π-acids (Section 10.2.2.2).

### 10.2.4.3 Oxo–imido complexes

[Re(NBu$^t$)$_2$(O)(Mes)] is prepared by reacting [Re(NBu$^t$)$_2$Cl(OH)$_2$] or "aged" [Re(NBu$^t$)$_2$Cl$_3$] with MgBr(Mes) and by reacting [Re(NBu$^t$)$_2$Cl$_2$(Mes)] with Ag$_2$O.[35] [Re(Ndip)$_2$(O)(CH$_2$Bu$^t$)] is formed when [Re(Ndip)$_2$Cl$_2$(CH$_2$Bu$^t$)] or [Re(Ndip)$_2${C(H)Bu$^t$}Cl] is treated with OH$^-$.[72,77] The analogous methyl derivative, [Re(Ndip)$_2$(O)Me], is synthesized from (**12**) and 2 equiv. of 2,6-diisopropylphenylisocyanate in DME.[71] It exists as a monomer in nonpolar solvents and as a dimer in polar solvents.

[Re(NBu$^t$)$_2$(Xyl)$_2$] reacts with 1 equiv. of NO to yield [Re(NBu$^t$)$_2$(O)(Xyl)], the key step in the reaction being insertion of NO into the Re–C bond. When an excess of NO is used, the dimers [{Re(NBu$^t$)(μ-O)(O)R}$_2$] (R = Mes or Xyl) are formed according to Equation (9).[35] An anologous rhenium(VII) dinuclear compound with oxo and imido bridges, [{Re(Ndip)(O)Me}$_2$(μ-Ndip)(μ-O)], is formed from (**12**) and 1.5 equiv. of 2,6-diisopropylphenylisocyanate in DME.[71]

$$[Re(NBu^t)_2R_2] + 3\,NO \xrightarrow[\substack{80\text{–}90\%}]{\text{hexane}} 1/2\,[\{Re(NBu^t)(\mu\text{-}O)(O)R\}_2] + N_2O + Bu^tN{=}NR \tag{9}$$

## 10.3 ALKYLIDENE/ALKYLIDYNE COMPLEXES

Numerous alkylidene and alkylidyne complexes have been prepared because of their potential use in catalytic alkene and alkyne metathesis reactions. In Sections 10.3.1 and 10.3.2 the synthesis and nonmetathesis reactivity of alkylidene/alkylidyne complexes are reviewed, and in Section 10.3.3, metathesis reactions of [Re(CBu$^t$){C(H)R$^1$}(OR$^2$)$_2$], [Re(CBu$^t$)(Ndip){OCMe(CF$_3$)$_2$}$_2$] and closely related complexes are discussed. Alkylidene intermediates have also been invoked or are presumed in metathesis and aldehyde alkenation reactions involving (**12**).

### 10.3.1 Rhenium(VII)

Schrock and co-workers, motivated by their successful synthesis of group 5 and 6 alkylidenes, and their use of group 6 alkylidenes in alkene metathesis reactions, were the first to prepare rhenium(VII) alkylidenes, initially using [Re(NBu$^t$)$_2$R$_3$] compounds as precursors.[75,76] For example, photolysis of [Re(NBu$^t$)$_2$(CH$_2$TMS)$_3$] through Pyrex glass yields [Re(NBu$^t$)$_2${C(H)TMS}(CH$_2$TMS)] as a yellow crystalline solid. Photolysis of the benzyl derivative [Re(NBu$^t$)$_2$(CH$_2$Ph)$_3$] gives a mixture of compounds including the unstable benzylidene complex, but photolysis of [Re(NBu$^t$)$_2$Me$_3$] does not give a methylidene complex. Photochemical activation is not required in the synthesis of related neopentylidene and neophylidene compounds;[75,76,78] they are spontaneously formed when [Re(NBu$^t$)$_2$Cl$_3$] or [Re(Naryl)$_2$Cl$_3$(py)] (aryl = dip or Xyl) are reacted with Grignard or dialkyl zinc reagents, presumably via [Re(NR$^2$)$_2$R$^1$$_3$] intermediates. The ease of α-hydrogen abstraction in the rhenium system appears to vary in the order CH$_2$Bu$^t$ ~ CH$_2$CMe$_2$Ph > CH$_2$TMS > CH$_2$Ph ≫ Me, as is found in other early transition metal systems. The reaction sequence shown in Scheme 7 is a general route to analogous alkoxide derivatives, [Re(Ndip)$_2${C(H)Bu$^t$}(OR)] (R = CH(CF$_3$)$_2$, CMe(CF$_3$)$_2$ or dip).[72,77]

$$[Re(Ndip)_2Cl_2(CH_2Bu^t)] \xrightarrow[-dbu \cdot HCl]{\text{dbu/ether}} [Re(Ndip)_2\{C(H)Bu^t\}Cl] \xrightarrow[\text{ether, }-40\,°C]{\text{LiOR}} [Re(Ndip)_2\{C(H)Bu^t\}(OR)]$$

**Scheme 7**

Relatively few rhenium(VII) oxo alkylidenes have been prepared. Compound (**19**) is formed on photolysis of [ReO$_2$(CH$_2$Bu$^t$)$_3$] in pyridine[70] and by acid- or alumina-catalyzed hydrolysis of [Re(Ndip)$_2${C(H)Bu$^t$}(CH$_2$Bu$^t$)].[78] Hydrolysis without the acid or alumina treatment, or conproportionation between the diimido and dioxo complexes both give the oxo–imido complex [Re(Ndip)-(O){C(H)Bu$^t$}(CH$_2$Bu$^t$)].

In the structure of (**19**) the alkylidene ligand orientation is such that the *p*–π donor orbital of the ligand (counting CR$_2$$^{2-}$) is in the plane bisecting the O–Re–O angle.[70] Proton NMR studies suggest that

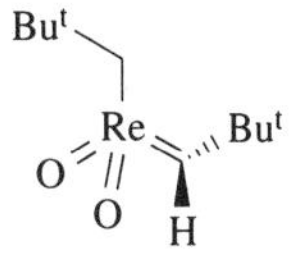

**(19)** 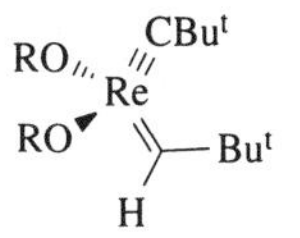      **(21)** R = CMe(CF$_3$)$_2$ 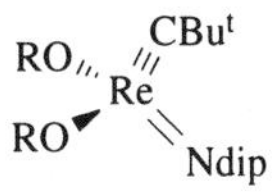      **(22)** R = CMe(CF$_3$)$_2$      **(24)**

**(23)** R = CH(CF$_3$)$_2$

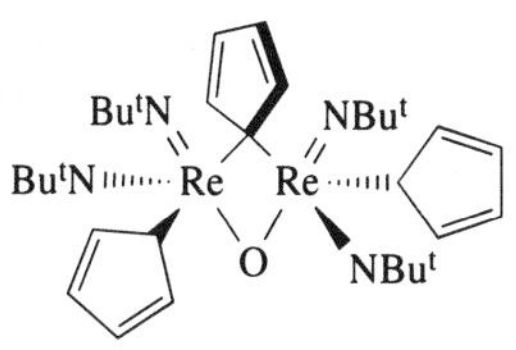

all of the related diimido alkylidenes have structures analogous to (**19**). The quinuclidene adduct of (**19**) has also been structurally characterized.[70] It resembles the trigonal-bipyramidal (tbp) Lewis-base adducts of (**12**) (Scheme 6).

[Re(NR$^1$)$_2$\{C(H)R$^2$\}R$^3$] (R$^1$ = Bu$^t$ or Xyl, R$^2$ = Bu$^t$, R$^3$ = CH$_2$Bu$^t$; R$^1$ = Xyl, R$^2$ = CMe$_2$Ph, R$^3$ = CH$_2$CMe$_2$Ph) complexes react with sources of HCl to yield chloro alkylidene–alkylidyne compounds (e.g., (**20**) via Equation (10)).[75,76,78,79] Remarkably, the reaction in Equation (10) can even be carried out in air using aqueous HCl, illustrating the stability of the alkylidene and alkylidyne ligands in (**20**). There is considerable evidence that reaction (10) and similar reactions proceed by a sequence of steps that include proton transfers from the alkylidene and alkyl ligands to imido and amido ligands, respectively (i.e., Re–C multiple bonds are formed rather than Re–N multiple bonds). In this regard, one of the presumed intermediates in the reaction to form the *t*-butylimido derivative, [Re\{N(H)Bu$^t$\}(CBu$^t$)Cl$_2$(CH$_2$Bu$^t$)], can be trapped as a Cl$^-$ adduct if the reaction is run in the presence of [NEt$_4$]Cl.[76]

$$[Re(NXyl)_2\{C(H)Bu^t\}(CH_2Bu^t)] + 3\,HCl \xrightarrow[85\%]{DME}$$

$$[\{Re(CBu^t)\{C(H)Bu^t\}Cl(\mu\text{-}Cl)(NH_2Xyl)\}_2] + [XylNH_3]Cl \quad (10)$$

**(20)**

Compound (**20**) reacts with pyridine, TMEDA, excess Bu$^t$NH$_2$ or 1,2-phenyldiamine to form the octahedral complexes [Re(CBu$^t$)\{C(H)Bu$^t$\}Cl$_2$L$_2$].[78,79] The analogue of (**20**) where the amine ligand is Bu$^t$NH$_2$ reacts directly with LiOBu$^t$ or LiO-TMS to give [Re(CBu$^t$)\{C(H)Bu$^t$\}(OR)$_2$],[76] with LiCH$_2$Bu$^t$ to yield [Re(CBu$^t$)\{C(H)Bu$^t$\}(CH$_2$Bu$^t$)$_2$],[76] and with TMS-I/py or TMSI/TMEDA to produce [Re(CBu$^t$)\{C(H)Bu$^t$\}I$_2$L$_2$].[75,76]

The most important starting material for alkene-metathesis-active rhenium(VII) compounds is ligand-free (**20**), [Re(CBu$^t$)\{C(H)Bu$^t$\}Cl$_2$]$_x$, which was first prepared by treating [Re(CBu$^t$)\{C(H)Bu$^t$\}Cl$_2$(1,2-(H$_2$N)$_2$C$_6$H$_4$)] with HCl.[79] A more convenient high-yield preparation is the reaction sequence shown in Scheme 8 starting from readily prepared [Re(NXyl)$_2$\{C(H)Bu$^t$\}(CH$_2$Bu$^t$)].[78] Salt metathesis reactions with alkali metal alkoxides give the important catalyst (**21**), as well as [Re(CBu$^t$)\{C(H)Bu$^t$\}(OR)$_2$] where R = Bu$^t$, CMe$_2$(CF$_3$), dip, or SiBu$^t_3$ as orange or yellow-orange solids or oils in nearly quantitative yields.[78,79] The alkoxide compounds prepared in this way have *syn*-alkylidene configurations as shown for (**21**), but in this and other cases *syn*- and *anti*-rotamers interconvert thermally or on exposure to light. In the solid state, the THF adduct of (**21**) has a face-capped tetrahedral geometry. Interestingly, the THF is not coordinated at a site alkene is expected to occupy prior to rhenacyclobutane formation in a metathesis process (viz., adjacent to a Re=C π-face).

$$[Re(NXyl)_2\{C(H)Bu^t\}(CH_2Bu^t)] \xrightarrow[\text{or alumina}]{H_2O,\ cat.\ H^+} (19) \xrightarrow{aq.\ HCl} [\{Re(CBu^t)\{C(H)Bu^t\}Cl_2\}_x]$$

**Scheme 8**

[NEt$_4$][Re(CBu$^t$)Cl$_4$\{N(H)dip\}] and [Re(CBu$^t$)(Ndip)Cl$_2$(DME)] are important starting materials for alkyne-metathesis-active rhenium(VII) compounds. The former is prepared in nearly quantitative yield by adding 3 equiv. of HCl to [Re(Ndip)$_2$\{C(H)Bu$^t$\}Cl] in ether to give [dipNH$_3$][Re(CBu$^t$)-Cl$_4$\{N(H)dip\}] (cf. Equation (10)), followed by cation exchange.[80,81] [Re(CBu$^t$)(Ndip)Cl$_2$(DME)] is prepared according to Scheme 9. The intermediate in Scheme 9, "Re(CBu$^t$)\{N(H)dip\}Cl$_3$," can be isolated as a THF or pyridine ligand adduct, and it dimerizes to [(Re(CBu$^t$)Cl$_2$(μ-Cl)\{N(H)dip\})$_2$] in the absence of ligand.[81]

[NEt$_4$][Re(CBu$^t$)\{N(H)dip\}Cl$_4$] and [Re(CBu$^t$)(Ndip)Cl$_2$(DME)] react with a variety of lithium alkoxides to give [Re(CBu$^t$)(Ndip)(OR)$_2$] compounds, including the important alkyne-metathesis-active compounds (**22**) and (**23**), the latter as a THF adduct, in high yields.[80,81]

$$[NEt_4][Re(CBu^t)\{N(H)dip\}Cl_4] \xrightarrow[-[NEt_4][ZnCl_3]]{ZnCl_2/DME} \text{"Re(CBu}^t)\{N(H)dip\}Cl_3(DME)\text{"} \xrightarrow[-[NHEt_3]Cl]{NEt_3/DME} [Re(CBu^t)(Ndip)Cl_2(DME)]$$

**Scheme 9**

Substituted phenols formally add to the Re≡C bond in [Re(CBu$^t$)(Ndip)Cl$_2$(DME)] to give [Re{C(H)Bu$^t$}(Ndip)Cl$_2$(O-aryl)] compounds (aryl = 2,6-Cl$_2$C$_6$H$_3$, Xyl, dip, 2,6-(MeO)$_2$C$_6$H$_3$), which in turn react with pyridine to give [Re(CBu$^t$)(Ndip)Cl$_2$(py)$_2$] by elimination of arylOH.[82] In contrast to the reactions that form (22), [Re(CBu$^t$)Cl$_2$(μ-Cl){N(H)dip}]$_2$ reacts with the electron-withdrawing aryl oxides C$_6$F$_5$O$^-$, 2,6-Cl$_2$C$_6$H$_3$O$^-$, and C$_6$Cl$_5$O$^-$ to give five-coordinate alkylidene compounds, [Re{C(H)Bu$^t$}(Ndip)(O-aryl)$_3$], and [Re(CBu$^t$){N(H)dip}Cl$_3$(THF)] reacts with KOC$_6$F$_5$ to give [Re{C(H)Bu$^t$}(Ndip)(OC$_6$F$_5$)$_3$(THF)].[82]

[Re(CBu$^t$)(CH$_2$Bu$^t$)$_3$X] (X = Cl or I) is prepared by adding HX to [Re(CBu$^t$){C(H)Bu$^t$}-(CH$_2$Bu$^t$)$_2$], and the triflate analogue, [Re(CBu$^t$)(CH$_2$Bu$^t$)$_3$(OTf)], is synthesized by reacting [Re(CBu$^t$)(CH$_2$Bu$^t$)$_3$Cl] with TMS-OTf.[76] [Re(C-TMS)(CH$_2$TMS)$_3$Cl], as well as [{Re(CH$_2$TMS)$_4$}$_2$(μ-N$_2$)] and [{Re(μ-C-TMS)(CH$_2$TMS)$_2$}$_2$],[3] is formed in the reaction of [ReCl$_4$(THF)$_2$] with 4 equiv. of Mg(CH$_2$TMS)Cl.[83] The unusual cyclopentadienylidene complex (24) is formed in low yield from [Re(NBu$^t$)$_3$Cl] and NaCp.[84] [Re(NBu$^t$)$_3$Cl] also reacts with 2 equiv. of the Wittig reagent Ph$_3$PCH$_2$ to give [PPh$_3$Me]Cl and [Re{C(H)PPh$_3$}(NBu$^t$)$_3$], a zwitterionic organometallic perrhenate analogue.[85]

Reaction (11) forms an interesting alternative class of rhenium(VII) alkylidynes (mq = the monoanion of mercaptoquinoline).[86] In the reaction, the rhenium hydride starting material is activated by abstraction of H$^-$. A terminal alkyne must be used initially or be generated in the reaction by isomerization because reaction (11) does not occur for PhC≡CPh. A proposed mechanism for Equation (11) involves a rhenium(VII) vinylidene intermediate, [ReH$_3$(=C=CHR)(mq)(PPh$_3$)$_2$]$^+$, and subsequent hydride transfer from rhenium to the β-carbon of the vinylidene. The alkylidyne–hydride products of (11) are reversibly deprotonated by NEt$_3$.

$$[ReH_4(mq)(PPh_3)_2] + R^1C≡CR^2 + [E]PF_6 \xrightarrow[>70\%]{CH_2Cl_2} [ReH_2(CCH_2R^3)(mq)(PPh_3)_3][PF_6] + EH \quad (11)$$

E = H or CPh$_3$
R$^1$ = R$^2$ = Et, R$^3$ = Bu$^n$; R$^1$ = H, R$^2$ = R$^3$ = Pr$^n$, Ph, *p*-Tol; R$^1$ = Me, R$^2$ = Et or Ph and R$^3$ = Pr$^n$ or CH$_2$Ph

Oxidation of [ReCp*(CBu$^t$)Cl$_2$] with Ag[SbF$_6$] gives [ReCp*(CBu$^t$)Cl$_2$]$^+$, a rare example of a rhenium(VII) alkylidyne cation, and oxidation of [ReCp*(CBu$^t$)Br$_2$] with bromine produces [ReCp*(CBu$^t$)Br$_3$].[87,88]

## 10.3.2 Rhenium(IV), (V), and (VI)

As mentioned in Section 10.3.1, (21) is an efficient alkene metathesis catalyst. When metathesis results in an alkylidene that is relatively small and has a heteroatom substituent, however, a remarkable Re=Re bond-forming reaction takes place.[89] Thus, (21) reacts with vinyl ethers to give [Re(CBu$^t$){C(H)OR}{OCMe(CF$_3$)$_2$}$_2$(THF)$_2$] (R = Et or TMS) in THF and THF-free analogues in hydrocarbons. The ethoxy-substituted alkylidene [Re(CBu$^t$){C(H)OEt}{OCMe(CF$_3$)$_2$}$_2$] decomposes in minutes at room temperature to give *cis/trans*-diethoxyethylene, *cis/trans*-3,3-dimethylbutenyl ethyl ether, and rhenium(V) dimer (25) in 30–40% yield. The *t*-butoxide dimer is prepared in a one-pot synthesis according to Equation (12). Also, (21) reacts with [Re(CBu$^t$){C(H)-OEt}{OCMe(CF$_3$)$_2$}$_2$(THF)$_2$] in 4 h at 85 °C to give *cis/trans*-EtOCH=CHBu$^t$ and (25) in 90% yield, and heating [Re(CBu$^t$){C(H)OEt}{OCMe(CF$_3$)$_2$}(THF)$_2$] in C$_6$D$_6$ gives (25) and *cis/trans*-1,2-diethoxyethylene. The formation of 1,2-diethoxyethylene in the latter reaction, and dimethylbutenyl ethyl ether in the former reaction and Equation (12), suggests that the dimers form via coupling of two alkylidene ligands, perhaps involving 1,3-dimetallacycle intermediates. The Re–Re bond distances in (25) and its *t*-butoxide derivative (0.238 and 0.240 nm) are consistent with double bonds. They are members of a small family of transition metal compounds with unsupported M–M double bonds (cf. Structure (5)).[2a]

$$[Re(CBu^t)\{C(H)Bu^t\}(OBu^t)_2] \xrightarrow[50-60\%]{\text{excess } CH_2=CHOEt} [\{Re(CBu^t)(OBu^t)_2\}_2] + CH_2=CHBu^t + EtO(H)C=CHBu^t \quad (12)$$

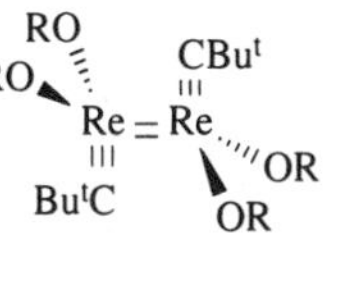

(25) R = CMe(CF₃)₂ rendered: **(25)** R = CMe(CF$_3$)$_2$ 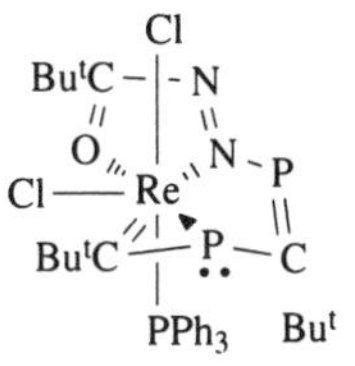    **(26)** 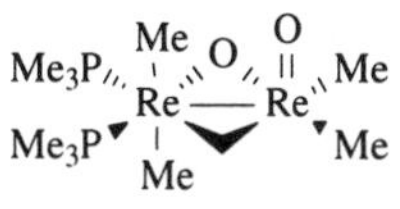    **(27)**

Compound (**21**) and its analogues shown in Scheme 10 also do not undergo simple metathesis reactions with ethene.[90] When the alkoxide is OBu$^t$, OCMe(CF$_3$)$_2$, or OCMe$_2$(CF$_3$), the reactions with ethene at low temperature give rhenacyclobutanes. Upon warming, the OCMe(CF$_3$)$_2$ rhenacyclobutane derivative reverts to starting material, and the OBu$^t$ and OCMe$_2$(CF$_3$) derivatives convert to unstable rhenium(V) rhenacyclopentene complexes. In contrast, [Re(CBu$^t$){C(H)SPh}{OCMe(CF$_3$)$_2$}$_2$(THF)$_2$] reacts with ethene to give the relatively stable rhenacyclopentene [Re{C(Bu$^t$)CH$_2$CH$_2$C(H)-SPh}{OCMe(CF$_3$)$_2$}$_2$], and no rhenacyclobutane intermediate is observed. Rhenacyclopentene formation in these complexes is proposed to occur by coupling of the less hindered rhenacyclobutane α-methylene and the neopentylidyne ligand, or by competitive cycloaddition of ethene to the parent alkyidene–alkylidyne complex. [Re{C(Bu$^t$)CH$_2$CH$_2$C(H)SPh}{OCMe(CF$_3$)$_2$}$_2$] forms monoadducts with PMe$_3$ and PMe$_2$Ph. The crystal structure of the PMe$_2$Ph adduct reveals that it is roughly trigonal bipyramidal with axial alkoxide and phosphine ligands and a dative interaction between the sulfur from the SPh ring substituent and rhenium.

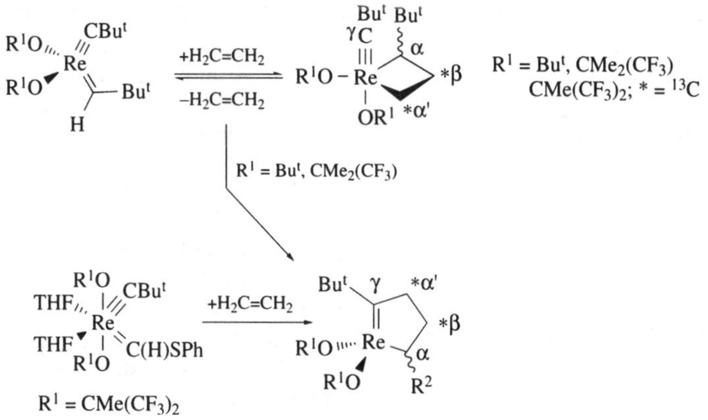

**Scheme 10**

The alkylidyne–hydride products formed by Equation (11) are reversibly deprotonated by NEt$_3$ yielding [ReH(CCH$_2$R$^3$)(mq)(PPh$_3$)$_2$] compounds.[86] The crystal structure of the pentylidyne derivative reveals that it has an octahedral geometry with *trans* phosphines and the pentylidyne *trans* to the nitrogen of the quinoline moiety. Another nontraditional high-oxidation-state compound with a Re–C multiple bond is (**26**), formed from the reaction of [ReCl$_2${N$_2$C(O)Ph}(PPh$_3$)$_2$] with Bu$^t$C≡P.[91] Also, *trans*-[Re{C=C(H)R}Cl(dppe)$_2$] and *trans*-[Re{η$^2$-H$_2$C=C=C(H)Ph}Cl(dppe)$_2$] react with HBF$_4$ to give *trans*-[Re(CCH$_2$R)X(dppe)$_2$][BF$_4$] (R = Bu$^t$ or Ph, X = Cl; R = Bu$^t$, X = F) and *trans*-[Re{=C(CH$_2$Ph)CH$_2$}(dppe)$_2$][BF$_4$], respectively.[92–5] Several closely related [Re{CN(H)R}X(dppe)$_2$]$^+$ complexes are prepared by β-protonation of coordinated isocyanides,[94,95] and *trans*-[Re{CN(H)R}Cl(CNR)$_2$(PMePh$_2$)$_2$][SbF$_6$] (R = Me, Bu$^t$) complexes are formed from [ReCl$_2$(CNR)$_3$(PMePh$_2$)$_2$]$^+$ under conditions that were expected to produce reductive coupling of the isocyanide ligands.[96] In both cases, the complexes are perhaps best considered aminocarbynes rather than alkylidynes.

Several intermediate and high-oxidation-state cyclopentadienyl alkylidyne complexes have been prepared. [ReCp*(O)(CH$_2$R)$_2$] (R = Ph or Bu$^t$) reacts with [TiCpX$_3$] (X = Cl, Br) to give Re$^{VI}$ [ReCp*(CBu$^t$)X$_2$] and diamagnetic Re$^{IV}$ [{ReCp*X}$_2$(μ-X)(μ-CPh)] (Re=Re).[87,88,97] In these reactions the titanium reagent serves to replace the oxo group with two chlorides, forming [TiCp(O)X]$_4$ in the process. Intermediate rhenium(V) alkylidenes are proposed. [ReCp*(CBu$^t$)Cl$_2$] reacts with TMS-X (X = Br, I) to yield the bromide and iodide derivatives, with Ag[SbF$_6$] to form [ReCp*(CBu$^t$)Cl$_2$]$^+$, and with PMe$_3$ to give [ReCp*(CBu$^t$)(PMe$_3$)$_2$]Cl and [Re(CBu$^t$)Cl$_2$(PMe$_3$)$_2$].[87,88]

A few intermediate-oxidation-state alkylidene/alkylidyne oxo complexes are known.[15] Unstable [Re(O)Me₃(PMe₃)] decomposes bimolecularly at room temperature to yield methane and (27), which has been structurally characterized. Extended Hückel calculations suggest that (27) is best described as a $d^3$–$d^1$ complex with a normal covalent Re–Re σ-bond and a lone pair on the ReMe₂(PMe₃)₂ center. Consistent with this description, (27) is readily oxidized by pyridine *N*-oxide to give diamagnetic [Me₂(O)Re(μ-CH₂)(μ-O)Re(O)Me₂], a structural analogue of (9). The methylidene protons in [Me₂(O)Re(μ-CH₂)(μ-O)Re(O)Me₂] exchange readily with deuterium from CD₃OD, and the methylidene is deprotonated by LiCH₂Buᵗ, forming [Me₂(O)Re(μ-CH)(μ-O)Re(O)Me₂]⁻, a rare example of an alkylidyne anion.

[Re{C(H)Me}(OAlCl₂)Et₂(PPh₃)₂] is said to be formed by α-hydrogen abstraction from AlCl₂Et and [ReOEt₃(PPh₃)₂],[16] and methylidene and (trimethylsilyl)methylidene complexes are said to be formed from [ReCl₅] and LiCH₂TMS.[98] In neither case were the proposed alkylidene complexes isolated.

### 10.3.3 Alkene and Alkyne Metathesis and Related Reactions of Rhenium(VII) Alkylidene/ Alkylidyne Complexes

#### *10.3.3.1 Alkene metathesis*

Compound (21) is an effective catalyst for the metathesis of internal and functionalized alkenes.[79,99] The activity of (21) with internal alkenes is characterized by a relatively slow induction period resulting from a slow metathesis reaction that replaces the bulky neopentylidene ligand with a smaller, more active alkylidene ligand. For example, reaction of (21) with 100 equiv. of *cis*-2-pentene in benzene gives a 1:2:1 equilibrium mixture of 2-butenes, 2-pentenes and 3-hexenes in 2.5 h. If another 100 equiv. of *cis*-2-pentene is added, equilibrium is reached in less than 30 min. Importantly, similar behavior and activity are observed for the functionalized alkene methyl oleate, but the rate is limited by ester coordination to the metal in competition with the alkene. Similarly, the metathesis reactions are slower in the presence of THF or DME. Intermediate alkylidene complexes in the reactions are observed by ¹H NMR, and in one example involving the reaction of (21) with 3-hexene, a TMEDA adduct of the expected propylidene complex has been isolated.

Metathesis reactions involving terminal alkenes have been used to prepare new isolable alkylidene complexes from (21).[99] Vinylferrocene reacts reversibly with (21) in noncoordinating solvent to give BuᵗHC=CH₂ and *anti*-[Re(CBuᵗ){C(H)(ferrocenyl)}{OCMe(CF₃)₂}₂], which is structurally charac- terized, but Re=CH(CH₂)₇Me and Re=CH(CH₂)₇CO₂Me complexes produced from the reaction of (21) with 1-decene and methyl 9-decenoate under similar conditions are unstable. In the presence of a few equivalents of DME, however, (21) reacts with a slight excess of propene, butene, or styrenes to yield the new alkylidene complexes [Re(CBuᵗ){C(H)R}{OCMe(CF₃)₂}₂(DME)] (R = Me, Et, Ph, C₆H₄-*p*- NMe₂). The addition of a near-stoichiometric amount of alkene in these reactions limits productive metathesis to give ethylene, which reacts nonmetathetically with alkylidenes (see Scheme 10), and the back-reaction with Buᵗ(H)C=CH₂ is relatively slow, presumably because of competitive DME coordination. Importantly, (21) reacts with functionalized terminal alkenes in the presence of THF to yield [Re(CBuᵗ){C(H)R}{OCMe(CF₃)₂}₂(THF)₂] (R = OEt, O-TMS, SPh, CHN(CH₂)₃C(O)). *Syn* isomers (as in (21)) are obtained in all of these reactions except for the case of 1-vinyl-2-pyrrolidinone, which gives exclusively the *anti* isomer. Mixtures of *syn* and *anti* isomers form when the *syn* isomers are exposed to light. Alkenes with oxygen, sulfur, or nitrogen bonded directly to the alkenic carbon react more rapidly with (21) in THF than nonfunctionalized alkenes because the functionalized alkenes are more nucleophilic.

Compound (21) and its derivatives are isoelectronic and isostructural with the well-behaved group 6 alkene metathesis catalysts [M{C(H)R²}(N-aryl)(OR¹)₂] (M = Mo and W). The metathesis activity of the rhenium catalysts is less than the group 6 compounds, however, and the rhenium compounds are more sensitive to the nature of the alkoxide ligand and are more tolerant of alkene heteroatom functional groups.[99] A theoretical examination of [Re(CH)(CH₂)(OH)₂] has appeared.[100]

#### *10.3.3.2 Alkyne metathesis*

Compounds (22) and (23) react with internal symmetric alkynes to give rhenacycles that are primarily of one type (type I, Scheme 11).[80,81] On the basis of NMR studies, which show two Cα resonances at very different chemical shifts (Δδ > 75 ppm), and the crystal structure of

$[\overline{Re\{C(Et)C(Et)C}(Et)\}(Ndip)\{OCMe(CF_3)_2\}_2]$, all of the type I rhenacycles are proposed to have the structure shown in Scheme 11. A type I rhenacycle, $[\overline{Re\{=C(Bu^i)C(H)=C}(Bu^i)\}(Ndip)\{OCH(CF_3)_2\}_2]$, is also formed when (23) reacts with the terminal alkyne $Bu^iC{\equiv}CH$. A different type of rhenacycle (type II) is proposed for the $Pr^i$ derivative $[\overline{Re\{C(Pr^i)C(Pr^i)C}(Pr^i)\}(Ndip)\{OCMe(CF_3)_2\}_2]$, formed from $Pr^iC{\equiv}CPr^i$ and (22). This compound gives rise to $C_\alpha$ resonances that have very similar chemical shifts ($\Delta\delta \approx 2$ ppm). Consistent with this, a poor quality crystal structure indicates the compound has a tbp structure with a nearly symmetrical ring, as shown in Scheme 11. Type I rhenacycles are stable to loss of alkyne and do not react further with internal alkynes, whereas the type II rhenacycle readily loses $Pr^iC{\equiv}CPr^i$.

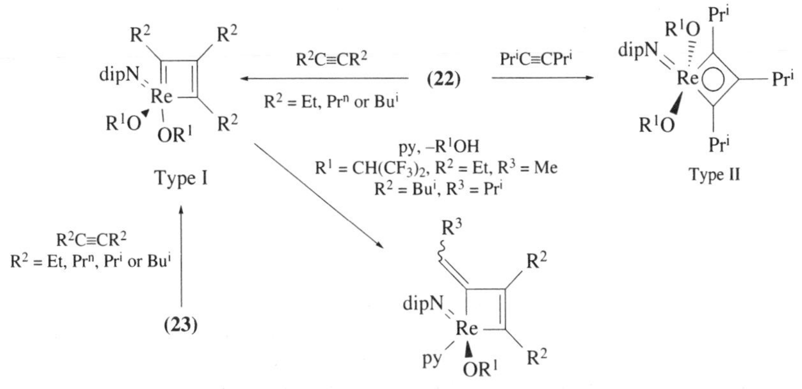

**Scheme 11**

Unsymmetrical bulky alkynes such as 3-heptyne, 4-nonyne, or 5-undecyne (20 equiv.) undergo rapid metathesis in the presence of (22) but the activity stops after about 5 min. In all cases the expected primary metathesis products and type I rhenacycles are observed in the product mixtures. In contrast, (22) will smoothly catalyze the co-metathesis $Pr^iCCPr^i + Bu^sCCBu^s \rightleftharpoons Pr^iCCBu^s$. In this reaction only type II rhenacycles are expected because of the bulky $Pr^i$ and $Bu^s$ ring substituents. On the basis of these results, it is postulated that type I rhenacycles are metathesis-inactive, and, furthermore, that type II rhenacycles are the active species for metathesis of internal alkynes, even in those cases when they cannot be observed directly.

The results of the metathesis experiments and their relationship to the type I and II rhenacycles have been discussed by Schrock and co-workers.[81] They assume that approach of alkyne to (22) and its derivatives takes place on either the C/N/O or C/O/O face of the pseudotetrahedron. Approach on one of the C/N/O faces of the catalyst leads irreversibly to only type I rhenacycles. Approach on the C/O/O face, on the other hand, leads to a rhenacycle that pseudorotates (via a type II rhenacyle) to a rhenacycle capable of productive metathesis by the microscopic reversal of the initial addition.

Reduction to rhenium(V) compounds with expulsion of ROH appears to be one fate of type I rhenacycles over a period of hours or days.[81] Scheme 11 shows examples of this type of reduction occuring in the presence of pyridine. The metallacycle $[\overline{Re\{C(Bu^i)=C(Bu^i)C}=C(H)Pr^i\}(Ndip)\text{-}\{OCMe(CF_3)_2\}(Bu^iCCBu^i)]$ is formed by a similar reduction process. $[Re\{=C(Me)C(Me)\text{-}=CMe\{OCMe(CF_3)_2\}\}(Ndip)\{OCMe(CF_3)_2\}(MeCCMe)]$ is obtained from (22) and 3 equiv. of 2-butyne by a somewhat different reduction involving a rhenacycle. In this case, the proposed reaction sequence consists of alkyne metathesis, intramolecular alkoxide ligand attack on Re–$C_\alpha$ in $[Re\{C(Me)C(Me)\text{-}C(Me)\}(Ndip)\{OCMe(CF_3)_2\}_2]$, and finally, capture of alkyne by the rhenium(V) alkylidene.

## 10.4 COMPLEXES WITH CYCLOPENTADIENYL AND OTHER CONJUGATED POLYENE LIGANDS

### 10.4.1 Monocyclopentadienyl Complexes

#### 10.4.1.1 Key starting materials

Most of the chemistry of intermediate- and high-oxidation-state monocyclopentadienyl complexes is derived from the remarkable rhenium(VII) compound (28) (Scheme 12). Compound (28) was first prepared by inadvertent oxygenation of [ReCp*(CO)$_3$] or [ReCp*(CO)(MeCN)($p$-N$_2$C$_6$H$_4$OMe)]$^+$.[101,102] It has since been prepared from [ReCp*(CO)$_3$] by photolysis in oxygenated THF (30–40% yield)[103] and by oxidation with H$_2$O$_2$ in refluxing benzene/H$_2$O,[104-6] [Mn$_2$O$_7$] in CCl$_4$,[107] H$_2$O$_2$ and a catalytic amount of (12) in ether, or H$_2$O$_2$ and (CF$_3$C(O))$_2$O in ether.[108] The C$_5$Me$_4$Et derivative of (28) has also been prepared from the tricarbonyl,[2f,105] and preparations of [Re($\eta$-C$_5$H$_4$R)O$_3$] (R = H or Me) from [Re$_2$O$_7$] and Zn(C$_5$H$_4$R)$_2$ have been published.[109]

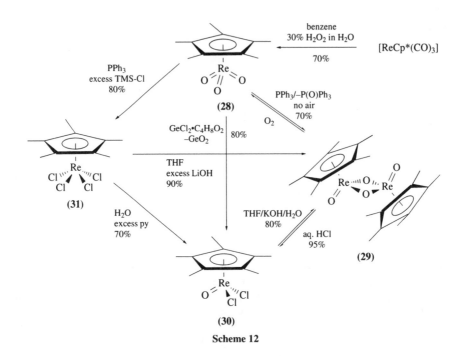

**Scheme 12**

Compound (28) is soluble in most common solvents except H$_2$O and hexane, sublimes at 40 °C (10$^{-2}$ torr), and crystallizes as yellow needles. It is an 18-electron compound with each oxo ligand contributing four electrons. This suggests an Re–O bond order of 2, and as might be expected on this basis, the Re–O bonds in (28) are quite reactive.

An electron diffraction study of (28) has been carried out.[107] A compilation of IR and $^{17}$O NMR data for rhenium(VII) Cp trioxo compounds, (12), and [ReO$_3$Ph] has also been presented.[109]

There are three synthetically important rhenium(V) compounds derived directly from (28) (Scheme 12). The diamagnetic dimer (29) is prepared from (28) by reduction with phosphines[106,110] or zirconium(III).[111] In the crystal structure of (29) the long Re–Re distance (0.314 nm) is consistent with no Re–Re bonding.[106] A theoretical analysis indicates that the four $d$ electrons reside in rhenium-based nonbonding MOs.[112]

The diamagnetic dichloride (30) is prepared from (28) and GeCl$_2$ or [ZrCp$_2$($\mu$-Cl)]$_2$.[106,111] Adding aqueous HCl to (29) yields (30) as well, and similar procedures are used to make the bromide and fluoride derivatives.[106,113-15] The iodide is prepared by photolysis of [ReCp*(CO)$_3$] and PhIO[113] or from (29) and [py·H]I.[106] Addition of KOH to (30) gives (29).[106,116] The unsubstituted Cp complex [ReCp(O)I$_2$] has been prepared serendipitously from [ReCp(CO)$_2$I]LnI (Ln = Yb or Sm) and atmospheric oxygen.[117]

The addition of 1 equiv. of PPh$_3$ and a large excess of TMS-Cl to (28) gives paramagnetic (31) in high yield.[2f,106,114,118] Compound (31) reacts with 1 equiv. of water to give (30), with LiOH to give (29),[106,116] and with PMe$_3$ to yield diamagnetic [ReCp*Cl$_4$(PMe$_3$)].[114] In addition to being a key starting material for the preparation of rhenium(V) compounds, (31) is also an important starting point for the synthesis of reduced rhenium complexes including [ReCp*Cl$_2$(R$^1$CCR$^2$)] and [ReCp*Cl$_2$(2,3-dimethylbutadiene)].[2]

### 10.4.1.2 [ReCp*(O)X(Y)] and related compounds

#### (i) Characterization, structural aspects, and electronic structure

A compendium of $^{17}$O NMR chemical shifts for (28), (29), and a variety of [ReCp*(O)X$_2$] and [ReCp*(O)X(Y)] compounds, where X and Y are selected halogens, alkyls, alkoxides, or thiolates, has been published.[119] The chemical shifts of the oxo ligands depend on the σ-donor ability of the X, Y ligands and roughly correlate with the electron density at rhenium.

Structures of the [ReCp*(O)X(Y)] compounds exhibit a distortion in the Cp* coordination, approaching $\eta^3$:$\eta^2$ coordination with the oxo group *trans* to the $\eta^2$ portion of the ring. The distortion from $\eta^5$ coordination has been attributed to a strong *trans* influence from the oxo group[2f] and, in a theoretical analysis, to oxo lone pair/Cp* π-orbital interactions.[112]

Formation of oxo-bridged compounds is sometimes observed for [ReCp*(O)X(Y)] compounds. The parent complex (30), for example, dimerizes to [ReCp*Cl$_2$(μ-O)]$_2$ in acetone–H$_2$O solution with concomitant Re–Re double bond formation ($d$(Re–Re) = 0.272 nm).[106] Similarly, Re–Re bonds are observed in [Cp*{$\overline{O(CH_2)_2O}$}Re(μ-O)$_2$Re(OReO$_3$)$_2$Cp*] (0.260 nm)[120] and the related compounds [Cp*(O)Re(μ-O)$_2$ReCl$_2$Cp*] (0.269 nm)[106] and [Cp*(O)Re(μ-O)$_2$Re(OReO$_3$)$_2$Cp*] (0.265 nm).[110,115] The Re–Re bonds in these complexes contrast with (29) where there is apparently no Re–Re bonding.[106] Related dinuclear compounds that have not been structurally characterized include [{ReCp*X$_2$(μ-O)}$_2$] (X = CN and NCO), [{ReCp*(μ-O)($\eta^2$-EO$_3$)}$_2$] (E = S, Se), [Cp*Cl$_2$Re(μ-O)$_2$ReCl(OReO$_4$)Cp*], and [Cp*(O)Re(μ-O)$_2$Re($\eta^2$-SO$_3$)Cp*].[116]

A theoretical analysis of [ReCp*(O)Me$_2$] indicates that it has an Re–O triple bond (i.e., the oxo group donates six electrons to the rhenium and overall there is an 18-electron count) and the two $d$ electrons reside in a nonbonding MO.[121]

#### (ii) Alkyls

The diamagnetic monoalkyls [ReCp*(O)ClR] (R = Ph, CH$_2$TMS) and dialkyls [ReCp*(O)R$_2$] (R = Me, CD$_3$, Et, Bu$^n$, CH$_2$Ph, Ph, $\eta^1$-CH$_2$CH=CH$_2$, CH$_2$Bu$^t$, CH$_2$TMS) are prepared by reaction of (30) with the appropriate amount of MgClR in THF.[87,88,122] When MgBr(CH$_2$CH=CH$_2$) is used in place of MgCl(CH$_2$CH=CH$_2$) there is oxo abstraction to form paramagnetic Re$^{IV}$ [ReCp*Br$_2$($\eta^3$-CH$_2$CH=CH$_2$)] instead of the di-$\eta^1$-allyl complex.[87,122] The reaction of (28) with AlMe$_3$ also produces [ReCp*(O)Me$_2$].[115]

[ReCp*(O)ClMe], prepared from [ReCp*Cl$_3$Me] and water, reacts with Grignard reagents to give the mixed alkyls [ReCp*(O)MeR] (R = Et, CH$_2$TMS, CH$_2$Bu$^t$).[87] The thermally robust four-membered rhenacycles [Re(CH$_2$CMe$_2$CH$_2$)Cp*O] and [Re(CH$_2$C$_6$H$_4$)Cp*(O)] are prepared by reacting (30) with BrMgCH$_2$CMe$_2$CH$_2$MgBr and [Mg(CH$_2$C$_6$H$_4$)]$_n$, respectively.[122]

Alkynes that are not too sterically demanding react with (28) over 16 h to give rhenapyran compounds according to Equation (13).[123] Short reaction times result in formation of Re$^{III}$–alkyne complexes. The methyl-substituted rhenapyran reacts with oxygen in hot toluene to give tetramethylfuran and (28), and with iodine at room temperature to yield tetramethylfuran and [ReCp*(O)I$_2$]. Reaction (13) is proposed to occur by addition of alkyne to an Re–O multiple bond followed by insertion of another alkyne into the Re–O single bond of the rhenacylooxabutene intermediate.

$$(28) + PPh_3 + R^1C\equiv CR^2 \xrightarrow[\text{16 h}]{\text{neat alkyne}}$$ (13)

$$R^1 = R^2 = Me, Et; R^1 = Ph, R^2 = H$$

*(iii) Alkoxides, thiolates, amides, pseudohalides, and related compounds*

Most compounds of this type are prepared straightforwardly from (**30**). For example, (**30**) reacts with NaOMe to give [ReCp*(O)(OMe)$_2$][106] and with 2 equiv. of RS$^-$ (R = Et, Bu$^t$, Ph, *o*-pyridyl) to yield [ReCp*(O)(SR)$_2$].[124] [ReCp*(O)(3,4-S$_2$C$_6$H$_3$-1-Me)] and [ReCp*(O){1,2-S(NH)C$_6$H$_4$}] are prepared similarly.[124,125] The hydroxide [{ReCp*(O)(μ-OH)}$_2$]$^{2+}$ is formed when HBF$_4$ is added to (**29**).[113] Compound (**28**) has also been used to prepare thiolates and alkoxides, reacting with PhSH to give [ReCp*(O)(SPh)$_2$], PhSSPh, and H$_2$O,[125] and with R$_3$Sn–SnR$_3$ to produce [ReCp*(O)(OSnR$_3$)$_2$] (R = Me or Bu$^n$).[126]

The five-membered rhenacylces [$\overline{\text{Re}\{\text{E}^1\text{(CH}_2\text{)}_2\text{E}^2\}}$Cp*(O)] (E$^1$ = E$^2$ = O, S, or NR; E$^1$ = O, E$^2$ = S; E$^1$ = O, E$^2$ = NR (R = H, Me, or Bu$^t$)) are synthesized from (**30**).[120,127] The glycolate and dithioglycolate are prepared using NaOCH$_2$CH$_2$OH and NaSCH$_2$CH$_2$SH, and the rest are synthesized by simply adding the ligand precursors directly to (**30**).

When the glycolate [$\overline{\text{Re}\{\text{O(CH}_2\text{)}_2\text{O}\}}$Cp*(O)] is heated under vacuum at 150 °C or refluxed in toluene it eliminates ethene quantitatively to form (**28**).[120] This elimination reaction, formally a retro [2 + 4] reaction, is the microscopic reverse of a proposed key step in ethene hydroxylation. The dithioglycolate is thermally more robust, requiring heating at 200 °C to eliminate ethene. [$\overline{\text{Re}\{\text{OC(H)(Me)C(H)(Me)O}\}}$-Cp*(O)], prepared from (2*R*,3*R*)-2,3-butanediol, eliminates *trans*-butene at temperatures above 150 °C, and [$\overline{\text{Re}\{\text{O(CH}_2\text{)}_3\text{O}\}}$Cp*(O)] decomposes above 0 °C, eliminating propene. The elimination of propene from [$\overline{\text{Re}\{\text{O(CH}_2\text{)}_3\text{O}\}}$Cp*(O)] requires a hydrogen shift, but the details of the mechanism are not known. In the presence of air, [$\overline{\text{Re}\{\text{O(CH}_2\text{)}_2\text{O}\}}$Cp*(O)] is thermally transformed to [Cp*{$\overline{\text{O(CH}_2\text{)}_2\text{O}}$}Re(μ-O)$_2$Re(OReO$_3$)$_2$Cp*].

Compound (**30**) undergoes salt metathesis reactions with a variety of silver salts, for example, Scheme 13.[116,128] Compounds prepared by similar methods include [{ReCp*X$_2$(μ-O)}$_2$], X = CN and NCO (O-ligated), [Cp*Cl$_2$Re(μ-O)$_2$ReCl(OReO$_4$)Cp*], [$\overline{\text{Re}\{\text{OC(O)CH(O)CHC(O)O}\}}$Cp*(O)], [Cp*(O)Re(μ-O)$_2$Re(η$^2$-SO$_3$)Cp*], and [{$\overline{\text{Re}\{\text{OC(O)C(O)O}\}}$Cp*(μ-Cl)}$_2$].

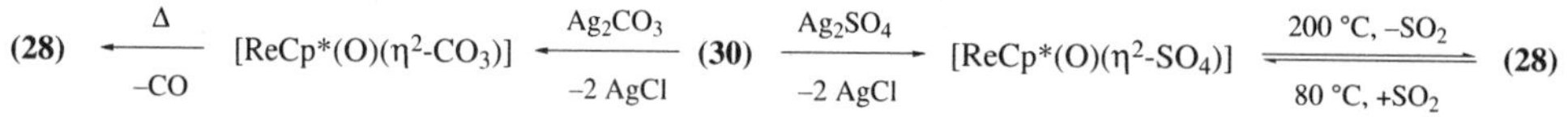

$$(\textbf{28}) \xleftarrow[-\text{CO}]{\Delta} [\text{ReCp*(O)(}\eta^2\text{-CO}_3\text{)}] \xleftarrow[-2\,\text{AgCl}]{\text{Ag}_2\text{CO}_3} (\textbf{30}) \xrightarrow[-2\,\text{AgCl}]{\text{Ag}_2\text{SO}_4} [\text{ReCp*(O)(}\eta^2\text{-SO}_4\text{)}] \xrightleftharpoons[80\,°\text{C, +SO}_2]{200\,°\text{C, -SO}_2} (\textbf{28})$$

**Scheme 13**

The rhenacycles [ReCp*(O){η$^2$-O$_2$C=CPh$_2$}], [ReCp*(O){η$^2$-O(NPh)C=O}],[115,129] and [$\overline{\text{Re}\{\text{OC(=O)CR}_2\text{O}\}}$Cp*(O)] (R = Ph, Me),[129-31] and the dimers [{ReCp*(μ-O)(η$^2$-EO$_3$)}$_2$] (E = S, Se),[116] are prepared from (**28**) or (**29**) by formal [2 + 2] and [2 + 4] cycloaddition reactions.

Compound (**29**) reacts with quinones to yield the rhenium(V) compounds [ReCp*(O)(O$_2$C$_{14}$H$_8$)], [ReCp*(O)(O$_2$C$_6$Cl$_4$)], and its dimer, [{ReCp*(O)(O$_2$C$_6$Cl$_4$)}$_2$], as well as [ReCp*(O$_2$C$_{14}$H$_8$)$_2$] and [ReCp*(O$_2$C$_6$Cl$_4$)$_2$].[132]

[Cp*(O)Re(μ-O)$_2$ReCl$_2$Cp*] is formed on heating (**30**) in toluene, stirring (**30**) with MgO/H$_2$O, or reacting (**30**) with 0.5 equiv. of (**29**).[106] Treatment of [Cp*(O)Re(μ-O)$_2$ReCl$_2$Cp*] with aqueous HCl in THF gives the dimer form of (**30**), [{ReCp*Cl$_2$(μ-O)}$_2$].

### 10.4.1.3 [ReCp*X$_4$] derivatives

*(i) Alkyls and related compounds*

With the exceptions of [ReCp*Me$_4$] and [Re(η-C$_5$Me$_4$R)Cl$_3$Me] (R = Me, Et), direct alkylations of (**31**) do not yield alkyls cleanly because of rhenium reduction. [ReCp*Me$_4$] is formed in 40–50% yield from reaction of (**31**) with 4 equiv. of [MgClMe].[133] Proton NMR spectra for [ReCp*Me$_4$] exhibit strongly temperature-dependent paramagnetic shifts. On this basis and a theoretical analysis, a singlet–triplet equilibrium is proposed. [ReCp*Me$_4$] slowly decomposes at room temperature and very rapidly at 68 °C, liberating 2 mol. equiv. of CH$_4$ and forming the proposed compound [{ReCp*(CH$_2$)$_2$}$_2$]. The monomethyl compounds [Re(η-C$_5$Me$_4$R)Cl$_3$Me] (R = Me, Et) have been prepared in high yield from (**31**) and SnMe$_4$.[134] When similar reactions with SnEt$_4$ and SnBu$^i_4$ are carried out, the diamagnetic rhenium(IV) dimer [ReCp*Cl$_2$(μ-Cl)]$_2$ is formed. [Re(η-C$_5$Me$_4$R)Cl$_3$Me] reacts with H$_2$O/py to form [Re(η-C$_5$Me$_4$R)(O)ClMe] (R = Me, Et)[87,134] and with PMe$_3$ to yield paramagnetic [Re(η-C$_5$Me$_4$R)Cl$_3$Me(PMe$_3$)].[134]

The oxo group in [ReCp*(O)R$_2$] compounds can be replaced with two chlorides by using [TiCpCl$_3$] as an oxo abstracting/chlorinating agent.[87,88,97] In the case of R = Me, [ReCp*Cl$_2$Me$_2$] is isolated from the reaction. When [ReCp*(O)R$_2$] (R = CH$_2$Bu$^t$ or CH$_2$TMS) is reacted the presumed [ReCp*Cl$_2$R$_2$] compounds are not stable, reacting further to give alkylidyne complexes (see Section 10.3.2). Alkyl ligands with β-hydrogens produce a different reaction course; when R = CHMe$_2$, the reaction gives [{ReCp*Cl(μ-Cl)}$_2$] (Re=Re) as shown in Scheme 14, and when R = Et, [{ReCp*Cl$_2$(μ-Cl)}$_2$] is isolated.

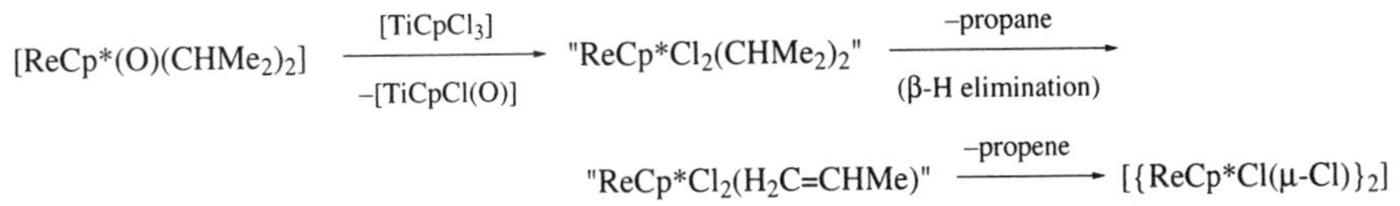

[ReCp*(O)(CHMe$_2$)$_2$] $\xrightarrow[\text{–[TiCpCl(O)]}]{\text{[TiCpCl}_3\text{]}}$ "ReCp*Cl$_2$(CHMe$_2$)$_2$" $\xrightarrow[\text{(β-H elimination)}]{\text{–propane}}$

"ReCp*Cl$_2$(H$_2$C=CHMe)" $\xrightarrow{\text{–propene}}$ [{ReCp*Cl(μ-Cl)}$_2$]

**Scheme 14**

In a rare example of an alkane elimination reaction resulting from the addition of acid to a high-oxidation-state rhenium alkyl, [ReCp*Me$_4$] reacts with triflic acid to give [ReCp*Me$_3$(OTf)] and methane.[135] The triflate shows temperature-dependent paramagnetism. Ammonia and hydrazine displace triflate from the complex to yield cationic adducts (Scheme 15). By analogy to [WCp*Me$_3$(η$^2$-N$_2$H$_4$)]$^+$, the diamagnetic hydrazine adduct probably has an η$^2$-N$_2$H$_4$ ligand with one nitrogen *trans* to the Cp* ring in a pseudooctahedral complex. The ammonia adduct, in contrast, has equivalent ammonia ligands detectable by $^1$H NMR even at −95 °C, suggesting that either a methyl ligand is *trans* to the Cp* ring or that an ammonia ligand is *trans* to Cp* and the ammonia ligands rapidly interconvert. Both the hydrazine and ammonia adducts are deprotonated to form monoamido complexes in high yields. In the latter case the deprotonation is accompanied by loss of the remaining ammonia ligand. Protonation of [ReCp*Me$_3$(NHNH$_2$)] with triflic acid reforms the hydrazine complex, and protonation of [ReCp*Me$_3$(NH$_2$)] forms a monoammonia adduct that exhibits temperature-dependent paramagnetism.

Upon reduction, [ReCp*Me$_3$(η$^2$-N$_2$H$_4$)]$^+$ yields ammonia (Scheme 15).[135] When the reduction is carried out using Zn/Hg or Na/Hg in the presence of 2,6-lutidine hydrochloride, 1.20 and 1.46 mol. equiv. of ammonia, respectively, are produced. In the absence of a proton source the yield is 1.00 equiv. of ammonia when the reducing agent is Zn/Hg. From these experiments, it is concluded that the N–N bond in $d^3$ [ReCp*Me$_3$(N$_2$H$_4$)] is cleaved on reduction but that the putative Re=NH species is not efficiently protonated and/or reduced to give the second equivalent of ammonia.

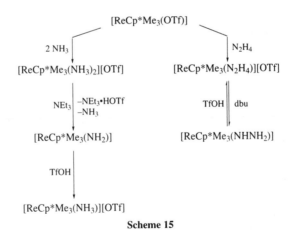

**Scheme 15**

[ReCp*Cl$_2$(μ-Cl)]$_2$ disproportionates upon reaction with 2-butyne to give Re$^{III}$ [ReCp*Cl$_2$(MeCCMe)] and (**31**), which further reacts with 2-butyne to yield the novel diamagnetic rhenium(V) compound (**32**) (Equation (14)).[136] The synthesis shown in Equation (14) is more straightforward, producing (**32**) in 90% yield.[137] Compound (**32**) (R$^1$ = Me, R$^3$ = H) reacts with LiOH in the presence of pyridine to form a rhenium(III) complex with an unusual 1,2,3-butatriene ligand.[138]

*(ii) Miscellaneous*

Compound (**31**) has been used to prepare the cyclohexyl xanthate and diethyldithiocarbamate derivatives [ReCp*Cl$_3$(S$_2$CR)],[124] the toluene-3,4-dithiolate [ReCp*(S$_2$C$_7$H$_6$)$_2$],[125] and the 2-amidothiophenolate [ReCp*{1,2-S(NH)C$_6$H$_4$}$_2$].[125] [ReCp*{1,2-S(NH)C$_6$H$_4$}$_2$] can also be prepared

$$(31) + R^1C{\equiv}CR^2 \xrightarrow[-HCl]{CH_2Cl_2,\ 40\ ^\circ C,\ 1\ h} \qquad (14)$$

$$R^1 = R^2 = Me,\ R^3 = H;\ R^1 = R^2 = Et,\ R^3 = Me$$
$$R^1 = Me,\ R^2 = Et,\ R^3 = Me;\ R^1 = Ph,\ R^2 = Me,\ R^3 = H$$

(32)

from [ReCp*(O){1,2-S(NH)C$_6$H$_4$}] or (**28**). TMS-Cl replaces the oxo group with chlorides in [ReCp*(O){o-S(NH)C$_6$H$_4$}] to give [ReCp*Cl$_2${o-S(NH)C$_6$H$_4$}]. Rhenium(V) *O,O*-catecholate-type chelate compounds, [ReCp*(O$_2$C$_6$Cl$_4$)$_2$] and [ReCp*(O$_2$C$_{14}$H$_8$)$_2$], are formed (along with other rhenium(V) compounds) when (**29**) reacts with tetrachloro-*o*-benzoquinone and 9,10-phenanthrene-quinone.[132]

### 10.4.1.4 Hydrides

The rhenium(VII) polyhydride (**33**) is synthesized in 75% yield according to Equation (15).[139-41] When LiAlD$_4$/MeOH and LiAlH$_4$/MeOD are used in the reaction, [ReCp*H$_2$D$_4$] and [ReCp*H$_4$D$_2$], respectively, are formed, suggesting that the lithium aluminum hydride replaces the oxo and chloride ligands.[141] The nature of the oxidation step in Equation (15) is not clear. An occasional side-product of reaction (15) is [{ReCp*H$_2$(μ-H)}$_2$].[139,140]

$$(30) \xrightarrow[\text{ether, }-95\ ^\circ C]{\text{excess LAH}} \xrightarrow[\text{ether, }-95\ ^\circ C]{\text{MeOH}} \qquad (15)$$

(33)

In solution (**33**) is fluxional, but at low temperature a binominal sextet with relative intensity 1 and a doublet with relative intensity 5 are observed in the hydride region of the $^1$H NMR spectrum, consistent with a pentagonal-bipyramidal structure. $T_1$ determinations suggest a classical hydride description is appropriate. Data from an electron diffraction study of (**33**) and an x-ray crystallographic study of the C$_5$Me$_4$Et derivative confirm that both compounds have pentagonal-bipyramidal structures with symmetrically π-bonded C$_5$ rings.[141] The hydrides are pushed down away from the pentagonal plane towards the unique hydride ligand ($\angle$H'–Re–H = 66–70°).

Thermolysis or photolysis of (**33**) gives reductive elimination of H$_2$ and formation of the dimer [{ReCp*H$_2$(μ-H)}$_2$], and photolysis in the presence of phosphine gives [ReCp*H$_2$(PMe$_2$R)$_2$] (R = Me or Ph).[140] Compound (**33**) reacts with I$_2$ to give [{ReCp*I$_2$(μ-I)}$_2$] and with CCl$_4$ to yield [{ReCp*Cl$_2$(μ-H)}$_2$].

Photolysis through quartz of [ReH$_7$(PPh$_3$)$_2$] and CpH in THF gives [ReCpH$_4$(PPh$_3$)], a compound previously prepared by photolysis of [ReH$_5$(PPh$_3$)$_2$] and CpH in THF.[142]

Oxidation of [ReCpH$_2${P(aryl)$_3$}$_2$] compounds (aryl = Ph, *p*-Tol, *p*-FC$_6$H$_4$, or *p*-MeOC$_6$H$_4$) gives the relatively stable rhenium(IV) 17-electron species [ReCpH$_2${P(aryl)$_3$}$_2$]$^+$.[143] [ReCp*H$_{2-x}$Cl$_x$(PMe$_3$)$_2$]$^+$ ($x = 0$, 1, or 2) compounds are produced similarly.[140] [ReCpH$_3$(PPh$_3$)$_2$]$^+$ is prepared by adding H$^+$ to [ReCpH$_2$(PPh$_3$)$_2$].[143,144]

### 10.4.1.5 Rhenium(IV) dimers

Compound (**31**) is reduced by 1 equiv. of SnBu$^n_4$ or an excess of aluminum in the presence of HgCl$_2$ to give [{ReCp*Cl$_2$(μ-Cl)}$_2$] in high yield (>80%).[134,137,145] The rhenium(IV) dimer reacts with Cl$_2$ to re-form (**31**) and with PMe$_3$ to give paramagnetic [ReCp*Cl$_3$(PMe$_3$)], and it can be further reduced (Al/HgCl$_2$ with heating) to [{ReCp*Cl(μ-Cl)}$_2$] (Re=Re). [{ReCp*Cl$_2$(μ-Cl)}$_2$] disproportionates upon reaction with 2-butyne to give [ReCp*Cl$_2$(MeCCMe)] and (**31**). [{ReCp*Cl$_2$(μ-Cl)}$_2$] is diamagnetic.

The crystal structure of the η-$C_5Me_4Et$ derivative reveals a Re–Re distance of 0.307 nm, consistent with a weak Re–Re single bond.[137]

The related compounds [{ReCp*$I_2$(μ-I)}$_2$], [{ReCp*$Cl_2$(μ-H)}$_2$], and [{ReCp*$H_2$(μ-H)}$_2$] have been prepared from (33) (see Section 10.4.1.4, and below, for syntheses).[140]

Treatment of (31) at low temperatures with excess LAH followed by methanol produces a separable mixture of (33) and [{ReCp*$H_2$(μ-H)}$_2$] in 36% yield.[140] The hydride dimer can also be prepared by thermal (200 °C) or photochemical (0 °C/hexane) reductive elimination of $H_2$ from (33). The hydride dimer shows fluxional NMR behavior at room temperature, but at −95 °C resonances for both the bridge and terminal hydrides are observed. The short Re–Re bond distance ($d$(Re–Re) = 0.245 nm) in [{Re(η-$C_5Me_4Et$)$H_2$(μ-H)}$_2$] has been interpreted as a triple bond in accordance with the 18-electron rule,[140] but a theoretical study suggests that the electronic configuration of the hydride is $(\sigma/\delta)^2(\pi)^2(\delta^*)^2$, which approximately translates to an Re–Re bond order of 1.[146]

[ReCp*(O)(RCCR)] (R = Me or Ph) can be oxidized to the rhenium(IV) dication [{ReCp*(μ-O)(RCCR)}$_2$][$BF_4$]$_2$.[147]

### 10.4.1.6 [{ReCp*(μ-O)$_2$}$_3$]$^{2+}$

In air-saturated solutions the reaction of (28) with $PPh_3$ produces (34) in 70–90% yield (cf. Scheme 12, reaction with $PPh_3$).[110,148] The crystal structure of (34) shows that it has virtual $D_{3h}$ symmetry with $d$(Re–Re) = 0.275 nm. The compound is reported to be diamagnetic, but this has been disputed on the basis of extended Hückel and SCF $X_\alpha$-SW calculations that suggest a $(1a_1')^2(1e')^2$ $d$-block configuration and an $^3A_2$ ground state.[149]

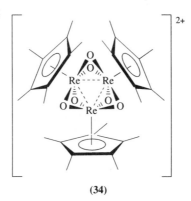

(34)

### 10.4.1.7 Imido complexes

The imido analogues of (30), [ReCp*(NR)$Cl_2$] (R = Me or $Bu^t$), are prepared by reacting (31) with $RNH_2$.[2f,71,127] The methyl imido derivative can also be prepared in high yield from (31) and 1 equiv. of MeN(TMS)$_2$.[127] The five-membered rhenacycles [Re{$E^1$(CH$_2$)$_2$E$^2$}Cp*(NMe)] ($E^1$ = $E^2$ = O, S, or NR; $E^1$ = O, $E^2$ = S; $E^1$ = O, $E^2$ = NR (R = H, Me, or $Bu^t$)) are prepared from [ReCp*(NMe)$Cl_2$] by the same methods used to prepare the analogous oxo compounds (see Section 10.4.1.2(iii)).

Reaction of 2 equiv. of (N-trimethylsilyl)-α-aminopicoline with (30) gives the metallacycle [Re{NCH$_2$-2-NC$_5$H$_4$}Cp*Cl(O-TMS)].[150] The compound has a square-pyramidal geometry with the chlorine ligand *trans* to the pyridine nitrogen. In the structure, the Re=N bond is longer and the Re–N(py) bond is shorter than normal, suggesting that the complex may have a [Re{N(H)=C(H)-2-NC$_5$H$_4$}] ring system (i.e., a $CH_2$ hydrogen has shifted to the "imido" nitrogen). Spectroscopic evidence, however, supports the imido formulation.

### 10.4.1.8 Sulfides and polysulfides

The rhenium(V) dimer [{ReCp*(O)(μ-S)}$_2$] is formed from (29) and $CS_2$ in toluene.[116] The structure is similar to (29) ($d$(Re–Re) = 0.366 nm).

Compound (30) reacts with [TiCp$_2$(S$_5$)] to give [ReCp*(O)(S$_4$)], which has a five-membered ReS$_4$ ring.[151] A similar reaction with $Na_2S_4$ gives a mixture of [ReCp*(O)(S$_4$)] and [ReCp*(S$_3$)(S$_4$)].[152] A better preparation of $[ReCp^*(S_3)(S_4)]$ is from (**30**) and excess $(NH_4)_2S_x$ ($x \approx 10$) in methanol. The crystal structure of $[ReCp^*(S_3)(S_4)]$ reveals that it has four- and five-membered $ReS_3$ and $ReS_4$ rings and is square pyramidal.[152] $[ReCp^*(S_3)(S_4)]$ reacts with excess $PPh_3$ to give $[\{ReCp^*(S_4)\}_2]$ (Re=Re), a dimer in which each monomer unit has one end of an $S_4^{2-}$ ligand terminally bonded to rhenium and the other bridging between the rhenium atoms.[152]

## 10.4.2 Biscyclopentadienyl Complexes

$[ReCp_2H]$ reacts with $Cl_2$ to form $[ReCp_2Cl_2]Cl$. Anion exchange using $Ag[SbF_6]$ in $SO_2$ gives $[ReCp_2Cl_2][SbF_6]$,[153] which can be oxidized to the unique rhenium(VII) 16-electron compound $[ReCp_2Cl_2]^{3+}$, as shown in Equation (16).[153]

$$[ReCp_2Cl_2][SbF_6] + 3\,SbF_5 \xrightarrow[96\%]{SO_2,\,25\,°C} [ReCp_2Cl_2][SbF_6]_3 + SbF_3 \qquad (16)$$

$[ReCp_2Me_2][PF_6]$ is formed by reacting $[ReCp_2H]$ with KOH/MeI and by adding MeI to $[ReCp_2Me]$, followed by anion exchange.[154] The dihydride $[ReCp^*_2H_2][BF_4]$ is prepared by protonating $[ReCp^*_2H]$ with $HBF_4$.[155] Photolysis of $[ReCp^*_2H_2][BF_4]$ gives rhenium(III) compounds.[155] The dihydride $[ReCp_2H_2][PF_6]$ reacts with diazomethane in RCN (R = Me, Ph) to give $[ReCp_2\{CR(NHMe)\}][PF_6]$.[154]

Protonation of $[ReCp_2Me]$ with HCl or $HBF_4$ gives $[ReCp_2H(Me)]X$.[156] $[ReCp_2H(Me)]^+$ exchanges the hydride and methyl protons, and it irreversibly eliminates methane to form $[ReCp_2Cl]$, both by intramolecular processes. Rate studies show that the site exchange occurs much more rapidly than the reductive elimination and the reductive elimination occurs with an inverse kinetic isotope effect ($k_H/k_D = 0.8 \pm 0.1$). The observation of an inverse kinetic isotope effect is consistent with a rapid preequilibrium involving the methyl hydride and a proposed methane σ-complex intermediate.

## 10.4.3 Cyclohexadienyl Complexes

Rhenium(V) $[Re(\eta\text{-}C_6H_7)H_3(PPh_3)_2][BF_4]$ and $[Re(\eta\text{-}2,4,6\text{-}exo\text{-}Me_3C_6H_4)H_3(PPh_3)_2][BF_4]$ are synthesized by protonating rhenium(III) dihydride precursors.[157] NMR studies suggest that the two phosphine ligands are *trans* in the pseudooctahedral complexes. The addition of MeOH to the trihydride cations induces loss of $H_2$ and formation of cationic rhenium(III) arene complexes.[158]

The reaction of $[Re(NBu^t)_2Cl_3]$ with excess MgBr(Mes) in THF gives a mixture of $[Re(NBu^t)_2(Mes)_2]$ and (**35**), which is isolated in 15% yield.[35]

(**35**)

## 10.5 REFERENCES

1. J. P. Collman, L. S. Hegedus, J. R. Norton and R. G. Finke, 'Principles and Applications of Organotransition Metal Chemistry', University Science Books, Mill Valley, CA, 1987, pp. 22–30.
2. (a) C. P. Casey, *Science*, 1993, **259**, 1552; (b) W. A. Herrmann, *Angew. Chem., Int. Ed. Engl.*, 1988, **27**, 1297; *Angew. Chem.*, 1988, **100**, 1269; (c) W. A. Herrmann, *Comments Inorg. Chem.*, 1988, **7**, 73; (d) W. A. Herrmann, *J. Organomet. Chem.*, 1990, **382**, 1; (e) W. A. Herrmann, *ibid.*, 1986, **300**, 111; (f) W. A. Herrmann, E. Herdtweck, M. Flöel, J. Kulpe, U. Küsthardt and J. Okuda, *Polyhedron*, 1987, **6**, 1165; (g) W. A. Herrmann and J. Okuda, *J. Mol. Catal.*, 1987, **41**, 109.
3. N. M. Boag and H. D. Kaesz, in 'COMC-I', vol. 4, p. 161.

4. (a) P. D. Savage, G. Wilkinson, M. Motevalli and M. B. Hursthouse, *J. Chem. Soc., Dalton Trans.*, 1988, 669; (b) J. Arnold, G. Wilkinson, B. Hussain and M. B. Hursthouse, *ibid.*, 1989, 2149.
5. (a) J. Arnold, G. Wilkinson, B. Hussain and M. B. Hursthouse, *Organometallics*, 1989, **8**, 415; (b) J. Arnold, G. Wilkinson, B. Hussain and M. B. Hursthouse, *J. Chem. Soc., Chem. Commun.*, 1988, 704.
6. H.-W. Swidersky, O. Kindel, F. Weller and K. Dehnicke, *Z. Anorg. Allg. Chem.*, 1990, **580**, 18.
7. H.-W. Swidersky, J. Pebler, K. Dehnicke and D. Fenske, *Z. Naturforsch., Teil B*, 1990, **45**, 1227.
8. E. Hey, F. Weller and K. Dehnicke, *Z. Anorg. Allg. Chem.*, 1984, **514**, 25.
9. D. Wolff von Gudenberg, I. Sens, U. Müller, B. Neumüller and K. Dehnicke, *Z. Anorg. Allg. Chem.*, 1992, **613**, 49.
10. (a) I. M. Gardiner, M. A. Bruck and D. E. Wigley, *Inorg. Chem.*, 1989, **28**, 1769; (b) I. M. Gardiner, M. A. Bruck, P. A. Wexler and D. E. Wigley, *ibid.*, 1989, **28**, 3688.
11. W. A. Herrmann, J. K. Felixberger, J. G. Kuchler and E. Herdtweck, *Z. Naturforsch., Teil B*, 1990, **45**, 876.
12. J. D. Allison and R. A. Walton, *J. Am. Chem. Soc.*, 1984, **106**, 163.
13. (a) S. Cai, D. M. Hoffman, D. Lappas and H.-G. Woo, *Organometallics*, 1987, **6**, 2273; (b) D. M. Hoffman, D. Lappas and D. A. Wierda, *Acta Crystallogr., Sect. C*, 1988, **44**, 1661.
14. D. M. Hoffman and D. A. Wierda, *Polyhedron*, 1989, **8**, 959.
15. D. M. Hoffman and D. A. Wierda, *J. Am. Chem. Soc.*, 1990, **112**, 7056.
16. W. K. Rybak and J. J. Ziólkowski, *J. Mol. Catal.*, 1987, **42**, 347.
17. D. Lappas, Ph.D. Thesis, Harvard University, 1990.
18. (a) D. M. Hoffman, J. C. Huffman, D. Lappas and D. A. Wierda, *Organometallics*, 1993, **12**, 4312; (b) D. Lappas, D. M. Hoffman, K. Folting and J. C. Huffman, *Angew. Chem., Int. Ed. Engl.*, 1988, **27**, 587; *Angew. Chem.*, 1988, **100**, 585.
19. W. A. Herrmann, J. G. Kuchler, G. Weichselbaumer, E. Herdtweck and P. Kiprof, *J. Organomet. Chem.*, 1989, **372**, 351.
20. J. Takacs, M. R. Cook, P. Kiprof, J. G. Kuchler and W. A. Herrmann, *Organometallics*, 1991, **10**, 316.
21. J. Takacs, P. Kiprof, J. G. Kuchler and W. A. Herrmann, *J. Organomet. Chem.*, 1989, **369**, C1.
22. J. Takacs, P. Kiprof, J. Riede and W. A. Herrmann, *Organometallics*, 1990, **9**, 782.
23. A. R. Middleton and G. Wilkinson, *J. Chem. Soc., Dalton Trans.*, 1981, 1898.
24. P. Stavropoulos, P. G. Edwards, G. Wilkinson, M. Motevalli, K. M. Abdul Malik and M. B. Hursthouse, *J. Chem. Soc., Dalton Trans.*, 1985, 2167.
25. P. Stavropoulos, G. Wilkinson, M. Motevalli and M. B. Hursthouse, *Polyhedron*, 1987, **6**, 1081.
26. J. K. Felixberger, J. G. Kuchler, E. Herdtweck, R. A. Paciello and W. A. Herrmann, *Angew. Chem., Int. Ed. Engl.*, 1988, **27**, 946; *Angew. Chem.*, 1988, **100**, 975.
27. S. Cai Ph.D., Thesis, Harvard University, 1990.
28. (a) S. Cai, D. M. Hoffman, J. C. Huffman, D. A. Wierda and H.-G. Woo, *Inorg. Chem.*, 1987, **26**, 3693. (b) S. Cai, D. M. Hoffman and D. A. Wierda, *Polyhedron*, 1990, **9**, 957.
29. P. Stavropoulos, P. G. Edwards, T. Behling, G. Wilkinson, M. Motevalli and M. B. Hursthouse, *J. Chem. Soc., Dalton Trans.*, 1987, 169.
30. W. A. Herrmann, R. W. Albach and J. Behm, *J. Chem. Soc., Chem. Commun.*, 1991, 367.
31. K. W. Chiu, W.-K. Wong, G. Wilkinson, A. M. R. Galas and M. B. Hursthouse, *Polyhedron*, 1982, **1**, 31.
32. D. S. Williams, M. H. Schofield and R. R. Schrock, *Organometallics*, 1993, **12**, 4560.
33. I. A. Weinstock, R. R. Schrock, D. S. Williams and W. E. Crowe, *Organometallics*, 1991, **10**, 1.
34. D. S. Williams and R. R. Schrock, *Organometallics*, 1993, **12**, 1148.
35. A. Gutierrez, G. Wilkinson, B. Hussain-Bates and M. B. Hursthouse, *Polyhedron*, 1990, **9**, 2081.
36. V. W.-W. Yam, K.-K. Tam and T.-F. Lai, *J. Chem. Soc., Dalton Trans.*, 1993, 651.
37. N. J. Hazel, J. A. K. Howard and J. L. Spencer, *J. Chem. Soc., Chem. Commun.*, 1984, 1663.
38. D. Baudry *et al.*, *J. Chem. Soc., Chem. Commun.*, 1985, 670.
39. X.-L. Luo and R. H. Crabtree, *J. Chem. Soc., Chem. Commun.*, 1990, 189.
40. E. Hey, F. Weller and K. Dehnicke, *Naturwissenschaften*, 1983, **70**, 41.
41. K. W. Chiu, R. A. Jones, G. Wilkinson, A. M. R. Galas and M. B. Hursthouse, *J. Chem. Soc., Dalton Trans.*, 1981, 2088.
42. C. J. Longley, P. D. Savage, G. Wilkinson, B. Hussain and M. B. Hursthouse, *Polyhedron*, 1988, **7**, 1079.
43. B. S. McGilligan, J. Arnold, G. Wilkinson, B. Hussain-Bates and M. B. Hursthouse, *J. Chem. Soc., Dalton Trans.*, 1990, 2465.
44. W. A. Herrmann, J. G. Kuchler, J. K. Felixberger, E. Herdtweck and W. Wagner, *Angew. Chem., Int. Ed. Engl.*, 1988, **27**, 394; *Angew. Chem.*, 1988, **100**, 420.
45. W. A. Herrmann, C. C. Romao, P. Kiprof, J. Behm, M. R. Cook and M. Taillefer, *J. Organomet. Chem.*, 1991, **413**, 11.
46. J. M. Huggins, D. R. Whitt and L. Lebioda, *J. Organomet. Chem.*, 1986, **312**, C15.
47. S. Cai, D. M. Hoffman and D. A. Wierda, *Organometallics*, 1988, **7**, 2069.
48. S. Cai, D. M. Hoffman and D. A. Wierda, *Inorg. Chem.*, 1989, **28**, 3784.
49. S. Cai, D. M. Hoffman and D. A. Wierda, *Inorg. Chem.*, 1991, **30**, 827.
50. W. A. Herrmann, P. Watzlowik and P. Kiprof, *Chem. Ber.*, 1991, **124**, 1101.
51. W. A. Herrmann, D. Marz, W. Wagner, J. Kuchler, G. Weichselbaumer and R. Fischer (Hoechst AG), *Ger. Offen. DE* 3 902 357 (1990) (*Chem. Abstr.*, 1991, **114**, 143 714u).
52. W. A. Herrmann, F. E. Kühn, R. W. Fischer, W. R. Thiel and C. C. Romao, *Inorg. Chem.*, 1992, **31**, 4431.
53. (a) W. A. Herrmann, C. C. Romao, R. W. Fischer, P. Kiprof and C. de Méric de Bellefon, *Angew. Chem., Int. Ed. Engl.*, 1991, **30**, 185; *Angew. Chem.*, 1991, **103**, 183; (b) W. A. Herrmann *et al.*, *Chem. Ber.*, 1993, **126**, 45.
54. W. A. Herrmann, M. Ladwig, P. Kiprof and J. Riede, *J. Organomet. Chem.*, 1989, **371**, C13.
55. C. de Méric de Bellefon, W. A. Herrmann, P. Kiprof and C. R. Whitaker, *Organometallics*, 1992, **11**, 1072.
56. W. A. Herrmann, G. Weichselbaumer and E. Herdtweck, *J. Organomet. Chem.*, 1989, **372**, 371.
57. W. A. Herrmann, J. G. Kuchler, P. Kiprof and J. Riede, *J. Organomet. Chem.*, 1990, **395**, 55.
58. H. Kunkely, T. Türk, C. Teixeira, C. de Méric de Bellefon, W. A. Herrmann and A. Vogler, *Organometallics*, 1991, **10**, 2090.
59. W. A. Herrmann and P. Watzlowik, *J. Organomet. Chem.*, 1992, **441**, 265.
60. W. A. Herrmann, R. W. Fischer and W. Scherer, *Adv. Mater.*, 1992, **4**, 653.
61. W. A. Herrmann, W. Wagner, U. N. Flessner, U. Volhardt and H. Komber, *Angew. Chem., Int. Ed. Engl.*, 1991, **30**, 1636; *Angew. Chem.*, 1991, **103**, 1704.

62. W. A. Herrmann, W. Wagner and U. Volhardt (Hoechst AG), *Ger. Offen.* 3 940 196 (1990) (*Chem. Abstr.*, 1990, **113**, 231 676t).
63. R. Buffon *et al.*, *J. Mol. Catal.*, 1992, **76**, 287.
64. R. Buffon, A. Choplin, M. Leconte, J.-M. Basset, R. Touroude and W. A. Herrmann, *J. Mol. Catal.*, 1992, **72**, L7.
65. J. R. McCoy and M. F. Farona, *J. Mol. Catal.*, 1991, **66**, 51.
66. W. A. Herrmann and M. Wang, *Angew. Chem., Int. Ed. Engl.*, 1991, **30**, 1641; *Angew. Chem.*, 1991, **103**, 1709.
67. W. A. Herrmann, R. W. Fischer and D. W. Marz, *Angew. Chem., Int. Ed. Engl.*, 1991, **30**, 1638; *Angew. Chem.*, 1991, **103**, 1706.
68. W. A. Herrmann, R. W. Fischer, W. Scherer and M. U. Rauch, *Angew. Chem., Int. Ed. Engl.*, 1993, **32**, 1157; *Angew. Chem.*, 1993, **105**, 1209.
69. S. Cai, D. M. Hoffman and D. A. Wierda, *J. Chem. Soc., Chem. Commun.*, 1988, 313.
70. S. Cai, D. M. Hoffman and D. A. Wierda, *J. Chem. Soc., Chem. Commun.*, 1988, 1489.
71. W. A. Herrmann *et al.*, *Organometallics*, 1990, **9**, 489.
72. A. D. Horton and R. R. Schrock, *Polyhedron*, 1988, **7**, 1841.
73. M. R. Cook, W. A. Herrmann, P. Kiprof and J. Takacs, *J. Chem. Soc., Dalton Trans.*, 1991, 797.
74. D. S. Williams, J. T. Anhaus, M. H. Schofield, R. R. Schrock and W. M. Davis, *J. Am. Chem. Soc.*, 1991, **113**, 5480.
75. D. S. Edwards and R. R. Schrock, *J. Am. Chem. Soc.*, 1982, **104**, 6806.
76. D. S. Edwards, L. V. Biondi, J. W. Ziller, M. R. Churchill and R. R. Schrock, *Organometallics*, 1983, **2**, 1505.
77. A. D. Horton, R. R. Schrock and J. H. Freudenberger, *Organometallics*, 1987, **6**, 893.
78. R. Toreki, R. R. Schrock and W. M. Davis, *J. Am. Chem. Soc.*, 1992, **114**, 3367.
79. R. Toreki and R. R. Schrock, *J. Am. Chem. Soc.*, 1990, **112**, 2448.
80. R. R. Schrock, I. A. Weinstock, A. D. Horton, A. H. Liu and M. H. Schofield, *J. Am. Chem. Soc.*, 1988, **110**, 2686.
81. I. A. Weinstock, R. R. Schrock and W. M. Davis, *J. Am. Chem. Soc.*, 1991, **113**, 135.
82. M. H. Schofield, R. R. Schrock and L. Y. Park, *Organometallics*, 1991, **10**, 1844.
83. P. D. Savage, G. Wilkinson, M. Motevalli and M. B. Hursthouse, *Polyhedron*, 1987, **6**, 1599.
84. (a) A. A. Danopoulos, G. Wilkinson and D. J. Williams, *J. Chem. Soc., Chem. Commun.*, 1991, 181; (b) V. Saboonchian, A. A. Danopoulos, A. Gutierrez, G. Wilkinson and D. J. Williams, *Polyhedron*, 1991, **10**, 2241.
85. J. Sundermeyer, K. Weber and O. Nürnberg, *J. Chem. Soc., Chem. Commun.*, 1992, 1631.
86. M. Leeaphon, P. E. Fanwick and R. A. Walton, *J. Am. Chem. Soc.*, 1992, **114**, 1890.
87. W. A. Herrmann, J. K. Felixberger, R. Anwander, E. Herdtweck, P. Kiprof and J. Riede, *Organometallics*, 1990, **9**, 1434.
88. J. K. Felixberger, P. Kiprof, E. Herdtweck, W. A. Herrmann, R. Jakobi and P. Gütlich, *Angew. Chem., Int. Ed. Engl.*, 1989, **28**, 334; *Angew. Chem.*, 1989, **101**, 346.
89. R. Toreki, R. R. Schrock and M. G. Vale, *J. Am. Chem. Soc.*, 1991, **113**, 3610.
90. G. A. Vaughan, R. Toreki, R. R. Schrock and W. M. Davis, *J. Am. Chem. Soc.*, 1993, **115**, 2980.
91. P. B. Hitchcock, M. F. Meidine, J. F. Nixon and A. J. L. Pombeiro, *J. Chem. Soc., Chem. Commun.*, 1991, 1031.
92. A. J. L. Pombeiro, A. Hills, D. L. Hughes and R. L. Richards, *J. Organomet. Chem.*, 1988, **352**, C5.
93. A. J. L. Pombeiro, D. L. Hughes, R. L. Richards, J. Silvestre and R. Hoffmann, *J. Chem. Soc., Chem. Commun.*, 1986, 1125.
94. A. J. L. Pombeiro and R. L. Richards, *Coord. Chem. Rev.*, 1990, **104**, 13.
95. A. J. L. Pombeiro, in 'Advances in Metal Carbene Chemistry', ed. U. Schubert, Kluwer, Dordrecht, 1989, p. 79.
96. (a) S. Warner and S. J. Lippard, *Organometallics*, 1989, **8**, 228; (b) R. N. Vrtis, Ch. P. Rao, S. Warner and S. J. Lippard, *J. Am. Chem. Soc.*, 1988, **110**, 2669.
97. W. A. Herrmann, P. Kiprof and J. K. Felixberger, *Chem. Ber.*, 1990, **123**, 1971.
98. I. L. Kershenbaum, I. A. Oreshkin, B. A. Dolgoplosk, E. I. Tinyakova and L. N. Grebenyak, *Dokl. Akad. Nauk. SSSR*, 1981, **256**, 1400 (*Dokl. Akad. Nauk. USSR (Engl. Transl.)*, 1981, 69).
99. R. Toreki, G. A. Vaughan, R. R. Schrock and W. M. Davis, *J. Am. Chem. Soc.*, 1993, **115**, 127.
100. T. R. Cundari and M. S. Gordon, *Organometallics*, 1992, **11**, 55.
101. W. A. Herrmann, R. Serrano and H. Bock, *Angew. Chem., Int. Ed. Engl.*, 1984, **23**, 383; *Angew. Chem.*, 1984, **96**, 364.
102. A. H. Klahn-Oliva and D. Sutton, *Organometallics*, 1984, **3**, 1313.
103. W. A. Herrmann, R. Serrano, A. Schäfer, U. Küsthardt, M. L. Ziegler and E. Guggolz, *J. Organomet. Chem.*, 1984, **272**, 55.
104. W. A. Herrmann, E. Voss and M. Flöel, *J. Organomet. Chem.*, 1985, **297**, C5.
105. J. Okuda, E. Herdtweck and W. A. Herrmann, *Inorg. Chem.*, 1988, **27**, 1254.
106. W. A. Herrmann, M. Flöel, J. Kulpe, J. K. Felixberger and E. Herdtweck, *J. Organomet. Chem.*, 1988, **355**, 297.
107. W. A. Herrmann *et al.*, *J. Am. Chem. Soc.*, 1991, **113**, 6527.
108. W. R. Thiel, R. W. Fischer and W. A. Herrmann, *J. Organomet. Chem.*, 1993, **459**, C9.
109. W. A. Herrmann, M. Taillefer, C. de Méric de Bellefon and J. Behm, *Inorg. Chem.*, 1991, **30**, 3247.
110. W. A. Herrmann, R. Serrano, U. Küsthardt, E. Guggolz, B. Nuber and M. L. Ziegler, *J. Organomet. Chem.*, 1985, **287**, 329.
111. W. A. Herrmann, T. Cuenca and U. Küsthardt, *J. Organomet. Chem.*, 1986, **309**, C15.
112. B. E. Bursten and R. H. Cayton, *Inorg. Chem.*, 1989, **28**, 2846.
113. W. A. Herrmann, U. Küsthardt, M. Flöel, J. Kulpe, E. Herdtweck and E. Voss, *J. Organomet. Chem.*, 1986, **314**, 151.
114. W. A. Herrmann, E. Voss, U. Küsthardt and E. Herdtweck, *J. Organomet. Chem.*, 1985, **294**, C37.
115. W. A. Herrmann, R. Serrano, U. Küsthardt, M. L. Ziegler, E. Guggolz and T. Zahn, *Angew. Chem., Int. Ed. Engl.*, 1984, **23**, 515; *Angew. Chem.*, 1984, **96**, 498.
116. W. A. Herrmann, K. A. Jung and E. Herdtweck, *Chem. Ber.*, 1989, **122**, 2041.
117. V. N. Kalinin *et. al.*, *Organomet. Chem. USSR*, 1992, **5**, 221 (*Metalloorg. Khim.*, 1992, **5**, 460).
118. R. A. Paciello, P. Kiprof, E. Herdtweck and W. A. Herrmann, *Inorg. Chem.*, 1989, **28**, 2890.
119. H.-J. Kneuper, P. Härter and W. A. Herrmann, *J. Organomet. Chem.*, 1988, **340**, 353.
120. W. A. Herrmann, D. Marz, E. Herdtweck, A. Schäfer, W. Wagner and H.-J. Kneuper, *Angew. Chem., Int. Ed. Engl.*, 1987, **26**, 462; *Angew. Chem.*, 1987, **99**, 462.
121. B. E. Bursten and R. H. Cayton, *Organometallics*, 1987, **6**, 2004.

122. (a) H. J. R. de Boer, B. J. J. van de Heisteeg, M. Flöel, W. A. Herrmann, O. S. Akkerman and F. Bickelhaupt, *Angew. Chem., Int. Ed. Engl.*, 1987, **26**, 73; *Angew. Chem.*, 1987, **99**, 88; (b) W. A. Herrmann, M. Flöel and E. Herdtweck, *J. Organomet. Chem.*, 1988, **358**, 321.

123. E. J. M. de Boer, J. de With and A. G. Orpen, *J. Am. Chem. Soc.*, 1986, **108**, 8271.

124. J. Takacs, P. Kiprof, G. Weichselbaumer and W. A. Herrmann, *Organometallics*, 1989, **8**, 2394.

125. J. Takacs, P. Kiprof and W. A. Herrmann, *Polyhedron*, 1990, **9**, 2211.

126. W. A. Herrmann and D. W. Marz, *J. Organomet. Chem.*, 1989, **362**, C5.

127. W. A. Herrmann, D. W. Marz and E. Herdtweck, *J. Organomet. Chem.*, 1990, **394**, 285.

128. W. A. Herrmann, K. A. Jung, A. Schäfer and H.-J. Kneuper, *Angew. Chem., Int. Ed. Engl.*, 1987, **26**, 464; *Angew. Chem.*, 1987, **99**, 464.

129. U. Küsthardt, W. A. Herrmann, M. L. Ziegler, T. Zahn and B. Nuber, *J. Organomet. Chem.*, 1986, **311**, 163.

130. W. A. Herrmann, U. Küsthardt, M. L. Ziegler and T. Zahn, *Angew. Chem., Int. Ed. Engl.*, 1985, **24**, 860; *Angew. Chem.*, 1985, **97**, 857.

131. W. A. Herrmann, U. Küsthardt, A. Schäfer and E. Herdtweck, *Angew. Chem., Int. Ed. Engl.*, 1986, **25**, 817; *Angew. Chem.*, 1986, **98**, 818.

132. W. A. Herrmann, U. Küsthardt and E. Herdtweck, *J. Organomet. Chem.*, 1985, **294**, C33.

133. M. Flöel, E. Herdtweck, W. Wagner, J. Kulpe, P. Härter and W. A. Herrmann, *Angew. Chem., Int. Ed. Engl.*, 1987, **26**, 787; *Angew. Chem.*, 1987, **99**, 787.

134. W. A. Herrmann, J. K. Felixberger, E. Herdtweck, A. Schäfer and J. Okuda, *Angew. Chem., Int. Ed. Engl.*, 1987, **26**, 466; *Angew. Chem.*, 1987, **99**, 466.

135. M. G. Vale and R. R. Schrock, *Organometallics*, 1991, **10**, 1662.

136. W. A. Herrmann, R. A. Fischer and E. Herdtweck, *Angew. Chem., Int. Ed. Engl.*, 1987, **26**, 1263; *Angew. Chem.*, 1987, **99**, 1286.

137. W. A. Herrmann, R. A. Fischer, J. K. Felixberger, R. A. Paciello, P. Kiprof and E. Herdtweck, *Z. Naturforsch., Teil B*, 1988, **43**, 1391.

138. R. A. Fischer, R. W. Fischer, W. A. Herrmann and E. Herdtweck, *Chem. Ber.*, 1989, **122**, 2035.

139. W. A. Herrmann and J. Okuda, *Angew. Chem., Int. Ed. Engl.*, 1986, **25**, 1092; *Angew. Chem.*, 1986, **98**, 1109.

140. W. A. Herrmann, H. G. Theiler, E. Herdtweck and P. Kiprof, *J. Organomet. Chem.*, 1989, **367**, 291.

141. W. A. Herrmann, H. G. Theiler, P. Kiprof, J. Tremmel and R. Blom, *J. Organomet. Chem.*, 1990, **395**, 69.

142. D. Baudry, J.-M. Cormier and M. Ephritikhine, *J. Organomet. Chem.*, 1992, **427**, 349.

143. M. R. Detty and W. D. Jones, *J. Am. Chem. Soc.*, 1987, **109**, 5666.

144. D. Baudry and M. Ephritikhine, *J. Chem. Soc., Chem. Commun.*, 1980, 249.

145. W. A. Herrmann, R. A. Fischer and E. Herdtweck, *J. Organomet. Chem.*, 1987, **329**, C1.

146. B. E. Bursten and R. H. Cayton, *Organometallics*, 1988, **7**, 1349.

147. W. A. Herrmann, R. A. Fischer, W. Amslinger and E. Herdtweck, *J. Organomet. Chem.*, 1989, **362**, 333.

148. W. A. Herrmann, R. Serrano, M. L. Ziegler, H. Pfisterer and B. Nuber, *Angew. Chem., Int. Ed. Engl.*, 1985, **24**, 50; *Angew. Chem.*, 1985, **97**, 50.

149. P. Hofmann and N. Rösch, *J. Chem. Soc., Chem. Commun.*, 1986, 843.

150. W. A. Herrmann, D. W. Marz and E. Herdtweck, *Z. Naturforsch., Teil B*, 1991, **46**, 747.

151. J. Kulpe, E. Herdtweck, G. Weichselbaumer and W. A. Herrmann, *J. Organomet. Chem.*, 1988, **348**, 369.

152. M. Herberhold, G.-X. Jin and W. Milius, *Angew. Chem., Int. Ed. Engl.*, 1993, **32**, 85; *Angew. Chem.*, 1993, **105**, 127.

153. P. Gowik, T. Klapötke and I. Tornieporth-Oetting, *Chem. Ber.*, 1989, **122**, 2273.

154. D. Baudry and M. Ephritikhine, *J. Organomet. Chem.*, 1980, **195**, 213.

155. F. G. N. Cloke, J. P. Day, J. C. Green, C. P. Morley and A. C. Swain, *J. Chem. Soc., Dalton Trans.*, 1991, 789.

156. G. L. Gould and D. M. Heinekey, *J. Am. Chem. Soc.*, 1989, **111**, 5502.

157. D. Baudry, P. Boydell and M. Ephritikhine, *J. Chem. Soc., Dalton Trans.*, 1986, 525.

158. D. Baudry, P. Boydell and M. Ephritikhine, *J. Chem. Soc., Dalton Trans.*, 1986, 531.

# Author Index

This Author Index comprises an alphabetical listing of the names of the authors cited in the text and the references listed at the end of each chapter in this volume.

Each entry consists of the author's name, followed by a list of numbers, for example

Templeton, J. L., 366, 385[233] (350, 366), 387[370] (363)

For each name, the page numbers for the citation in the reference list are given, followed by the reference number in superscript and the page number(s) in parentheses of where that reference is cited in the text. Where a name is referred to in text only, the page number of the citation appears with no superscript number. References cited both in the text and in the tables are included.

Although much effort has gone into eliminating inaccuracies resulting from the use of different combinations of initials by the same author, the use by some journals of only one initial, and different spellings of the same name as a result of the transliteration processes, the accuracy of some entries may have been affected by these factors.

# Subject Index

JOHN NEWTON

*David John (Services), Slough, UK*

This Subject Index contains individual entries to the text pages of Volume 6. The index covers general types of organometallic compound, specific organometallic compounds, general and specific organic compounds where their synthesis or use involves organometallic compounds, types of reaction (insertion, oxidative addition, etc.), spectroscopic techniques (NMR, IR, etc.), and topics involving organometallic compounds.

Because authors may have approached similar topics from different viewpoints, index entries to those topics may not always appear under the same headings. Both synonyms and alternatives should therefore be considered to obtain all the entries on a particular topic. Commonly used synonyms include alkyne/acetylene, compound/complex, preparation/synthesis, etc. Entries where the oxidative state of a metal has been specified occur after all the entries for the unspecified oxidation state, and the same or similar compounds may occur under both types of heading. Thus $Cr(C_6H_6)_2$ occurs under Chromium, bis($\eta$-benzene) and again under Chromium(0), bis($\eta$-benzene). Similar ligands may also occur in different entries. Thus a carbene–metal complex may occur under Carbene complexes, Carbene ligands, or Carbenes, as well as under the specific metal. Individual organometallic compounds may also be listed in the Cumulative Formula Index in Volume 14.